Antimicrobial and Anti-infective Activity of Natural Products

Antimicrobial and Anti-infective Activity of Natural Products

Editors

Valério Monteiro-Neto
Elizabeth S. Fernandes

MDPI
Basel • Beijing • Wuhan • Barcelona • Belgrade • Novi Sad • Cluj • Manchester

Editors

Valério Monteiro-Neto
Department of Pathology
Federal University of
Maranhão
São Luís
Brazil

Elizabeth S. Fernandes
Instituto de Pesquisa Pelé
Pequeno Príncipe
Faculdades Pequeno Príncipe
Curitiba
Brazil

Editorial Office
MDPI
St. Alban-Anlage 66
4052 Basel, Switzerland

This is a reprint of articles from the Special Issue published online in the open access journal *Antibiotics* (ISSN 2079-6382) (available at: www.mdpi.com/journal/antibiotics/special_issues/Anti_Natural).

For citation purposes, cite each article independently as indicated on the article page online and as indicated below:

Lastname, A.A.; Lastname, B.B. Article Title. *Journal Name* **Year**, *Volume Number*, Page Range.

ISBN 978-3-0365-8725-7 (Hbk)
ISBN 978-3-0365-8724-0 (PDF)
doi.org/10.3390/books978-3-0365-8724-0

Contents

About the Editors

Valério Monteiro-Neto

Valério Monteiro-Neto received his Ph.D. in Biology of Host-Pathogen Interactions from the University of São Paulo, Brazil (2000). From 2005 to 2006, he was a postdoctoral fellow at the Laboratory of Microbiology at the Health Sciences Center at the University of Arizona (USA). He is currently an Associate Professor of Microbiology at the Federal University of Maranhão (Brazil) and Head of the Department of Pathology. His research interests mainly involve bacterial pathogenicity, bioprospecting antimicrobial activity, and the immunomodulatory activity of plant products and probiotics.

Elizabeth S. Fernandes

Elizabeth S. Fernandes is a Brazilian pharmacologist whose research is focused on the search for novel natural and synthetic compounds with anti-inflammatory, analgesic, and/or antimicrobial activities. She is especially interested in those able to modulate host-pathogen interactions and the connections between painful and inflammatory molecules. She is currently a Professor of Pharmacology of the Post-graduate Program in Biotechnology Applied to Child and Adolescent Health at Faculdades Pequeno Príncipe and leads her research laboratory at the Instituto de Pesquisa Pelé Pequeno Príncipe, both part of the Pequeno Príncipe Hospital Complex, located in South Brazil.

Preface

Antimicrobial drugs represent one of the great scientific advances in the medical field. However, a dramatic increase in antimicrobial resistance has compromised the effectiveness of these drugs. Thus, there is an increased need to search for new antimicrobial compounds. A growing area of research in this regard is the investigation of antimicrobial activities in natural products from different sources, including plants, microorganisms, and animals. Another important area of research in combating resistant microorganisms is the investigation of compounds that neutralize or inactivate bacterial resistance mechanisms—for example, compounds that inhibit efflux pumps or betalactamases. Furthermore, there are molecules that can interfere with bacterial virulence properties. The main advantage claimed for antivirulence therapy is the reduced selective pressure on the pathogen and, consequently, the lower possibility of the emergence of resistance.

This reprint, organized with the recent publications of the Special Issue Antimicrobial and Anti-infective Activity of Natural Products in the journal *Antibiotics*, aimed to report recent advances in the discovery of new antimicrobial compounds from natural products, mechanisms of action of pure compounds, in silico evidence of antimicrobial activity, synergistic associations with antibiotics, compounds with antiviral activity or capable of neutralizing bacterial resistance to antibiotics, as well as in vivo studies to demonstrate the effectiveness of compounds.

Valério Monteiro-Neto and Elizabeth S. Fernandes
Editors

antibiotics

Editorial

Antimicrobial and Anti-Infective Activity of Natural Products—Gaining Knowledge from Novel Studies

Elizabeth S. Fernandes [1,2,*], Isabella F. da Silva Figueiredo [1], Cinara R. A. V. Monteiro [3] and Valério Monteiro-Neto [3]

1 Programa de Pós-Graduação em Biotecnologia Aplicada à Saúde da Criança e do Adolescente, Faculdades Pequeno Príncipe, Av. Iguaçu No 333, Curitiba 80230-020, PR, Brazil; bellaafigueiredo@hotmail.com
2 Instituto de Pesquisa Pelé Pequeno Príncipe, Av. Silva Jardim No 1632, Curitiba 80240-020, PR, Brazil
3 Programa de Pós-Graduação em Ciências da Saúde, Universidade Federal do Maranhão, Av. dos Portugueses No 1966, Cidade Universitária Dom Delgado, São Luís 65085-040, MA, Brazil; cinaraaragao@hotmail.com (C.R.A.V.M.); valerio.monteiro@ufma.br (V.M.-N.)
* Correspondence: elizabeth.fernandes@pelepequenoprincipe.org.br

Abstract: Despite advances in the development of antimicrobial drugs in the last centuries, antimicrobial resistance has consistently raised in the last decades, compromising their effectiveness. Novel antimicrobial compounds, especially from natural sources, including plants, microorganisms, and animals, have since become a growing area of research. In this context, studies covering the investigation of their ability to combat resistant microorganisms, either by neutralization or inactivation of pathogen resistance mechanisms and virulence properties, have gained attention. Herein, a collection of 19 manuscripts focused on the antimicrobial and anti-infective activity of natural products, including their mechanisms of action, in silico evidence of antimicrobial activity, synergistic associations with antibiotics, and other aspects, will be discussed.

Citation: Fernandes, E.S.; da Silva Figueiredo, I.F.; Monteiro, C.R.A.V.; Monteiro-Neto, V. Antimicrobial and Anti-Infective Activity of Natural Products—Gaining Knowledge from Novel Studies. *Antibiotics* **2023**, *12*, 1051. https://doi.org/10.3390/antibiotics12061051

Received: 31 May 2023
Accepted: 9 June 2023
Published: 15 June 2023

Of the published papers, 15 are original research studies, three are reviews (one is a systematic review of the literature), and one is a brief report. Kopel and collaborators [1] reviewed and discussed the in vitro models and clinical trials for assessing the antimicrobial activities of phytochemicals. They highlighted the great diversity of compounds, such as alkaloids, organosulfur compounds, phenols, coumarins, and terpenes found in plants, as well as their broad spectrum of antimicrobial activity and resistance. Amongst the novel compounds discussed are the alkaloids sanguinarine, found in *Sanguinaria canadensis*, and tomatidine, found in solanaceous plants, such as tomatoes and potatoes. These were shown to interfere with bacterial cell division and cytokinesis through disruption of the plasma membrane, increased cell permeabilization, and inhibition of protein synthesis; enhancement of reactive oxygen species generation and reduction in ergosterol formation by fungi were also discussed as mechanisms of action of the alkaloids. They also commented on the ability of these compounds, as well as of other alkaloids, to potentiate the action of commercial antimicrobials, such as vancomycin, streptomycin, and ciprofloxacin. Organosulfur compounds, such as allicin from raw garlic, isothiocyanates, including sulforaphane, and allyl, benzyl, and phenethyl isothiocyanates were all suggested to act against a range of microorganisms, presenting antimicrobial and anti-virulence activities, which may be equivalent or even stronger than common antibiotics. Their mechanisms of action varied from disruption of cell walls, interference with essential biochemical pathways, and inhibition of enzymes, amongst others. Phenolic compounds, such as flavonoids (e.g., galangin, kaempferol quercetin, catechins, etc.) act on bacterial enzymes and toxins, disturbing cytoplasmic membranes, preventing the formation of biofilms, alone or in synergy with wide spectrum antibiotics. All the above-described mechanisms of action were first demonstrated by means of in vitro assays, such as microdilution and agar well diffusion methods, biofilm and hyphae formation techniques, and biochemical studies. Promising results on the in vitro antimicrobial actions of natural compounds can evolve to clinical trials to assess the effects of plant extract-based mouthwashes, gels, creams, suppositories, as well as other formulations for systemic usage, for a variety of diseases. Finally, the authors highlighted the need for further randomized clinical trials with large populations of subjects to evaluate the efficacy of natural products, in addition to the importance of regulations and quality standards for developing efficient and affordable phytochemicals as alternative to antibiotics.

Another review by Araújo et al. [2] introduced surface-active compounds—biomolecules produced by microorganisms (e.g., *Acinetobacter* sp., *Bacillus* sp., *Candida* sp., *Lactobacillus* sp., *Pseudomonas* sp., *Serratia* sp.)—able to interact with surfaces and hydrophobic or hydrophilic interfaces. Biosurfactants and emulsifiers are included in this group of compounds, and their actions as antimicrobials, modulators of virulence factors, anticancer, and wound healing agents were discussed. Biosurfactants can be classified as glycolipids, lipopeptides, lipoproteins or fatty acids, and phospholipid polymers. Their chemical compositions vary according with the producer microorganism—which can be found in different types of water and land, and even in extreme conditions of contaminants, temperatures, pH values, and salinity levels. The most abundant are glycolipids and lipopeptides. Bioemulsifiers comprehend composites of heteropolysaccharides, lipopolysaccharides, proteins, glycoproteins, or lipoproteins; and they can be isolated from contaminated soil, mangroves, seawater, freshwater, and human skin. These surface-active compounds have applications in different industrial sectors, such as pharmaceutical, textile, agriculture, cosmetics, personal care, and food industries, as well as for environmental purposes in soil remediation, hydrocarbon degradation, and oil recovery. Their antimicrobial and antibiofilm actions are due to the ability of breaking the outer and inner membranes of pathogens, reducing replication, inhibition of cell adhesion, and blocking of the invasion of host cells by harmful microorganisms. Surface-active compounds are also able to counteract inflammation by reducing neutrophil migration and inflammatory mediator release, as well as by modulating lymphocyte populations. Additional anti-cancer properties include apoptosis and impairment of the replication of cancer cells, whilst pro-healing effects may occur via numerous pathways that regulate the synthesis of collagenases, metalloproteinases, growth factors, and cytokines; this results in promoting fibroblast and epithelial cell proliferation, re-epithelialization, collagen deposition, and thus, faster healing. The authors emphasized the suitability of biosurfactants and bioemulsifiers as candidates for future biotechnological, biomedical, and pharmaceutical applications.

A systematic review by De La Hoz-Romo and collaborators [3] explored the potential of marine actinobacteria-derived compounds for the treatment of acne vulgaris, a multifactorial disease, which has been associated with microbial dysbiosis and inflammation of the pilosebaceous unit. *Cutibacterium acnes* is the most abundant microorganism of the pilosebaceous unit, followed by *Staphylococcus epidermidis*; however, these may become opportunistic pathogens involved in skin dysbiosis. In this context, antimicrobials are important tools for disease management. Marine actinobacteria represent an interesting source of diverse compounds, with antibacterial, antibiofilm, anticoagulant, antiviral, and antibacterial effects. In their manuscript, the authors performed a systematic analysis of metabolites and extracts produced by marine actinobacteria with antimicrobial, anti-biofilm, and quorum-sensing inhibition activities (quorum quenching, QQ), as therapeutic alternatives treatment of acne vulgaris, some skin diseases, and infectious diseases. They also classified them in clusters and associated these with their corresponding biosynthetic genes, considering their structure–activity relationship. They found that most of the studies on the anti-infective activity of marine actinobacteria-derived compounds were from China, India, and Egypt. Their sources were marine sediments, sponges, and other marine invertebrates, such as sea squirts, corals, echinoderm-derived organisms, mollusks, and jellyfish, as well as marine algae, water, mangroves, seagrasses, and fishes. The genuses most reported were *Streptomyces*, *Nocardiopsis*, *Micromonospora*, *Salinospora*, and *Verruscosispora*. These bacterial crude extracts and compounds were effective against *Staphylococcus* sp., inhibiting its growth and biofilm formation. Amongst the compounds identified in the bacterial crude extracts (most of them from *Streptomyces* sp.) were flavonoids, citreamicins, anthracyclines, chromomycins, napyradiomycins, marinomycins, and kokumarin. Then, the strategies to maximize the anti-infective activity and yield of metabolites were discussed by the authors; they found that starch and yeast extract peptone are the best sources of carbon and nitrogen, respectively, to maximize the production of anti-infective compound from cultured marine actinobacteria. Finally, the compounds were stratified into compounds with antibacterial activity and compounds with QQ activity; different biosynthetic gene clusters were identified and associated with the compounds. The results emphasize the antimicrobial potential of these marine bacterial compounds and the need for novel studies in the field.

Four of the original research studies investigated novel antifungals derived from plants. Mendonça et al. [4] demonstrated, by using in silico and in vitro analysis, that the ethyl acetate fraction obtained from the leaves of *P. granatum* and its isolated compound, galloyl-hexahydroxydiphenoyl-glucose (a hydrolysable tannin), are effective against *Candida albicans* and *C. glabrata*, inhibiting their growth and biofilm formation. Both the fraction and the compound were able to potentiate the effects of fluconazole—a fungistatic usually used to treat yeast infections. *P. granatum* fraction, as well as galloyl-hexahydroxydiphenoyl-glucose, reduced phospholipase production by *Candida* sp., an important virulence factor of fungi. Finally, the study highlights the importance of *P. granatum* as a source of antimicrobials and the need for further studies with the tannin identified in the study. In another report by Motta et al. [5], *C. albicans* and *C. glabrata* were incubated with the hydroalcoholic extract from the leaves of *Vismia guianensis*, a native Brazilian plant. The extract showed antifungal and fungicidal activities against *Candida* sp., inhibiting fungi growth and adhesion, as well as disrupting both early and mature biofilms. Fourteen compounds were identified in the extract, with a predominance of anthraquinones, flavonoids, and vismione D. In silico analysis predicted a high probability for vismione D to act as an antifungal and an anti-inflammatory. On the other hand, kaempferol and quercetin exhibited the highest predictive value as anti-inflammatories, and quercetin and catechin had the highest values as antioxidant compounds. By molecular docking, the authors determined possible interactions between *V. guianensis* compounds and *C. albicans* CaCYP51—a cytochrome P450 enzyme required for the biosynthesis of sterols. Vismione D presented the highest predicted affinity in relation to CaCYP51. The interesting results indicate the need for further investigations with *V. guianensis* and its compounds, especially in regards of *Candida*-induced diseases. Silva and colleagues [6] assessed the antifungal potential of sulforaphane, an isothiocyanate found in cruciferous plants, with previous suggested antibacterial actions. The authors conducted both in silico analysis and in vitro assays; as expected, in silico analysis confirmed its antimicrobial properties, which include antiparasitic, antifungal, and antibacterial actions, as well as mechanisms ranging from alterations of membrane composition to inhibition of fungi RNA. The compound was estimated to present moderate risk for mutagenic, tumorigenic, and reproductive tract deleterious effects. The analysis also indicated that sulforaphane has a greater potential to permeate the skin than fluconazole, suggesting its potential use to treat superficial mycosis. In vitro assays with *Candida* sp. ATCC strains and clinical isolates from the oral cavity and vaginal smears demonstrated the ability of the compound to inhibit fungi growth, as well as hyphae and biofilm formation. Additional tests revealed that sulforaphane can potentiate fluconazole antifungal actions when assessed at sub-inhibitory concentrations. Similar results on hyphae and biofilm formation by *C. albicans* were seen when the compounds were combined. These effects were related to their ability to interfere with the expression of hyphae growth-related and biofilm formation-related genes by *C. albicans*, although this may vary according with strains and isolates. The authors conclude that the association of sulforaphane and fluconazole may be beneficial to treat *Candida* spp. infections and may trigger fewer adverse reactions caused by these compounds. They also discuss the importance of additional research to address the effects of their combination on mixed populations of microorganisms, including those found in the oral cavity and the vagina. In another study, Lecsombun et al. [7] investigated the antifungal activity of the essential oil and ethanolic extract of *Coleus amboinicus* against *Microsporum canis*—a zoophilic dermatophyte found in domestic cats and dogs, which can also cause disease in humans. The activity of the essential oil and extract of *C. amboinicus* was assessed in vitro against clinical isolates obtained from feline samples, classified as weak, moderate, and strong biofilm-producers. In order to evaluate their effects on biofilm formation, minimum inhibitory concentrations (MICs) were determined and, then, the microorganisms were incubated with MIC and 2 × MIC for 96 h. Both the essential oil and the extract inhibited biofilm formation by *M. canis*. Similar inhibitory effects on growth and biofilm formation were observed when these preparations were assessed against *C. albicans* and *Trichophyton rubrum*. Chemical analysis revealed the presence of 18 compounds in the essential oil, including carvacrol (major compound), p-cymene, and γ-terpinene, which are all monoterpenes. The flavonoids caffeic acid, rosmarinic acid, and apigenin were detected in *C. amboinicus* extract. These results highlight the antifungal potential of *C. amboinicus* formulations for treating zoonotic infections caused by *M. canis*.

The antimicrobial and anti-parasitic actions of extracts and compounds obtained from fungi and bacteria were analysed in three of the of the original research studies. The first, performed by Mazzone and collaborators [8], explored the in vitro activity of natural products from *Paraboeremia selaginellae*—an endophytic fungus against *Toxoplasma gondii*. They isolated eight natural products from the crude extract of *P. selaginellae* found in the leaves of *Philodendron monstera*—an ornamental plant native to the Americas; six of them inhibited *T. gondii* proliferation with low or no effects on other microrganisms. Possible toxic actions of these compounds were assessed in human Hs27 fibroblast, THP-1, Huh-7, and Hek 293 cell lines, with three of them (biphenyl ether derivatives) showing no toxicity when tested at ≤100 μM. The results indicate the promising potential of endophytic fungi as sources for novel anti-toxoplasma compounds. In another investigation, Diyaolu et al. [9] isolated and identified a strain of *Aspergillus fumigatus* (UIAU-3F) from soil samples collected from the River Oyun in Kwara State in Nigeria. *A. fumigatus* was cultured in different conditions, resulting in three fungus extracts prepared and subjected to comparative metabolomics. The extract obtained from the fungus cultured in rice medium produced the greatest diversity of metabolites. Following its fractioning into two compounds, fumitremorgin C and pseurotin D were isolated. Molecular docking was performed for the compounds against two enzymes, cruzain and l-lactate dehydrogenase, from *Trypanosoma cruzi* and *Plasmodium falciparum*, respectively. The compounds were found to interact with both enzymes; their docking was comparable to the docking observed for benznidazole (an anti-trypanosoma) and chloroquine (an anti-plasmodium). When tested in vitro, fumitremorgin C presented lower efficacy than benznidazole and chloroquine, but it displayed significantly higher activities against *T. cruzi* and *P. falciparum* in comparison to pseurotin D. The lack of effects of pseurotin D was suggested to be due to its low cell permeability, interstitial hypertension, and/or metabolic degradation. This was the first study of the anti-parasitic effects of fumitremorgin C and pseurotin D, and it highlights the need for further research, especially with fumitremorgin C, to determine mechanisms of action and potential use in vivo. In the third study, Mondal and Thomas [10] isolated and characterized novel Actinomycetes found in marine sediments, and they assessed its antimicrobial activity against fish pathogens. Sixteen (16) different Actinomycete colonies were obtained from marine sediment samples from the coast of Digha in India. The isolates were then tested against two fish pathogens—*Aeromonas hydrophila* and *Vibrio parahemolyticus*. Two of them presented antibacterial activity, with *Beijerinckia fluminensis* VIT01 being the most potent. Of note, this study reported, for the first time, the isolation of *B. fluminensis* from marine sediments. This bacterium was found to be able to hydrolyze starch, gelatin, and casein, grow in salt concentrations as high as 7%, and survive in temperatures from 28–40 °C. Fourier-transform infrared spectroscopy of the crude bacterial extract revealed 14 vibrational bands, corresponding to eight groups of compounds, including thiol, carboxylic acid, isothiocyanate, tertiary alcohol, amine, alkene, 1,2-disubstituted, and fluoro and halo compounds. Gas chromatography-mass spectrometry (GC/MS) detected 18 different major compounds; of those, N, N-Dimethylheptanamide, Glycine,N-Octyl-, Ethyl Ester, Glycyl-L-Proline, Actinomycin C2, (S)-3,4-Dimethylpentanol, and 7-Tetradecene were suggested to be antibacterial metabolites. The data indicate these compounds could be used as alternative treatments for infections in aquaculture.

Eight of the original research studies investigated the antibacterial effects of plant extracts, fractions, and compounds. The study by Figueiredo et al. [11] looked at the protective effects of cinnamaldehyde—a compound known for its antimicrobial and immunomodulatory actions, in mice systemically injected with a pathogenic *Escherichia coli* (strain F5, which was isolated from the blood stream). The compound, given by gavage, at 60 mg/kg 2 h after infection, reduced mortality by 40% over five days of observation. Cinnamaldehyde was effective against *E. coli* in vitro, but not in vivo; however, septic mice treated with the compound exhibited less tissue haemorrhage, inflammation, and damage in different organs, especially in the lungs in comparison with vehicle-treated mice. Cinnamaldehyde partially reverted leukocyte counts in the peritoneal and bone marrow of septic mice. A stimulatory effect was also observed in *E. coli*-infected mice treated with the compound; however, the same animals presented with reduced circulating and peritoneal levels of inflammatory mediators (cytokines and chemokines). The study reinforces previous data on the ability of cinnamaldehyde to act as a modulator of the immune/inflammatory response in sepsis. Moraes-Neto et al. [12] investigated the antimycobacterial and anti-inflammatory activities of the ethyl acetate fraction of *Bixa orellana* leaves and one of its active compound—ellagic acid. In silico analysis indicated anti-inflammatory, antioxidant, antibacterial, antimycobacterial, and hepatoprotective activities for ellagic acid. Additionally, the compound was predicted to be better absorbed by the intestine than clarithromycin, and to not be mutagenic, tumorigenic, irritant, or harmful to the reproductive system. Molecular docking revealed that ellagic acid can bind to COX_2 and to mycobacterial dihydrofolate reductase, indicating its potential mechanisms of anti-inflammatory and antimicrobial activities. Both ellagic acid and *B. orellana* ethyl acetate fraction were effective against *Mycobacterium abscessus* in vitro and protected infected *Tenebrio molitor* larvae, reducing their mortality when tested at MIC, $2 \times$ MIC, and $4 \times$ MIC. The anti-inflammatory effects of *B. orellana* fraction and ellagic acid were confirmed in a mouse model of paw inflammation induced by carragenan, in which they both reduced oedema formation. The results support the anti-inflammatory and antimicrobial potential of *B. orellana* and highlight the need for further investigations on the mechanisms of action of the plant fraction and its active compound and for broader studies with a higher number of mycobacteria strains.

Protocatechuic acid (PCA) is a plant-derived phenolic acid with known antimicrobial activity and ability to enhance antibiotic action; although a promising molecule, its low bioavailability and fast metabolism/excretion has limited its clinical usage [13]. Miklasińska-Majdanik et al. [14] used an esterified PCA—protocatechuic acid ethyl ester to gain knowledge on its antimicrobial actions against staphylococcal strains, alone or in combination with erythromycin. Twelve strains were tested in microdilution and checkerboard assays; protocatechuic acid ethyl ester reduced erythromycin MIC values by $\geq$50% when incubated with nine of them. Although these are preliminary results, they show a clear and potent anti-staphylococcal effect for protocatechuic acid ethyl ester in vitro. In another study, da Silva and colleagues [15] assessed the actions of solasodine—a natural steroidal alkaloid found in plants of the *Solanum* genus—against *S. aureus*, *P. aeruginosa*, and *E. coli*. They also analysed the in vitro combination effects of solasodine with gentamicin, ciprofloxacin, norfloxacin, or ampicillin, as well as the inhibitory actions of the compound on NorA and MepA multidrug efflux pumps. From the eight bacterial strains tested, solasodine had no effects on four of them. The compound was able to potentiate both gentamicin and norfloxacin activities when tested against multidrug resistant strains of *S. aureus*, *P. aeruginosa*, and *E. coli*. Ampicillin, on the other hand, was only potentiated by solasodine when incubated with *S. aureus*. The synergistic effects between solasodine and ciprofloxacin were found to be related to inhibition of the MePA efflux pump. An antagonistic effect between the compound and ampicillin or norfloxacin was also observed for some of the bacterial strains of *S. aureus* tested. The authors suggest the different effects noted may depend on the mechanism of action of the antibiotics, or even antibiotic chelation by solasodine. Nonetheless, future studies on the combination effects of plant alkaloids with antibiotics represent an interesting tool for treating bacterial infections whilst reducing resistance.

Another original report by Gazwi and collaborators showed the antimicrobial (antifungal and antibacterial), antioxidant, and cytotoxic properties of the hydroalcoholic extract from the red flowers of *Malvaviscus arboreus* Cav.—a tropical and subtropical perennial deciduous shrub native to Central and South America [16]. Phytochemical analysis of the extract by GC-MS detected 21 compounds; 11-octadecenoic acid methyl ester, 9,12-octadecadienoic acid (Z, Z)-2-hydroxy-1-(hydroxymethyl) ethyl ester, 9,12-Octadecadienoic acid (Z, Z)-methyl ester, exadecenoic acid methyl ester, hexadecanoic acid, and oleic acid were the predominant compounds. Additionally, 13 phenolic compounds were identified by high-performance liquid chromatography: naringin, hesperidin, kaempferol, luteolin, apigenin, catechin, caffeic acid, cinnamic acid, gallic acid, syringic acid, benzoic acid, and ellagic acid. Proton nuclear magnetic resonance revealed the presence of oxygenated saturated and unsaturated hydrocarbon compounds, such as saturated and unsaturated fatty acids in the extract. Different compounds detected were previously reported as antimicrobials and antioxidants. In accordance, *M. arboreus* acted as an efficient free radical scavenger and presented a strong antibacterial activity against *Vibrio damsela* and moderate activity against *V. fluvialis* and *S. typhimurium*. An antifungal property was also noted against *Aspergillus terreus*, *A. fumigatus*, and *A. flavus*. The extract exhibited a long-lasting action on *V. damsel* and was able to potentiate amoxicillin/clavulanic acid antibacterial effects on this microorganism. *M. arboreus* triggered apoptosis (via caspase 3/7 activation), autophagy, and cell cycle arrest in hepatocellular carcinoma HepG2 cells. Overall, the data indicate a promising usage of compounds from the hydroalcoholic extract from the red flowers of *M. arboreus* as anticancer, antioxidant, and antimicrobial therapies. In the study by Almutairi [17], the antimicrobial effects of frankincense (*Boswellia sacra*) oil, its interaction with imipenem and gentamicin against methicillin-resistant *S. aureus*, and multidrug-resistant *P. aeruginosa* were determined in vitro and in vivo. GC/MS analysis revealed the presence of 40 constituents. Poor antibacterial effects and no interactions with the tested antibiotics were observed when the oil was incubated with *S. aureus* or *P. aeruginosa*. No interactions between the antibiotics and the frankincense oil were observed in vivo, although the oil alone greatly attenuated bacterial load and tissue damage in the lungs of rats with *S. aureus*-induced pneumonia. The lack of potential benefits from the combination between frankincense oil and the tested antibiotics is discouraging; however, the in vivo results allow us to suggest that the oil, per se, may act as a modulator of inflammatory responses in infectious diseases. Another interesting study by Pérez-Delgado [18] explored the in vitro antibacterial activity of *Apis mellifera*, venom collected from bees of Lambayeque, northern Peru, against *E. coli*, *Pseudomonas aeruginosa*, and *Staphylococcus aureus* reference strains. Three peptides of low molecular weights (5 kDa, 6 kDa, and 7 kDa) were separated from the crude bee venom. A strong and moderate antimicrobial effect was noted against *E. coli* and *S. aureus*, respectively, whilst no effects were observed for the peptide fractions when incubated with *P. aeruginosa*. At the MIC found for *E. coli*, the fraction did not promote haemolysis, although this was noted for higher concentrations. Additionally, no antioxidant actions were detected for *A. mellifera* fraction; this was attributed to the lack of melittin—a polypeptide with anti-inflammatory and antioxidant actions [19]—in its composition. The study indicates the importance of other peptides as antimicrobials.

The beneficial role of another plant-derived preparation was investigated against *Bipolaris sorokiniana*—a fungi known to infect wheat crops. Its effects were also tested against bacteria resistant to antibiotics (*E. coli*, *Klebsiella pneumonia* and *S. aureus*). The study by Alqahtani et al. [20] demonstrated the effects of silver nanoparticles, containing

Mangifera indica leaf extract. An important antibacterial effect was noted for the nanoparticle in vitro. When applied to diseased wheat, the nanoparticles markedly increased reduced sugar levels and total phenol content. The same plants also exhibited higher activation of the enzymes superoxide dismutase, catalase, peroxidase, and glutathione reductase, indicating protection against oxidative stress damage. The nanoparticles also enhanced the activity of phenylalanine ammonia lyase, an enzyme involved in the synthesis of secondary metabolites. Reduction in disease severity was only observed for the highest concentrations of nanoparticles ($\geq$30 ppm). Overall, the application of silver nanoparticles on wheat crops causes may prevent *B. sorokiniana* infections.

Finally, the brief report by Heuser and colleagues [21] looked at the bactericidal activity of sodium bituminosulfonate (a substance derived from sulfur-rich oil shale) against methicillin-susceptible and methicillin-resistant clinical isolates of *S. aureus*. All 40 isolates tested were susceptible to sodium bituminosulfonate, which demonstrated a rapid bactericidal action in vitro. The results suggest this compound is a promising alternative to the available topical antibiotics.

Taken together, the manuscripts discussed herein unveiled the potential of different natural preparations and compounds from plants and microorganisms to treating infections of importance to humans, animals, and wheat. Some of them also predicted side-effects and cellular targets for the natural compounds. All these studies presented interesting highlights regarding the need for further investigations in order to uncover more in-depth information on their additional properties, safety, pharmacokinetic, and pharmacodynamic profiles.

Author Contributions: The editorial board members E.S.F. and V.M.-N. contributed equally to editing this collection. They also wrote the original draft, reviewed and edited this editorial, and were responsible for funding acquisition, I.F.d.S.F. and C.R.A.V.M. equally contributed to the interpretation and summarization of the Special Issue information. All authors have read and agreed to the published version of the manuscript.

Funding: This work was supported by the Conselho Nacional de Desenvolvimento Científico e Tecnológico (CNPq; grant numbers 305676/2019-9 and 408053/2018-6, to E.S.F.; and CNPq grant number 315072/2020-2 to V.M.-N.), Instituto de Pesquisa Pelé Pequeno Príncipe, INCT-INOVAMED, FAPEMA/CAPES—ACT-05691/21, and ACT 01784-21.

Acknowledgments: The Special Issue editorial committee would like to thank all the contributors to this collection, authors and reviewers, for their availability and dedicated time to the editorial process.

Conflicts of Interest: The authors declare no conflict of interest.

References

1. Kopel, J.; McDonald, J.; Hamood, A. An Assessment of the In Vitro Models and Clinical Trials Related to the Antimicrobial Activities of Phytochemicals. *Antibiotics* **2022**, *11*, 1838. [CrossRef]
2. Araujo, J.; Monteiro, J.; Silva, D.; Alencar, A.; Silva, K.; Coelho, L.; Pacheco, W.; Silva, D.; Silva, M.; Silva, L.; et al. Surface-Active Compounds Produced by Microorganisms: Promising Molecules for the Development of Antimicrobial, Anti-Inflammatory, and Healing Agents. *Antibiotics* **2022**, *11*, 1106. [CrossRef]
3. De La Hoz-Romo, M.C.; Diaz, L.; Villamil, L. Marine Actinobacteria a New Source of Antibacterial Metabolites to Treat Acne Vulgaris Disease-A Systematic Literature Review. *Antibiotics* **2022**, *11*, 965. [CrossRef]
4. Mendonca, A.M.S.; Monteiro, C.A.; Moraes-Neto, R.N.; Monteiro, A.S.; Mondego-Oliveira, R.; Nascimento, C.E.C.; da Silva, L.C.N.; Lima-Neto, L.G.; Carvalho, R.C.; de Sousa, E.M. Ethyl Acetate Fraction of Punica granatum and Its Galloyl-HHDP-Glucose Compound, Alone or in Combination with Fluconazole, Have Antifungal and Antivirulence Properties against *Candida* spp. *Antibiotics* **2022**, *11*, 265. [CrossRef] [PubMed]
5. Motta, E.P.; Farias, J.R.; Costa, A.; Silva, A.F.D.; Oliveira Lopes, A.J.; Cartagenes, M.; Nicolete, R.; Abreu, A.G.; Fernandes, E.S.; Nascimento, F.R.F.; et al. The Anti-Virulence Effect of Vismia guianensis against Candida albicans and Candida glabrata. *Antibiotics* **2022**, *11*, 1834. [CrossRef]
6. Silva, B.L.R.; Simao, G.; Campos, C.D.L.; Monteiro, C.; Bueno, L.R.; Ortis, G.B.; Mendes, S.J.F.; Moreira, I.V.; Maria-Ferreira, D.; Sousa, E.M.; et al. In Silico and In Vitro Analysis of Sulforaphane Anti-Candida Activity. *Antibiotics* **2022**, *11*, 1842. [CrossRef]

7. Leesombun, A.; Thanapakdeechaikul, K.; Suwannawiang, J.; Mukto, P.; Sungpradit, S.; Bangphoomi, N.; Changbunjong, T.; Thongjuy, O.; Weluwanarak, T.; Boonmasawai, S. Effects of Coleus amboinicus L. Essential Oil and Ethanolic Extracts on Planktonic Cells and Biofilm Formation of Microsporum canis Isolated from Feline Dermatophytosis. *Antibiotics* **2022**, *11*, 1734. [CrossRef] [PubMed]
8. Mazzone, F.; Simons, V.E.; van Geelen, L.; Frank, M.; Mandi, A.; Kurtan, T.; Pfeffer, K.; Kalscheuer, R. In Vitro Biological Activity of Natural Products from the Endophytic Fungus Paraboeremia selaginellae against Toxoplasma gondii. *Antibiotics* **2022**, *11*, 1176. [CrossRef] [PubMed]
9. Diyaolu, O.A.; Preet, G.; Fagbemi, A.A.; Annang, F.; Perez-Moreno, G.; Bosch-Navarrete, C.; Adebisi, O.O.; Oluwabusola, E.T.; Milne, B.F.; Jaspars, M.; et al. Antiparasitic Activities of Compounds Isolated from Aspergillus fumigatus Strain Discovered in Northcentral Nigeria. *Antibiotics* **2023**, *12*, 109. [CrossRef] [PubMed]
10. Mondal, H.; Thomas, J. Isolation and Characterization of a Novel Actinomycete Isolated from Marine Sediments and Its Antibacterial Activity against Fish Pathogens. *Antibiotics* **2022**, *11*, 1546. [CrossRef] [PubMed]
11. Figueiredo, I.F.S.; Araujo, L.G.; Assuncao, R.G.; Dutra, I.L.; Nascimento, J.R.; Rego, F.S.; Rolim, C.S.; Alves, L.S.R.; Frazao, M.A.; Cadete, S.F.; et al. Cinnamaldehyde Increases the Survival of Mice Submitted to Sepsis Induced by Extraintestinal Pathogenic Escherichia coli. *Antibiotics* **2022**, *11*, 364. [CrossRef] [PubMed]
12. Moraes-Neto, R.N.; Coutinho, G.G.; Ataide, A.C.S.; de Oliveira Rezende, A.; Nascimento, C.E.C.; de Albuquerque, R.P.; da Rocha, C.Q.; Rego, A.S.; de Sousa Cartagenes, M.D.S.; Abreu-Silva, A.L.; et al. Ethyl Acetate Fraction of Bixa orellana and Its Component Ellagic Acid Exert Antibacterial and Anti-Inflammatory Properties against Mycobacterium abscessus subsp. massiliense. *Antibiotics* **2022**, *11*, 817. [CrossRef]
13. Kakkar, S.; Bais, S. A review on protocatechuic Acid and its pharmacological potential. *ISRN Pharmacol.* **2014**, *2014*, 952943. [CrossRef] [PubMed]
14. Miklasinska-Majdanik, M.; Kepa, M.; Kulczak, M.; Ochwat, M.; Wasik, T.J. The Array of Antibacterial Action of Protocatechuic Acid Ethyl Ester and Erythromycin on Staphylococcal Strains. *Antibiotics* **2022**, *11*, 848. [CrossRef] [PubMed]
15. da Silva, A.R.P.; Costa, M.D.S.; Araujo, N.J.S.; de Freitas, T.S.; de Almeida, R.S.; Barbosa Filho, J.M.; Tavares, J.F.; de Souza, E.O.; de Farias, P.A.M.; Pinheiro, J.C.A.; et al. Potentiation of Antibiotic Action and Efflux Pump Inhibitory Effect on *Staphylococcus aureus* Strains by Solasodine. *Antibiotics* **2022**, *11*, 1309. [CrossRef] [PubMed]
16. Gazwi, H.S.S.; Shoeib, N.A.; Mahmoud, M.E.; Soltan, O.I.A.; Hamed, M.M.; Ragab, A.E. Phytochemical Profile of the Ethanol Extract of Malvaviscus arboreus Red Flower and Investigation of the Antioxidant, Antimicrobial, and Cytotoxic Activities. *Antibiotics* **2022**, *11*, 1652. [CrossRef]
17. Almutairi, M.B.F.; Alrouji, M.; Almuhanna, Y.; Asad, M.; Joseph, B. In-Vitro and In-Vivo Antibacterial Effects of Frankincense Oil and Its Interaction with Some Antibiotics against Multidrug-Resistant Pathogens. *Antibiotics* **2022**, *11*, 1591. [CrossRef]
18. Perez-Delgado, O.; Espinoza-Culupu, A.O.; Lopez-Lopez, E. Antimicrobial Activity of Apis mellifera Bee Venom Collected in Northern Peru. *Antibiotics* **2023**, *12*, 779. [CrossRef]
19. Khalil, A.; Elesawy, B.H.; Ali, T.M.; Ahmed, O.M. Bee Venom: From Venom to Drug. *Molecules* **2021**, *26*, 4941. [CrossRef]
20. Alqahtani, Y.S.; Bahafi, A.; Mirajkar, K.K.; Basavaraju, R.R.; Mitra, S.; Shailaja, S.; More, S.S.; Muddapur, U.M.; Khan, A.A.; Sudarshan, P.R.; et al. In Vitro Antibacterial Activity of Green Synthesized Silver Nanoparticles Using Mangifera indica Aqueous Leaf Extract against Multidrug-Resistant Pathogens. *Antibiotics* **2022**, *11*, 1503. [CrossRef]
21. Heuser, E.; Becker, K.; Idelevich, E.A. Bactericidal Activity of Sodium Bituminosulfonate against *Staphylococcus aureus*. *Antibiotics* **2022**, *11*, 896. [CrossRef] [PubMed]

antibiotics

MDPI

Article

Antimicrobial Activity of *Apis mellifera* Bee Venom Collected in Northern Peru

Orlando Pérez-Delgado [1,*], Abraham Omar Espinoza-Culupú [1] and Elmer López-López [2]

1 Health Science Research Laboratory, Universidad Señor de Sipán, Chiclayo 14001, Peru
2 Faculty of Health Sciences, Universidad Señor de Sipán, Chiclayo 14001, Peru
* Correspondence: perezdelgado@crece.uss.edu.pe

Abstract: Due to the emergence of microorganisms resistant to antibiotics and the failure of antibiotic therapies, there is an urgent need to search for new therapeutic options, as well as new molecules with antimicrobial potential. The objective of the present study was to evaluate the in vitro antibacterial activity of *Apis mellifera* venom collected in the beekeeping areas of the city of Lambayeque in northern Peru against *Escherichia coli*, *Pseudomonas aeruginosa,* and *Staphylococcus aureus*. Bee venom extraction was performed by electrical impulses and separated using the Amicon ultra centrifugal filter. Subsequently, the fractions were quantified by spectrometric 280 nm and evaluated under denaturant conditions in SDS-PAGE. The fractions were pitted against *Escherichia coli* ATCC 25922, *Staphylococcus aureus* ATCC 29213, and *Pseudomonas aeruginosa* ATCC 27853. A purified fraction (PF) of the venom of *A. mellifera* and three low molecular weight bands of 7 KDa, 6 KDa, and 5 KDa were identified that showed activity against *E. coli* with a MIC of 6.88 μg/mL, while for *P. aeruginosa* and *S. aureus,* it did not present a MIC. No hemolytic activity at a concentration lower than 15.6 μg/mL and no antioxidant activity. The venom of A. mellifera contains a potential presence of peptides and a predilection of antibacterial activity against *E. coli.*

Keywords: venom; antimicrobial activity; fraction; bee; protein

Citation: Pérez-Delgado, O.; Espinoza-Culupú, A.O.; López-López, E. Antimicrobial Activity of *Apis mellifera* Bee Venom Collected in Northern Peru. *Antibiotics* **2023**, *12*, 779. https://doi.org/10.3390/antibiotics12040779

Academic Editors: Valério Monteiro-Neto and Elizabeth S. Fernandes

Received: 8 February 2023
Revised: 10 April 2023
Accepted: 16 April 2023
Published: 19 April 2023

1. Introduction

World Health Organization (WHO) published the first global surveillance report on antibiotic resistance, showing that five out of six regions had more than 50% resistance to third-generation cephalosporins and fluoroquinolones in *Escherichia coli* and methicillin resistance in *Staphylococcus aureus* in hospital settings [1].

S. aureus is a human commensal that can cause systemic infections in the host; this requires evading the immune response and the ability to proliferate in different niches in the host; currently, the infection by staphylococci in the face of immune mediators and the disease is not well known [2]. However, the main agent of bacteremia and infective endocarditis (IE), as well as osteoarticular, skin, and soft tissue infections, pleuropulmonary infections [3], and even the appearance of methicillin-resistant *S. aureus* (MRSA), which is a therapeutic problem in patients [4]. Staphylococcal infection has also been reported from hosts or carriers of asymptomatic nasopharyngeal bacteria, even with certain risk factors such as passive smoking and a large family [5]. The results of certain studies have determined that *S. aureus* has generated resistance against ampicillin, penicillin, rifampicin, clindamycin, oxacillin, and erythromycin [6]. Variable susceptibility to levofloxacin, ciprofloxacin, gentamicin, tetracycline, and sulfamethoxazole-trimethoprim has also been shown [7], and patterns have also been shown regarding the mecA, rpoB, blaZ, ermB, tetM, and nuc genes [6,8]. This resistance has been acquired through different mechanisms, the most frequent being reduced membrane permeability, excessive production of β-lactamase, and acquiring resistance genes or gene mutations [9].

Likewise, *Pseudomonas aeruginosa* can cause nosocomial outbreaks related to its resistance and virulence properties [10], being a producer of β-lactamases and multiresistant to a wide range of antimicrobials such as penicillin, cephalosporin, cephamycin, and carbapenem [11]. In addition, 35 resistomes (antimicrobial resistance genes) have been identified that confer resistance to 10 different antibiotics (including four classes of beta-lactams) and 214 virulence factor genes [12], in addition to the susceptibility of *P. aeruginosa* to carbapenems, piperacillin-tazobactam, and amikacin has undergone alterations before and during COVID-19 [13]. There are phenotypic studies that *P. aeruginosa*, as producers of extended-spectrum beta-lactamase (ESBL) and metallo-beta-lactamases (MBL), also present genes associated with biofilm formation and virulence, such as toxA and lasB [14]. Even using polymerase chain reaction (PCR) techniques and molecular markers such as Random Amplified Polymorphic DNA (RAPD), they have identified strains resistant to imipenem, Ticarcillin + Clavulanate, Piperacillin, and Ticarcillin + Clavulanate, these strains being isolated from swimming pools [15].

E. coli is responsible for a large number of virulent variants associated with human diseases, such as urinary tract infection (UTI) with a resistance rate of >55% to first to fourth-generation cephalosporins [16], neonatal and traveler's diarrhea [17], and multiresistant isolates (MDRs) the most prevalent genes being CTX-M-1, followed by NDM-1 for Betalactamases and the genes ermB and aac(6′)-Ib for resistance to macrolides and aminoglycosides [18]. *E. coli* is frequently discharged into the environment through feces, including wastewater, and is considered an indicator of fecal contamination. Many strains can carry resistance genes [19]. According to isolates, they have usually been reported to be sensitive to netilmicin, gentamicin, chloramphenicol, pipemidic acid, nalidixic acid, ciprofloxacin, amoxicillin/clavulanic acid, and nitrofurantoin, as well as increased susceptibility to cefotaxime, ceftriaxone, and aztreonam [20]. Uncomplicated UTI isolates have been found to have a higher susceptibility than complicated UTI isolates to amoxicillin, amoxicillin/clavulanic acid, and ciprofloxacin [21].

Antimicrobial resistance has been reported in gram-negative and gram-positive bacteria, with reports of up to 96.2% for *Pseudomonas* spp. and 66.7% for *E. coli* [22], with isolates reported in heart disease intensive care units [23] and in bloodstream infections [24] antimicrobial-resistant bacteria are also considered the most frequent uropathogens [25]. There are reports of MRSA in up to 50% of patients [26], with resistance profiles for cefoxitin, chloramphenicol, clindamycin, and gentamicin [27].

The failure of antibiotics, including the latest generation, to counteract superbugs highlights the need to search for new molecules with antimicrobial potential to control the global problem of antimicrobial resistance. One group of these new molecules are peptides that are antimicrobial (AMP) and are promising molecules for combating antimicrobial resistance (AMR) [28]. AMP has been found the most in the venoms of different organisms, such as scorpions [29–31], snakes [32], spiders [33], and bees [34], among other venoms.

The composition of bee venom is very variable, having Peptides: Melitin (the main component of the venom), Apamin, Mast Cell-Degranulating Peptide (MCD), Tertiapin, Secapin, and Its Isoforms, Adolapin, Procamine, and Minimine; Polypeptides: Api m 6, Cardiopep, Icarapin, and Major Royal Jelly Proteins; Enzymes: Phospholipase A2 (PLA2), Hyaluronidase, Acid Phosphatase, and Dipeptidylpeptidase IV; Serine Proteases [35].

Bees are insects found on all continents, many of these species have yet to be described and are an exciting source for the study and search for new molecules with antimicrobial properties. There are experimental and clinical reports on *Apis mellifera* venom and its anti-inflammatory, antimicrobial, and anticancer effects; the components present in the venom, such as proteins, vary from a summer season compared to a winter season [36–38], in addition, have shown different therapeutic properties against oxidative stress induced by beta-amyloid [39–41]. For Parkinson's disease, the neuroprotective potential of bee venom against oxidative stress induced by rotenone (pesticide) has been demonstrated in a mouse model, including preventing the decrease in dopamine and also restoring locomotor activity in mice [42,43]. For Lyme disease, the melittin present in the venom showed in vitro antibacterial effects against the causative agent *Borrelia burgdorferi* [44] and even had significant antibacterial effects against *E. coli*, *S. aureus*, and *Salmonella typhyimurium* [45]. Melittin also exhibited antibacterial activity against MRSA strains [46], with antimicrobial potential against agents that cause dental caries, with antifungal capacity including suppression of biofilm formation [47,48]. Its significant antiviral potential has also been demonstrated in in vitro and in vivo assays on different enveloped (Influenza A) and non-enveloped (enterovirus-71) viruses [49]. In addition, phospholipase A2 (PLA2) can also block the replication of the virus, being shown to be responsible for the inhibition of HIV replication [50]. The present study aimed to evaluate the in vitro antibacterial activity of *Apis mellifera* venom collected in the beekeeping areas of the city of Lambayeque in northern Peru against *Escherichia coli*, *Pseudomonas aeruginosa*, and *Staphylococcus aureus*.

2. Results

As seen in Figure 1, 15% SDS-PAGE-Tricine of the purified fraction (PF) of crude venom from *A. mellifera* yielded 3 low molecular weight bands, i.e., 7 kDa, 6 kDa, and 5 kDa.

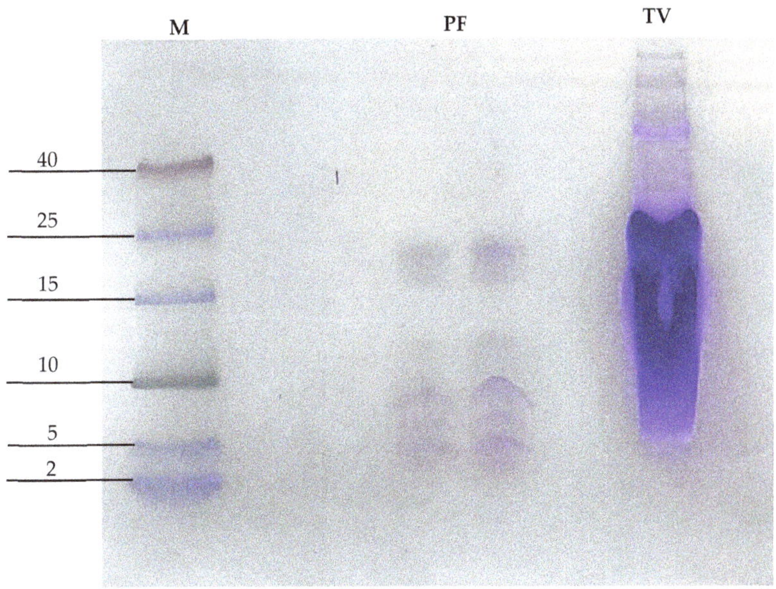

Figure 1. SDS polyacrylamide gel electrophoresis with Tricine (SDS-PAGE-Tricine) of the purified fraction; M denoted the marker lane (molecular weight 2–40 kDa) (molecular weight marker), PF denotes the protein fractions, and TV denotes total venom.

The PF of *A. mellifera* venom had a minimum inhibitory concentration (MIC) of 6.88 μg/mL ($p < 0.05$) for *E. coli*. For the *S. aureus* strain, at higher concentrations, the venom exhibited antibacterial activity. For *P. aeruginosa*, no antibacterial activity was observed (Figure 2).

Concentrations of the PF of *A. mellifera* venom above 125 μg/mL in erythrocyte suspension produced more than 50% hemolysis. At a concentration of 31.25 μg /mL, less than 10% hemolysis occurred, but at 7.8 μg /mL PF, no erythrocyte lysis was evidenced (Figure 3).

The PF of the *A. mellifera* venom showed no antioxidant activity was observed at the concentrations evaluated.

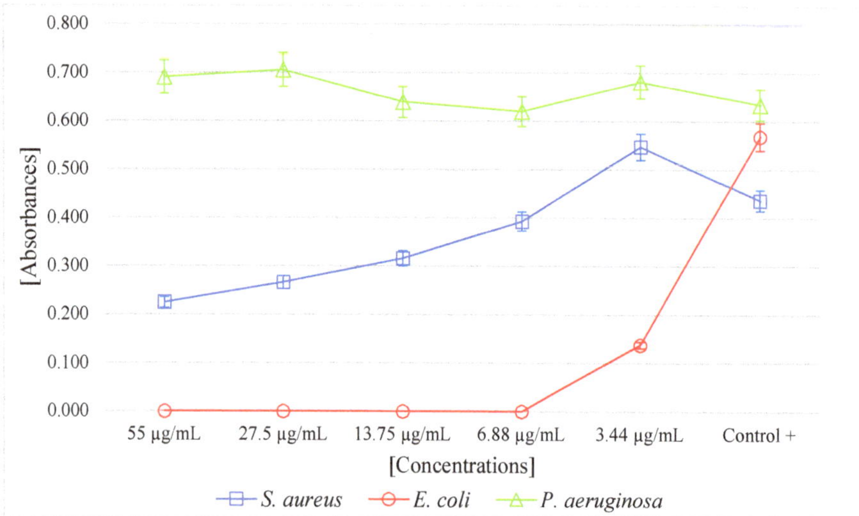

Figure 2. Microbial growth of *S. aureus*, *E. coli*, and *P. aeruginosa* incubated with the purified fraction of *Apis mellifera* venom.

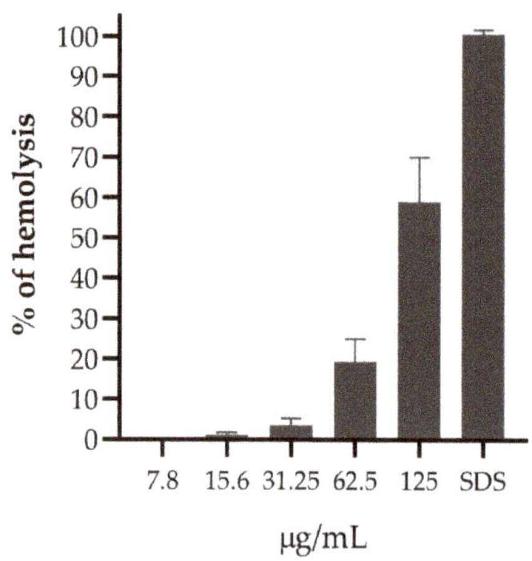

Figure 3. Hemolytic activity of the purified fraction of *Apis mellifera* venom.

3. Discussion

The purified fraction (PF) of the *A. mellifera* venom revealed the presence of peptides using the SDS-PAGE-Tricine technique and found three peptides of 7 kDa, 6 kDa, and 5 kDa. The chemical composition of the venom of *A. mellifera* is highly variable; such as melittin (3 kDa), apamin (2 kDa), and cecropin (4 kDa), enzymes, such as phospholipase A2 (19 kDa) and hyaluronidase (38 kDa), biologically active amines, such as histamine and epinephrine, as well as peptides not reported [39,51,52] and this suggests that many of these components may contribute their anti-inflammatory, antifungal, antiviral, healing and analgesic properties [36,53–55].

Our results were coherent with other studies; the PF of *A. mellifera* venom collected from Íllimo showed antibacterial activity against *E. coli*, but the same was not observed for *P. aeruginosa* despite being a gram-negative bacterium. For *S. aureus*, as the concentration of the venom fraction increased, bacterial growth was affected. Interestingly, the results of other studies of the antibacterial activity of the crude venom showed highly variable MICs against *E. coli* and for *S. aureus* [56], including also demonstrating the inhibitory effect through a viability assay at a temperature of 25 °C against *E. coli* and *P. putida*, causing membrane permeability and loss of ATP, showing no effect against *P. fluorescens* [57], as well as the crude bee venom extract in a region of Egypt significantly inhibited the growth of *E. coli* ATCC8739 and *S. aureus* ATCC 6538P [58], in addition, the action of crude bee venom from Iran demonstrated inhibition through the Kirby–Bauer method against *E. coli* and *S. aureus*, but not against *P. aeruginosa* [59].

Other studies have shown the existence of antimicrobial peptides in the venom; it follows that the interaction against the cell envelope of the bacteria is due to the attraction between the positively charged venom peptides and the phospholipids, causing a rupture or instability of the venom membrane, in addition to forming pores; however, this mechanism requires a certain concentration threshold [60]. Direct insertion of melittin leads to pore formation, whereas the parallel conformation is inactive and prevents other melittin molecules from being inserted, thus, preventing pore formation [61]. However, melittin has a molecular weight of 3 KDa; in our study, we found three peptides in the range of 5 KDa to 7 KDa; this finding demonstrates that melittin is not the only peptide present in bee venom with antibacterial activity.

The PF of *A. mellifera* venom at a concentration lower than 15.6 μg/mL demonstrated low hemolytic activity. Few studies have revealed the hemolytic activity of bee venom in Peru; on the contrary, in other latitudes, they have revealed that melittin has not presented significant hemolytic activity below a concentration of 0.25 μg/mL [62], the hemolytic action was also demonstrated against erythrocytes of different species, with variable sensitivity to bee venom pools, with sheep erythrocytes being the most resistant to hemolytic action compared to equine erythrocytes, including humans erythrocytes showed good resistance to hemolytic action, it follows that hemolysis can be increased by the action of phospholipase 2 (PLA2) after the action of melittin [63].

The PF of the venom of *A. mellifera* has not shown antioxidant capacity, but in other studies, they demonstrated antioxidant capacity. Curiously, they worked with the total venom or apitoxin, having the capacity to inhibit the free radical DPPH (2,2-diphenyl-1-picrylhydrazine) between 60% and 75% of antioxidant activity [64]. In the same sense, demonstrated with the venom of *A. mellifera syriaca* eliminating DPPH radicals between 50 to 65% [65]. When analyzing the venom of four bee species, *A. dorsata*, *A. mellifera*, *A. florea*, and *A. cerena*, they showed that *A. dorsata* contained the highest amount of melittin; they also revealed that the extract of *A. dorsata* had the highest antioxidant activity from the DPPH and ABTS (3-ethylbenzothiazoline-6-sulfonic acid) assays, including melittin alone, revealed very poor antioxidant activity among all bee venom extracts [66], this suggests that in our study, of the peptides present in the venom PF, melittin was not present.

The bioactive components present in the venom have generated much interest in medicine through the different species of the *Apis* genus, and their application in in vitro antimicrobial activity [67], their cytotoxic action against cancer cells [68], even the synergistic effect of the venom with some antibiotics such as Cephotax, Cefepime, and Tavanic has been revealed [69]. Through Transmission Electron Microscopy, the deformation of the cell wall was appreciated, resulting in the destruction of the cell wall, changes in the permeability of the membrane, leakage of cell content, inactivation of metabolic activity, and finally, cell death [57], as the inhibitory effect on F1-F0—ATPase has also been demonstrated [70].

4. Materials and Methods

4.1. Bee Venom Samples

The venom was obtained from Africanized bees *A. mellifera* (Linnaeus, 1758) in hives from the Cruz Verde town center of the Ilimo District located at latitude 6°28′26″ and longitude 79°50′34″ (Lambayeque); an electrical impulse of 3 volts with an electrical intensity of 0.004 A was passed through a collecting box (beeWhisper 6.0; Model 2020), without harming the specimens (Figure 4a,b) [71].

Figure 4. (**a**) beeWhisper 6.0. collector box; (**b**) electrostimulation of the collector box; (**c**) recovery of *A. mellifera* dry venom; (**d**) storage of the venom.

The venom was collected on a glass plate and allowed to dry; it was then transferred into a 50-mL Falcon tube (Figure 4c,d). Then, the bee venom was resuspended in sterile deionized water and centrifuged at 10,000× *g* at 4 °C for 10 min to remove insoluble materials. The supernatant was collected and stored in 2 mL microtubes at −20 °C.

4.2. Fraction Concentration and Electrophoresis

The fractions were collected and quantified via absorbance at 280 nm (Navi UV/vis Nano spectrophotometer, Seongnam-si, Republic of Korea) using the formula [mg/mL] = (1.56 × Abs 280 nm) − (0.76 × Abs 260 nm) [72].

Crude fractions of bee venom were collected and concentrated using the Amicon ultra centrifugal filter (Merck Millipore, Cork, Ireland) with Cutoff from 3 kDa to 100 kDa [73], quantified via absorbance, and 50 µg protein from the FP were evaluated on a gel Tricine-SDS-PAGE (15%) under denaturing conditions [74] with a voltage of 100 volts and stained with Coomassie blue [75].

4.3. Antimicrobial Activity Test

The MIC values of the fraction were determined using the broth microdilution method in 96-well plates [76] against the strains of *E. coli* ATCC 25922, *S. aureus* ATCC 29213, and *P. aeruginosa* ATCC 27853; 50 µL of bacterial solution containing 5 × 10^4 CFU/mL was placed in each well, then 50 µL of different concentrations of the fraction (55 µg/mL to 3.44 µg/mL) were added and incubated at 37 °C for 24 h. The positive control was broth plus inoculum, and the negative control was only broth. Growth of the positive control was determined by a growth button of ≥2 mm or defined turbidity. Finally, the plates were read by absorbance at 630 nm (SmartReader 96—Accuris) to determine the minimum inhibitory concentration (MIC). All assays were performed in triplicate.

4.4. Evaluation of Hemolytic Activity

The hemolytic test was evaluated following the protocol described by Oddo [77], red blood cells washed with PBS (137 mM NaCl, 2.7 mM KCl, 10 mM Na_2HPO_4, 1.8 mM KH_2PO_4) and resuspended at a concentration of 0.5% and incubated for 1 h with different concentrations of the fractions and then centrifuged at 10,000× *g* for 10 min, 60 µL of supernatant was transferred to a 96-well polypropylene plate, and the absorbance was read at 405 nm. The results were normalized with the positive controls of hemolysis (0.25% SDS) and negative controls (PBS). Assays were performed in triplicate.

4.5. Evaluation of Antioxidant Activity

20 µL of different concentrations of the fraction (55 µg/mL to 3.44 µg/mL) were added with 380 µL of ABTS radical in ethanol, incubated at room temperature protected from light for 30 min, then the absorbance of the mixture was measured at 734 nm [66]. To calculate the % decoloration, the following equation was used: % decoloration = $[(C - S)/C] \times 100$, where C is the absorbance of the control, and S is the absorbance of the problem sample. Trolox was used as a positive control. The experiments were done in triplicate.

4.6. Statistic Analysis

The MegaStat add-in for Excel was used to determine the antibacterial activity of the purified fraction of *Apis mellifera* venom. Analysis of variance (ANOVA) was performed at a significance level of 5%.

5. Conclusions

In summary, *A. mellifera* bee venom contained peptides with weights of 7 kDa, 6 kDa, and 5 kDa and exhibited antibacterial activity against *E. coli* ATCC 25922 at a concentration of 6.88 µg/mL.

Author Contributions: Writing—preparation of the original draft, methodology, O.P.-D.; formal analysis, A.O.E.-C.; data curation, E.L.-L. All authors have read and agreed to the published version of the manuscript.

Funding: This work was financially supported by the Universidad Señor de Sipán with RD No. 074-2020/PD-USS. The funders had no role in the collection, analysis, and interpretation of the data.

Institutional Review Board Statement: Not applicable.

Informed Consent Statement: Not applicable.

Data Availability Statement: All data related to the manuscript were available in the main manuscript.

Acknowledgments: We are grateful for the assistance of Señor de Sipán University for the technical, translation, and logistical support provided for this research. We thank Luis Orlando Sánchez Suclupe, for his support with the hives for the extraction of bee venom, Karl Weiss Cruz Verde—Íllimo Bee-keeping Association. We thank the Thematic Network for the development of antiviral and antimicrobial peptides for multiresistant strains (RED BUDEPAV-AM) and for participating as an exhibitor at the international symposium on antimicrobial and antiviral peptides.

Conflicts of Interest: The authors declare that there are no conflict of interest with respect to the publication of this article.

References

1. World Health Organization. *Antimicrobial Resistance: Global Report on Surveillance*; World Health Organization: Geneva, Switzerland, 2014; ISBN 978-92-4-156474-8.
2. Pollitt, E.J.G.; Szkuta, P.T.; Burns, N.; Foster, S.J. Staphylococcus Aureus Infection Dynamics. *PLoS Pathog.* **2018**, *14*, e1007112. [CrossRef]

3. Tong, S.Y.C.; Davis, J.S.; Eichenberger, E.; Holland, T.L.; Fowler, V.G. Staphylococcus Aureus Infections: Epidemiology, Pathophysiology, Clinical Manifestations, and Management. *Clin. Microbiol. Rev.* **2015**, *28*, 603–661. [CrossRef]
4. Safdari, H.; Aryan, E.; Sadeghian, H.; Shams, S.F.; Aganj, M. Frequency of Methicillin-Resistant Staphylococcus Aureus (MRSA) in Nose and Cellular Phone of Medical and Non-Medical Personnel of Emergency Departments of Ghaem Hospital in Mashhad City. *Clin. Epidemiol. Glob. Health* **2020**, *8*, 1043–1046. [CrossRef]
5. Belayhun, C.; Tilahun, M.; Seid, A.; Shibabaw, A.; Sharew, B.; Belete, M.A.; Demsiss, W. Asymptomatic Nasopharyngeal Bacterial Carriage, Multi-Drug Resistance Pattern and Associated Factors among Primary School Children at Debre Berhan Town, North Shewa, Ethiopia. *Ann. Clin. Microbiol. Antimicrob.* **2023**, *22*, 9. [CrossRef]
6. Akanbi, O.E.; Njom, H.A.; Fri, J.; Otigbu, A.C.; Clarke, A.M. Antimicrobial Susceptibility of Staphylococcus Aureus Isolated from Recreational Waters and Beach Sand in Eastern Cape Province of South Africa. *Int. J. Environ. Res. Public Health* **2017**, *14*, 1001. [CrossRef]
7. Qodrati, M.; SeyedAlinaghi, S.; Dehghan Manshadi, S.A.; Abdollahi, A.; Dadras, O. Antimicrobial Susceptibility Testing of Staphylococcus Aureus Isolates from Patients at a Tertiary Hospital in Tehran, Iran, 2018–2019. *Eur. J. Med. Res.* **2022**, *27*, 152. [CrossRef]
8. Nikbakht, M.; Ahangarzadeh Rezaee, M.; Hasani, A.; Nahaei, M.R.; Sadeghi, J.; Jedari Seifi, S. Molecular Characterization and Antimicrobial Susceptibility Patterns of Methicillin-Resistant Staphylococcus Aureus Isolates in Tabriz, Northwest of Iran. *Arch. Pediatr. Infect. Dis.* 2017, *in press*. [CrossRef]
9. Guo, Y.; Song, G.; Sun, M.; Wang, J.; Wang, Y. Prevalence and Therapies of Antibiotic-Resistance in *Staphylococcus aureus*. *Front. Cell. Infect. Microbiol.* **2020**, *10*, 107. [CrossRef]
10. Hu, P.; Chen, J.; Chen, Y.; Zhou, T.; Xu, X.; Pei, X. Molecular Epidemiology, Resistance, and Virulence Properties of *Pseudomonas aeruginosa* Cross-Colonization Clonal Isolates in the Non-Outbreak Setting. *Infect. Genet. Evol.* **2017**, *55*, 288–296. [CrossRef]
11. Nasser, M.; Gayen, S.; Kharat, A.S. Prevalence of β-Lactamase and Antibiotic-Resistant *Pseudomonas aeruginosa* in the Arab Region. *J. Glob. Antimicrob. Resist.* **2020**, *22*, 152–160. [CrossRef]
12. Hoque, M.N.; Jahan, M.I.; Hossain, M.A.; Sultana, M. Genomic Diversity and Molecular Epidemiology of a Multidrug-Resistant *Pseudomonas aeruginosa* DMC30b Isolated from a Hospitalized Burn Patient in Bangladesh. *J. Glob. Antimicrob. Resist.* **2022**, *31*, 110–118. [CrossRef]
13. Coșeriu, R.L.; Vintilă, C.; Mare, A.D.; Ciurea, C.N.; Togănel, R.O.; Cighir, A.; Simion, A.; Man, A. Epidemiology, Evolution of Antimicrobial Profile and Genomic Fingerprints of *Pseudomonas aeruginosa* before and during COVID-19: Transition from Resistance to Susceptibility. *Life* **2022**, *12*, 2049. [CrossRef]
14. Asadpour, L. Antimicrobial Resistance, Biofilm-Forming Ability and Virulence Potential of *Pseudomonas aeruginosa* Isolated from Burn Patients in Northern Iran. *J. Glob. Antimicrob. Resist.* **2018**, *13*, 214–220. [CrossRef]
15. Schiavano, G.F.; Carloni, E.; Andreoni, F.; Magi, S.; Chironna, M.; Brandi, G.; Amagliani, G. Prevalence and Antibiotic Resistance of *Pseudomonas aeruginosa* in Water Samples in Central Italy and Molecular Characterization of OprD in Imipenem Resistant Isolates. *PLoS ONE* **2017**, *12*, e0189172. [CrossRef]
16. Alotaibi, B.S.; Tantry, B.A.; Farhana, A.; Alammar, M.A.; Shah, N.N.; Mohammed, A.H.; Wani, F.; Bandy, A. Resistance Pattern in Mostly Gram-Negative Bacteria Causing Urinary Tract Infections. *Infect. Disord. Drug Targets* **2023**, *23*, 56–64. [CrossRef]
17. Umpiérrez, A.; Ernst, D.; Fernández, M.; Oliver, M.; Casaux, M.L.; Caffarena, R.D.; Schild, C.; Giannitti, F.; Fraga, M.; Zunino, P. Virulence Genes of *Escherichia coli* in Diarrheic and Healthy Calves. *Rev. Argent. Microbiol.* **2021**, *53*, 34–38. [CrossRef]
18. Sarjana Safain, K.; Bhuyan, G.S.; Hassan Hasib, S.; Islam, M.S.; Mahmud-Un-Nabi, M.A.; Sultana, R.; Tasnim, S.; Noor, F.A.; Sarker, S.K.; Islam, M.T.; et al. Genotypic and Phenotypic Profiles of Antibiotic-resistant Bacteria Isolated from Hospitalised Patients in Bangladesh. *Trop. Med. Int. Health* **2021**, *26*, 720–729. [CrossRef]
19. Jang, J.; Hur, H.-G.; Sadowsky, M.J.; Byappanahalli, M.N.; Yan, T.; Ishii, S. Environmental *Escherichia coli*.: Ecology and Public Health Implications—A Review. *J. Appl. Microbiol.* **2017**, *123*, 570–581. [CrossRef]
20. Edmond, T.; Yehouenou, L.C.; Malick, Z.F.; Arsene, K.A.; Rene, K.K.; Diouara, A.A.M.; Tonde, I.; Bankole, H.S.; Wilfried, B.K.; Marius, E.A.; et al. Antimicrobial Susceptibility of Community Acquired *Escherichia coli* in Urinary Tract Infections (UTI) in Benin for Eleven Years (2005–2015). *Am. J. Infect. Dis.* **2017**, *13*, 21–27. [CrossRef]
21. Grados, M.C.; Thuissard, I.J.; Alós, J.-I. Stratification by Demographic and Clinical Data of the Antibiotic Susceptibility of Escherichia Coli from Urinary Tract Infections of the Community. *Atención Primaria* **2019**, *51*, 494–498. [CrossRef]
22. Amanati, A.; Sajedianfard, S.; Khajeh, S.; Ghasempour, S.; Mehrangiz, S.; Nematolahi, S.; Shahhosein, Z. Bloodstream Infections in Adult Patients with Malignancy, Epidemiology, Microbiology, and Risk Factors Associated with Mortality and Multi-Drug Resistance. *BMC Infect. Dis.* **2021**, *21*, 636. [CrossRef]
23. Mahmoudi, S.; Mahzari, M.; Banar, M.; Pourakbari, B.; Haghi Ashtiani, M.T.; Mohammadi, M.; Keshavarz Valian, S.; Mamishi, S. Antimicrobial Resistance Patterns of Gram-Negative Bacteria Isolated from Bloodstream Infections in an Iranian Referral Paediatric Hospital: A 5.5-Year Study. *J. Glob. Antimicrob. Resist.* **2017**, *11*, 17–22. [CrossRef]
24. Shi, N.; Kang, J.; Wang, S.; Song, Y.; Yin, D.; Li, X.; Guo, Q.; Duan, J.; Zhang, S. Bacteriological Profile and Antimicrobial Susceptibility Patterns of Gram-Negative Bloodstream Infection and Risk Factors Associated with Mortality and Drug Resistance: A Retrospective Study from Shanxi, China. *Infect. Drug Resist.* **2022**, *15*, 3561–3578. [CrossRef]

25. Esposito, S.; Maglietta, G.; Di Costanzo, M.; Ceccoli, M.; Vergine, G.; La Scola, C.; Malaventura, C.; Falcioni, A.; Iacono, A.; Crisafi, A.; et al. Retrospective 8-Year Study on the Antibiotic Resistance of Uropathogens in Children Hospitalised for Urinary Tract Infection in the Emilia-Romagna Region, Italy. *Antibiotics* **2021**, *10*, 1207. [CrossRef]
26. Boschetti, G.; Sgarabotto, D.; Meloni, M.; Bruseghin, M.; Whisstock, C.; Marin, M.; Ninkovic, S.; Pinfi, M.; Brocco, E. Antimicrobial Resistance Patterns in Diabetic Foot Infections, an Epidemiological Study in Northeastern Italy. *Antibiotics* **2021**, *10*, 1241. [CrossRef]
27. Obakiro, S.B.; Kiyimba, K.; Paasi, G.; Napyo, A.; Anthierens, S.; Waako, P.; Royen, P.V.; Iramiot, J.S.; Goossens, H.; Kostyanev, T. Prevalence of Antibiotic-Resistant Bacteria among Patients in Two Tertiary Hospitals in Eastern Uganda. *J. Glob. Antimicrob. Resist.* **2021**, *25*, 82–86. [CrossRef]
28. Mahlapuu, M.; Håkansson, J.; Ringstad, L.; Björn, C. Antimicrobial Peptides: An Emerging Category of Therapeutic Agents. *Front. Cell. Infect. Microbiol.* **2016**, *6*, 194. [CrossRef]
29. Tawfik, M.M.; Bertelsen, M.; Abdel-Rahman, M.A.; Strong, P.N.; Miller, K. Scorpion Venom Antimicrobial Peptides Induce Siderophore Biosynthesis and Oxidative Stress Responses in *Escherichia coli*. *mSphere* **2021**, *6*, e00267-21. [CrossRef]
30. Pérez-Delgado, O.; Rincon-Cortés, C.A.; Vega-Castro, N.A.; Reyes-Montaño, E.A.; Gómez-Garzón, M. Purificación Parcial de Péptidos Del Veneno de Escorpión Hadruroides Charcasus (Karsch, 1879) Con Actividad Antimicrobiana. *Bionat. Lat. Am. J. Biotechnol. Life Sci.* **2021**, *6*, 1917–1923. [CrossRef]
31. Zhao, Z.; Ma, Y.; Dai, C.; Zhao, R.; Li, S.; Wu, Y.; Cao, Z.; Li, W. Imcroporin, a New Cationic Antimicrobial Peptide from the Venom of the Scorpion *Isometrus maculates*. *Antimicrob. Agents Chemother.* **2009**, *53*, 3472–3477. [CrossRef]
32. de Barros, E.; Gonçalves, R.M.; Cardoso, M.H.; Santos, N.C.; Franco, O.L.; Cândido, E.S. Snake Venom Cathelicidins as Natural Antimicrobial Peptides. *Front. Pharmacol.* **2019**, *10*, 1415. [CrossRef]
33. Lee, B.; Shin, M.K.; Yoo, J.S.; Jang, W.; Sung, J.-S. Identifying Novel Antimicrobial Peptides from Venom Gland of Spider Pardosa Astrigera by Deep Multi-Task Learning. *Front. Microbiol.* **2022**, *13*, 971503. [CrossRef]
34. Ko, S.J.; Park, E.; Asandei, A.; Choi, J.-Y.; Lee, S.-C.; Seo, C.H.; Luchian, T.; Park, Y. Bee Venom-Derived Antimicrobial Peptide Melectin Has Broad-Spectrum Potency, Cell Selectivity, and Salt-Resistant Properties. *Sci. Rep.* **2020**, *10*, 10145. [CrossRef]
35. Abd El-Wahed, A.A.; Khalifa, S.A.M.; Sheikh, B.Y.; Farag, M.A.; Saeed, A.; Larik, F.A.; Koca-Caliskan, U.; AlAjmi, M.F.; Hassan, M.; Wahabi, H.A.; et al. Bee Venom Composition: From Chemistry to Biological Activity. In *Studies in Natural Products Chemistry*; Elsevier: Amsterdam, The Netherlands, 2019; Volume 60, pp. 459–484, ISBN 978-0-444-64181-6.
36. Amin, M.A.; Abdel-Raheem, I.T. Accelerated Wound Healing and Anti-Inflammatory Effects of Physically Cross Linked Polyvinyl Alcohol–Chitosan Hydrogel Containing Honey Bee Venom in Diabetic Rats. *Arch. Pharm. Res.* **2014**, *37*, 1016–1031. [CrossRef]
37. Ward, R.; Coffey, M.; Kavanagh, K. Proteomic Analysis of Summer and Winter *Apis mellifera* Workers Shows Reduced Protein Abundance in Winter Samples. *J. Insect Physiol.* **2022**, *139*, 104397. [CrossRef]
38. Kwon, N.-Y.; Sung, S.-H.; Sung, H.-K.; Park, J.-K. Anticancer Activity of Bee Venom Components against Breast Cancer. *Toxins* **2022**, *14*, 460. [CrossRef]
39. Kim, H.; Park, S.-Y.; Lee, G. Potential Therapeutic Applications of Bee Venom on Skin Disease and Its Mechanisms: A Literature Review. *Toxins* **2019**, *11*, 374. [CrossRef]
40. Hegazi, A.; Abdou, A.M.; El-Moez, S.I.A.; Allah, F.A. Evaluation of the Antibacterial Activity of Bee Venom from Different Sources. *World Appl. Sci. J.* **2014**, *30*, 266–270.
41. Nguyen, C.D.; Yoo, J.; Hwang, S.-Y.; Cho, S.-Y.; Kim, M.; Jang, H.; No, K.O.; Shin, J.C.; Kim, J.-H.; Lee, G. Bee Venom Activates the Nrf2/HO-1 and TrkB/CREB/BDNF Pathways in Neuronal Cell Responses against Oxidative Stress Induced by Aβ1–42. *Int. J. Mol. Sci.* **2022**, *23*, 1193. [CrossRef]
42. Tanner, C.M.; Kamel, F.; Ross, G.W.; Hoppin, J.A.; Goldman, S.M.; Korell, M.; Marras, C.; Bhudhikanok, G.S.; Kasten, M.; Chade, A.R.; et al. Rotenone, Paraquat, and Parkinson's Disease. *Environ. Health Perspect.* **2011**, *119*, 866–872. [CrossRef]
43. Khalil, W.K.B.; Assaf, N.; ElShebiney, S.A.; Salem, N.A. Neuroprotective Effects of Bee Venom Acupuncture Therapy against Rotenone-Induced Oxidative Stress and Apoptosis. *Neurochem. Int.* **2015**, *80*, 79–86. [CrossRef] [PubMed]
44. Socarras, K.; Theophilus, P.; Torres, J.; Gupta, K.; Sapi, E. Antimicrobial Activity of Bee Venom and Melittin against Borrelia Burgdorferi. *Antibiotics* **2017**, *6*, 31. [CrossRef]
45. Zolfagharian, H.; Mohajeri, M.; Babaie, M. Bee Venom (*Apis mellifera*) an Effective Potential Alternative to Gentamicin for Specific Bacteria Strains: Bee Venom an Effective Potential for Bacteria. *J. Pharmacopunct.* **2016**, *19*, 225–230. [CrossRef]
46. Han, S.; Kim, J.; Hong, I.; Woo, S.; Kim, S.; Jang, H.; Pak, S. Antibacterial Activity and Antibiotic-Enhancing Effects of Honeybee Venom against Methicillin-Resistant Staphylococcus Aureus. *Molecules* **2016**, *21*, 79. [CrossRef]
47. Leandro, L.F.; Mendes, C.A.; Casemiro, L.A.; Vinholis, A.H.C.; Cunha, W.R.; de Almeida, R.; Martins, C.H.G. Antimicrobial Activity of Apitoxin, Melittin and Phospholipase A2 of Honey Bee (*Apis mellifera*) Venom against Oral Pathogens. *Ann. Acad. Bras. Ciênc.* **2015**, *87*, 147–155. [CrossRef] [PubMed]
48. El-Didamony, S.E.; Kalaba, M.H.; El-Fakharany, E.M.; Sultan, M.H.; Sharaf, M.H. Antifungal and Antibiofilm Activities of Bee Venom Loaded on Chitosan Nanoparticles: A Novel Approach for Combating Fungal Human Pathogens. *World J. Microbiol. Biotechnol.* **2022**, *38*, 244. [CrossRef]
49. Uddin, M.B.; Lee, B.-H.; Nikapitiya, C.; Kim, J.-H.; Kim, T.-H.; Lee, H.-C.; Kim, C.G.; Lee, J.-S.; Kim, C.-J. Inhibitory Effects of Bee Venom and Its Components against Viruses In Vitro and In Vivo. *J. Microbiol.* **2016**, *54*, 853–866. [CrossRef] [PubMed]

50. Fenard, D.; Lambeau, G.; Valentin, E.; Lefebvre, J.C.; Lazdunski, M.; Doglio, A. Secreted Phospholipases A(2), a New Class of HIV Inhibitors That Block Virus Entry into Host Cells. *J. Clin. Investig.* **1999**, *104*, 611–618. [CrossRef]
51. de Brito, J.C.M.; Bastos, E.M.A.F.; Heneine, L.G.D.; de Souza Figueiredo, K.C. Fractionation of *Apis mellifera* Venom by Means of Ultrafiltration: Removal of Phospholipase A_2. *Braz. J. Chem. Eng.* **2018**, *35*, 229–236. [CrossRef]
52. El-Seedi, H.; Abd El-Wahed, A.; Yosri, N.; Musharraf, S.G.; Chen, L.; Moustafa, M.; Zou, X.; Al-Mousawi, S.; Guo, Z.; Khatib, A.; et al. Antimicrobial Properties of *Apis mellifera*'s Bee Venom. *Toxins* **2020**, *12*, 451. [CrossRef] [PubMed]
53. Memariani, H.; Memariani, M.; Moravvej, H.; Shahidi-Dadras, M. Melittin: A Venom-Derived Peptide with Promising Anti-Viral Properties. *Eur. J. Clin. Microbiol. Infect. Dis.* **2020**, *39*, 5–17. [CrossRef] [PubMed]
54. Kurek-Górecka, A.; Komosinska-Vassev, K.; Rzepecka-Stojko, A.; Olczyk, P. Bee Venom in Wound Healing. *Molecules* **2020**, *26*, 148. [CrossRef] [PubMed]
55. Huh, J.-E.; Seo, B.-K.; Lee, J.-W.; Park, Y.-C.; Baek, Y.-H. Analgesic Effects of Diluted Bee Venom Acupuncture Mediated by δ-Opioid and A2-Adrenergic Receptors in Osteoarthritic Rats. *Altern. Ther. Health Med.* **2018**, *24*, 28–35. [PubMed]
56. Maitip, J.; Mookhploy, W.; Khorndork, S.; Chantawannakul, P. Comparative Study of Antimicrobial Properties of Bee Venom Extracts and Melittins of Honey Bees. *Antibiotics* **2021**, *10*, 1503. [CrossRef]
57. Haktanir, I.; Masoura, M.; Mantzouridou, F.T.; Gkatzionis, K. Mechanism of Antimicrobial Activity of Honeybee (*Apis mellifera*) Venom on Gram-Negative Bacteria: *Escherichia coli* and *Pseudomonas* spp. *AMB Expr.* **2021**, *11*, 54. [CrossRef]
58. Bakhiet, E.K.; Hussien, H.A.M.; Elshehaby, M. *Apis mellifera* Venom Inhibits Bacterial and Fungal Pathogens In Vitro. *Pak. J. Biol. Sci.* **2022**, *25*, 875–884. [CrossRef] [PubMed]
59. Babaie, M.; Ghaem panah, A.; Mehrabi, Z.; Mollaei, A.; Sima Khalilifard, B. Partial Purification and Characterization of Antimicrobial Effects from Snake (*Echis carinatus*), Scorpion (*Mesosobuthus epues*) and Bee (*Apis mellifera*) Venoms. *Iran. J. Med. Microbiol.* **2020**, *14*, 460–477. [CrossRef]
60. Pucca, M.B.; Cerni, F.A.; Oliveira, I.S.; Jenkins, T.P.; Argemí, L.; Sørensen, C.V.; Ahmadi, S.; Barbosa, J.E.; Laustsen, A.H. Bee Updated: Current Knowledge on Bee Venom and Bee Envenoming Therapy. *Front. Immunol.* **2019**, *10*, 2090. [CrossRef]
61. van den Bogaart, G.; Guzmán, J.V.; Mika, J.T.; Poolman, B. On the Mechanism of Pore Formation by Melittin. *J. Biol. Chem.* **2008**, *283*, 33854–33857. [CrossRef]
62. Zarrinnahad, H.; Mahmoodzadeh, A.; Hamidi, M.P.; Mahdavi, M.; Moradi, A.; Bagheri, K.P.; Shahbazzadeh, D. Apoptotic Effect of Melittin Purified from Iranian Honey Bee Venom on Human Cervical Cancer HeLa Cell Line. *Int. J. Pept. Res. Ther.* **2018**, *24*, 563–570. [CrossRef]
63. de Roodt, A.R.; Lanari, L.C.; Lago, N.R.; Bustillo, S.; Litwin, S.; Morón-Goñi, F.; Gould, E.G.; van Grootheest, J.H.; Dokmetjian, J.C.; Dolab, J.A.; et al. Toxicological Study of Bee Venom (*Apis mellifera* Mellifera) from Different Regions of the Province of Buenos Aires, Argentina. *Toxicon* **2020**, *188*, 27–38. [CrossRef] [PubMed]
64. Viana, G.A.; Freitas, C.I.A.; Almeida, J.G.L.d.; Medeiros, G.V.D.d.; Teófilo, T.d.S.; Rodrigues, V.H.V.; Coelho, W.A.C.; Batista, J.S. Antioxidant, Genotoxic, Antigenotoxic, and Antineoplastic Activities of Apitoxin Produced by *Apis mellifera* in Northeast, Brazil. *Cienc. Rural* **2021**, *51*, e20200545. [CrossRef]
65. Frangieh, J.; Salma, Y.; Haddad, K.; Mattei, C.; Legros, C.; Fajloun, Z.; El Obeid, D. First Characterization of the Venom from *Apis mellifera* Syriaca, a Honeybee from the Middle East Region. *Toxins* **2019**, *11*, 191. [CrossRef] [PubMed]
66. Somwongin, S.; Chantawannakul, P.; Chaiyana, W. Antioxidant Activity and Irritation Property of Venoms from Apis Species. *Toxicon* **2018**, *145*, 32–39. [CrossRef]
67. Tanuwidjaja, I.; Svečnjak, L.; Gugić, D.; Levanić, M.; Jurić, S.; Vinceković, M.; Mrkonjić Fuka, M. Chemical Profiling and Antimicrobial Properties of Honey Bee (*Apis mellifera* L.) Venom. *Molecules* **2021**, *26*, 3049. [CrossRef]
68. Yaacoub, C.; Rifi, M.; El-Obeid, D.; Mawlawi, H.; Sabatier, J.-M.; Coutard, B.; Fajloun, Z. The Cytotoxic Effect of *Apis mellifera* Venom with a Synergistic Potential of Its Two Main Components—Melittin and PLA2—On Colon Cancer HCT116 Cell Lines. *Molecules* **2021**, *26*, 2264. [CrossRef]
69. Kamel, A.; Suleiman, W.; Elfeky, A.; El-Sherbiny, G.; Elhaw, M. Characterization of Bee Venom and Its Synergistic Effect Combating Antibiotic Resistance of *Pseudomonas aeruginosa*. *Egypt. J. Chem.* **2021**, *65*, 1–2. [CrossRef]
70. Nehme, H.; Ayde, H.; El Obeid, D.; Sabatier, J.M.; Fajloun, Z. Potential Inhibitory Effect of *Apis mellifera*'s Venom and of Its Two Main Components—Melittin and PLA2—On *Escherichia coli* F1F0-ATPase. *Antibiotics* **2020**, *9*, 824. [CrossRef]
71. Sobral, F.; Sampaio, A.; Falcão, S.; Queiroz, M.J.R.P.; Calhelha, R.C.; Vilas-Boas, M.; Ferreira, I.C.F.R. Chemical Characterization, Antioxidant, Anti-Inflammatory and Cytotoxic Properties of Bee Venom Collected in Northeast Portugal. *Food Chem. Toxicol.* **2016**, *94*, 172–177. [CrossRef]
72. Noble, J.E. Quantification of Protein Concentration Using UV Absorbance and Coomassie Dyes. In *Methods in Enzymology*; Elsevier: Amsterdam, The Netherlands, 2014; Volume 536, pp. 17–26, ISBN 978-0-12-420070-8.
73. Pérez-Delgado, O.; Espinoza-Vergara, M.A.; Castro-Vega, N.A.; Reyes-Montaño, E.A. Evaluación Preliminar de Actividad Antibacteriana in Vitro Del Veneno de Escorpión Hadruroides Charcasus (Karsch, 1879) Contra *Pseudomonas aeruginosa* y *Staphylococcus aureus*. *Rev. Cuerpo Med. HNAAA* **2019**, *12*, 6–12. [CrossRef]
74. Schägger, H.; von Jagow, G. Tricine-Sodium Dodecyl Sulfate-Polyacrylamide Gel Electrophoresis for the Separation of Proteins in the Range from 1 to 100 KDa. *Anal. Biochem.* **1987**, *166*, 368–379. [CrossRef] [PubMed]
75. Brunelle, J.L.; Green, R. Coomassie Blue Staining. In *Methods in Enzymology*; Elsevier: Amsterdam, The Netherlands, 2014; Volume 541, pp. 161–167, ISBN 978-0-12-420119-4.

76. *Approved Standard M07*; Methods for Dilution Antimicrobial Susceptibility Tests for Bacteria That Grow Aerobically, 9th ed. Clinical and Laboratory Standards Institute: Wayne, PA, USA, 2012; Volume 32, ISBN 1-56238-784-7.
77. Oddo, A.; Hansen, P.R. Hemolytic Activity of Antimicrobial Peptides. In *Antimicrobial Peptides*; Hansen, P.R., Ed.; Methods in Molecular Biology; Springer: New York, NY, USA, 2017; Volume 1548, pp. 427–435, ISBN 978-1-4939-6735-3.

antibiotics

MDPI

Article

Antiparasitic Activities of Compounds Isolated from *Aspergillus fumigatus* Strain Discovered in Northcentral Nigeria

Oluwatofunmilayo A. Diyaolu [1,*], Gagan Preet [1], Adeshola A. Fagbemi [2], Frederick Annang [3], Guiomar Pérez-Moreno [4], Cristina Bosch-Navarrete [4], Olusoji O. Adebisi [5], Emmanuel T. Oluwabusola [1], Bruce F. Milne [1,6], Marcel Jaspars [1] and Rainer Ebel [1]

1 Marine Biodiscovery Centre, Department of Chemistry, University of Aberdeen, Aberdeen AB24 3UE, UK
2 Department of Pharmaceutical Chemistry, Faculty of Pharmacy, Lead City University, Ibadan 200005, Nigeria
3 Fundación MEDINA, Parque Tecnológico de Ciencias de la Salud, Avenida del Conocimiento 34, Armilla, 18016 Granada, Spain
4 Institut de Parasitiologia Biomedicina "Lopez-Neyra", Consejo Superior de Investigaciones Cientificas (CSIC) Avda. Del Conocimiento 17, Armilla, 18016 Granada, Spain
5 School of Biosciences, Aston University Birmingham, Birmingham B4 7ET, UK
6 CFisUC, Department of Physics, University of Coimbra, Rua Larga, 3004-516 Coimbra, Portugal
* Correspondence: r01oad17@abdn.ac.uk

Abstract: In this study, we explored a fungal strain UIAU-3F identified as *Aspergillus fumigatus* isolated from soil samples collected from the River Oyun in Kwara State, Nigeria. In order to explore its chemical diversity, the fungal strain UIAU-3F was cultured in three different fermentation media, which resulted in different chemical profiles, evidenced by LC-ESI-MS-based metabolomics and multivariate analysis. The methanolic extract afforded two known compounds, fumitremorgin C (**1**) and pseurotin D (**2**). The in vitro antiparasitic assays of **1** against *Trypanosoma cruzi* and *Plasmodium falciparum* showed moderate activity with IC_{50} values of 9.6 μM and 2.3 μM, respectively, while **2** displayed IC_{50} values > 50 μM. Molecular docking analysis was performed on major protein targets to better understand the potential mechanism of the antitrypanosomal and antiplasmodial activities of the two known compounds.

Keywords: *Aspergillus fumigatus*; OSMAC; metabolomics; molecular docking; antitrypanosomal; antiplasmodial; in silico molecular docking

Citation: Diyaolu, O.A.; Preet, G.; Fagbemi, A.A.; Annang, F.; Pérez-Moreno, G.; Bosch-Navarrete, C.; Adebisi, O.O.; Oluwabusola, E.T.; Milne, B.F.; Jaspars, M.; et al. Antiparasitic Activities of Compounds Isolated from *Aspergillus fumigatus* Strain Discovered in Northcentral Nigeria. *Antibiotics* **2023**, *12*, 109. https://doi.org/10.3390/antibiotics12010109

Academic Editors: Valério Monteiro-Neto and Elizabeth S. Fernandes

Received: 24 November 2022
Revised: 29 December 2022
Accepted: 30 December 2022
Published: 6 January 2023

1. Introduction

Microbes continue to make a tremendous impact on natural product drug discovery. They produce many primary metabolites, such as amino acids, nucleotides, vitamins, and organic acids [1]. In addition, they produce secondary metabolites, which constitute a substantial part of the market's pharmaceuticals today, to the extent that microbial natural products have become a principal source of drug lead compounds [2]. Various infectious diseases emerge worldwide, requiring a constant and broadened search for newer, more efficient bioactive molecules [2].

Aspergillus is a genus of all-pervasive fungi that are beneficially and morbidly important. The members of the genus *Aspergillus* are urbane and ubiquitous constituents of different habitats [3]; this is because they can colonise a broad range of substrates. A wide range of secondary metabolites has been isolated from this genus [4]. These metabolites have been linked to 24 biosynthetic families. Examples include alkaloids, benzoquinones, flavonoids, phenols, anthraquinones, steroids, terpenoids, tetralones, and xanthones [5]. Some bioactivities of compounds isolated from *Aspergillus fumigatus* have been studied and reported in the literature, including antibacterial [6], antinematodal [7,8], antiprotozoal [9,10], antifungal [11], antiviral [12], anticholinesterase [13], and cytotoxic activities [14].

Parasitic diseases concern health and human life, particularly in the developing world. On a global scale, parasitic diseases are responsible for almost one million deaths yearly [15]. They remain the significant killers of children in low-income nations [16–18]. In 2019, the World Health Organisation (WHO) tagged malaria as a significant parasitic disease, causing 409,000 deaths, most of whom were African children below the age of five [19]. However, it is thought that many cases are undiagnosed, and therefore unpublished, so the actual figure may well be higher. In tropical regions, the combination of warm and humid climates with exponential population growth and poverty contributes to the rise in disseminating parasitic infections [20], usually mediated by vectors such as mosquitoes and ticks. Protozoans belonging to the *Plasmodium* group are the causative agents for malaria and cause worldwide deaths [16,17]. Antiparasitic drugs currently in use suffer from several drawbacks, as they tend to be expensive, their side effects may be severe, they may be challenging to administer, or drug resistance towards them may be rapidly expanding [21–23].

For instance, African Trypanosomiasis, one of the neglected tropical diseases caused by *Trypanosoma brucei*, is transmitted by the tsetse fly (*Glossina* species) found exclusively in sub-Saharan Africa [24]. The parasite, which comprises two physically identical subspecies, affects people in two different ways. While the *T. rhodesiense* strain produces a more severe acute African trypanosomiasis in eastern and southern Africa, the *T. gambiense* strain causes a slowly progressive disease in western and central Africa [25]. An estimated more than 60 million people in sub-Saharan Africa have been affected [7].

However, the ability of *Aspergillus fumigatus* to biosynthesise bioactive natural products could result in previously unexplored therapeutic options to fight these parasites. Consequently, we decided to extend our investigation of *Aspergillus fumigatus* by screening the isolated compounds for their potential antimalarial and antitrypanosomal activities using in silico molecular docking and in vitro studies. Furthermore, we studied the effect of altering the growth conditions of the fungus on their metabolic profiles using LC-ESI-MS-based metabolomics followed by multivariate analysis. Herein, this work presents for the first time the antiplasmodial and trypanocidal activities of fumitremorgin C (**1**) and pseurotin D (**2**). To identify the potential connections between the binding model and the antiparasitic properties, we employed molecular docking to fit compounds **1** and **2** into the active site of the target enzyme.

2. Results and Discussion

2.1. Characterisation and Identification of Strain UIAU-3F

The identification of UIAU-3F was conducted based on the morphology and phylogenetic analysis. Incubation was performed at 28 °C for 10 days, upon which strain UIAU-3F formed colonies on ISP2 agar plates displaying characteristic hyphal structures (Figure 1A). The ITS gene region of the ribosomal DNA of the strain was amplified by PCR- and sequenced. The phylogenetic tree (Figure 1B) constructed from the ITS gene sequence indicated that UIAU-3F belonged to the genus of *Aspergillus* with the highest similarity to *A. fumigatus* (100%, accession number: EF66998).

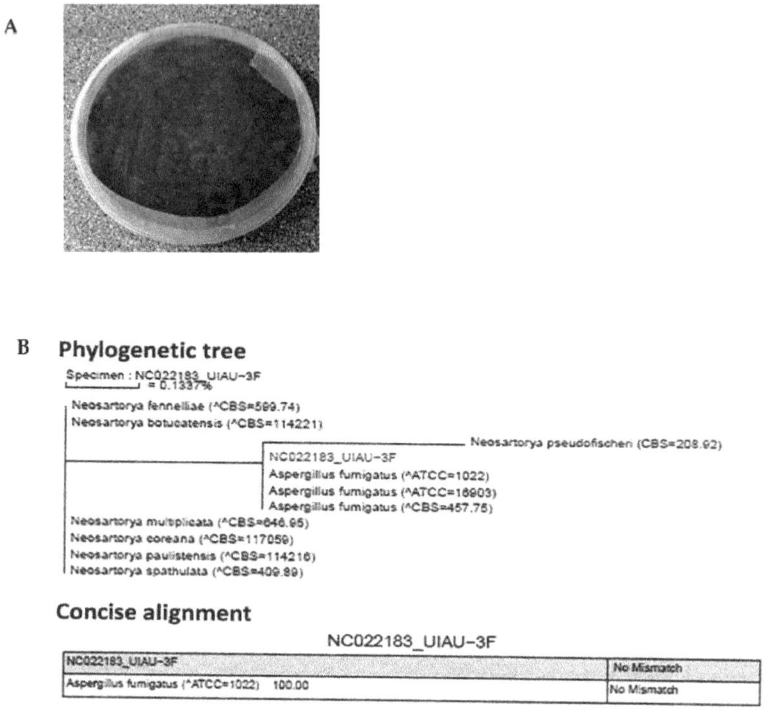

Figure 1. (**A**) Colony characteristics of *Aspergillus fumigatus* UIAU-3F grown on ISP2 agar at 28 °C for 7 days. (**B**) Phylogenetic tree of *Aspergillus fumigatus* UIAU-3F.

2.2. Non-Targeted Multivariate Analysis

After fungal identification and culturing of the strain in the different growth media, the extracts obtained were subjected to comparative metabolomics. Supervised statistical PCA, PLS-DA, permutation tests, and heat maps were used to compare the presence or absence and relative abundance of parent ions in each extract. Using metaboAnalyst software, the peak lists generated from filtered positive and negative mode mass spectral data were analysed. Ions present in solvent and media blanks were eliminated from the analysis to avoid uninformative skewing of the results. We detected 770 features between all three extracts, with each feature representing a distinctive combination of the m/z value and chromatogram peak characteristics. ANOVA was used to compare and determine if there was a statistically significant difference between the three fungal extracts (ISP2, ISP2-S, and RICE). The p-value threshold was set to 0.05 (default). Metabolites with a p-value below the threshold are shown in red and are deemed statistically significant. Of the metabolites predominantly originating from the RICE extract, 433 are statistically significant and 337 are insignificant (Figure 2A).

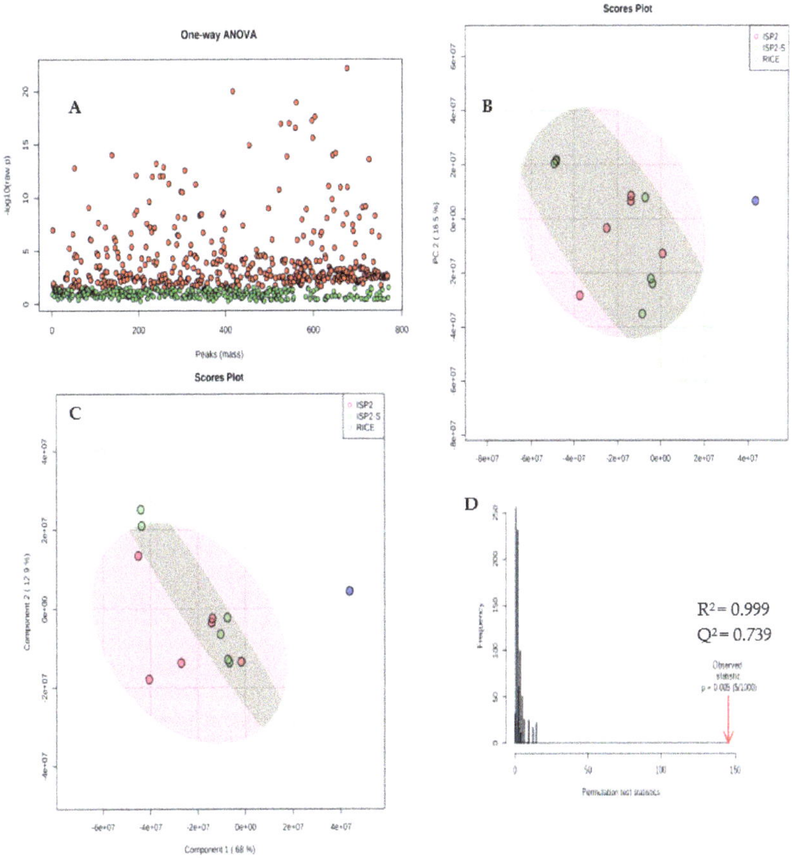

Figure 2. (**A**) Multivariate analysis. ANOVA results show statistically significant metabolites in red. (**B**) Principal Component Analysis (PCA) score plot. Green circles correspond to ISP2−S extract; pink circles correspond to ISP2 extract; blue circles correspond to RICE extract. (**C**) Partial least squares discriminant analysis PLS−DA model. Colour coding as before. (**D**) A permutation test plot.

The PCA score plot enables the visualisation of how the three extracts are related to one another. Extracts are clustered according to their similarity, with the more distinct sample groups showing more separation. The score plot comprising six replicates of the RICE, ISP2, and ISP2-S extracts in positive and negative mode ionisations was acquired, as shown in Figure 2B. Metabolites obtained from the RICE media (RICE 1–6) were separated from metabolites obtained from ISP2 (1–6) and ISP2-S medium by PC1. However, metabolites from ISP2 positively correlate with ISP2-S; this may be because of the close similarity in media culturing conditions and therefore suggests similarities in their secondary metabolites. The first two components, PC1 and PC2, explained 68.2% and 16.5% of the total variance, respectively.

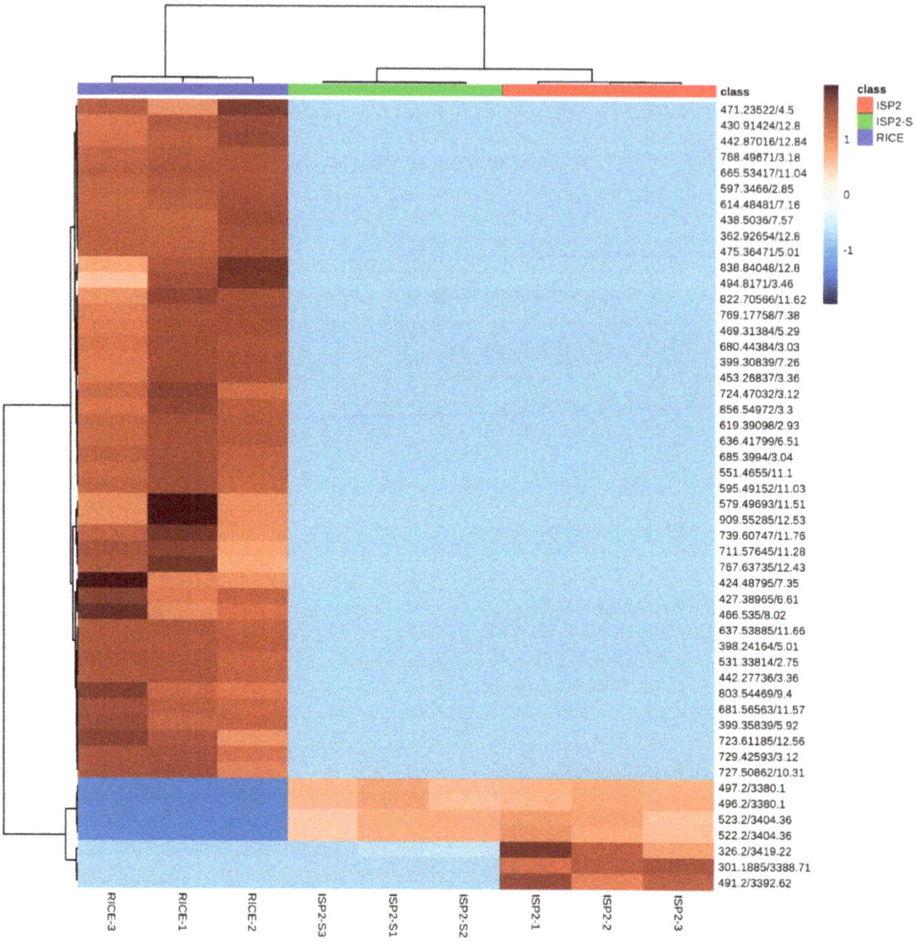

Figure 3. Hierarchical clustering of detected ions in ISP2 extract (red), ISP2-S extract (green), and RICE extract (blue). Heatmap displays relative abundance (low, blue; high, red) of respective features.

PLS–DA sharpens the separation between the extracts as it can perform both classification and metabolite selection. To identify the metabolites that contribute to the different metabolomic patterns, the PLS–DA model was applied (Figure 2C). As PLS–DA analysis is prone to overfitting, it is essential to validate PLS–DA analysis by conducting cross-validation or a permutation test. Cross-validation of Q^2 and R^2 is used to select the optimal number of compounds used in the PLS–DA model for classification. This model obtained an R^2 of 0.993 and a Q^2 of 0.739. The R^2 suggests that the model can be considered to have a substantial predictive ability and the Q^2 obtained is close to the R^2, indicating that the data fit the model (Figure 2D).

A heatmap was generated by visualising the intensities of the metabolites in order to determine the abundance of the secondary metabolomes in the three different extracts (Figure 3). The samples were conducted in triplicate. Multiple red bands indicate a rich secondary metabolome with a high diversity of metabolites, while deep blue bands exemplify a more limited set of secondary metabolites. The RICE culture extract showed the richest metabolome of the three extracts, as is evident from the heatmap display (Figure 3).

Metabolites Identification

The loadings plot generated (Figure 4) enables the identification of the metabolites within the three extracts responsible for the driving patterns seen in the score plot. Metabolites displaying similar information are grouped, indicating a correlation between the ISP2 and ISP2-S extracts. Compounds labelled **1–5** in the loading plot can be found in both ISP2 and ISP2-S extracts. An inverse correlation is displayed between the RICE and ISP2 extracts, as metabolites of the RICE extract are positioned on the opposite side of the plot of origin and farther away (compounds **6–10**).

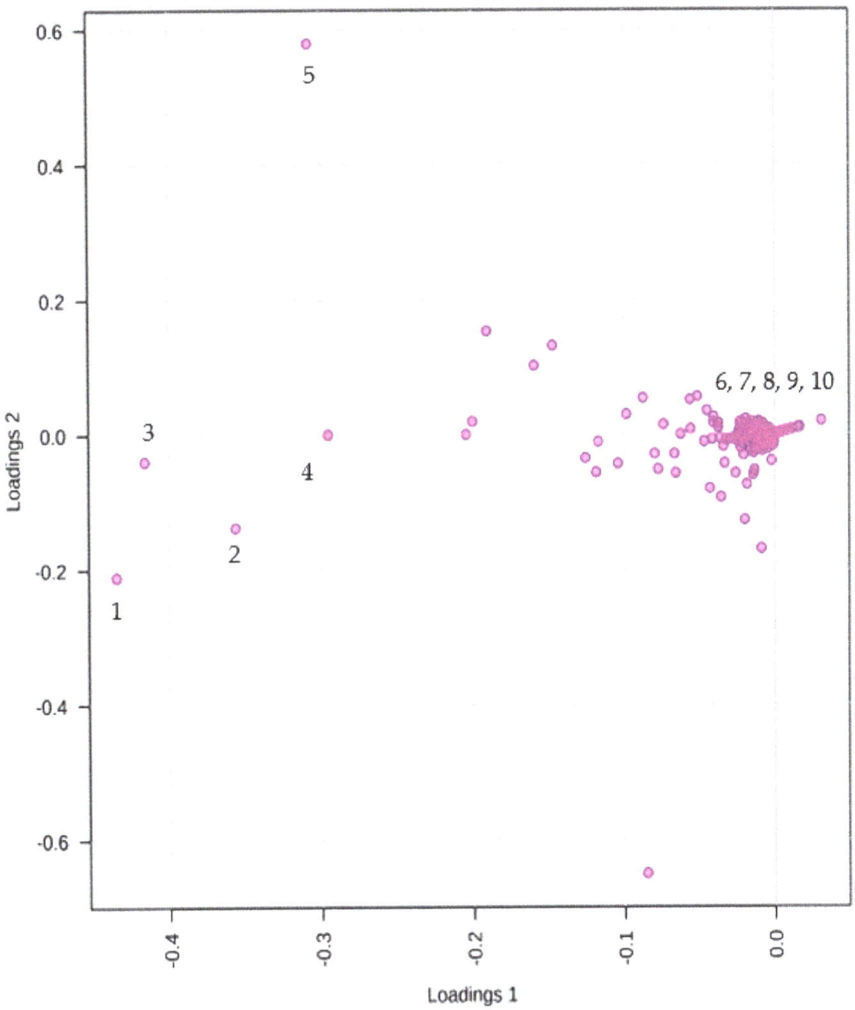

Figure 4. Principal Component Analysis (PCA) loadings score plot to display discriminant meta-bolites.

Potential biomarkers were initially identified by their molecular weights and fragmentation patterns using mass spectrometry. The precise mass of each differential metabolite was searched using different databases (Reaxys, NPAtlas, PubChem) to confirm its identity, possible molecular formula, and chemical composition (Table 1).

Table 1. Significant differential metabolites identified (p-value < 0.05 VIP).

Metabolite	$[M+H]^+$	t_R (min)	MF	Compound	Source	Ref
1	480.1595	4.5	$C_{27}H_{33}N_3O_5$	Fumitremorgen B	ISP2, ISP2-S	[26]
2	496.2338	3.8	$C_{27}H_{33}N_3O_6$	Spirotryprostatin C	ISP2, ISP2-S	[27]
3	528.5353	7.7	$C_{27}H_{33}N_3O_8$	Spirotryprostatin E	ISP2, ISP2-S	[28]
4	524.3704	5.6	$C_{33}H_{49}NO_4$	Anthcolorin E	ISP2, ISP2-S	[29]
5	522.3404	3.3	$C_{33}H_{47}NO_4$	Anthcolorin H	ISP2, ISP2-S	[29]
6	478.5882	7.1	$C_{28}H_{34}N_2O_5$	Fumitremorgen B derivative	RICE	[30]
7	431.4423	6.9	$C_{22}H_{25}NO_8$	Pseurotin D	RICE	[31]
8	381.4316	6.2	$C_{21}H_{23}N_3O_4$	Cyclotryprostatin C	RICE	[28]
9	379.4151	6.4	$C_{21}H_{21}N_3O_4$	Cyclotryprostatin D	RICE	[32]
10	511.6752	4.5	$C_{27}H_{33}N_3O_7$	Verruculogen	RICE	[32]

t_R—retention time; MF—molecular formula; Ref—reference.

2.3. Structural Elucidation of Isolated Compounds

The fungal UIAU-3F extract obtained from the RICE fermentation medium was analysed using HR-LCMS and ^1H NMR spectroscopy. This extract was selected for fractionation and subsequent purification based on the diversity of metabolites dereplicated in the crude extract (Table 1) and its larger production mass compared to the ISP2 and ISP2-S extracts (Figure 4). The ^1H NMR spectrum showed a wide range of interesting signals that indicated the chemical diversity of this extract. The UIAU-3F extract was fractionated and purified, as described in Section 3. Two known compounds, fumitremorgin C (**1**) [33] and pseurotin D (**2**) [34], were isolated and identified by comparing their experimental NMR and HRESIMS data (Figures S2–S6 and S8–S12) with the published literatures.

2.4. Determination of Antitrypanosomal and Antiplasmodium Activity Using Molecular Docking

In order to ascertain the antitrypanosomal and antiplasmodial activities of compounds **1** and **2**, molecular docking studies were conducted. Two enzymes were used in the docking study: Cruzain, the principal papain-like cysteine protease of *Trypanosoma cruzi*, and l-lactate dehydrogenase from *Plasmodium falciparum*. Cruzain is crucial for the survival and the multiplication of the parasite *Trypanosoma cruzi* [33], while l-lactate dehydrogenase is considered a potential molecular target for antimalarials due to the parasite's dependence on glycolysis for energy production [34]. To gain an insight into the differences in binding between the compounds and these proteins, rigid receptor docking was performed.

Docking poses were analysed and compared to benznidazole and chloroquine standards for antitrypanosomal and antiplasmodial activities, respectively. The first docking was performed on the crystal structure of cruzain from *Trypanosoma cruzi* (PDB: 3I06) [35]. In contrast, the second docking was performed on the crystal structure of L-lactate dehydrogenase from *Plasmodium falciparum* (PDB: 1LDG).

Compounds **1** and **2** were subjected to docking analysis, and the specificities of their interaction with these targets were identified, as shown in Figure 5 Based on binding energies and interacting residues, the best-docked complexes were obtained. Docking poses were analysed and compared to the standards. In both the two molecular docking studies, **1** and **2** docked very well compared to the standards (Table 2 and Figure 5).

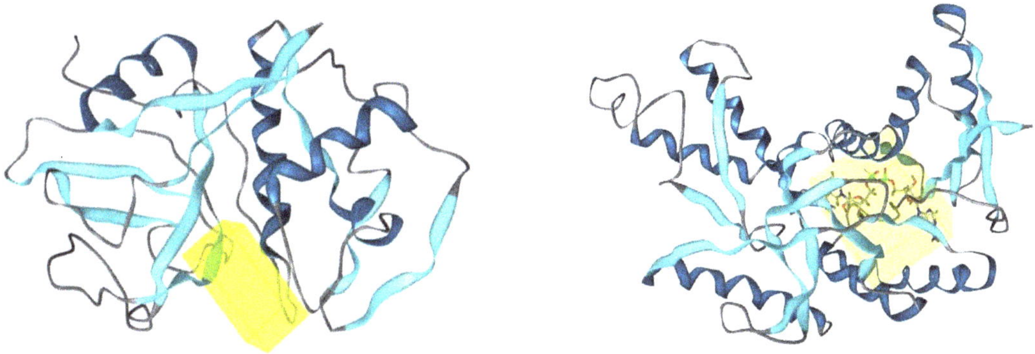

(a) (b)

Figure 5. (**a**) Binding site (yellow colour) of cruzain from *Trypanosoma cruzi.* (**b**) Binding site (yellow colour) l-lactate dehydrogenase from *Plasmodium falciparum.*

Table 2. Docking analysis of two ligands on two different protein receptors with respect to benznidazole and chloroquine standard.

Compounds	Docking Score (-) (kcal mol^{-1}) (PDB: 3I06) Cruzain from *Trypanosoma cruzi*	Docking Score (-) (kcal mol^{-1}) (PDB: 1LDG) l-lactate Dehydrogenase from *Plasmodium falciparum*
1	7.1	7.5
2	6.8	8.9
Benznidazole (standard) [a]	5.6	-
Chloroquine (standard) [b]	-	6.3

(Standard) [a] for antitrypanosomal activity. (Standard) [b] for antiplasmodium activity.

Ligplots in Figure 6 show significant amino acid residue interactions with the two ligands. In both cases, Trp184(A) was found to be involved in hydrophobic interactions. Compound **1** interacts with cruzain, thus acting as a hydrogen bond donator (HBD) at the receptor site interacting region involving residue Gly20A. Cruzain's residue Trp184(A) was also involved in aromatic interaction with the indole moiety of the structure. The interaction of **2** with cruzain involves both hydrogen bond acceptor (HBA) interaction with Trp184A and HBD with His162A.

Ligplots in Figure 7 show significant amino acid residue interactions with the two ligands. In both cases, Thr97(A), Ile31(A), Thr101(A), Ala98(A), and Ile54(A) were found to be involved in hydrophobic interactions. The interaction of compound **2** with l-lactate dehydrogenase involves both HBA interaction with Ile31(A) and HBD with Gly99(A).

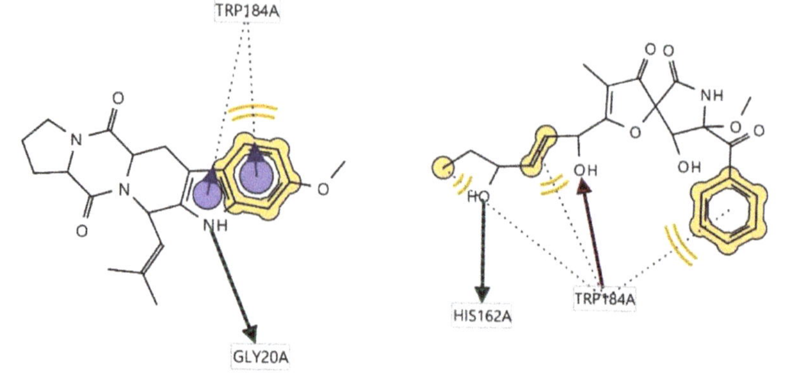

(**a**) (**b**)

Figure 6. Ligplots showing interacting residues of cruzain complex with compounds **1** (**a**) and **2** (**b**). Red arrow dotted lines, HBAs; green arrow dotted lines, HBDs; Yellow lines, H; Purple arrows, AR interactions.

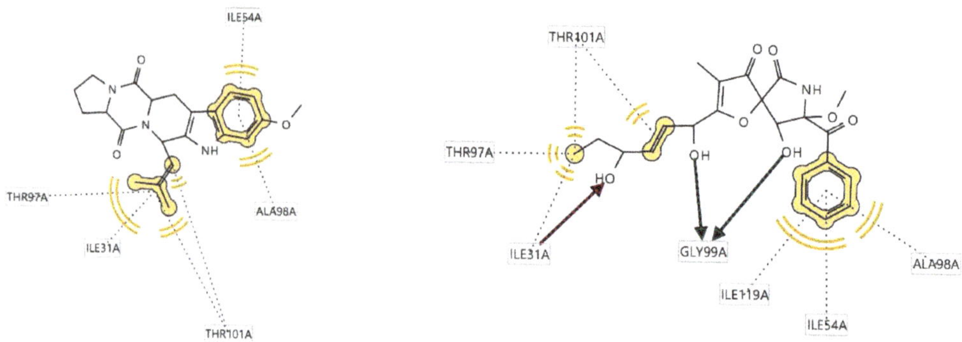

(**a**) (**b**)

Figure 7. Ligplots showing interacting residues of l-lactate dehydrogenase complex with compounds **1** (**a**) and **2** (**b**). Red arrow dotted lines, HBAs; green arrow dotted lines, HBDs; yellow lines, H interactions.

2.5. In Vitro Antiparasitic Activity

The compounds were tested using the in vitro β-B-galactosidase-transgenic *T. cruzi* and the *P. falciparum* 3D7 lactase dehydrogenase assay. Although compound **2** showed activity in the in silico studies, it displayed only moderate antitrypanosomal and antiplasmodial activities in vitro with IC_{50} values greater than 50 μM compared to the standards (Table 3). This may be due to several factors, such as low cell permeability to the parasite cells, interstitial hypertension, and metabolic degradation. However, **1** displayed significantly higher activities than **2** against *T. cruzi* and *P. falciparum*, with IC_{50} values of 9.6 μM and 2.3 μM (Figure 8). The IC_{50} values are higher than the standard: 2.6 μM for benznidazole and 0.017 μM for chloroquine.

Table 3. In vitro antitrypanosomal and antimalarial activities of **1** and **2**.

	IC_{50} (μM)	
Compounds	**Antitrypanosomal Activity** ***Trypanosoma cruzi*** **C2C4 Strain**	**Antimalarial Activity** ***Plasmodium falciparum*** **3D7 Strain**
1	9.6	2.3
2	>50	>50
Benznidazole [a]	2.6	NA
Chloroquine [b]	NA	0.017

[a] Benznidazole is the standard for the *T. cruzi* C2C4 strain. [b] Chloroquine is the standard for the *P. falciparum* 3D7 strain. NA, not applicable.

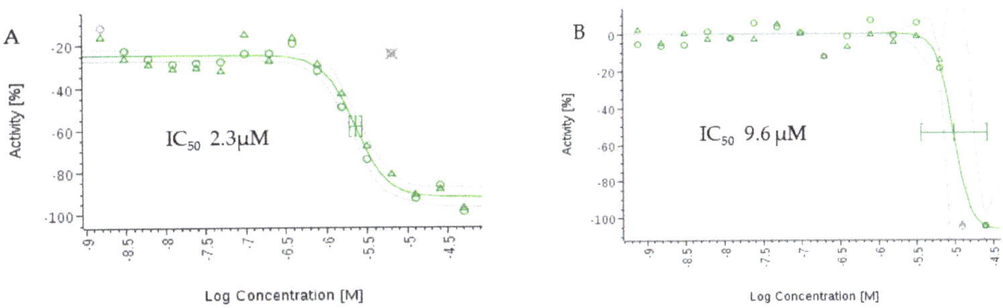

Figure 8. (**A**) Dose−dependent IC_{50} curve of **1** for *P. falciparum* 3D7. (**B**) Dose−dependent IC_{50} curve of **1** for *T. cruzi* Tulahuen.

Compound **1** showed moderate antiparasitic activity against *T. cruzi* Tulahuen C4 and *P. falciparum*; to our knowledge, this is the first account of the inhibitory activity of this compound against these two tropical parasites. Watts et al. suggest that the peroxide ring present in the fused spiro-pentacyclic diketopiperazines is not the bioactive pharmacophore. They screened the diketopiperazines verruculogen TR-2, fumitremorgin B, 12,13-dihydroxyfumitremorgin C, and cyclotryprostatin A isolated from *Aspergillus fumigatus* for their trypanocidal potential by testing for inhibitory activity against a panel of cysteine proteases. Fumitremorgin B and 12,13-dihydroxyfumitremorgin C exhibited IC_{50} values of 0.2 μM and 7.4 μM, respectively. The abundant enzyme isoform in the parasite is suggested to be rhodesain (also referred to as trypanopain or brucipain) [21], although it is believed that members of the rhodesain family of enzymes are not crucial for the sustenance of the parasite. However, they play essential roles in the second stage of host infection, parasite migration across the blood–brain barrier [36]. We have evaluated **1** and **2** against the major protein targets to determine whether the compounds have an affinity for the essential receptors connected to antiparasitic mechanisms. Molecular docking analysis indicated that these compounds gave potential binding consistent with antitrypanosomal and antiplasmodial activities.

3. Material and Methods

3.1. General Procedure

Bruker Avance III 400 and 600 MHz (equipped with a liquid nitrogen-cooled Prodigy cryoprobe) spectrometers were used to record NMR spectra at the University of Aberdeen, UK. Spectra were acquired at 25 °C and chemical shifts were referenced to the residual solvent peaks: 3.31 and 49.1 ppm (CD_3OD), 2.50 and 39.52 ppm (DMSO-d_6), and 4.78 ppm (D_2O). An edited Heteronuclear Single Quantum Coherence (HSQC) experiment was used to determine the multiplicities of the peaks in the ^{13}C NMR spectrum. Deuterated solvents were acquired from Cambridge Isotope Laboratories, Inc., UK. High-resolution electrospray mass spectra were obtained using a Bruker maXis II electrospray ionisation quadrupole time-of-flight (ESI-qToF) mass spectrometer coupled with an Agilent reversed-phase HPLC system. Solvents for HPLC, i.e., acetonitrile and methanol, were obtained from Rathburn Chemicals Ltd., Scotland, UK, while water was obtained from a Milli Q water system.

3.2. *Aspergillus fumigatus* Collection

Soil samples were collected from the River Oyun in Kwara State, Nigeria. The fungal strain, UIAU-3F, was isolated from the soil sediment. The collection site, with coordinates (8°27′57.8412″ N 4°40′3.4428″ E) (Figure 9), in Ilorin Kwara State, Nigeria, was highly contaminated with heavy metals [37,38]. The strains were isolated by Dr Olusoji Olusegun Adebisi from the University of Ilorin, Kwara State, and stored at the university's microbiology department with the voucher specimen number UIL-0022.

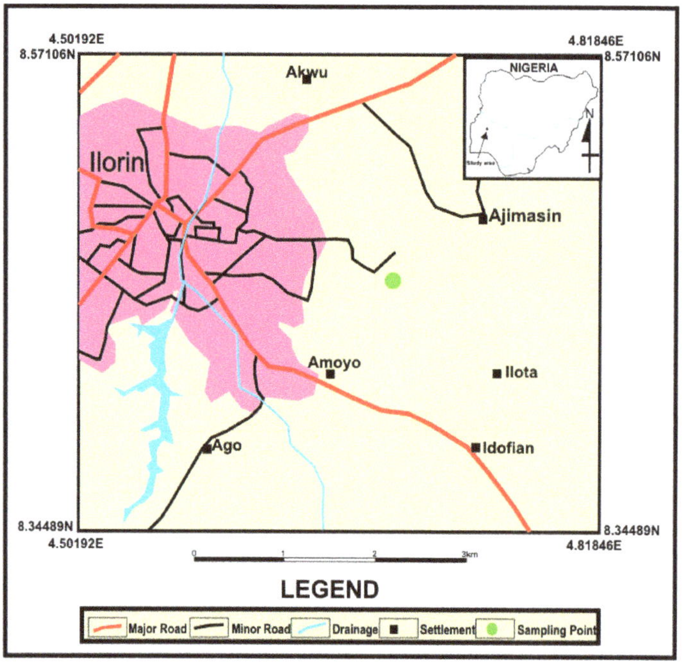

Figure 9. Map showing sample collection site.

3.3. Chemical Reagents

The high analytical grade chemicals and other reagents used in this study were all acquired from Fisher Scientific UK Ltd. in Leicestershire, UK.

3.4. DNA Isolation and Molecular Identification

The pure strain was preserved in duplicates in 25% aqueous glycerol stock and stored at −80 °C at the Marine Biodiscovery Centre, University of Aberdeen. Subsequently, it was inoculated in 100 mL of ISP2 liquid medium for 5 days; the pellet was collected in a 2 mL microcentrifuge tube and the supernatant was discarded. The DNA was extracted using the QIAGEN-DNeasy UltraClean Microbial Kit following the manufacturer's quick start protocol. Whole-genome sequencing of strain UIAU-3F was conducted by the National Collection of Industrial, Food and Marine Bacteria (NCIMB), Bucksburn, Aberdeen.

For assembly, the information collected from the single SMRT sequencing cell was used. To create the final genomic sequence, the raw reads were improved upon and assembled using the hierarchical genome assembly method (HGAP) [39].

3.5. Fermentation and Extraction

The isolated *Aspergillus fumigatus* strain was fermented under three different culturing conditions: solid rice medium (RICE), liquid ISP2 (malt extract 10 g L^{-1}, yeast extract 4 g L^{-1}, and glucose 4 g L^{-1}) medium with continuous shaking (ISP2-S), and static liquid ISP2 medium. The solid medium fermentation (RICE medium) was performed using 100 g of commercial rice and 110 mL water in a 1 L Erlenmeyer conical flask autoclaved for 20 min at 121 °C. After cooling, 2 mL of the fungal spore suspension was inoculated into the flask and incubated at 28 °C under static conditions for 30 days. Liquid fermentations were performed in 2 L Erlenmeyer flasks containing 1.5 L ISP2 medium. After inoculation, they were allowed to grow for 10 days at 30 °C while shaking at 150 rpm. The mycelium was removed from the liquid culture by filtration and soaked in 100% methanol for 24 h, while the liquid fungal medium was treated with the HP 20 resin (30 g L^{-1}). The combined methanol extract from the mycelium and the liquid fungal medium was evaporated until dryness under reduced pressure, before 500 mL of EtOAc was added to the solid rice media, and EtOAc extract was filtered and dried with a vacuum rotary evaporator (BÜCHI R-114, Switzerland) at 40 °C.

3.6. Metabolic Profiling and Multivariate Statistical Analysis

A final concentration of 0.1 mg mL^{-1} of crude extracts was prepared using 100% methanol, centrifuged and injected into a Phenomenex Kinetex XB-C18 (2.6 μm, 100 × 2.1 mm) column. Samples were analysed using a Bruker maXis II electrospray ionisation quadrupole time-of-flight (ESI-qToF) mass spectrometer coupled with an Agilent 1290 UHPLC system. The elution conditions were 5% MeCN + 0.1% FA to 100% MeCN + 0.1% FA in 15 min. The Bruker maXis II had a mass range of 100–2000 m/z and was equipped with the positive and negative mode conditions. The internal settings included 4.5 kV capillary voltage, 4.5 bar nebuliser gas, 12.0 L min^{-1} dry gas, and a dry temperature of 250 °C. MS/MS fragmentation experiments were adjusted under Auto MS/MS scan mode with no stepwise collision. A mass of m/z 248.96 was used for the negative mode and an m/z of 226 for positive mode calibration. The external reference lock mass (sodium formate) was infused at a constant flow of 0.18 mL h^{-1}.

Using MSConvert software, the HPLC-MS/MS data acquired from the crude extracts were converted from data analysis (.d) to.mzXML file format. The mzXML files were further processed using MZmine software [40]. The following parameter modules were used: mass detection (RT 2.5–35 min, centroid); chromatogram builder (MS level 1; minimum height 1.0 × 104; minimum time 0.5 min; m/z tolerance 10 ppm); deconvolution of the spectra (Savitzky–Golay algorithm); isotopic peaks grouper (m/z tolerance 10 ppm; retention time (RT) tolerance 0.1 min); duplicate peak filtering; smoothing; data alignment (join aligner; m/z tolerance 10 ppm; RT tolerance 0.5 min); gap-fling (intensity tolerance 20%; m/z tolerance 10 ppm; RT tolerance 0.5 min); and peak filtering range (0.00–0.5 min). After processing the data, peaks in several samples with the same RT and m/z were assumed to come from the same component. Furthermore, MetaboAnalyst 4.0 [41] was used for multivariate statistical analysis. The resulting peak intensity table was uploaded to MetaboAnalyst and, during the data integrity check, any missing values were replaced with small values. Prior to being auto-scaled and normalised to the median, the data were log-transformed. Heat maps of chemical profiles were constructed using hierarchical clustering and Principal Component Analysis (PCA).

3.7. Isolation of Metabolites

The ISP2-S crude extract (1.15 g) was sequentially fractionated using a modified Kupchan liquid–liquid partitioning method [42]. The crude extract was partitioned between an aqueous (water fraction; 0.30 g) and organic (dichloromethane fraction; 0.62 g) solvent layer. This afforded two fractions (A and D). Fraction D was then purified by solid-phase extraction (SPE) with the aid of a 10 g pre-packed Phenomenex strata C18-E (55 µm, 70 Å) column as the stationary phase. Aqueous methanol (25%, 50%, 100%, and 100% + 0.1% TFA) was the mobile phase. The fraction FD-100% MeOH (32.3 mg) was purified by HPLC and a Phenomenex C18 Luna phenyl-hexyl (250 × 10 mm 10 µ micron) column. The mobile phase was made of methanol, water, and trifluoroacetic acid (TFA) (0.05%). A linear gradient solvent from 60:40 H_2O/MeOH to 0:100 MeOH ran for 30 min with a flow rate of 1.8 mL min^{-1} to afford compounds **1** (4.42 mg, t_R = 13.9 min) and **2** (3.80 mg, t_R = 15.6 min). Optical rotations of both compounds were recorded using a PerkinElmer model 241 polarimeter and compared to that obtained from the literature.

3.8. Molecular Docking

The crystallographic protein structures used for docking were PDB: 3I06 [43], which is cruzain from *Trypanosoma cruzi*, and PDB: 1LDG [44] of l-lactate dehydrogenase from *Plasmodium falciparum*. These were obtained from the Protein Data Bank and used to perform docking simulations. The receptor sites were predicted using LigandScout (Inte: Ligand) Advanced software [35], which identifies putative binding pockets by creating a grid surface and calculating the buriedness value of each grid point on the surface. The resulting pocket grid consists of several clusters of grid points, rendered using an iso surface connecting the grid points. The iso surface represents an empty space suitable for creating a pocket.

Molecular docking analysis was performed using Autodock Vina v.1.2.0 (The Scripps Research Institute, La Jolla, CA, USA) docking software [45]. All ligands and protein structures were prepared using the Dock Prep tool with default parameters in Chimera 1.16 [46]. The net charges were set to neutral for all ligands. The box centre and size coordinates in Angstrom units for PDB: 3I06 were −7.1 × −33.4 × −0.6 and 11.2 × 13.3 × 16.7 and for PDB: 1LDG, 32.9 × 25.7 × 34.6 and 20.4 × 17.8 × 22.0 around the active site. The search parameters used for production runs were: number of binding modes = 10, exhaustiveness = 32, and maximum energy difference = 3 kcal mol^{-1}. The results were tested for convergence at exhaustiveness 8, 16, 24, and 32 keeping all the parameters fixed. LigandScout (Inte: Ligand) Advanced software [35] was used to visualise and calculate protein–ligand interactions.

3.9. In Vitro Antitrypanosomal Activity

The procedure used for the in vitro -β-Galactosidase transgenic *T. cruzi* experiment was previously described by Annang and colleagues [47]. It entails *T. cruzi* Tulahuen C4 strain expression of the galactosidase gene (LacZ), with L6 rat skeletal muscle cells serving as the host cells. At 37 °C and 5% CO_2, they were grown in RPMI-1640 that had been supplemented with 10% iFBs, 2 mM l-glutamine, 100 U mL^{-1} penicillin, and 100 g mL^{-1} streptomycin. To test the compounds, *T. cruzi* amastigote-infected L6 cell culture (2103 infected L6 cells per well) was dispersed into 384-well assay plates (1, 2). The initial concentration was 25 M, with each molecule examined in triplicate in 16-point dose–response curves (1/2 serial dilution). The plates were incubated for 96 h at 37 °C. Following the addition, 1.5 L of 100 M CRPG and 0.1% NP40 were further incubated at 37 °C for 4 h in the dark. The absorbance at 585 nm was measured using an EnVision plate reader (PerkinElmer, Waltham, MA, USA). The in-plate was used to normalise the test. Benznidazole at 10 µg mL^{-1} was used as the negative control and 0.167% DMSO as the positive control.

3.10. Plasmodium falciparum 3D7 Lactase Dehydrogenase In-Vitro Assay

The *P. falciparum* 3D7 strain parasites were cultured in a freshly collected type O positive (O+) human erythrocyte. The assay was executed using a previously described standardised method by Annang et al. [48]. After 72 h of incubating the parasites with the test compounds (**1**, **2**), the synthetic cofactor APAD+, which is specific to the parasite LDH enzyme [49], was used to quantify the proportions of parasite viability by measuring the activity of this intracellular enzyme that is released upon lysis of the parasites. The assay was performed in triplicate using a 16 points dose–response curve (1/2 serial dilution) with starting concentrations of 50 μM in 384-well plates. Each plate had 100 nM of chloroquine as a negative control and a parasite culture media as a positive growth control. LDH activity was assessed after the plates had been incubated for 72 h, frozen for 4 h, and then thawed at room temperature for 1 h. To perform this, 70 mL of a freshly made solution containing 100 mM Tris-HCL at a pH of 8.0, 143 mM sodium l-lactate, 143 mM APAD, 178.75 mM NBT, 1 g mL^{-1} diaphorase, and 0.7% Tween 20 was poured into the plates. After a 10 min incubation period at room temperature and a gentle shake of the plates to ensure uniformity, the absorbance was measured at 650 nm. The absorbance in this test was measured using the EnVision plate reader (PerkinElmer, Waltham, MA, USA).

3.11. Statistical Analysis

All experiments were performed in triplicate and results were expressed as the mean ± SEM. The statistical significance of the means differences was confirmed by analysis of variance (ANOVA) with Duncan's post hoc tests, with *p*-values < 0.05 considered statistically significant.

4. Conclusions

We investigated the metabolomic diversity and the antiparasitic activity of fumitremorgin C (**1**) and pseurotin D (**2**) isolated from *Aspergillus fumigatus* found in soil samples collected from the River Oyun in Kwara State, Nigeria. Fermentation of the fungus under different growth conditions (OSMAC approach) stimulated their metabolite production, as evidenced by the LC-HRMS metabolomic profiling. In vitro studies revealed **1** as the most active against *Trypanosoma cruzi* and *Plasmodium falciparum*, with IC_{50} values of 9.6 and 2.3 μM, respectively. Molecular docking supported this observation by predicting binding modes and affinities in line with the experimental activity profile. As a result, we feel this compound warrants further investigation, including in vivo and mechanism of action studies.

Supplementary Materials: The following supporting information can be downloaded at: https://www.mdpi.com/article/10.3390/antibiotics12010109/s1, Figure S1: Structure of fumitremorgin C (**1**); Figure S2: ESI MS (positive ionisation mode) of fumitremorgin C (**1**); Figure S3: 1H NMR spectrum of fumitremorgin C (**1**) at 600 MHz in CD3OD; Figure S4: HSQC spectrum of fumitremorgin C (**1**) at 600 MHz in CD3OD; Figure S5: COSY spectrum of fumitremorgin C (**1**) at 600 MHz in CD3OD; Figure S6: HMBC spectrum of fumitremorgin C (**1**) at 600 MHz in CD3OD; Figure S7: Structure of pseurotin D (**2**); Figure S8: ESI MS (positive ionisation mode) of pseurotin D (**2**); Figure S9: 1H NMR spectrum of pseurotin D (**2**) at 600 MHz in CD3OD; Figure S10: HSQC spectrum of pseurotin D (**2**) at 600 MHz in CD3OD; Figure S11: COSY spectrum of pseurotin D (**2**) at 600 MHz in CD3OD; Figure S12: HMBC spectrum of pseurotin D (**2**) at 600 MHz in CD3OD; Table S1: NMR data for fumitremorgin C (**1**); Table S2: NMR data for pseurotin D (**2**).

Author Contributions: Conceptualisation, O.A.D.; methodology, O.A.D., E.T.O. and F.A.; funding acquisition, O.A.D.; data curation, O.A.D.; formal analysis, O.A.D.; writing—original draft preparation, O.A.D.; writing—review and editing, O.A.D., E.T.O., M.J. and R.E.; raw material acquisition, O.O.A.; data analysis, O.A.D., G.P., E.T.O., G.P.-M., C.B.-N., F.A., A.A.F. and B.F.M.; supervision, M.J. and R.E. All authors have read and agreed to the published version of the manuscript.

Funding: The Schlumberger Foundation Faculty for the Future Scholarship Program funded this research and the article processing charge. B.F.M. acknowledges national funds from the Portuguese FCT–Fundação para a Ciência e a Tecnologia, I.P., within the projects UIDB/04564/2020 and UIDP/04564/2020.

Institutional Review Board Statement: Not applicable.

Informed Consent Statement: Not applicable.

Data Availability Statement: Not applicable.

Acknowledgments: The authors are grateful to the Schlumberger Foundation Faculty for the Future board for their financial support in conducting the present study.

Conflicts of Interest: The authors declare no conflict of interest.

References

1. Klich, M.A. Biogeography of Aspergillus Species in Soil and Litter. *Mycologia* **2002**, *94*, 21–27. [CrossRef] [PubMed]
2. Lević, J.; Gošič-dondo, S.N.; Ivanović, D.; Stanković, S.; Krnjaja, V.; Bočarov-Stančić, A.; Stepanić, A. An outbreak of Aspergillus species in response to environmental conditions in Serbia. *Pestic. Phytomed./Pestic. Fitomed.* **2013**, *7*, 28.
3. Frisvad, J.C.; Rank, C.; Nielsen, K.F.; Larsen, T.O. Metabolomics of Aspergillus fumigatus. *Med. Mycol.* **2009**, *47* (Suppl. S1), S53–S71. [CrossRef] [PubMed]
4. Vadlapudi, V.; Borah, N.; Yellusani, K.R.; Gade, S.; Reddy, P.; Rajamanikyam, M.; Vempati, L.N.S.; Gubbala, S.P.; Chopra, P.; Upadhyayula, S.M.; et al. Aspergillus Secondary Metabolite Database, a resource to understand the Secondary metabolome of *Aspergillus genus. Sci. Rep.* **2017**, *7*, 7325. [CrossRef] [PubMed]
5. Liu, J.-Y.; Song, Y.; Zhang, Z.; Wang, L.; Guo, Z.; Zou, W.; Tan, R. *Aspergillus fumigatus* CY018, an endophytic fungus in Cynodon dactylon as a versatile producer of new and bioactive metabolites. *J. Biotechnol.* **2004**, *114*, 279–287. [CrossRef]
6. Waksman, S.A.; Geiger, W.B. The Nature of the Antibiotic Substances Produced by *Aspergillus fumigatus*. *J. Bacteriol.* **1944**, *47*, 391–397. [CrossRef]
7. Hayashi, A.; Fujioka, S.; Nukina, M.; Kawano, T.; Shimada, A.; Kimura, Y. Fumiquinones A and B, nematicidal quinones produced by *Aspergillus fumigatus*. *Biosci. Biotechnol. Biochem.* **2007**, *71*, 1697–1702. [CrossRef]
8. Parker, G.F.; Jenner, P.C. Distribution of trypacidin in cultures of *Aspergillus fumigatus*. *Appl. Microbiol.* **1968**, *16*, 1251–1252. [CrossRef]
9. Balan, J.; Ebringer, L.; Němec, P. Trypacidin a new antiprotozoal antibiotic. *Die Nat.* **1964**, *51*, 227. [CrossRef]
10. Van Middlesworth, F.; Dufresne, C.; Wincott, F.E.; Mosley, R.T.; Wilson, K.E. Determination of the relative and absolute stereochemistry of sphingofungins A, B, C and D. *Tetrahedron Lett.* **1992**, *33*, 297–300.
11. Lindequist, U.; Lesnau, A.; Teuscher, E.; Pilgrim, H. The antiviral action of ergosterol peroxide. *Die Pharm.* **1989**, *44*, 579–580.
12. Tomoda, H.; Kim, Y.K.; Nishida, H.; Masuma, R.; Omura, S. Pyripyropenes, novel inhibitors of acyl-CoA: Cholesterol acyltransferase produced by *Aspergillus fumigatus* I. Production, isolation, and biological properties. *J. Antibiot.* **1994**, *47*, 148–153. [CrossRef]
13. Tomoda, H.; Ōmura, S. Pyripyropenes, highly potent inhibitors of acyl-CoA: Cholesterol acyltransferase produced by *Aspergillus fumigatus*. *J. Antibiot.* **1993**, *46*, 1168–1169.
14. Bhatti, A.B.; Usman, M.; Kandi, V. Current Scenario of HIV/AIDS, Treatment Options, and Major Challenges with Compliance to Antiretroviral Therapy. *Cureus* **2016**, *8*, e515. [CrossRef]
15. May, J.F. World Population Policies: Their Origin, Evolution, and Impact. *Can. Stud. Popul.* **2012**, *39*, 125.
16. Roberts, D.; Matthews, G. Risk factors of malaria in children under the age of five years old in Uganda. *Malar. J.* **2016**, *15*, 246. [CrossRef]
17. Afoakwah, C.; Deng, X.; Onur, I. Malaria infection among children under-five: The use of large-scale interventions in Ghana. *BMC Public Health* **2018**, *18*, 536. [CrossRef]
18. Iskander, D. *The Power of Parasites: Malaria as (un) Conscious Strategy*; Springer Nature: Singapore, 2021.
19. Crompton, D.W.; Savioli, L. Intestinal parasitic infections and urbanization. *Bull. World Health Organ.* **1993**, *71*, 1–7.
20. Peniche, A.G.; Renslo, A.R.; Melby, P.C.; Travi, B.L. Antileishmanial Activity of Disulfiram and Thiuram Disulfide Analogs in an Ex Vivo Model System Is Selectively Enhanced by the Addition of Divalent Metal Ions. *Antimicrob. Agents Chemother.* **2015**, *59*, 6463–6470. [CrossRef]
21. Watts, K.R.; Ratnam, J.; Ang, K.H.; Tenney, K.; Compton, J.E.; McKerrow, J.; Crews, P. Assessing the trypanocidal potential of natural and semi-synthetic diketopiperazines from two deep water marine-derived fungi. *Bioorganic Med. Chem.* **2010**, *18*, 2566–2574. [CrossRef]
22. Tagboto, S.; Townson, S. Antiparasitic properties of medicinal plants and other naturally occurring products. *Adv. Parasitol.* **2001**, *50*, 199–295. [CrossRef] [PubMed]
23. Wink, M. Medicinal Plants: A Source of Anti-Parasitic Secondary Metabolites. *Molecules* **2012**, *17*, 12771–12791. [CrossRef] [PubMed]
24. Simarro, P.P.; Cecchi, G.; Paone, M.; Franco, J.R.; Diarra, A.; Ruiz, A.J.; Fèvre, E.M.; Courtin, F.; Mattioli, R.C.; Jannin, J.G. The Atlas of human African trypanosomiasis: A contribution to global mapping of neglected tropical diseases. *Int. J. Health Geogr.* **2010**, *9*, 57. [CrossRef] [PubMed]
25. Kourbeli, V.; Chontzopoulou, E.; Moschovou, K.; Pavlos, D.; Mavromoustakos, T.; Papanastasiou, I.P. An overview on target-based drug design against kinetoplastid protozoan infections: Human African trypanosomiasis, Chagas disease and leishmaniases. *Molecules* **2021**, *26*, 4629. [CrossRef] [PubMed]

26. Tian, D.; Gou, X.; Jia, J.; Wei, J.; Zheng, M.; Ding, W.; Bi, H.; Wu, B.; Tang, J. New diketopiperazine alkaloid and polyketides from marine-derived fungus *Penicillium* sp. TW58-16 with antibacterial activity against *Helicobacter pylori*. *Fitoterapia* **2022**, *156*, 105095. [CrossRef] [PubMed]
27. Zhang, R.; Wang, H.; Chen, B.; Dai, H.; Sun, J.; Han, J.; Liu, H. Discovery of Anti-MRSA Secondary Metabolites from a Marine-Derived Fungus *Aspergillus fumigatus*. *Mar. Drugs* **2022**, *20*, 302. [CrossRef]
28. Zhang, H.; Liu, R.; Yang, J.; Li, H.; Zhou, F. Bioactive Alkaloids of Aspergillus fumigatus, an Endophytic Fungus from *Astragalus membranaceus*. *Chem. Nat. Compd.* **2017**, *53*, 802–805. [CrossRef]
29. Nakanishi, K.; Doi, M.; Usami, Y.; Amagata, T.; Minoura, K.; Tanaka, R.; Numata, A.; Yamada, T. Anthcolorins A–F, novel cytotoxic metabolites from a sea urchin-derived *Aspergillus versicolor*. *Tetrahedron* **2013**, *69*, 4617–4623. [CrossRef]
30. Steyn, P.S.; Vleggaar, R.; Rabie, C.J. Alkaloids from *Aspergillus caespitosus*. *Phytochemistry* **1981**, *20*, 538–539. [CrossRef]
31. Yan, L.-H.; Li, X.-M.; Chi, L.-P.; Li, X.; Wang, B.-G. Six New Antimicrobial Metabolites from the Deep-Sea Sediment-Derived Fungus *Aspergillus fumigatus* SD-406. *Mar. Drugs* **2021**, *20*, 4. [CrossRef]
32. Ma, Y.-M.; Liang, X.-A.; Zhang, H.-C.; Liu, R. Cytotoxic and Antibiotic Cyclic Pentapeptide from an Endophytic *Aspergillus tamarii* of *Ficus carica*. *J. Agric. Food Chem.* **2016**, *64*, 3789–3793. [CrossRef]
33. Sajid, M.; Robertson, S.A.; Brinen, L.S.; McKerrow, J.H. Cruzain—The path from target validation to the clinic. In *Cysteine Proteases of Pathogenic Organisms*; Robinson, M.W., Dalton, J.P., Eds.; Springer: Berlin/Heidelberg, Germany, 2011.
34. Egan, T.J.; Ncokazi, K.K. Quinoline antimalarials decrease the rate of β-hematin formation. *J. Inorg. Biochem.* **2005**, *99*, 1532–1539. [CrossRef]
35. Wolber, G.; Langer, T. LigandScout: 3-D pharmacophores derived from protein-bound ligands and their use as virtual screening filters. *J. Chem. Inf. Model.* **2005**, *45*, 160–169. [CrossRef]
36. Wilson, G.; Bryan, J.; Cranston, K.; Kitzes, J.; Nederbragt, L.; Teal, T.K. Good enough practices in scientific computing. *PLoS Comput. Biol.* **2017**, *13*, e1005510. [CrossRef]
37. Ogunkunle, C.O.; Mustapha, K.; Oyedeji, S.; Fatoba, P.O. Assessment of metallic pollution status of surface water and aquatic macrophytes of earthen dams in Ilorin, north-central of Nigeria as indicators of environmental health. *J. King Saud Univ.-Sci.* **2016**, *28*, 324–331. [CrossRef]
38. Okoro, H.K.; Jimoh, H.A. Speciation and determination of priority metals in sediments of Oyun River, Ilorin, Kwara, Nigeria. *Bull. Chem. Soc. Ethiop.* **2016**, *30*, 199–208. [CrossRef]
39. Chin, C.S.; Alexander, D.H.; Marks, P.; Klammer, A.A.; Drake, J.; Heiner, C.; Clum, A.; Copeland, A.; Huddleston, J.; Eichler, E.E.; et al. Nonhybrid, finished microbial genome assemblies from long-read SMRT sequencing data. *Nat. Methods* **2013**, *10*, 563–569. [CrossRef]
40. Adusumilli, R.; Mallick, P. Data conversion with ProteoWizard msConvert. In *Proteomics*; Humana Press: New York, NY, USA, 2017; pp. 339–368.
41. Chong, J.; Soufan, O.; Li, C.; Caraus, I.; Li, S.; Bourque, G.; Wishart, D.S.; Xia, J. MetaboAnalyst 4.0: Towards more transparent and integrative metabolomics analysis. *Nucleic Acids Res.* **2018**, *46*, W486–W494. [CrossRef]
42. Brown, K.S., Jr.; Kupchan, S.M. A convenient separation of alkaloid mixtures by partition chromatography, using an indicator in the stationary phase. *J. Chromatogr. A* **1962**, *9*, 71–80. [CrossRef]
43. Mott, B.T.; Ferreira, R.S.; Simeonov, A.; Jadhav, A.; Ang, K.K.; Leister, W.; Shen, M.; Silveira, J.T.; Doyle, P.S.; Arkin, M.R.; et al. Identification and optimization of inhibitors of trypanosomal cysteine proteases: Cruzain, rhodesain, and TbCatB. *J. Med. Chem.* **2010**, *53*, 52–60. [CrossRef]
44. Dunn, C.R.; Banfield, M.J.; Barker, J.J.; Higham, C.W.; Moreton, K.M.; Turgut-Balik, D.; Brady, R.L.; Holbrook, J.J. The structure of lactate dehydrogenase from *Plasmodium falciparum* reveals a new target for anti-malarial design. *Nat. Struct. Biol.* **1996**, *3*. [CrossRef] [PubMed]
45. Eberhardt, J.; Santos-Martins, D.; Tillack, A.F.; Forli, S. AutoDock Vina 1.2. 0: New docking methods, expanded force field, and python bindings. *J. Chem. Inf. Model.* **2021**, *61*, 3891–3898. [CrossRef] [PubMed]
46. Pettersen, E.F.; Goddard, T.D.; Huang, C.C.; Couch, G.S.; Greenblatt, D.M.; Meng, E.C.; Ferrin, T.E. UCSF Chimera—A visualization system for exploratory research and analysis. *J. Comput. Chem.* **2004**, *25*, 1605–1612. [CrossRef] [PubMed]
47. Annang, F.; Pérez-Moreno, G.; García-Hernández, R.; Cordon-Obras, C.; Martín, J.; Tormo, J.R.; Rodríguez, L.; De Pedro, N.; Gómez-Pérez, V.; Valente, M.; et al. High-throughput screening platform for natural product–based drug discovery against 3 neglected tropical diseases: Human african trypanosomiasis, leishmaniasis, and chagas disease. *J. Biomol. Screen.* **2015**, *20*, 82–91. [CrossRef]
48. Annang, F.; Perez-Victoria, I.; Perez-Moreno, G.; Domingo, E.; Gonzalez, I.; Tormo, J.R.; Martin, J.; Ruiz-Perez, L.M.; Genilloud, O.; Gonzalez-Pacanowska, D.; et al. MDN-0185, an antiplasmodial polycyclic xanthone isolated from *Micromonospora* sp. CA-256353. *J. Nat. Prod.* **2018**, *81*, 1687–1691. [CrossRef]
49. Ignatushchenko, M.V.; Riscoe, M.; Winter, R.W. Xanthones as antimalarial agents: Stage specificity. *Am. J. Trop. Med. Hyg.* **2000**, *62*, 77–81. [CrossRef]

antibiotics

Article

In Silico and In Vitro Analysis of Sulforaphane Anti-*Candida* Activity

Bruna L. R. Silva [1,†], Gisele Simão [2,3,†], Carmem D. L. Campos [4,†], Cinara R. A. V. Monteiro [4], Laryssa R. Bueno [2,3], Gabriel B. Ortis [2,3], Saulo J. F. Mendes [5], Israel Viegas Moreira [4], Daniele Maria-Ferreira [2,3], Eduardo M. Sousa [1,5], Flávia C. B. Vidal [6], Cristina de Andrade Monteiro [7], Valério Monteiro-Neto [1,4,*] and Elizabeth S. Fernandes [2,3,*]

1 Programa de Pós-Graduação em Biodiversidade e Biotecnologia da Rede BIONORTE, Universidade Federal do Maranhão, São Luís 65085-040, MA, Brazil
2 Programa de Pós-Graduação em Biotecnologia Aplicada à Saúde da Criança e do Adolescente, Faculdades Pequeno Príncipe, Av. Iguaçu No 333, Curitiba 80230-020, PR, Brazil
3 Instituto de Pesquisa Pelé Pequeno Príncipe, Av. Silva Jardim No 1632, Curitiba 80240-020, PR, Brazil
4 Programa de Pós-Graduação em Ciências da Saúde, Universidade Federal do Maranhão, Av. dos Portugueses No 1966, Cidade Universitária Dom Delgado, São Luís 65085-040, MA, Brazil
5 Universidade CEUMA, Graduation Programme, Rua Josué Montello No 1, Renascença II, São Luís 65075-120, MA, Brazil
6 Departamento de Morfologia, Universidade Federal do Maranhão, Av. dos Portugueses No 1966, Cidade Universitária Dom Delgado, São Luís 65085-040, MA, Brazil
7 Departamento Acadêmico de Biologia, Instituto Federal do Maranhão, Av. Getúlio Vargas No 2158/2159, Monte Castelo, São Luís 65030-005, MA, Brazil
* Correspondence: valerio.monteiro@ufma.br (V.M.-N.); elizabeth.fernandes@pelepequenoprincipe.org.br (E.S.F.)
† These authors equally contributed to this work.

Citation: Silva, B.L.R.; Simão, G.; Campos, C.D.L.; Monteiro, C.R.A.V.; Bueno, L.R.; Ortis, G.B.; Mendes, S.J.F.; Moreira, I.V.; Maria-Ferreira, D.; Sousa, E.M.; et al. In Silico and In Vitro Analysis of Sulforaphane Anti-*Candida* Activity. *Antibiotics* **2022**, *11*, 1842. https://doi.org/10.3390/antibiotics11121842

Academic Editors: Carlos M. Franco and Marc Maresca

Received: 31 October 2022
Accepted: 15 December 2022
Published: 19 December 2022

Abstract: Oropharyngeal candidiasis/candidosis is a common and recurrent opportunistic fungal infection. Fluconazole (FLZ), one of the most used and effective antifungal agents, has been associated with a rise of resistant *Candida* species in immunocompromised patients undergoing prophylactic therapy. Sulforaphane (SFN), a compound from cruciferous vegetables, is an antimicrobial with yet controversial activities and mechanisms on fungi. Herein, the in silico and antifungal activities of SFN against *C. albicans* were investigated. In silico analyzes for the prediction of the biological activities and oral bioavailability of SFN, its possible toxicity and pharmacokinetic parameters, as well as the estimates of its gastrointestinal absorption, permeability to the blood-brain barrier and skin, and similarities to drugs, were performed by using different software. SFN in vitro anti-*Candida* activities alone and in combination with fluconazole (FLZ) were determined by the broth microdilution method and the checkerboard, biofilm and hyphae formation tests. Amongst the identified probable biological activities of SFN, nine indicated an antimicrobial potential. SFN was predicted to be highly absorbable by the gastrointestinal tract, to present good oral availability, and not to be irritant and/or hepatotoxic. SFN presented antifungal activity against *C. albicans* and prevented both biofilm and hyphae formation by this microorganism. SFN was additive/synergistic to FLZ. Overall, the data highlights the anti-*Candida* activity of SFN and its potential to be used as an adjuvant therapy to FLZ in clinical settings.

Keywords: sulforaphane; antifungal activity; *Candida albicans*; in silico analysis

1. Introduction

Oropharyngeal candidiasis/candidosis (OPC) is the most common, prevalent, and recurrent opportunistic fungal infection [1] and occurs in approximately 90% of HIV+ patients [2]. Opportunistic infections can be caused by different *Candida* spp., such as *C. tropicalis*, *C. glabrata*, *C. parapsilosis*, *C. krusei* and *C. dubliniensis* [3,4]; however, *C. albicans* is described as the most frequent species associated with oral opportunistic infections [5,6].

The spectrum of *Candida* spp. infection is diverse, as these pathogens can cause from asymptomatic or mild oral disease [7] to OPC, esophagitis, vulvovaginitis, onychomycosis, cutaneous candidiasis, and even evolve into invasive or systemic diseases [8]. Although highly active antiretroviral therapy has improved the prognosis of AIDS, contributing to the decline of most opportunistic infections, HIV+ patients continue to experience significant morbidity associated with *Candida*-induced infections [9,10].

Several antifungal agents are available for OPC treatment [11]. The most frequently used azole derivatives are itraconazole and fluconazole (FLZ) (both for oral administration) due to their low cost and toxicity [12]. FLZ is considered the most effective antifungal agent in OPC [13]. However, recent studies have reported isolates of resistant *Candida* species (for review, see: [14]) in immunocompromised patients [15–17], and associated this event with the increased use of azoles, especially FLZ, in prophylactic therapy [18], limiting therapy success [19,20]. In this context, the search for new antifungal compounds and also therapies able to inhibit *Candida* spp. adhesion, yeast-hyphae transition, and biofilm formation, with low toxicity, are necessary and urgent.

Isothiocyanates (ITCs) are secondary metabolites generated by the hydrolysis of glucosinolates, which are found in cruciferous plants of the Brassicaceae family [21]. ITCs attract attention due to their antibacterial [22–24], antifungal [25], antiviral [26], anticancer [27], and chemopreventive actions [28,29] associated with moderate toxicity to normal human cells.

One of the main and promising ITCs is sulforaphane (SFN: $C_6H_{11}NOS_2$), a natural phytochemical derived from cruciferous vegetables such as broccoli, cabbage, and cauliflower, with several cytoprotective effects in human cells [30,31]. SFN is anti-inflammatory [32], antioxidant [33], anti-angiogenic [34], anticancer [35,36] and neuroprotective [37,38] depending on the concentration or dose and duration of exposure. In this context, SFN benefits to human health have been investigated in different clinical trials for airway, gastrointestinal and metabolic diseases, as well as autism and cancer (for review, see: [39]).

SFN antimicrobial potential has also been investigated, especially against bacteria, with little information on its antifungal potential and mechanisms. SFN is antimicrobial against *Helicobacter pylori* [40,41] and inhibits a range of Gram (−) bacteria, including *Escherichia coli* and *Pseudomonas aeruginosa*, and Gram (+) bacteria, such as *Enterococcus faecalis*, and *Staphylococcus aureus* [42,43]. Although the antibacterial effects of SFN have been consistent across these studies, its antifungal activity is yet controversial. In fact, some studies showed that *C. albicans* is resistant to SFN [43], whilst others demonstrated that *C. albicans* and *Aspergillus niger* strains are sensitive to this compound [44–46]. In order to gain knowledge regarding the antifungal activity of SFN, the in silico and in vitro activities of this compound were assessed against *C. albicans* ATCC 90028 and oral isolates from HIV+ patients. In addition, the ability of SFN to potentiate the effects of FLZ against *C. albicans* was investigated.

2. Results

2.1. *In Silico Analysis*

2.1.1. Analysis of Biological Activity

The analysis of probable biological activities indicated that SFN has a >30% (Pa > 0.3) probability of presenting 147 activities. Of these, 30 have a moderate probability (Pa > 0.5) of occurrence and eight have a high probability of occurrence (Pa > 0.7). Of the total activities identified, nine were antimicrobial (Table 1). The biological activities of FLZ were also analyzed. Of the total evaluated, 56 activities with >30% (Pa > 0.3) occurrence were identified. Of those, only eight had a high probability of occurrence (Pa > 0.7). Amongst the biological activities with >30% probability of occurrence, seven were antimicrobial (Table 1). Additional non-antimicrobial biological activities of SFN and FLZ are described in Supplementary Table S2.

Table 1. In silico identification of the antimicrobial activities of sulforaphane (SFN) and fluconazole (FLZ).

SFN			FLZ		
Antimicrobial Activities	**Pa**	**Pi**	**Antimicrobial Activities**	**Pa**	**Pi**
Anti-*Helicobacter pylori*	0.742	0.002	Lanosterol 14 alpha demethylase inhibitor	0.846	0.001
Yeast RNA Inhibitor	0.444	0.019	Steroid synthesis inhibitor	0.744	0.001
Glycoprotein-phosphatidylinositol inhibitor	0.516	0.093	Antifungal	0.726	0.008
Antiparasitic	0.441	0.023	Phospholipid translation ATPase Inhibitor	0.480	0.069
Omptin inhibitor	0.469	0.091	NADPH-cytochrome-c2 reductase inhibitor	0.366	0.134
Phospholipid translation ATPase Inhibitor	0.363	0.140	Sugar-phosphatase inhibitor	0.356	0.148
Endopeptidase So inhibitor	0.327	0.101	Cell wall synthesis inhibitor	0.351	0.002
Mannose isomerase inhibitor	0.302	0.071			
P. gingivalis TPR protease inhibitor	0.301	0.108			

Pa: probability of a compound being active; Pi: probability of a compound being inactive.

2.1.2. Estimated Oral Bioavailability and Expected Toxicity

To predict the oral bioavailability of SFN, its partition coefficient (water/oil; iLogP), molecular weight (MW), total polar surface area (TPSA), number of hydrogen bond donors (nHBD), and number of hydrogen bond acceptors (nHBA) values were analyzed. Our results demonstrate that SFN fits the criteria to have good estimated oral bioavailability (iLogP = 2.11; molecular weight = 177.29 g/mol; TPSA = 29.43 Å^2; nHBD = 2; and nHBA = 0) (Table 2). For comparison, FLZ exhibited iLogP = 0.41; molecular weight = 306.27 g/mol; TPSA = 81.65 Å^2; nHBD = 1; and nHBA = 7 (Table 2).

Table 2. In silico estimations of the oral bioavailability, toxic effects, absorption, solubility, and drug-likeness score of sulforaphane in comparison with fluconazole.

Estimated Oral Bioavailability	**SFN**	**FLZ**
iLogP	2.11	0.41
MW (g/mol)	177.29	306.27
TPSA	29.43 Å^2	81.65 Å^2
nHBD	2	1
nHBA	0	7
Predicted toxic effects		
Mutagenic effects	Moderate	No
Tumorigenic effects	Moderate	No
Irritant effects	None	No
Hepatotoxicity	None	No
Effects on reproduction	Moderate	No
LD_{50}	1000 mg/kg	1410 mg/kg
Toxicity class	4	4
Estimated absorption		
GI absorption	High	High
BBB permeability	No	No
Log K_p	−6.38 cm/s	−7.92 cm/s

Table 2. *Cont.*

Estimated Oral Bioavailability	SFN	FLZ
Predicted solubility and drug-likeness and score		
Log S	−2.10	−2.17
DL	−6.47	1.99
DS	0.25	0.87

BBB: blood-brain barrier; DL: drug-likeness; DS: drug-score; GI: gastrointestinal absorption; iLogP: partition coefficient water: oil–lipophilicity index; LD_{50}: lethal dose 50%; Log Kp: skin permeation index; Log S: solubility; MW: molecular weight; nHBA: number of hydrogen bond acceptors; nHBD: number of hydrogen bond donors; TPSA: total polar surface area.

Table 2 also shows the expected toxic effects of SFN compared to those of FLZ. Predictive analyses suggested that SFN is not irritating or hepatotoxic but has moderate mutagenic, tumorigenic, and deleterious effects on reproduction. FLZ did not affect these parameters. The estimated LD50 was 1000 and 1410 mg/kg for SFN and FLZ, respectively (Table 2). SFN and FLZ toxicity scores were equal to 4.0, indicating they are classified as harmful if ingested at their LD50 (Table 2).

Predictive gastrointestinal absorption, permeability through the BBB, and skin permeation (log Kp in centimetres (cm)/s) are shown in Table 2. Both SFN and FLZ were suggested to be highly absorbed by the gastrointestinal tract and unable to cross the BBB. The estimated values of Log Kp were −6.38 cm/s and −7.92 cm/s for SFN and FLZ, respectively (Table 2). Both compounds were expected to be water-soluble, with Log S of −2.10 and −2.17 for SFN and FLZ, respectively. The similarity of drugs to SFN was estimated at −6.47 and 1.99 for FLZ. Drug scores were 0.25 for SFN and 0.87 for FLZ.

2.2. In Vitro Analysis

2.2.1. In Vitro Antifungal Activity

MICs and MFCs against *C. albicans* were determined for SFN in comparison with FLZ (Table 3). MIC values for SFN were equal to 30 µg/mL for all strains (ATCC 90028, Oral 38 HIV^+ and Oral 40 HIV^+) except for Oral 37 HIV^+ (60 µg/mL), whilst the MFC values ranged from 30 to 240 µg/mL (Table 3). MIC and MFC values varied for FLZ. When assessed against the ATCC strain, FLZ MIC and MFC values were 1 and 8 µg/mL, respectively (Table 3). On the other hand, the MIC values observed for the *C. albicans* clinical isolates were ≥4 µg/mL (Table 3). Also, the MFC values for FLZ were ≥64 µg/mL when assessed against these microorganisms (Table 3). The antifungal effects of SFN were also investigated in non-*Candida albicans* species (*C. parapsilosis*, *C. krusei*, and *C. glabrata)* in comparison to FLZ. In these fungi, SFN MIC and MFC values ranged from 1.87–60 µg/mL, whilst FLZ MIC values varied from 4–16 µg/mL (Supplementary Table S1). FLZ MFC values were >64 µg/mL for Oral 22 *C. krusei* and *C. glabrata* ATCC 2001; the other assessed non-*Candida albicans* species presented MFC values ≤64 µg/mL (Supplementary Table S1).

Table 3. Minimum Inhibitory Concentration (MIC; µg/mL) and Minimum Fungicidal Concentration (MFC; µg/mL) values of sulforaphane (SFN) compared to fluconazole (FLZ) against *C. albicans* ATCC 90028 and *C. albicans* clinical isolates (Oral 37 HIV^+, Oral 38 HIV^+, and Oral 40 HIV^+).

Strain	MIC (µg/mL)		MFC (µg/mL)	
	SFN	FLZ	SFN	FLZ
ATCC 90028	30	1	60	8
Oral 37 HIV^+	60	16	240	>64
Oral 38 HIV^+	30	4	30	64
Oral 40 HIV^+	30	8	60	64

2.2.2. In Vitro Evaluation of the Combined Effects of SFN and FLZ

ATCC 90028 and the isolates Oral 40 HIV^+ and Oral 37 HIV^+ were selected for further studies. A checkerboard microdilution method was used to assess the combined effects of SFN and FLZ. Their interaction was calculated and expressed as mean (Table 4) and individual FICIs (Table 5). As demonstrated in Table 4, the mean FICI values obtained indicate an indifferent effect for the combination of SFN and FLZ when assessed against *C. albicans* spp.

Table 4. Mean fractional inhibitory concentration indexes (FICIs) for all tested combinations between sulforaphane (SFN) and fluconazole (FLZ) against *C. albicans* ATCC 90028 and *C. albicans* clinical isolates (Oral 40 HIV$^+$ and Oral 37 HIV$^+$).

Strains	**Mean FICI (μg/mL)**	**Interaction**
C. albicans ATCC 90028	2.197	Indifferent
C. albicans Oral 37 HIV$^+$	1.412	Indifferent
C. albicans Oral 40 HIV$^+$	1.359	Indifferent

Synergism: ΣFICI ≤ 0.5; additive 0.5 > Σ FICI ≤ 1; indifference 1 > Σ FICI ≤ 4.0; and Antagonism: Σ FICI > 4.0.

Table 5. Individual fractional inhibitory concentration indexes (FICIs) for different sub-inhibitory concentrations of sulforaphane (SFN; 0.12–30 μg/mL; MIC/256-MIC/2) and fluconazole (FLZ; MIC/2: 0.5 μg/mL for ATCC 90028, 1 μg/mL for Oral 40 HIV$^+$, and 4 μg/mL for Oral 37 HIV$^+$).

Strain	**FIC (μg/mL) at an SFN Concentration (μg/mL) of:**								
	FLZ (μg/mL)	**MIC/256**	**MIC/128**	**MIC/64**	**MIC/32**	**MIC/16**	**MIC/8**	**MIC/4**	**MIC/2**
ATCC 90028	**MIC/2**	0.504	0.508	0.515	0.531	0.562	0.625	0.75	1
Oral 37 HIV$^+$	**MIC/4**	-	-	0.265	0.281	0.312	0.375	0.5	0.75
Oral 40 HIV$^+$	**MIC/4**	0.253	0.25	0.265	0.281	0.312	0.375	0.5	0.75

Synergism: ΣFICI ≤ 0.5; additive 0.5 > Σ FICI ≤ 1; indifference 1 > Σ FICI ≤ 4.0; and Antagonism: Σ FICI > 4.0.

However, when considering the individual FICI values for each tested concentration of the compounds, we found eight additive combinations between them against *C. albicans* ATCC 90028, and seven synergistic and one additive combinations against the Oral 40 HIV$^+$ isolate (Table 5). SFN exhibited the best additive effects when combined at concentrations ≤1.88 μg/mL (<MIC/16) for ATCC 90028 with FLZ at 0.5 μg/mL. On the other hand, SFN showed synergistic effects for the clinical isolate Oral 40 HIV+ at concentrations ≤7.5 μg/mL (MIC/4) with FLZ at 1.0 μg/mL. A similar effect was observed for Oral 37 HIV$^+$, as the combinations between SFN at concentrations ≤15 μg/mL (MIC/4) and FLZ at MIC/4 (4 μg/mL) were synergistic.

2.2.3. Effects on Hyphae Formation

Hyphae are important structures of *Candida* spp. biofilms and are essential to successful fungal colonization and invasion of the host, meaning that the greater the ability of a fungi to adhere to the host, grow hyphae and form biofilms, the higher the chances of persistent and severe infections [47,48]. Therefore, the individual and combined anti-hyphae effects of sub-inhibitory concentrations of the compounds were evaluated. As *C. albicans* oral isolates 37 HIV$^+$ and 40 HIV$^+$ presented similar mean and individual FICIs when combined with FLZ, the effects of the compounds on hyphae growth were investigated in *C. albicans* ATCC 90028 and oral isolate 40 HIV$^+$. *C. albicans* ATCC 90028 was inhibited by SFN (MIC/2 and MIC/4) and FLZ (MIC/2) (Figure 1a–c). When assessed individually for each compound, hyphae inhibition was similar for SFN and FLZ at MIC/2 (~80%) (Figure 1c). When combined, the best effects were observed for SFN at MIC/256 *plus* FLZ at MIC/4 (73.4%). The other combinations exhibited inhibitions ranging from 52.5–70% (Figure 1c). When incubated alone with the oral isolate 40 HIV$^+$, SFN (MIC/2) displayed the best inhibitory effect (100%) in comparison with FLZ (Figure 2a–c). Inhibitions of similar magnitude were observed for the following combinations: SFN MIC/256 *plus* FLZ MIC/4 and SFN MIC/64 *plus* FLZ MIC/2 (62.3%) (Figure 2c).

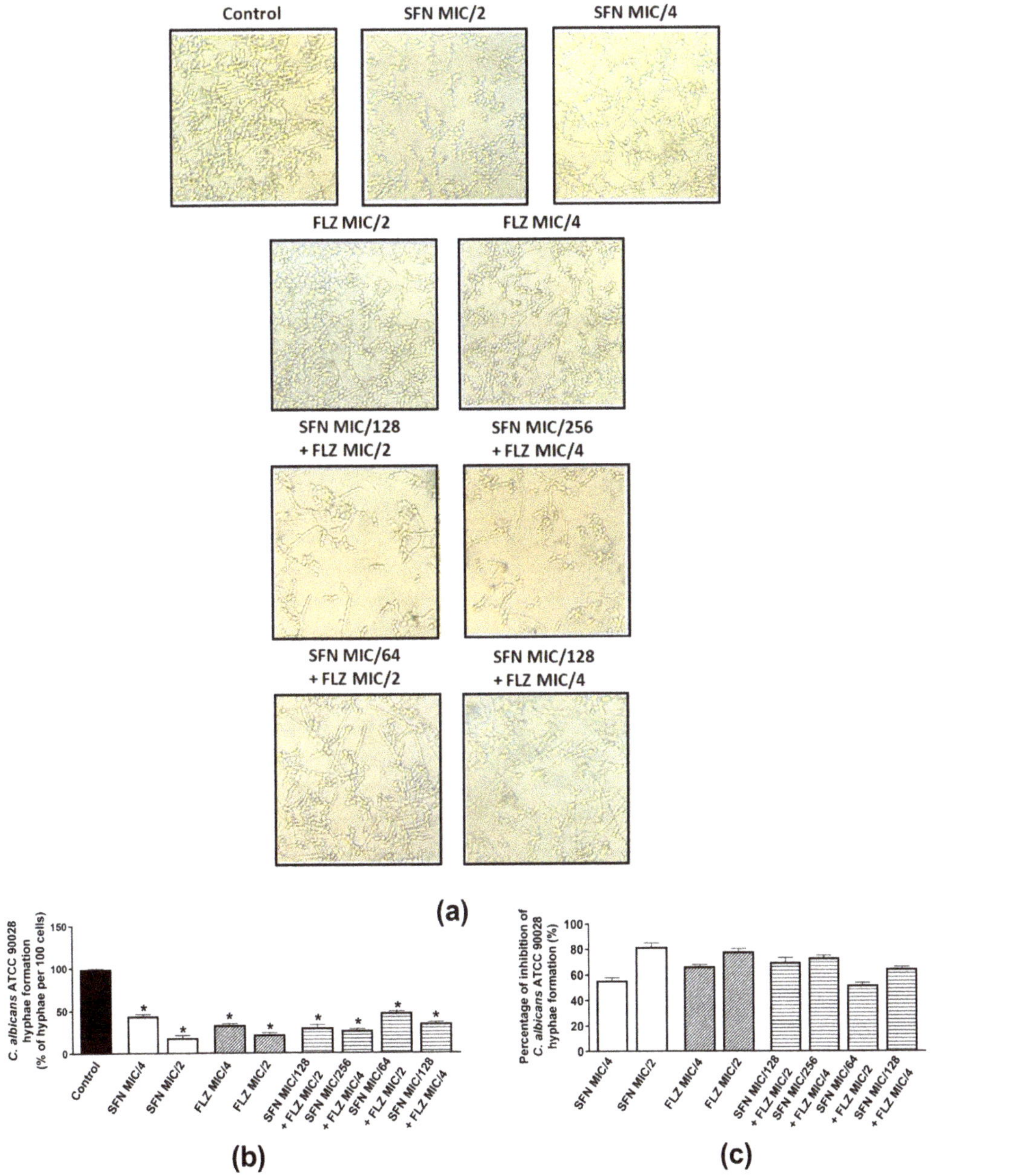

Figure 1. Effects of sulforaphane (SFN) and fluconazole (FLZ) on hyphae formation by *C. albicans* ATCC 90028. (**a**) Representative images of hyphae growth by *C. albicans* ATCC 90028 incubated with either FLZ or SFN at MIC/2 and MIC/4. The same panel represents the combined effects of SFN (MIC/64 to MIC/256) and FLZ (MIC/2 and MIC/4). (**b**) Percentage of hyphae *per* 100 cells of *C. albicans* ATCC 90028 and (**c**) percentage (%) of inhibition of *C. albicans* ATCC90028 hyphae formation. The control consisted of inoculum plus phosphate-buffered saline (PBS) containing 10% fetal bovine serum (FBS). * $p < 0.05$ differs from the control group.

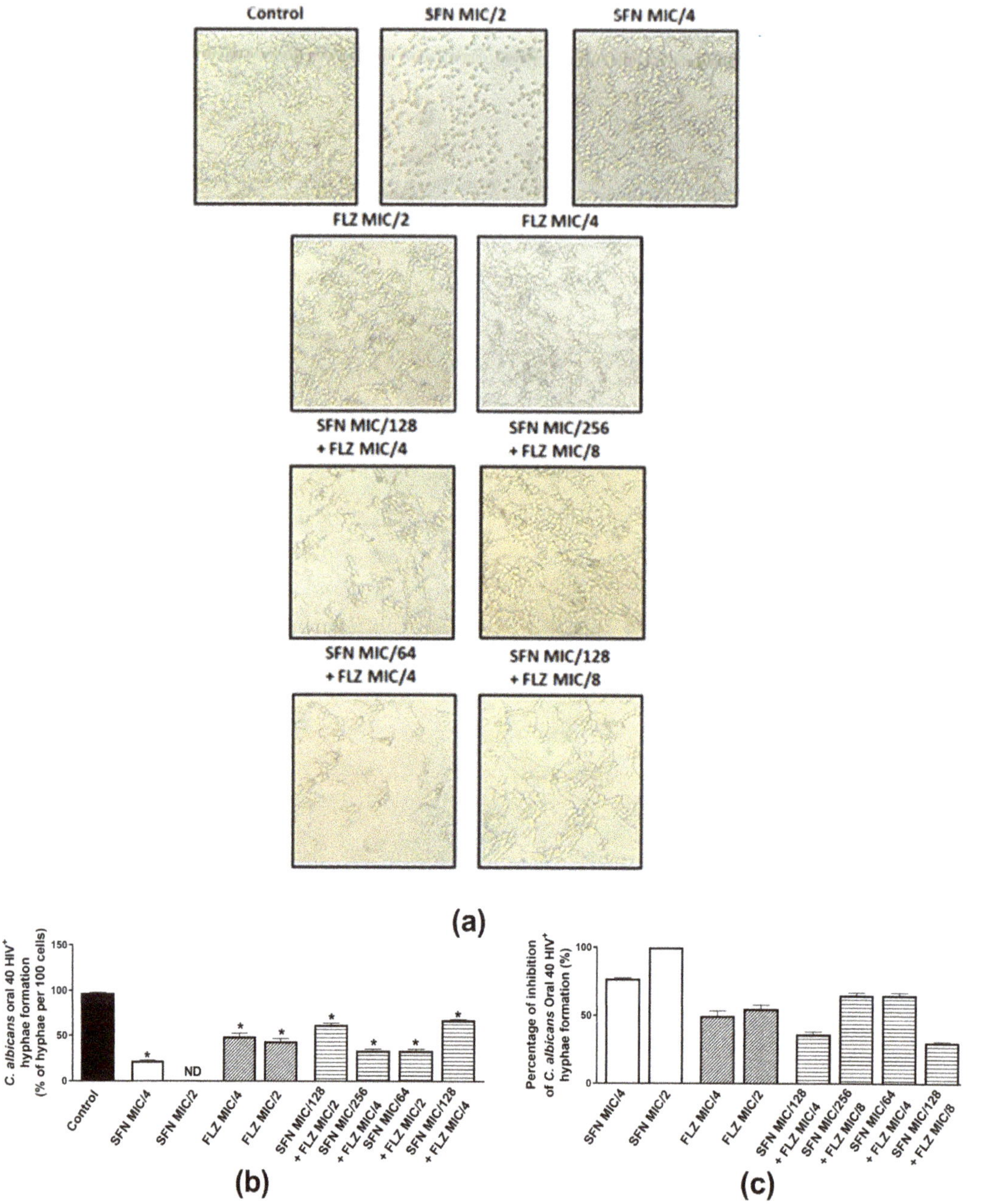

Figure 2. Effects of sulforaphane (SFN) and fluconazole (FLZ) on hyphae formation by *C. albicans* oral isolate 40 HIV$^+$. (**a**) Representative images of hyphae growth by Oral 40 HIV$^+$ incubated with either FLZ or SFN at MIC/2 and MIC/4. The same panel represents the combined effects of SFN (MIC/64 to MIC/256) and FLZ (MIC/4 and MIC/8). (**b**) Percentage of hyphae *per* 100 cells of Oral 40 HIV$^+$ and (**c**) percentage (%) of inhibition of Oral 40 HIV$^+$ hyphae formation. The control consisted of inoculum plus phosphate-buffered saline (PBS) containing 10% fetal bovine serum (FBS). * $p < 0.05$ differs from the control group.

2.2.4. Effects on Biofilm Formation

The effects of sub-inhibitory concentrations of SFN and FLZ on biofilm formation by *C. albicans* ATCC 90028 and *C. albicans* clinical isolate (Oral 40 HIV$^+$) were evaluated alone and in combination at the same concentrations tested in the hyphae growth assays. Only SFN (MIC/ 2 and MIC/4) was able to impair biofilm formation by the ATCC strain (Figure 3a). The best inhibition was noted for SFN at MIC/2 (31.3%). Both FLZ and SFN per se, significantly reduced biofilm formation by the Oral 40 HIV$^+$ isolate (Figure 3b). Whilst FLZ attenuated this response at both tested concentrations (MIC/2 and MIC/4; 16%), SFN only reduced biofilm formation by Oral 40 HIV$^+$ at MIC/2 (13.3%) (Figure 3b).

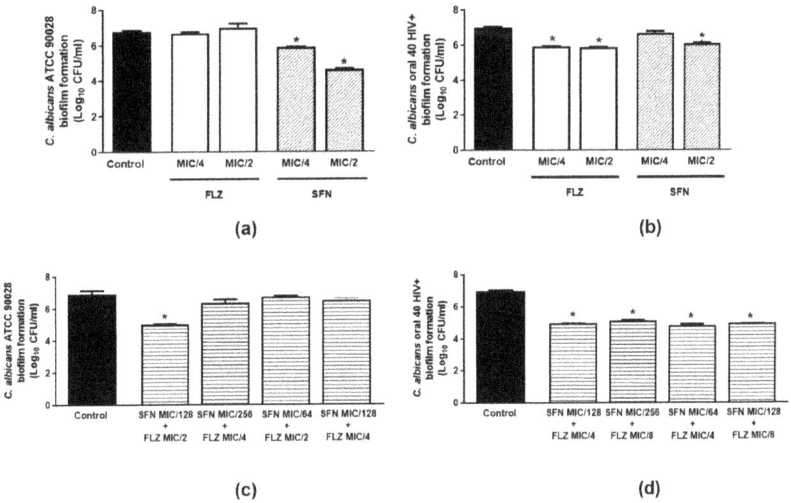

Figure 3. Effects of sulforaphane (SFN) and fluconazole (FLZ) on biofilm formation by *C. albicans* spp. SFN and FLZ were tested at MIC/2 and MIC/4 against (**a**) *C. albicans* ATCC 90028 and (**b**) *C. albicans* clinical isolate (Oral 40 HIV$^+$). The combined effects of SFN (MIC/64 to MIC/256) with FLZ (MIC/2–MIC/8) against (**c**) *C. albicans* ATCC 90028 and (**d**) *C. albicans* clinical isolate (Oral 40 HIV+) were investigated over 24 h. The control consisted of inoculum plus culture medium. Data were obtained from three independent experiments and are expressed as Log_{10} CFU/mL.* $p < 0.05$ differs from the control group.

We next assessed the combined effects of SFN (MIC/64 to MIC/264) with FLZ (MIC/2 to MIC/8) against the ATCC strain and the Oral 40 HIV$^+$ isolate. Only the combined incubation of SFN MIC/128 with FLZ MIC/2 was able to attenuate biofilm formation (27.2%) by *C. albicans* ATCC 90028 (Figure 3c). On the other hand, all tested combinations decreased the ability of the Oral 40 HIV$^+$ isolate to form biofilm (28.9%; Figure 3d).

2.2.5. Effects on mRNA Expression of Hyphae Growth- and Biofilm Formation-Related Genes

As pronounced inhibitory effects on hyphae growth and biofilm formation were observed for SFN and FLZ (MIC/2 and MIC/4), the expression of *C. albicans* virulence genes was analyzed. Figure 4a–c demonstrates the effects of SFN or FLZ on the mRNA expression of *ASL1*, *EFG1* and *RAS1* in *C. albicans* ATCC 90028. The relative expression of all three genes to *18S* rRNA was significantly impaired by all tested concentrations of the drugs in this microorganism strain. On the other hand, whilst FLZ at MIC/4 enhanced *ASL1*, *EFG1* and *RAS1* mRNA levels in *C. albicans* Oral 40 HIV$^+$, SFN-tested concentrations had no effects on the isolate (Figure 4d–f).

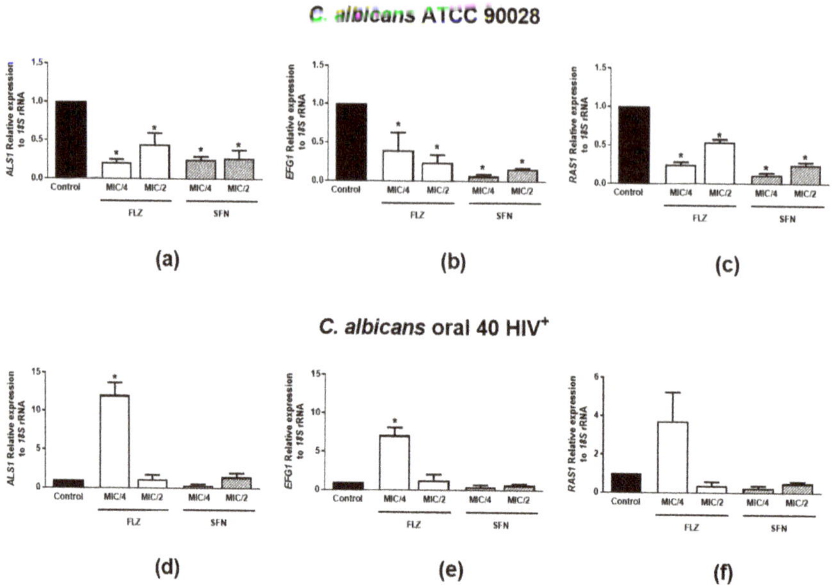

Figure 4. Effects of sulforaphane (SFN) and fluconazole (FLZ) on mRNA expression of hyphae growth- and biofilm formation-related genes by *C. albicans* spp. SFN and FLZ were tested at MIC/2 and MIC/4 against *C. albicans* ATCC 90028 and the oral isolate 40 HIV^{+}. The relative expressions of *ALS1* (**a**,**d**), *EFG1* (**b**,**e**) and *RAS1* (**c**,**f**) normalized to *18S* rRNA were quantified 24 h following incubation with either FLZ or SFN. The control consisted of inoculum plus culture medium. Data were obtained from three independent experiments and are depicted as relative expression of mRNA. * $p < 0.05$ differs from the control group.

3. Discussion

FLZ is considered a low-cost drug with little toxicity [12], which is readily absorbed with high bioavailability by oral route [49]. Despite its effectiveness, recent studies have reported the isolation of resistant *Candida* species in immunocompromised patients [15–17] and associated this with the increased use of azoles, especially FLZ, in prophylactic therapy [18], limiting treatment success [19,20]. This surge in FLZ-resistant fungi has indicated the need for new compounds able not only to kill but also to inhibit *Candida* spp. adhesion, yeast-hyphae transition, and biofilm formation whilst displaying few toxic effects.

SFN antibacterial effects have been widely shown [40–43,46]. However, controversial data has been found in regard to its antifungal activity, with varying results on fungi sensitivity to this compound [40–43,46]. Herein, we investigated SFN in silico properties in comparison with FLZ—a first-line medication used for the treatment of *C. albicans* infections. We also determined its antifungal actions against *C. albicans*, alone and in combination with FLZ.

In silico analysis confirmed SFN antimicrobial properties, which include antiparasitic, antifungal and antibacterial actions, and mechanisms ranging from alterations of membrane composition to inhibition of fungi RNA. SFN presented an estimated DS of 0.25 with a potential moderate risk for mutagenic, tumorigenic and reproduction tract deleterious effects. These estimates are controversial, considering that SFN has been demonstrated as a cytoprotective compound able to attenuate mutagenesis and tumorigenesis and prevent abnormalities in the reproductive system [35,36,50–53]. SFN was also predicted to be lethal when ingested at doses higher than its LD_{50} (1000 mg/kg). On the contrary, although potentially harmful at doses higher than 1410 mg/kg, FLZ was not suggested to be deleterious as a tumorigenic, mutagenic, hepatotoxic or irritant agent, and neither to affect reproduction.

Prediction of skin permeation (log Kp) indicated that SFN is more likely to be absorbable by skin layers than FLZ, indicating a potential use for SFN alone or in combination with FLZ to treat skin diseases, including infections. Both compounds were considered as highly absorbed by the GI tract but not through the BBB; this later observation could be interpreted as a low chance of undesirable actions for SFN and FLZ on the central nervous system when fungal infections in this tissue are absent. However, their predicted chemical properties (iLogP, MW, TPSA, nHBD and nHBA values) [47] suggest that FLZ is less likely to penetrate the BBB. In fact, previous research indicates that SFN [54–56] may cross the BBB, and data obtained from studies with FLZ suggest its ability to enter the brain through the BBB; this may be enhanced in infectious diseases, which affect barrier permeability [57–59]. These data allow us to hypothesize that the association of low doses of FLZ and SFN may be beneficial to treat superficial mycosis, such as those affecting the skin and/or mucosa. Also, the in silico analysis corroborated previous findings showing SFN as an anti-oxidant with anti-cancer activities [33,35].

Therefore, we initially assessed the antimicrobial effects of SFN and FLZ alone or in combination. FLZ, MIC and MFC values indicated this drug is more potent than SFN against *Candida* spp. Also, eight additive combinations were observed for these compounds when tested against *C. albicans* ATCC 90028, whilst seven synergistic combinations were observed against the clinical isolate (Oral 40 HIV$^+$), with the best effect seen when SFN was combined at concentrations ≤MIC/16 with FLZ at MIC/2 for the ATCC strain and ≤MIC/8 with FLZ/4 for this clinical isolate. These results confirmed SFN antimicrobial actions and showed, for the first time to our knowledge, its ability to potentiate FLZ actions when at sub-inhibitory concentrations.

Interestingly, both FLZ and SFN reduced hyphae formation by *C. albicans* at sub-inhibitory concentrations, and this effect was greatest for SFN (MIC/2). When incubated together, a greater anti-hyphae action was observed for these compounds when assessed against the ATCC strain and the clinical isolate Oral 40 HIV$^+$. Indeed, hyphae growth was attenuated by more than 60%. Of note, hyphae growth is an important component of mature *Candida*-induced biofilms [60], with young hyphae already present during the intermediate phase of biofilm formation (12–30 h) [61]. In addition to a fundamental role in biofilm formation, hyphae growth is a key process in invasive infections by *C. albicans*, meaning this fungal structure is vital to fungal pathogenicity [47].

Therefore, the capacity of sub-inhibitory concentrations of SFN or FLZ to markedly halt this process is a relevant finding. Thus, we next assessed the effects of sub-inhibitory concentrations of SFN and FLZ alone or in combination on *Candida*-induced biofilm formation. SFN greatly reduced biofilm formation by *Candida* spp. at the tested concentrations, whilst FLZ was only effective against the Oral 40 HIV$^+$ isolate. Similarly to that observed for hyphae growth, a more pronounced anti-biofilm action was noted when the compounds were combined.

Interestingly, FLZ at 1 μg/mL—the MIC observed for the ATCC strain and MIC/8 for the Oral 40 HIV$^+$ isolate tested herein, was previously shown to prevent hyphae growth without affecting biofilm-formation by resistant *Candida* spp., following 48 h incubation [48]. Another study demonstrated that a higher concentration of FLZ (512 μg/mL) is needed to attenuate biofilm formation by *Candida* spp. [62] 48 h-post incubation. These studies differ from the observed in our study in regards to the effects of FLZ on biofilm formation following 24 h incubation. Discrepancies between the studies may be due to differences in the incubation periods with the drug. Nonetheless, we show for the first time that the combination of low concentrations of SFN with sub-inhibitory concentrations of FLZ has anti-biofilm and anti-hyphae effects on *C. albicans*. Of note, SFN was previously suggested as an inhibitor of quorum sensing in bacteria [63]—a "machinery" also important for *Candida* spp. virulence [64]. Therefore, it is possible that its antifungal activity is related to the regulation of quorum-sensing genes in fungi. To assess the ability of SFN to alter the expression of genes involved in *C. albicans* quorum sensing and virulence [65–67], in comparison with FLZ, the relative levels of *ALS1*, *EFG1* and *RAS1* were investigated against the ATCC 90028 strain. Both compounds inhibited the expression of all three genes when tested at MIC/2 and MIC/4, indicating that their inhibitory effects on biofilm formation and hyphae growth by *C. albicans* are associated with their ability to down-regulate these genes. Although regulation of *EGF1* by FLZ was found to be attenuated across studies, controversial data has been reported for *ALS1* and *RAS1* mRNA expressions in *Candida* spp. [68,69], suggesting the up- or down-regulation of these genes by the compound is strain-dependent. This is supported by the data presented herein, as SFN and FLZ had different actions in the expression of virulence genes when assessed against ATCC 90028 and Oral 40 HIV^+. Nevertheless, SFN's ability to regulate these pathways in *Candida* spp. is a novel finding which deserves attention in further studies. Also, future investigations on the effects of SFN alone and in combination with FLZ on mixed populations of microorganisms, such as those found in the oral cavity and the vagina, are worthy of being pursued.

4. Materials and Methods

4.1. In Silico Analysis

4.1.1. Prediction of Biological Activities

The possible biological activities of SFN were evaluated using the online Prediction of Activity Spectra for Substances (PASS). The PASS computational tool calculates the probability of a given organic molecule being active (Pa) or inactive (Pi) on a biological target by comparing their structure to organic molecules with defined biological properties (www.way2drug.com/passonline) [70,71]. For comparison, the biological activities of FLZ were also assessed.

4.1.2. Prediction of Oral Bioavailability

The SwissADME web tool (http://www.swissadme.ch/index.php# accessed on 15 September 2021) [72] was used to predict the theoretical oral bioavailability of SFN, considering Lipinski's "Rule of Five" [73] as previously described [74]. The oral bioavailability of FLZ was evaluated for comparison.

4.1.3. Estimations of Toxicity and Pharmacokinetic Characteristics

To analyze the possible toxic effects and pharmacokinetic parameters (absorption, distribution, metabolism, and excretion) of SFN, the Osiris (www.organic-chemistry.org/prog/peo/drugScore.html accessed on 15 September 2021) [70] and SwissADME (http://www.swissadme.ch/index.php# accessed on 15 September 2021) [72] web tools were used. The pharmacokinetic parameters and toxicity were predicted by comparing the chemical structures of SFN with a database containing commercially available drugs and compounds. The toxic effects were classified as mutagenic, tumorigenic, irritating, and affecting the reproductive system.

Toxic doses, given as LD50 values in mg/kg of body weight, were estimated using the ProTox web tool (http://tox.charite.de/protox_II/index.php?site=compound_input accessed on 15 September 2021) [75]. The estimated gastrointestinal absorption, permeability through the blood-brain barrier, and the skin (log Kp in centimeters (cm)/s) of SFN were evaluated by the SwissADME program (http://www.swissadme.ch/index.php# accessed on 15 September 2021) [72]. Also, the probability of SFN becoming a commercial drug ("drug score") was calculated using the OSIRIS software (www.organic-chemistry.org/prog/peo/drugScore.html accessed on 15 September 2021).

All parameters were compared to those of FLZ.

4.2. In Vitro Analysis

4.2.1. Microorganisms

The reference strains *C. albicans* ATCC 90028, *C. krusei* ATCC 6258 and *C. glabrata* ATCC 2001 (American Type Culture Collection) were kindly provided by the Faculdade de Odontologia de Araraquara, Universidade do Estado de São Paulo. Three *C. albicans* clinical isolates from the oral cavity of HIV+ patients (Oral 37 HIV+, Oral 38 HIV$^+$, and Oral 40 HIV$^+$), the isolates from the oral cavity (Oral 01 HIV$^+$ *C. parapsilosis* and Oral 22 *C. krusei*), and vaginal smears of patients with vulvovaginal candidiasis (Vaginal 11 *C. glabrata* and Vaginal 14 *C. glabrata*) were donated by the microorganism collection sector of the Applied Microbiology Laboratory of Universidade CEUMA (UNICEUMA).

4.2.2. Inoculum Preparation

The strains were reactivated in Sabouraud Dextrose Agar (SDA, Kasvi, Italy) for 24 h at 37 °C. The fungal inoculum was prepared in phosphate-buffered saline (PBS; Sigma–Aldrich, Gillingham, UK; pH 7.0) to achieve a cell density of 1×10^6 colony forming units (CFU)/mL on a spectrophotometer (at 530 nm), equivalent to a turbidimetric scale McFarland's score of 0.5 [76].

4.2.3. Determination of Minimum Inhibitory (MIC) and Fungicidal (MFC) Concentrations

FLZ and L-SFN were purchased from Sigma–Aldrich (UK). The Minimum Inhibitory (MIC) and Fungicidal Concentrations (MFC) of SFN and FLZ were determined by the microdilution broth assay, as previously described in the document M27-A4 of the Clinical and Laboratory Standards Institute (CLSI, 2008). SFN (5000 µg/mL) was dissolved in 50% dimethylsulfoxide (DMSO; Sigma–Aldrich, UK) and then diluted in RPMI 1640 medium (Roswell Park Memorial Institute, Sigma–Aldrich, UK), containing glutamine, no bicarbonate, and buffered with sulfonic morpholino propane acid (MOPS; Sigma–Aldrich, UK) to different concentrations (0.117–60 µg/mL). FLZ was diluted in RPMI-MOPS to 0.03125–64 µg/mL. RPMI 1640 containing wells plus inoculum were used as negative control in the growth assays, and 1% DMSO (v/v) was used as vehicle control. Wells containing only RPMI 1640 medium were used for sterility control.

The fungal inoculum was prepared at 1×10^3 CFU/mL in RPMI 1640 medium. Then, 100 µL of the inoculum was incubated with SFN, FLZ or vehicle for 48 h at 37 °C. After the incubation period, fungal growth was analyzed visually. MIC was defined as the lowest concentration of SFN or FLZ at which no visible growth was detected. For the determination of MFC, 10 µL of the wells with concentrations >MIC were seeded on Sabouraud Dextrose Agar (SDA; Kasvi, Italy) and incubated at 37 °C for 24 h. The fungicidal activity was defined as the one in which no growth of colonies was observed.

4.2.4. Combined Effects of SFN with FLZ on Fungal Survival

The anti-*Candida* effects of SFN were evaluated against the ATCC 90028 strain and the oral isolates 37 HIV$^+$ and 40 HIV$^+$, in combination with FLZ, by using the checkerboard assay [77]. FLZ was tested at concentrations ranging from 0.0625 to 64 µg/mL and SFN at concentrations ranging from 0.117 to 240 µg/mL.

One hundred µL of the fungal inoculum prepared at 1×10^3 CFU/mL was incubated with 50 µL of SFN and 50 µL of FLZ in microplates with different combinations of drug concentrations, at 37 °C, for 48 h. The antifungal activity was assessed as described for the determination of MICs. After incubation, SFN-FLZ interactions were determined by the Fractional Inhibitory Concentration Index (ΣFICI): FICI = (MICFLZ+SFN/MICFLZ) + (MICSFN+FLZ/MICSFN), where:

MICFLZ+SFN: MIC of FLZ when in combination with SFN;

MICFLZ: MIC of FLZ;

MICSFN+FLZ: MIC of SFN when in combination with FLZ;

MICSFN: MIC of SFN;

The FICIs were calculated for all possible combinations of different concentrations for the same isolate at which no visible growth of the microorganism was observed, and the final result was expressed as mean of the FICIs. The interaction between these drugs was classified as: Synergism: $\Sigma FICI \leq 0.5$; additive $0.5 > \Sigma\ FICI \leq 1$; indifference $1 > \Sigma\ FICI \leq 4.0$; and Antagonism: $\Sigma\ FICI > 4.0$; as previously described [78].

4.2.5. Biofilm Formation Assay

Biofilm formation was assessed as previously described [79]. *C. albicans* (ATCC 90028 and Oral 40 HIV$^+$) were inoculated in nitrogen-based yeast medium (Yeast Nitrogen Base, YNB; Sigma–Aldrich, UK) supplemented with 50 mM glucose at 37 °C for 18 h. For this, uniform yeast cells in the exponential growth phase were centrifuged, washed twice with PBS, and adjusted to contain 1 × 10^6 CFU/mL in PBS (0.5 of McFarland scale, at 530 nm). Aliquots of this suspension (100 µL) were transferred to a 96-well polystyrene plate and incubated for 90 min at 37 °C to allow the initial adhesion of the yeasts. After the adhesion phase, the wells were washed three times with PBS in order to remove non-adherent cells. Then, 200 µL of YNB containing 100 mM glucose and different concentrations of SFN or FLZ alone or in combination were added to the wells and incubated at 37 °C for 24 h. SFN and FLZ were individually tested at MIC/2 and MIC/4 against *C. albicans* ATCC 90028 and *C. albicans* clinical isolate (Oral 40 HIV$^+$). The combined effects of SFN (MIC/64 to MIC/256) with FLZ (MIC/2 to MIC/8) against *C. albicans* were also investigated. Wells containing vehicle (1% DMSO v/v) in culture medium were used as positive controls for biofilm formation. As sterility control, YNB containing 100 mM glucose was used.

After the incubation period, the resulting biofilms were washed with PBS and collected by scraping off the bottom of each well. The biofilms were resuspended in 100 µL PBS. To quantify the yeast cells, the microdrop technique was used. For this, each biofilm sample was serially diluted, and an aliquot of each dilution (10 µL) was seeded onto SDA plates and incubated at 37 °C for 24 h. After this period, colonies were counted, and the results expressed as viable cells in CFU/mL.

4.2.6. Hyphae Formation Test

The effects of SFN on hyphae formation were assessed as previously described [80,81], with modifications. *C albicans* strains (1 × 10^6 CFU/mL; 1mL) were incubated either in RPMI 1640 medium or PBS containing 10% foetal bovine serum (FBS) supplemented with 1% DMSO (vehicle controls), FLZ or SFN (MIC/2 and MIC/4). Their combined effects were also assessed (MIC/64 to MIC/256 for SFN; and MIC/2 to MIC/8 for FLZ) against *C. albicans*. Samples were incubated under agitation (100 rpm) on glass slides placed at the bottom of each well of 24-wells plates, at 37 °C, for 48 h. After the incubation period, the slides were collected, prepared and photographed under a bright-field microscope. The number of hyphae and yeasts were counted, and the results expressed as the percentage of hyphae *per* 100 cells of *C. albicans*. Also, the percentage of inhibition of hyphae formation was calculated.

4.2.7. mRNA Expression of Hyphae Growth- and Biofilm Formation-Related Genes

The effects of SFN and FLZ on the expression of *C. albicans* virulence genes (*agglutinin-like sequence 1—ALS1*; *enhanced filamentous growth protein 1—EFG1*; and *Ras-like protein 1—RAS1*) [60–62]. *C. albicans* ATCC 90028 or the Oral 40 HIV+ isolate (1 × 10^6 CFU/mL; 1 mL) was submitted to the biofilm formation assay in the presence and absence of sub-inhibitory concentrations of FLZ or SFN (MIC/2 and MIC/4) for 24 h. The tested concentrations of the compounds were the same used in the biofilm formation and hyphae growth assays. Vehicle (1% DMSO; v/v)-treated wells were used as positive controls for biofilm formation.

After incubation, the cells were scraped from the bottom of each well, and the resulting suspensions, transferred to 1.5 mL tubes and centrifuged at 5000× *g*, 4 °C, for 5 min. For purification of total RNA (as previously described [82], the cell pellets were resuspended in TRIzol (Invitrogen®) and added of 600 µL of silica beads (0.2 mm diameter). Then, each sample was homogenized twice for 2.5 min followed by 2.5 min breaks on ice. The resulting suspensions were placed on ice and sonicated for 1 min at 15% sonication amplitude in continuous mode. The samples were centrifuged (10,000× *g*, at 4 °C, for 10 min), and the aqueous phases collected and transferred to new 1.5 mL tubes added of 200 µL of RNase-free chloroform. Samples were vortexed for 15 s, let to rest for 10 min, and centrifuged (10,000× *g*, at 4 °C, for 10 min). Three hundred µL of the supernatant of each sample were then mixed with 500 μL of RNase-free chloroform (Sigma–Aldrich, UK), vortexed for 15 s, and centrifuged (10,000× *g*, at 4 °C, for 10 min). The resulting aqueous phase was added of 500 μL of RNase-free isopropanol (Thermofisher, São Paulo, Brazil), vortexed for 15 s, and centrifuged (10,000× *g*, at 4 °C, for 10 min). The supernatants were discarded, and the pellets added of 1 mL of RNase-free 75% ethanol and centrifuged at 10,000× *g* for 5 min at room temperature. The supernatant was again discarded, and the pellets dried for 15 min for removal of residual ethanol. Samples were then resuspended in 50 μL of nuclease-free water (Thermofisher, Brazil). RNA quantities in each sample were measured by NanoDrop (Thermofisher, Brazil); samples were then kept at −80 °C to further analyse.

The total RNA (5 μg) of each sample was reverse transcribed to cDNA by using the PCR GoScript™ Reverse Transcriptase kit (Promega Corporation, USA) according to manufacturer's instructions. The resulting cDNAs were kept at −20 °C to further analyse. One μg of cDNA was amplified by qPCR on a StepOne qPCR system (Thermofisher; Brazil; hold: 10 min at 95 °C; cycling: 40 cycles: 10 s at 95 °C, 20 s at 60 °C, and 15 s at 72 °C; melt: 68–90 °C) using SYBR Green PCR Master mix (Thermofisher; Brazil). All primers for the virulence genes *ALS1* (Reverse: ATGATTCAAAGCGTCGTTC; Forward: TTGGGTTGGTCCTTAGATGG), *EFG1* (Reverse: TTGTTGTCCTGCTGTCTGTC; Forward: TATGCCCCAGCAAACAACTG) and *Ras1* (Reverse: GTCTTTCCATTTCTAAATCAC; Forward: TATGCCCCAGCAAACAACTG), and for the housekeeping gene *18S* rRNA (Reverse: TGCAACAACTTTAATATACGC; Forward: AATTACCCAATCCCGACAC), were purchased from Integrated DNA Technologies (Iowa, USA) and used at 500 nM. Results are depicted as relative expression of each virulence gene, normalized to *18S* rRNA and calculated by the $2^{-\Delta\Delta CT}$ method [83].

4.3. Statistical Analysis

The results are presented as the mean ± standard error of the mean (SEM). All experiments were carried out in triplicate and were obtained from three independent experiments. Significant differences among groups were determined by using one-way analysis of variance (ANOVA), followed by Bonferroni. The p-value < 0.05 was statistically significant.

5. Conclusions

Overall, the data gathered herein demonstrates that SFN is an anti-*Candida* compound once it presents antifungal and fungicidal activities. SFN also inhibits virulence factors (biofilm and hyphae growth) essential for the fungi's ability to colonize and invade host tissues by down-regulating hyphae growth- and biofilm development-associated genes in this pathogen. Its suggested ability to cross the BBB indicates that SFN alone could be used as an alternative therapy for both deep and superficial fungal infections. Also, the marked additive/synergistic effects observed for the combination of low concentrations of SFN with sub-inhibitory concentrations of FLZ, in regards to cell survival, biofilm formation and hyphae growth, indicate this association may be an interesting approach to managing *Candida* spp. infections with possible attenuation of adverse reactions caused by these compounds.

Supplementary Materials: The following supporting information can be downloaded at: https://www.mdpi.com/article/10.3390/antibiotics11121842/s1, Table S1: Minimum Inhibitory Concentration (MIC; μg/mL) and Minimum Fungicidal Concentration (MFC; μg/mL) values of sulforaphane (SFN) compared to fluconazole (FLZ) against non-Candida albicans strains and clinical isolates; Table S2: In silico identification of additional non-antimicrobial activities of sulforaphane (SFN) and fluconazole (FLZ).

Author Contributions: Conceptualization, C.d.A.M., V.M.-N. and E.S.F.; methodology, B.L.R.S., C.D.L.C., G.S., L.R.B., G.B.O., S.J.F.M., I.V.M., C.d.A.M., E.M.S. and F.C.B.V.; formal analysis, B.L.R.S., C.D.L.C., S.J.F.M., C.d.A.M. and D.M.-F.; data curation, C.d.A.M., V.M.-N. and E.S.F.; writing—original draft preparation, B.L.R.S., C.D.L.C., C.R.A.V.M., D.M.-F.; writing—review and editing, C.d.A.M., V.M.-N. and E.S.F.; supervision, C.d.A.M., V.M.-N. and E.S.F.; project administration,

C.d.A.M., V.M.-N. and E.S.F.; funding acquisition, E.S.F. All authors have read and agreed to the published version of the manuscript.

Funding: This work was supported by the Conselho Nacional de Desenvolvimento Científico e Tecnológico (CNPq; Brazil; grant numbers: 305676/2019-9 and 408053/2018-6 to E.S.F and 315072/2020-2 to V.M.-N.), the Instituto de Pesquisa Pelé Pequeno Príncipe (Brazil), and INCT-INOVAMED (Brazil). B.L.R.S. is a PhD student and V.M.-N. (grant number ACT-05691/21) who received a grant from Fundação de Amparo à Pesquisa e ao Desenvolvimento Científico e Tecnológico do Maranhão (FAPEMA, Brazil). G.S. is a MSc and, C.D.L.C. and L.R.B. are PhD students, all receiving grants from the Coordenação de Aperfeiçoamento de Pessoal de Nível Superior (CAPES; finance code 001;Brazil).

Institutional Review Board Statement: Not applicable.

Informed Consent Statement: No applicable.

Data Availability Statement: Data will be available upon request to B.L.R.S., C.D.L.C. and S.J.F.M.

Conflicts of Interest: The authors declare no conflict of interest.

References

1. Spalanzani, R.N.; Mattos, K.; Marques, L.I.; Barros, P.F.D.; Pereira, P.I.P.; Paniago, A.M.M.; Mendes, R.P.; Chang, M.R. Clinical and laboratorial features of oral candidiasis in HIV-positive patients. *Rev. Soc. Bras. Med. Trop.* **2018**, *51*, 352–356. [CrossRef] [PubMed]
2. Fidel, P.L., Jr. Candida-host interactions in HIV disease: Implications for oropharyngeal candidiasis. *Adv. Dent. Res.* **2011**, *23*, 45–49. [CrossRef] [PubMed]
3. Das, P.P.; Saikia, L.; Nath, R.; Phukan, S.K. Species distribution & antifungal susceptibility pattern of oropharyngeal Candida isolates from human immunodeficiency virus infected individuals. *Indian J. Med. Res.* **2016**, *143*, 495–501. [CrossRef] [PubMed]
4. Khedri, S.; Santos, A.L.S.; Roudbary, M.; Hadighi, R.; Falahati, M.; Farahyar, S.; Khoshmirsafa, M.; Kalantari, S. Iranian HIV/AIDS patients with oropharyngeal candidiasis: Identification, prevalence and antifungal susceptibility of Candida species. *Lett. Appl. Microbiol.* **2018**, *67*, 392–399. [CrossRef] [PubMed]
5. Hosain Pour, A.; Salari, S.; Ghasemi Nejad Almani, P. Oropharyngeal candidiasis in HIV/AIDS patients and non-HIV subjects in the Southeast of Iran. *Curr. Med. Mycol.* **2018**, *4*, 1–6. [CrossRef]
6. Salari, S.; Khosravi, A.R.; Mousavi, S.A.; Nikbakht-Brojeni, G.H. Mechanisms of resistance to fluconazole in Candida albicans clinical isolates from Iranian HIV-infected patients with oropharyngeal candidiasis. *J. Mycol. Med.* **2016**, *26*, 35–41. [CrossRef]
7. Hellstein, J.W.; Marek, C.L. Candidiasis: Red and White Manifestations in the Oral Cavity. *Head Neck Pathol.* **2019**, *13*, 25–32. [CrossRef]
8. Patil, S.; Majumdar, B.; Sarode, S.C.; Sarode, G.S.; Awan, K.H. Oropharyngeal Candidosis in HIV-Infected Patients-An Update. *Front. Microbiol.* **2018**, *9*, 980. [CrossRef]
9. Goulart, L.S.; Souza, W.W.R.; Vieira, C.A.; Lima, J.S.; Olinda, R.A.; Araujo, C. Oral colonization by Candida species in HIV-positive patients: Association and antifungal susceptibility study. *Einstein* **2018**, *16*, eAO4224. [CrossRef]
10. Hu, L.; He, C.; Zhao, C.; Chen, X.; Hua, H.; Yan, Z. Characterization of oral candidiasis and the Candida species profile in patients with oral mucosal diseases. *Microb. Pathog.* **2019**, *134*, 103575. [CrossRef]
11. Lortholary, O.; Petrikkos, G.; Akova, M.; Arendrup, M.C.; Arikan-Akdagli, S.; Bassetti, M.; Bille, J.; Calandra, T.; Castagnola, E.; Cornely, O.A.; et al. ESCMID* guideline for the diagnosis and management of Candida diseases 2012: Patients with HIV infection or AIDS. *Clin. Microbiol. Infect.* **2012**, *18*, 68–77. [CrossRef] [PubMed]
12. Pfaller, M.A.; Diekema, D.J.; Gibbs, D.L.; Newell, V.A.; Ellis, D.; Tullio, V.; Rodloff, A.; Fu, W.; Ling, T.A.; Global Antifungal Surveillance Group. Results from the ARTEMIS DISK Global Antifungal Surveillance Study, 1997 to 2007: A 10.5-year analysis of susceptibilities of Candida Species to fluconazole and voriconazole as determined by CLSI standardized disk diffusion. *J. Clin. Microbiol.* **2010**, *48*, 1366–1377. [CrossRef] [PubMed]
13. Rajadurai, S.G.; Maharajan, M.K.; Veettil, S.K.; Gopinath, D. Comparative Efficacy of Antifungal Agents Used in the Treatment of Oropharyngeal Candidiasis among HIV-Infected Adults: A Systematic Review and Network Meta-Analysis. *J. Fungi* **2021**, *7*, 637. [CrossRef] [PubMed]
14. Murphy, S.E.; Bicanic, T. Drug Resistance and Novel Therapeutic Approaches in Invasive Candidiasis. *Front. Cell Infect. Microbiol.* **2021**, *11*, 759408. [CrossRef]
15. Badiee, P.; Badali, H.; Boekhout, T.; Diba, K.; Moghadam, A.G.; Hossaini Nasab, A.; Jafarian, H.; Mohammadi, R.; Mirhendi, H.; Najafzadeh, M.J.; et al. Antifungal susceptibility testing of Candida species isolated from the immunocompromised patients admitted to ten university hospitals in Iran: Comparison of colonizing and infecting isolates. *BMC Infect. Dis.* **2017**, *17*, 727. [CrossRef]
16. Dehghani Nazhvani, A.; Haddadi, P.; Badiee, P.; Malekhoseini, S.A.; Jafarian, H. Antifungal Effects of Common Mouthwashes on Candida Strains Colonized in the Oral Cavities of Liver Transplant Recipients in South Iran in 2014. *Hepat. Mon.* **2016**, *16*, e31245. [CrossRef]

17. Shivaswamy, U.; Sumana, M.N. Antifungal Resistance of Candida Species Isolated from HIV Patients in a Tertiary Care Hospital, Mysuru, Karnataka. *Indian J. Dermatol.* **2020**, *65*, 423–425. [CrossRef]
18. Pfaller, M.A.; Rhomberg, P.R.; Messer, S.A.; Jones, R.N.; Castanheira, M. Isavuconazole, micafungin, and 8 comparator antifungal agents' susceptibility profiles for common and uncommon opportunistic fungi collected in 2013: Temporal analysis of antifungal drug resistance using CLSI species-specific clinical breakpoints and proposed epidemiological cutoff values. *Diagn. Microbiol. Infect. Dis.* **2015**, *82*, 303–313. [CrossRef]
19. Sandai, D.; Tabana, Y.M.; Ouweini, A.E.; Ayodeji, I.O. Resistance of Candida albicans Biofilms to Drugs and the Host Immune System. *Jundishapur J. Microbiol.* **2016**, *9*, e37385. [CrossRef]
20. Taff, H.T.; Mitchell, K.F.; Edward, J.A.; Andes, D.R. Mechanisms of Candida biofilm drug resistance. *Future Microbiol.* **2013**, *8*, 1325–1337. [CrossRef]
21. De Figueiredo, S.M.; Filho, S.A.; Nogueira-Machado, J.A.; Caligiorne, R.B. The anti-oxidant properties of isothiocyanates: A review. *Recent Pat. Endocr. Metab. Immune Drug Discov.* **2013**, *7*, 213–225. [CrossRef] [PubMed]
22. Dufour, V.; Stahl, M.; Baysse, C. The antibacterial properties of isothiocyanates. *Microbiology* **2015**, *161*, 229–243. [CrossRef] [PubMed]
23. Kurepina, N.; Kreiswirth, B.N.; Mustaev, A. Growth-inhibitory activity of natural and synthetic isothiocyanates against representative human microbial pathogens. *J. Appl. Microbiol.* **2013**, *115*, 943–954. [CrossRef] [PubMed]
24. Nowicki, D.; Rodzik, O.; Herman-Antosiewicz, A.; Szalewska-Palasz, A. Isothiocyanates as effective agents against enterohemorrhagic Escherichia coli: Insight to the mode of action. *Sci. Rep.* **2016**, *6*, 22263. [CrossRef] [PubMed]
25. Plaszko, T.; Szucs, Z.; Vasas, G.; Gonda, S. Effects of Glucosinolate-Derived Isothiocyanates on Fungi: A Comprehensive Review on Direct Effects, Mechanisms, Structure-Activity Relationship Data and Possible Agricultural Applications. *J. Fungi* **2021**, *7*, 539. [CrossRef]
26. Romeo, L.; Iori, R.; Rollin, P.; Bramanti, P.; Mazzon, E. Isothiocyanates: An Overview of Their Antimicrobial Activity against Human Infections. *Molecules* **2018**, *23*, 624. [CrossRef]
27. Gupta, P.; Kim, B.; Kim, S.H.; Srivastava, S.K. Molecular targets of isothiocyanates in cancer: Recent advances. *Mol. Nutr. Food Res.* **2014**, *58*, 1685–1707. [CrossRef]
28. Prawan, A.; Saw, C.L.; Khor, T.O.; Keum, Y.S.; Yu, S.; Hu, L.; Kong, A.N. Anti-NF-kappaB and anti-inflammatory activities of synthetic isothiocyanates: Effect of chemical structures and cellular signaling. *Chem. Biol. Interact.* **2009**, *179*, 202–211. [CrossRef]
29. Wagner, A.E.; Terschluesen, A.M.; Rimbach, G. Health promoting effects of brassica-derived phytochemicals: From chemopreventive and anti-inflammatory activities to epigenetic regulation. *Oxid. Med. Cell. Longev.* **2013**, *2013*, 964539. [CrossRef]
30. Cheung, K.L.; Kong, A.N. Molecular targets of dietary phenethyl isothiocyanate and sulforaphane for cancer chemoprevention. *AAPS J.* **2010**, *12*, 87–97. [CrossRef]
31. Houghton, C.A.; Fassett, R.G.; Coombes, J.S. Sulforaphane and Other Nutrigenomic Nrf2 Activators: Can the Clinician's Expectation Be Matched by the Reality? *Oxid. Med. Cell. Longev.* **2016**, *2016*, 7857186. [CrossRef] [PubMed]
32. Qi, T.; Xu, F.; Yan, X.; Li, S.; Li, H. Sulforaphane exerts anti-inflammatory effects against lipopolysaccharide-induced acute lung injury in mice through the Nrf2/ARE pathway. *Int. J. Mol. Med.* **2016**, *37*, 182–188. [CrossRef] [PubMed]
33. De Oliveira, M.R.; de Bittencourt Brasil, F.; Furstenau, C.R. Sulforaphane Promotes Mitochondrial Protection in SH-SY5Y Cells Exposed to Hydrogen Peroxide by an Nrf2-Dependent Mechanism. *Mol. Neurobiol.* **2018**, *55*, 4777–4787. [CrossRef] [PubMed]
34. Liu, P.; Atkinson, S.J.; Akbareian, S.E.; Zhou, Z.; Munsterberg, A.; Robinson, S.D.; Bao, Y. Sulforaphane exerts anti-angiogenesis effects against hepatocellular carcinoma through inhibition of STAT3/HIF-1alpha/VEGF signalling. *Sci. Rep.* **2017**, *7*, 12651. [CrossRef]
35. Russo, M.; Spagnuolo, C.; Russo, G.L.; Skalicka-Wozniak, K.; Daglia, M.; Sobarzo-Sanchez, E.; Nabavi, S.F.; Nabavi, S.M. Nrf2 targeting by sulforaphane: A potential therapy for cancer treatment. *Crit. Rev. Food Sci. Nutr.* **2018**, *58*, 1391–1405. [CrossRef]
36. Su, X.; Jiang, X.; Meng, L.; Dong, X.; Shen, Y.; Xin, Y. Anticancer Activity of Sulforaphane: The Epigenetic Mechanisms and the Nrf2 Signaling Pathway. *Oxid. Med. Cell. Longev.* **2018**, *2018*, 5438179. [CrossRef]
37. Angeloni, C.; Malaguti, M.; Rizzo, B.; Barbalace, M.C.; Fabbri, D.; Hrelia, S. Neuroprotective effect of sulforaphane against methylglyoxal cytotoxicity. *Chem. Res. Toxicol.* **2015**, *28*, 1234–1245. [CrossRef]
38. Tarozzi, A.; Angeloni, C.; Malaguti, M.; Morroni, F.; Hrelia, S.; Hrelia, P. Sulforaphane as a potential protective phytochemical against neurodegenerative diseases. *Oxid. Med. Cell. Longev.* **2013**, *2013*, 415078. [CrossRef]
39. Houghton, C.A. Sulforaphane: Its "Coming of Age" as a Clinically Relevant Nutraceutical in the Prevention and Treatment of Chronic Disease. *Oxid. Med. Cell. Longev.* **2019**, *2019*, 2716870. [CrossRef]
40. Haristoy, X.; Angioi-Duprez, K.; Duprez, A.; Lozniewski, A. Efficacy of sulforaphane in eradicating Helicobacter pylori in human gastric xenografts implanted in nude mice. *Antimicrob. Agents Chemother.* **2003**, *47*, 3982–3984. [CrossRef]
41. Yanaka, A.; Fahey, J.W.; Fukumoto, A.; Nakayama, M.; Inoue, S.; Zhang, S.; Tauchi, M.; Suzuki, H.; Hyodo, I.; Yamamoto, M. Dietary sulforaphane-rich broccoli sprouts reduce colonization and attenuate gastritis in Helicobacter pylori-infected mice and humans. *Cancer Prev. Res.* **2009**, *2*, 353–360. [CrossRef] [PubMed]
42. Aires, A.; Mota, V.R.; Saavedra, M.J.; Rosa, E.A.; Bennett, R.N. The antimicrobial effects of glucosinolates and their respective enzymatic hydrolysis products on bacteria isolated from the human intestinal tract. *J. Appl. Microbiol.* **2009**, *106*, 2086–2095. [CrossRef] [PubMed]

43. Johansson, N.L.; Pavia, C.S.; Chiao, J.W. Growth inhibition of a spectrum of bacterial and fungal pathogens by sulforaphane, an isothiocyanate product found in broccoli and other cruciferous vegetables. *Planta Med.* **2008**, *74*, 747–750. [CrossRef] [PubMed]
44. Cierpial, T.; Kielbasinski, P.; Kwiatkowska, M.; Lyzwa, P.; Lubelska, K.; Kuran, D.; Dabrowska, A.; Kruszewska, H.; Mielczarek, L.; Chilmonczyk, Z.; et al. Fluoroaryl analogs of sulforaphane—A group of compounds of anticancer and antimicrobial activity. *Bioorg. Chem.* **2020**, *94*, 103454. [CrossRef]
45. Devi, J.R.; Thangam, E.B. Studies on antioxidant and antimicrobial activities of purified sulforaphane from *Brassica oleraceae* var. *rubra*. *J. Pharm. Res.* **2012**, *5*, 3582–3584.
46. Murata, W.; Yamaguchi, Y.; Fujita, K.I.; Yamauchi, K.; Tanaka, T.; Ogita, A. Enhancement of paraben-fungicidal activity by sulforaphane, a cruciferous vegetable-derived isothiocyanate, via membrane structural damage in *Saccharomyces cerevisiae*. *Lett. Appl. Microbiol.* **2019**, *69*, 403–410. [CrossRef] [PubMed]
47. Desai, J.V. Candida albicans Hyphae: From Growth Initiation to Invasion. *J. Fungi* **2018**, *4*, 10. [CrossRef]
48. Yong, J.; Zu, R.; Huang, X.; Ge, Y.; Li, Y. Synergistic Effect of Berberine Hydrochloride and Fluconazole against *Candida albicans* Resistant Isolates. *Front. Microbiol.* **2020**, *11*, 1498. [CrossRef]
49. Nett, J.E.; Andes, D.R. Antifungal agents: Spectrum of activity, pharmacology, and clinical indications. *Infect. Dis. Clin.* **2016**, *30*, 51–83. [CrossRef]
50. Abel, E.L.; Boulware, S.; Fields, T.; McIvor, E.; Powell, K.L.; DiGiovanni, J.; Vasquez, K.M.; MacLeod, M.C. Sulforaphane induces phase II detoxication enzymes in mouse skin and prevents mutagenesis induced by a mustard gas analog. *Toxicol. Appl. Pharmacol.* **2013**, *266*, 439–442. [CrossRef]
51. Huo, L.; Su, Y.; Xu, G.; Zhai, L.; Zhao, J. Sulforaphane Protects the Male Reproductive System of Mice from Obesity-Induced Damage: Involvement of Oxidative Stress and Autophagy. *Int. J. Environ. Res. Public Health* **2019**, *16*, 3759. [CrossRef]
52. Sestili, P.; Paolillo, M.; Lenzi, M.; Colombo, E.; Vallorani, L.; Casadei, L.; Martinelli, C.; Fimognari, C. Sulforaphane induces DNA single strand breaks in cultured human cells. *Mutat. Res.* **2010**, *689*, 65–73. [CrossRef] [PubMed]
53. Wei, Y.; Chang, L.; Liu, G.; Wang, X.; Yang, Y.; Hashimoto, K. Long-lasting beneficial effects of maternal intake of sulforaphane glucosinolate on gut microbiota in adult offspring. *J. Nutr. Biochem.* **2022**, *109*, 109098. [CrossRef] [PubMed]
54. Clarke, J.D.; Hsu, A.; Williams, D.E.; Dashwood, R.H.; Stevens, J.F.; Yamamoto, M.; Ho, E. Metabolism and tissue distribution of sulforaphane in Nrf2 knockout and wild-type mice. *Pharm. Res.* **2011**, *28*, 3171–3179. [CrossRef] [PubMed]
55. Jazwa, A.; Rojo, A.I.; Innamorato, N.G.; Hesse, M.; Fernandez-Ruiz, J.; Cuadrado, A. Pharmacological targeting of the transcription factor Nrf2 at the basal ganglia provides disease modifying therapy for experimental parkinsonism. *Antioxid. Redox Signal.* **2011**, *14*, 2347–2360. [CrossRef]
56. Pajouhesh, H.; Lenz, G.R. Medicinal chemical properties of successful central nervous system drugs. *NeuroRx* **2005**, *2*, 541–553. [CrossRef]
57. Debruyne, D. Clinical pharmacokinetics of fluconazole in superficial and systemic mycoses. *Clin. Pharmacokinet.* **1997**, *33*, 52–77. [CrossRef]
58. Pappas, P.G.; Chetchotisakd, P.; Larsen, R.A.; Manosuthi, W.; Morris, M.I.; Anekthananon, T.; Sungkanuparph, S.; Supparatpinyo, K.; Nolen, T.L.; Zimmer, L.O.; et al. A phase II randomized trial of amphotericin B alone or combined with fluconazole in the treatment of HIV-associated cryptococcal meningitis. *Clin. Infect. Dis.* **2009**, *48*, 1775–1783. [CrossRef]
59. Thaler, F.; Bernard, B.; Tod, M.; Jedynak, C.P.; Petitjean, O.; Derome, P.; Loirat, P. Fluconazole penetration in cerebral parenchyma in humans at steady state. *Antimicrob. Agents Chemother.* **1995**, *39*, 1154–1156. [CrossRef]
60. Chandra, J.; Kuhn, D.M.; Mukherjee, P.K.; Hoyer, L.L.; McCormick, T.; Ghannoum, M.A. Biofilm formation by the fungal pathogen Candida albicans: Development, architecture, and drug resistance. *J. Bacteriol.* **2001**, *183*, 5385–5394. [CrossRef]
61. Cavalheiro, M.; Teixeira, M.C. Candida Biofilms: Threats, Challenges, and Promising Strategies. *Front. Med.* **2018**, *5*, 28. [CrossRef] [PubMed]
62. Raut, J.S.; Bansode, B.S.; Jadhav, A.K.; Karuppayil, S.M. Activity of Allyl Isothiocyanate and Its Synergy with Fluconazole against *Candida albicans* Biofilms. *J. Microbiol. Biotechnol.* **2017**, *27*, 685–693. [CrossRef] [PubMed]
63. Ganin, H.; Rayo, J.; Amara, N.; Levy, N.; Krief, P.; Meijler, M.M. Sulforaphane and erucin, natural isothiocyanates from broccoli, inhibit bacterial quorum sensing. *MedChemComm* **2013**, *4*, 175–179. [CrossRef]
64. Grainha, T.R.R.; Jorge, P.; Perez-Perez, M.; Perez Rodriguez, G.; Pereira, M.; Lourenco, A.M.G. Exploring anti-quorum sensing and anti-virulence based strategies to fight Candida albicans infections: An in silico approach. *FEMS Yeast Res.* **2018**, *18*, foy022. [CrossRef] [PubMed]
65. Feng, Q.; Summers, E.; Guo, B.; Fink, G. Ras signaling is required for serum-induced hyphal differentiation in Candida albicans. *J. Bacteriol.* **1999**, *181*, 6339–6346. [CrossRef]
66. Roudbarmohammadi, S.; Roudbary, M.; Bakhshi, B.; Katiraee, F.; Mohammadi, R.; Falahati, M. ALS1 and ALS3 gene expression and biofilm formation in Candida albicans isolated from vulvovaginal candidiasis. *Adv. Biomed. Res.* **2016**, *5*, 105. [CrossRef]
67. Sohn, K.; Urban, C.; Brunner, H.; Rupp, S. EFG1 is a major regulator of cell wall dynamics in Candida albicans as revealed by DNA microarrays. *Mol. Microbiol.* **2003**, *47*, 89–102. [CrossRef]
68. Liu, X.; Li, T.; Wang, D.; Yang, Y.; Sun, W.; Liu, J.; Sun, S. Synergistic Antifungal Effect of Fluconazole Combined with Licofelone against Resistant *Candida albicans*. *Front. Microbiol.* **2017**, *8*, 2101. [CrossRef]
69. Wu, J.; Wu, D.; Zhao, Y.; Si, Y.; Mei, L.; Shao, J.; Wang, T.; Yan, G.; Wang, C. Sodium New Houttuyfonate Inhibits Candida albicans Biofilm Formation by Inhibiting the Ras1-cAMP-Efg1 Pathway Revealed by RNA-seq. *Front. Microbiol.* **2020**, *11*, 2075. [CrossRef]

70. Ferreira, S.B.; Dantas, T.B.; de Figueredo Silva, D.; Ferreira, P.B.; de Melo, T.R.; de Oliveira Lima, E. In Silico and In Vitro Investigation of the Antifungal Activity of Isoeugenol against Penicillium citrinum. *Curr. Top. Med. Chem.* **2018**, *18*, 2186–2196. [CrossRef]
71. Khurana, N.; Ishar, M.P.; Gajbhiye, A.; Goel, R.K. PASS assisted prediction and pharmacological evaluation of novel nicotinic analogs for nootropic activity in mice. *Eur. J. Pharmacol.* **2011**, *662*, 22–30. [CrossRef] [PubMed]
72. Daina, A.; Michielin, O.; Zoete, V. SwissADME: A free web tool to evaluate pharmacokinetics, drug-likeness and medicinal chemistry friendliness of small molecules. *Sci. Rep.* **2017**, *7*, 42717. [CrossRef] [PubMed]
73. Lipinski, C.A. Lead- and drug-like compounds: The rule-of-five revolution. *Drug Discov. Today Technol.* **2004**, *1*, 337–341. [CrossRef] [PubMed]
74. Monteiro-Neto, V.; de Souza, C.D.; Gonzaga, L.F.; da Silveira, B.C.; Sousa, N.C.F.; Pontes, J.P.; Santos, D.M.; Martins, W.C.; Pessoa, J.F.V.; Carvalho Junior, A.R.; et al. Cuminaldehyde potentiates the antimicrobial actions of ciprofloxacin against *Staphylococcus aureus* and *Escherichia coli*. *PLoS ONE* **2020**, *15*, e0232987. [CrossRef] [PubMed]
75. Drwal, M.N.; Banerjee, P.; Dunkel, M.; Wettig, M.R.; Preissner, R. ProTox: A web server for the in silico prediction of rodent oral toxicity. *Nucleic Acids Res.* **2014**, *42*, W53–W58. [CrossRef]
76. CLSI. *Reference Method for Broth Dilution Antifungal Susceptibility Testing of Yeasts: Approved Standard. M27-A3*, 3rd ed.; CLSI: Wayne, PA, USA, 2008.
77. Santos, J.R.; Gouveia, L.F.; Taylor, E.L.; Resende-Stoianoff, M.A.; Pianetti, G.A.; Cesar, I.C.; Santos, D.A. Dynamic interaction between fluconazole and amphotericin B against *Cryptococcus gattii*. *Antimicrob. Agents Chemother.* **2012**, *56*, 2553–2558. [CrossRef]
78. Gómez-López, A.; Cuenca-Estrella, M.; Mellado, E.; Rodríguez-Tudela, J.L. In vitro evaluation of combination of terbinafine with itraconazole or amphotericin B against *Zygomycota*. *Diagn. Microbiol. Infect. Dis.* **2003**, *45*, 199–202. [CrossRef]
79. Atiencia-Carrera, M.B.; Cabezas-Mera, F.S.; Vizuete, K.; Debut, A.; Tejera, E.; Machado, A. Evaluation of the biofilm life cycle between *Candida albicans* and *Candida tropicalis*. *Front. Cell. Infect. Microbiol.* **2022**, *12*, 953168. [CrossRef]
80. Ha, K.C.; White, T.C. Effects of azole antifungal drugs on the transition from yeast cells to hyphae in susceptible and resistant isolates of the pathogenic yeast *Candida albicans*. *Antimicrob. Agents Chemother.* **1999**, *43*, 763–768. [CrossRef]
81. Tsang, P.W.; Bandara, H.M.; Fong, W.P. Purpurin suppresses *Candida albicans* biofilm formation and hyphal development. *PLoS ONE* **2012**, *7*, e50866. [CrossRef]
82. Roth, R.; Madhani, H.D.; Garcia, J.F. Total RNA Isolation and Quantification of Specific RNAs in Fission Yeast. *Methods Mol. Biol.* **2018**, *1721*, 63–72. [CrossRef] [PubMed]
83. Livak, K.J.; Schmittgen, T.D. Analysis of relative gene expression data using real-time quantitative PCR and the 2(-Delta Delta C(T)) Method. *Methods* **2001**, *25*, 402–408. [CrossRef] [PubMed]

antibiotics

MDPI

Review

An Assessment of the In Vitro Models and Clinical Trials Related to the Antimicrobial Activities of Phytochemicals

Jonathan Kopel [1], Julianna McDonald [2] and Abdul Hamood [3,4,*]

1 School of Medicine, Texas Tech University Health Sciences Center, Lubbock, TX 79430, USA
2 Texas Tech University, Lubbock, TX 79430, USA
3 Department of Immunology and Molecular Microbiology, Texas Tech University Health Sciences Center, Lubbock, TX 79430, USA
4 Department of Surgery, Texas Tech University Health Sciences Center, Lubbock, TX 79430, USA
* Correspondence: abdul.hamood@ttuhsc.edu

Abstract: An increased number antibiotic-resistant bacteria have emerged with the rise in antibiotic use worldwide. As such, there has been a growing interest in investigating novel antibiotics against antibiotic-resistant bacteria. Due to the extensive history of using plants for medicinal purposes, scientists and medical professionals have turned to plants as potential alternatives to common antibiotic treatments. Unlike other antibiotics in use, plant-based antibiotics have the innate ability to eliminate a broad spectrum of microorganisms through phytochemical defenses, including compounds such as alkaloids, organosulfur compounds, phenols, coumarins, and terpenes. In recent years, these antimicrobial compounds have been refined through extraction methods and tested against antibiotic-resistant strains of Gram-negative and Gram-positive bacteria. The results of the experiments demonstrated that plant extracts successfully inhibited bacteria independently or in combination with other antimicrobial products. In this review, we examine the use of plant-based antibiotics for their utilization against antibiotic-resistant bacterial infections. In addition, we examine recent clinical trials utilizing phytochemicals for the treatment of several microbial infections.

Keywords: antibiotic resistance; alkaloids; multidrug-resistant; MRSA; organosulfur compounds; phenolic compounds; phytochemicals; plant-based extracts; Pseudomonas aeruginosa; *Staphylococcus aureus*

Citation: Kopel, J.; McDonald, J.; Hamood, A. An Assessment of the In Vitro Models and Clinical Trials Related to the Antimicrobial Activities of Phytochemicals. *Antibiotics* **2022**, *11*, 1838. https://doi.org/10.3390/antibiotics11121838

Academic Editors: Valério Monteiro-Neto and Elizabeth S. Fernandes

Received: 18 November 2022
Accepted: 16 December 2022
Published: 17 December 2022

1. Introduction

The discovery of penicillin led to a cascade of medical innovations that enhanced the treatment of bacterial infections [1,2]. However, as the discovery and utility of antibiotics increased, their ability to successfully inhibit bacterial infections decreased due to the overuse of antibiotics [1,3]. Microbial infections and antibiotic resistance are a major problem leading to the deaths of millions of patients each year. The development of resistance has made the currently available antibacterial medications less effective [4]. This threat is further compounded by the growing recognition of biofilm formations among several bacterial species, which have made certain antibiotics ineffective for severe infections [5,6]. Therefore, new medicinal treatments that could restrict the growth of antibiotic resistance (e.g., bacterial pathogens) are required. Several methods have been proposed in recent years to combat antibiotic resistance. One of the suggested methods for achieving this includes combining antibiotics with other compounds, such as plant phytochemicals, to restore their antibacterial activities (Figure 1) [7,8]. This review explores the antimicrobial activities of several phytochemicals (e.g., alkaloids, organosulfur compounds, phenols, coumarins, and terpenes) against antibiotic-resistant bacteria.

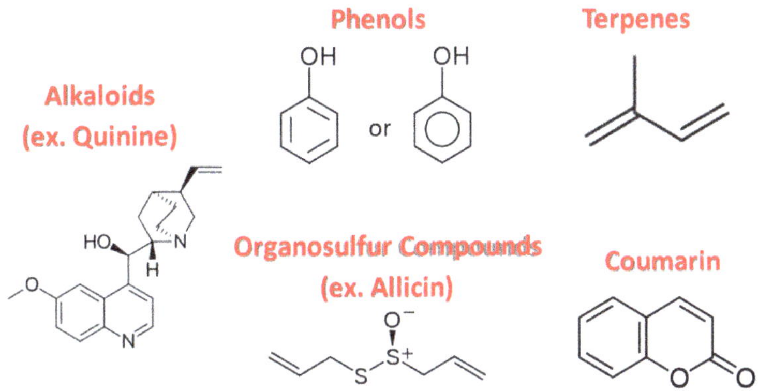

Figure 1. Structures of major phytochemical families.

2. Mechanisms of Antibacterial Activity and Resistance

An antibiotic's activity is attributed to two primary processes that interfere with the production or operation of an essential bacterial function and/or evade the established antibacterial resistance mechanisms [9,10]. Most antibiotics were discovered in soil samples that contained compounds that are able to eradicate bacteria. In addition, antibiotics such as streptomycin, erythromycin, tetracycline, vancomycin, penicillins, and cephalosporins were harvested from fungi and filamentous bacteria [9,10]. In contrast, the second- and third-generation beta-lactams of the penicillin and cephalosporin families were made using semisynthetic modifications, whereas the second-generation erythromycins, clarithromycin, and azithromycin were made through complete synthesis [9,10].

In general, the primary targets for antibacterial agents include bacterial protein biosynthesis, bacterial cell-wall biosynthesis, bacterial cell membranes, bacterial DNA replication and repair, and metabolic pathways [9,10]. An antibiotic's usefulness is limited when it has been shown to be an effective antibacterial agent and is widely used therapeutically. Without proper restrictions and surveillance, bacterial resistance manifests over a prolonged period of use. Bacteria have several methods for resisting antibiotics. Some bacteria have an intrinsic resistance to one or more antimicrobial agent classes. In most cases, bacteria develop resistance to antibiotics, primarily through several mechanisms, including the activation of efflux pumps, destruction through hydrolytic enzymes, the modification of antibiotic structures, and the alteration of target structures [9,10]. In addition, plasmids are another method by which antibacterial resistance propagates through subsequent bacterial generations. As such, there is a growing need to develop novel antibacterials that can overcome the inevitable production of antibiotic-resistant bacteria.

One method for circumventing antibiotic resistance has been the use of plant phytochemicals. For centuries, plant products have been utilized to treat infections across several cultures [11–13]. The revitalized interest in utilizing natural products or creating synthetic versions of natural products grew in response to antibiotic resistance [11–14]. In recent years, phytochemicals have demonstrated strong antimicrobial activities in conjunction with standard antibiotic regimens [11,15,16]. Currently, alkaloids, organosulfur compounds, and phenolic compounds extracted from plants have shown efficacy against microbes, particularly multidrug-resistant bacteria. As such, there is growing interest in using plant extracts to help address antibiotic resistance [13,17,18]. In this review, we examine the use of plant-based antibiotics for their utilization against antibiotic-resistant bacterial infections.

2.1. Alkaloids

Alkaloids comprise a major class of phytochemicals that are synthesized as secondary metabolites with low molecular weights and nitrogen contents [17]. Several antibacterial medications have been derived from alkaloids. For example, medications such as linezolid and trimethoprim contain alkaloids as their primary structures. Alkaloids exhibit antibacterial activities against a wide range of bacterial infections [17]. Alkaloids include a large family of chemicals with diverse heterocyclic structures, which are generally further classified by their carbon skeleton structures [11,13,17–20]. Several plant families contain alkaloid compounds with antimicrobial properties against *S. aureus*, *P. aeruginosa*, and other pathogenic bacteria [19].

Specifically, several recent studies showed that the alkaloid sanguinarine, a benzophenanthridine alkaloid found in *Sanguinaria canadensis* and other poppy fumaria species, has strong antibacterial, antifungal, and anti-inflammatory properties [17,18]. Sanguinarine was identified as having broad-spectrum antimicrobial properties that interfere with bacterial cell division and cytokinesis [19]. Specifically, sanguinarine inhibits the replication of methicillin-resistant *Staphylococcus aureus* (MRSA) through the disruption of the plasma membrane by interfering with the protein FtsZ's ability to assemble into filaments, thereby inhibiting bacterial fission [19,21–23]. When used in combination, sanguinarine enhanced the antimicrobial activity of vancomycin and streptomycin. The chelating EDTA, which disrupts the permeability of the bacterial cell wall, increases the amount of sanguinarine and streptomycin that enter the bacterial cell, thereby boosting their antimicrobial activities. Sanguinarine also demonstrated effectiveness against eight phytopathogenic fungi, including *Magnaporthe oryzae*, *Fusarium oxysporum*, *Fusarium graminearum*, and *Botrytis cinerea* [21]. Specifically, in vitro studies of fungi treated with sanguinarine showed that the alkaloid eliminated fungi by increasing the production of reactive oxygen species, which was linked to changes in the nuclear morphology and the redox potential of mitochondrial membranes [21]. In addition, sanguinarine has antimicrobial properties against pathogenic bacteria found in soil, including *Agrobacterium tumefaciens*, *Pseudomonas lachrymans*, and *Xanthomonas vesicatoria* [21].

In addition to sanguinarine, another alkaloid, tomatidine, showed strong antimicrobial activities against *S. aureus* strains as well as other Gram-negative and Gram-positive bacterial species [17–19,24]. Tomatidine was first isolated from solanaceous plants, such as tomatoes and potatoes. Later studies demonstrated the ability of tomatidine to disrupt the activity of ATP synthase in several bacterial species [17,18,23,24]. Furthermore, tomatidine is an aminoglycoside potentiator against *S. aureus* strains that are both sensitive and resistant to aminoglycosides [25]. Subsequent experiments showed that tomatidine is also effective against *Listeria* and *Bacillus* bacterial species [26]. The exact mechanism behind tomatidine's synergy with aminoglycosides is not fully understood [17,19,23,24]. However, it is believed that both aminoglycosides and tomatidine, when used together, reach their respective intracellular targets by increasing cell permeabilization [26]. In addition, tomatidine may also inhibit the formation of macromolecules within the bacterial target by blocking important steps in protein synthesis. Beyond its antimicrobial activity, tomatidine also inhibits fungal species, such as *Saccharomyces cerevisiae*, by blocking the formation of ergosterol via the inhibition of C-24 sterol methyltransferase and C-24 sterol reductase [27–30]. Other alkaloid compounds have demonstrated synergistic activities with antibiotics.

The alkaloid piperine potentiates the effect of ciprofloxacin against *S. aureus* [31]. Similar results were obtained when piperine and gentamicin were administered together to treat MRSA infections [32]. The mechanism behind this phenomenon remains to be fully understood. Piperine also exhibited a potent antibacterial activity against *Mycobacterium tuberculosis* and *Mycobacterium smegmatis* [33–35]. In addition, piperine improves the therapeutic effectiveness of rifampicin in immunocompromised patients infected with *M. tuberculosis* [32,35].

The isoquinoline alkaloid berberine is an effective plant alkaloid against a broad range of viruses, protozoa, fungi, and bacteria [35]. Although the mechanism is not fully elucidated yet, current experimental studies suggest that berberine's antimicrobial activity is related to disrupting bacterial cell walls, particularly in MRSA [35]. In addition to disrupting bacterial cell walls, berberine may interfere with bacterial division, protein synthesis, and biofilm development [35]. Further studies suggested that berberine's activity is related to DNA intercalation and the targeting of RNA polymerase, gyrase, and topoisomerase IV enzymes [36–39].

Overall, alkaloids have strong antimicrobial potentials that make them an attractive alternative to traditional antibiotics.

2.2. Organosulfur Compounds

Organosulfur compounds, or sulfur-containing compounds, constitute another class of phytochemicals that are considered secondary metabolites [11,13,17,18]. Originally, organosulfur compounds were found in two different families of plants that exhibited antimicrobial qualities: the *Alliaceae* and the *Cruciferae* (*Brassicaceae*) families [13]. One such organosulfur compound from the *Alliaceae* family is allicin or diallyl thiosulfinate [13,18].

Allicin, a volatile component derived from raw garlic, was initially credited with the antibacterial properties of garlic. Since then, allicin has been the subject of many studies aimed at investigating its potential inhibitory effects against *S. aureus, Escherichia coli, and Candida albicans* [40,41]. The inhibitor effects of allicin are equivalent to, if not stronger than, several common antibiotics (e.g., kanamycin, tetracycline, and penicillin) [42]. In contrast to these antibiotics, allicin targets a broad spectrum of microorganisms, including bacteria, yeasts, fungi, and parasites [42]. Allicin's antibacterial mechanism has been linked to the specific targeting of bacterial thiol-containing proteins and enzymes, thereby inhibiting essential metabolic pathways [43]. Specifically, allicin inhibits the growth of microorganisms through the natural reaction that occurs between its –S(O)–S-group and the –SH groups in bacterial and fungal proteins [42]. The thiosulfinate's oxygen atom, which acts as an electron-withdrawing agent, forms an electrophilic sulfur center that reacts readily with thiol groups, thereby contributing to allicin's reactivity. It has also been reported that the addition of beta-mercaptoethanol, which breaks disulfide bonds, inhibits the interaction between allicin and cysteine [42]. This result suggests that the disulfide bonds that form between the sulfhydryl groups of bacterial proteins and allicin play a vital role in its antimicrobial activity. Due to its broad-spectrum antimicrobial activity, allicin will likely be a practical solution to treat multidrug-resistant bacteria [13,18].

Other organosulfates, such as glucosinolates and isothiocyanates, inhibit a wide range of pathogenic bacteria [13,18]. Isothiocyanates are volatile organosulfur compounds produced when the enzyme myrosinase reacts with plant glucosinolates [13,18]. Myrosinase hydrolyzes glucosinolates into active substances such as nitriles, thiocyanates, and isothiocyanates. Isothiocyanates have strong inhibitory effects on several pathogenic bacteria by disrupting their cell walls [13,18]. The in vitro antibacterial efficacy of isothiocyanates against bacterial pathogens has been the subject of several investigations, but little is known about their in vivo antimicrobial properties. Most studies have been focused on sulforaphane's ability to fight *Helicobacter pylori* bacteria. *H. pylori* produces a urease enzyme, which hydrolyzes urea in ammonia and carbon dioxide, thereby neutralizing the gastric acid surrounding the bacteria [44]. In addition, several virulence factors produce by *H. pylori* cause excessive inflammation in the gastric mucosa [44]. Sulforaphane was found to inactivate urease and eliminate *H. pylori* infections [44]. Dufour et al. demonstrated that sulforaphane was particularly effective against several clinical isolates of *H. pylori*, many of which were resistant to common antibiotics [45]. Dufour et al. also hypothesized that the isothiocyanate's antimicrobial action is related to its reactivity with proteins that disrupt essential biochemical pathways within *H. pylori*. Isothiocyanates attack sulfhydryl groups at their individual thiol-containing amino acids, such as cysteine [45]. Isothiocyanates are also known to block the ATP binding sites of bacterial P-ATPase.

Another family of organosulfur compounds, allyl isothiocyanates (AITCs), have also shown strong bacteriostatic and bactericidal activities against *E. coli* and *S. aureus* [46]. Allyl ITCs synergize with streptomycin against *E. coli* and *P. aeruginosa* and lower the minimum inhibitory concentration (MIC) values of erythromycin against *S. pyogenes* [47,48]. The antibacterial properties of AITC have been attributed to several different mechanisms, such as weakening cell walls and releasing reactive cellular metabolites [49,50].

Prati et al. previously reported that benzyl isothiocyanate (BITC) is bactericidal against several MRSA clinical isolates [51]. This strong BITC antibacterial activity appears to be influenced by its lipophilic and electrophilic chemical composition, which allows it to penetrate bacterial outer membranes and disrupt their plasma membranes. Phenethyl isothiocyanate (PEITC) has demonstrated antibacterial activity against several bacteria obtained from the human digestive tract (e.g., *Enterococcus* spp., *Enterobacteriaceae* spp., *Lactobacillus* spp., *Bifidobacterium* spp., *Bacteroides* spp., and *Clostridium* spp.) [52–54]. Besides its antibacterial effect, PEITC is effective against several fungal species. It accomplishes this antifungal effect through reducing the rate of oxygen consumption, increasing the production of reactive oxygen species, and depolarizing the mitochondrial membrane [52–54]. Overall, phytochemicals such as organosulfur compounds are effective in inhibiting different pathogenic bacteria.

2.3. Phenolic Compounds

Phenolic compounds are a diverse class of substances found in many foods, such as fruits, vegetables, tea, wine, and honey [11,13,17–19]. Phenolic compounds are grouped into several groups, including phenolic acids, flavonoids, and non-flavonoids [55]. Chemically, phenolic compounds are aromatic in their structures and contain numerous hydroxyl groups. These hydroxyl groups donate electrons or hydrogen atoms to neutralize free radicals and other reactive oxygen species [55]. As a result, the antimicrobial activities of phenolic compounds include inhibiting efflux pumps and cell wall biosynthesis as well as inhibiting key bacterial enzymes such as urease and dihydrofolate reductase. Phenolic compounds are, therefore, contenders for future investigations and clinical trials due to their effective antimicrobial activities. Two common groups of phenolic compounds include flavonoids and non-flavonoids [17,19].

2.4. Flavonoids

Flavonoids have demonstrated antimicrobial abilities against both Gram-negative and Gram-positive pathogens [17]. The most effective antibacterial phenolic compounds include flavanols, flavonols, and phenolic acids. These compounds exhibit antibacterial activities through a variety of mechanisms, including inhibiting bacterial enzymes and toxins, disrupting cytoplasmic membranes, preventing the formation of biofilms, and working synergistically with a wide spectrum of antibiotics [56].

Previous studies revealed that the hydroxylation and lipophilic substituents of the flavonoid ring enhance its antibacterial activity, whereas the substitution of a methoxy, acetyl, or fluoride group has the opposite effect [57–64]. The hydroxyl groups on flavonoid rings inhibit bacterial enzymes involved in cellular respiration and disrupt bacterial membranes [14,57–59,65–67]. Similarly, lengthy aliphatic chain substitutions increase the hydrophobicity of flavonoids, which increases their interactions with antibiotics. These interactions facilitate the movement of antibiotics across the bacterial cell wall to inhibit their intracellular targets. In general, flavonoid compounds exhibit a broad range of antibiotic activities through a number of modifications to their ring structures [14,57–59,65–67].

One flavonoid, galangin, possess a potent antibacterial activity against *S. aureus* species by targeting their bacterial cell walls [68]. Cushnie et al. showed that incubating *S. aureus* bacteria with galangin reduced the number of *S. aureus* colonies by almost 15,000-fold. Interestingly, it was discovered that there was an increase in potassium loss from the *S. aureus* cytoplasm when incubated with galangin. Cushine et al. investigated the mechanism of potassium release by *S. aureus* during incubation with galangin using two difference compounds: novobiocin and penicillin G. Penicillin G, which is known to disrupt cell membranes, was used as a positive control, while novobiocin, which does not target the cell membrane, was used as a negative control. *Staph aureus* bacterial cells were then individually incubated with penicillin G or novobiocin to confirm that the increase in potassium loss was due to cell wall disruption. Novobiocin did not increase potassium release, whereas penicillin G increased potassium release. The study, therefore, showed that galangin eliminates bacteria by targeting their cell walls and inducing cell lysis [69].

Other phenols, such kaempferol and quercetin, demonstrated synergistic effects with rifampicin against MRSA strains. When combined with rifampicin, kaempferol and quercetin inhibited MRSA beta-lactamase enzymes, which increased the inhibition of bacterial growth by 57.8% and 75.8%, respectively. Similarly, kaempferol and quercetin synergize and enhance ciprofloxacin's activity against several bacterial topoisomerases [70,71].

Another well-known subset of flavonoids, known as flavanols, includes compounds such as catechin, epicatechin, epigallocatechin, epicatechin gallate, and epigallocatechin gallate; these compound exhibit both bacteriostatic and bactericidal activities. According to several studies, the ability of flavonols to attach to the lipid bilayer of bacterial plasma membranes is strongly associated with their antimicrobial activity [72–78]. The demonstrated inhibitory effect of the flavonol alkyl gallate against several *S. aureus* variants is due to its ability to decrease the production of several *S. aureus* virulence factors, such as coagulase or alpha-toxin [79]. In addition, flavonol interferes with biofilm formation by *S. aureus* [79]. Other flavanols, such as (−)-epicatechin gallate and (−)-epigallocatechin gallate, promote the aggregation of staphylococcal cell walls, which renders them more susceptible to beta-lactam antibiotics [80–83].

2.5. Non-Flavonoids

In addition to flavonoids, non-flavonoid compounds show a broad range of antimicrobial activities against several microorganisms. Some common non-flavonoids include stilbenes, coumarins, phenolic acids, and tannins. A recent study investigating sugarcane bagasse extract found it to be effective against several *S. aureus* strains by altering their membrane permeability. Specifically, the study found that there was more conductivity in the extract-exposed strains compared to the control strains, suggesting that sugarcane polyphenol extract may influence the integrity of bacterial membranes, leading to cellular electrolyte leakage [84]. Zhao et al. also showed that, following incubation with a subinhibitory dose of non-flavonoid polyphenols, phenolic acids also alter the shapes of bacterial cells. Scanning and transmission electron microscopy analyses revealed that *S. aureus* cells that had been exposed to the sugarcane bagasse extract displayed uneven surface wrinkles as well as fragmentation, adhesions, and the aggregation of cellular debris. These alterations suggested that the sugarcane bagasse extract severely damaged the outer cell walls of *S. aureus* cells, causing cytoplasmic components to seep out [85].

3. Clinical Trial Assessment of Phytochemicals against Microbes

In addition to in vivo and in vitro studies, several clinical trials have examined the efficacy of phytochemicals as antimicrobial agents (Table 1). The most prevalent sterols in plants are beta-sitosterols. Unlike other phytosterols, beta-sitosterols are not produced endogenously and can only be obtained from the diet [86]. A study by Donald et al. evaluated the use of the phytochemical beta-sitosterol against pulmonary tuberculosis [87]. Despite only differing from cholesterol by one ethyl group in the side chain, beta-sitosterol has several biological effects, such as boosting the proliferation of peripheral blood lymphocytes by increasing interleukin-2 and interferon-gamma production [88–91]. Given this observation, Donald et al. examined whether beta-sitosterol may be used individually or in conjunction with current antibiotics against tuberculosis, which currently includes a six-month regimen of isoniazid, rifampicin, pyrazindamide, and ethambutol [87]. Approximately a quarter of tuberculosis patients in resource-poor countries are unable to complete the current antibiotic regimen for tuberculosis. Therefore, assessing the efficacy of beta-sitosterol against pulmonary tuberculosis would provide these countries a readily available and tolerable treatment alternative.

Table 1. Phytochemical clinical trials against bacteria.

Author	Type of Study	Subjects	Phytochemical	Purpose	Results
Donald et al. [87]	Randomized Control Trial	Human	Beta-sitosterol	Treating *Mycobacterium tuberculosis* infection	• Improved weight gain, lymphocyte counts, radio- graphical findings on chest X-ray • Similar efficacy with standard antibiotics against M. tuberculosis
Ahmed et al. [92]	Clinical Trial	Human	Silymarin	Treating hepatitis C infection	• Decreased liver function tests • Improved blood counts and oxidative stress markers • Decreased viral load of HCV • Increased sex hormones
Rahim et al. [93]	Clinical Trial	Human—In Vitro	Alpha-Viniferin	Treating *S. aureus* from the nasal passages	• Reduced *S. aureus* levels in the nasal passage • Maintained nasal flora • High potency against *S. aureus*
Nawarathne et al. [94]	Clinical Trial	Human—In Vitro	*Nigella sativa L.* extract	Treating the *Propionibacterium acnes* infection	• All three formulations inhibited the growth of *S. aureus* and *P. acnes* • Very stable under different conditions (e.g., color, odor, homogeneity, washability, consistency, and pH)
Ferrazzano et al. [95]	Randomized Control Trial	Human—In Vitro	*Plantago lanceolata*	Reducing oral streptococci and lactobacilli bacterial species	• Potent antimicrobial against streptococci • Well tolerated by patients
Kerdar et al. [96]	Randomized Control Trial	Human	*Scrophularia striata*	Treating periodontitis due to *Streptococcus mutans*	• Improved plaque index, pocket depth, and bleeding on probing • Decreased the number of Streptococcus mutans in the long term
Mergia et al. [97]	Randomized Control Trial	Swiss Albino Murine Model	*Verbascum sinaiticum*	Treating *Trypanosoma* brucei species	• Improved mean survival and body weight • Lowered parasite load • Low toxicity to murine model
Askari et al. [98]	Randomized Control Trial	Human Subjects	Myrtle and oak gall	Treating bacterial vaginosis	• Reducing vaginal discharge and pH • Improved disease recurrence • Effective against mixed vaginitis
Karumathil et al. [99]	Randomized Control Trial	In Vitro Keratinocytes	Trans-cinnamaldehyde and Eugenol	Treating *Acinetobacter baumannii* wound infections	• Reduced *A. baumannii* adhesion and invasion • Reduced biofilm formation • Decreased transcription of biofilm production genes

Donald et al. used a blinded randomized placebo-controlled trial to assess the treatment duration for hospitalized pulmonary tuberculosis with positive sputum cultures of *Mycobacterium tuberculosis* at the South African National Tuberculosis Association in Cape Town, South Africa [87]. In addition, the patients' chest radiography, weight gain, Matoux test responses, hematological studies, and liver function tests were performed routinely throughout their treatment course. For a total period of six months, a total of 23 patients received 20 mg of beta-sitosterol, while 24 patients in the placebo group received an inactive ingredient known as talcum and the standard antibiotic therapy for hospitalized pulmonary tuberculosis patients [87]. At the beginning of the study, there were no significant differences in patient characteristics, such as age, sex, or health comorbidities, in the treatment and placebo groups. Donald et al. also included hospitalized pulmonary tuberculosis patients who had *M. tuberculosis* samples that were sensitive to the current antibiotic regimen against *M. tuberculosis*. After one month of treatment, 11 patients in the beta-sitosterol (58%) and placebo groups (61%) had positive sputum cultures for *M. tuberculosis*. At two months, only two patients, or 11% in each group, were still positive [87]. Following the start of the antibiotic treatment for pulmonary tuberculosis, the majority of the sputum cultures were expected to be negative, along with a radiographic improvement, by two months.

By the end of the study, Donald et al. reported three patients in the beta-sitosterol group and one patient in the placebo group with no signs of radiographic improvement at six months, despite the negative sputum. In addition, there were no significant differences in the baseline values for hemoglobin, hematocrit, neutrophils, globulin, creatinine, and urea at the beginning of the study [87]. However, weight gain was higher in the beta-sitsterol group compared to the placebo group (8.9 kg vs. 6.1 kg). Furthermore, the lymphocyte and eosinophil counts were higher in the beta-sitosterol groups compared to the placebo group. There were also differences in the monocyte counts, platelet counts, and sedimentation rates between the groups and time points [87]. Overall, Donald et al.'s study showed that there was improved weight gain and a higher immune response in pulmonary tuberculosis patients receiving beta-sitosterol. The efficacy was similar to the current antibiotic treatments for tuberculosis. The main limitation of the study was the low sample sizes for the treatment and placebo groups. In addition, it remains to be seen whether similar results would be detected in different patient populations.

Similar to pulmonary tuberculosis, Ahmed et al. examined the use of the flavonolignan silymarin for the treatment of hepatitis C (HCV) [92]. More than 185 million people around the globe have been infected by HCV, which has increased the number of patients who develop chronic liver failure and hepatocellular carcinoma [100–102]. Interferon monotherapy was the initial course of treatment before viruses were discovered. This drug had unpleasant side effects and was only moderately effective. The discovery of pegylated interferons, the addition of ribavirin, and antivirals were just a few of the methods that improved the overall efficacy [103]. Since 1997, a weekly infusion of PEG-IFN and ribavirin has improved the effectiveness and cure rate of the treatment [103]. With the simultaneous injection of ribavirin and PEG-IFN-alpha, a persistent virological response in 40–50% of HCV-infected people has been documented [103]. Given the severe side effects of HCV medications, there is a need to find effective and tolerable medications for HCV [103]. Silymarin consists of a combination of flavonolignans or phytochemicals that were extracted from the seeds and fruits of the *Silybum marianum* plant [104–108]. Three phytochemicals make up silymarin: silidianin, silicristin, and silybin [104–108]. The most potent and active phytochemical, silybin, is thought to be primarily responsible for silymarin's purported health advantages [104–108]. Numerous pharmacological effects of silymarin have been noted, but its antiretroviral effects stand out. Silymarin has previously been demonstrated to be safe in human patients at large doses (>1500 mg/day) [104–108]. Therefore, Ahmed et al. investigated the effectiveness of Silymarin in treating HCV infection [92].

A total of 30 patients were randomized into control and the treatment groups, each containing a total of 15 patients. Only antiretrovirals (sofosbuvir and ribavirin; 400 mg/800 mg each/day) were given to the control group. The treatment group received adjunct medication, including antiretrovirals (400/800 mg/day) and silymarin (400 mg/day), during an 8-week period. Ahmed et al. showed that silymarin significantly improved the blood parameters in treated patients when combined with sofosbuvir and ribavirin compared to the control group [92]. When compared to the control group, sofosbuvir/ribavirin and silymarin adjunct therapy in the treatment group increased the production of neutrophils, white blood cells, platelet counts, red blood cells, and hemoglobin [92]. Based on their findings, Ahmed et al. suggested that the silymarin adjuvant has a positive impact on the hematological parameters of HCV patients [92]. The levels of liver markers, such as aspartate transaminase (AST), alanine transaminase (ALT), and bilirubin, were lower in the treatment group. In addition to reducing the latent viral load, the adjunct therapy showed a positive impact on hematological indices and oxidative markers compared to the control group.

In addition, the study showed that when used as an adjuvant therapy with sofosbuvir/ribavirin, silymarin had a positive impact on the hormonal levels of both male and female HCV patients. In contrast to the control group, the adjunct therapy showed increased testosterone levels in male patients, which decreased in the control group. Progesterone levels stayed the same in both the treatment and control groups of male patients [92]. The serum levels of LH and FSH in female patients were checked and found to be higher in both the control and treatment groups. This demonstrates that both medications and adjuncts have positive or ameliorating effects. The progesterone levels in the treatment group, which were shown to be lower in the control group, tended to normalize with silymarin, although the testosterone levels in the female group remained nearly constant [92].

Overall, Ahmed et al.'s study showed that the viral RNA from infected HCV patients was successfully reduced by the sofosbuvir/ribavirin and silymarin treatment. Since the viral load decreased in both groups, comparing the effects of silymarin adjunct therapy on viral quantification was not possible [92]. However, the study showed that sofosbuvir/ribavirin and new-generation antivirals are sufficient to completely remove HCV viral RNA [92]. To examine silymarin's role in correcting HCV RNA levels, Ahmed et al. suggested further studies to examine silymarin's impact on the eradication of the HCV virus over a shorter period.

Another clinical application of phytochemicals has been the prevention of community-acquired or hospital-acquired infections as well as the treatment of antibiotic-resistant bacteria [109]. One of the most prevalent opportunistic pathogens in the world is *Staphylococcus aureus*. The anterior nares are the main niche for *S. aureus* and act as a reservoir for the transmission of the disease, even if the axilla, throat, and perineum are necessary reservoirs [109]. Several serious illnesses, such as endocarditis, pneumonia, bacteremia, and chronic osteomyelitis, can be brought on by *S. aureus* nasal colonization [110]. Due to the resistance to a wide array of therapeutically important antibiotics and a dearth of novel treatments, *S. aureus* infections have emerged as a substantial global concern [111]. Given that large portions of the world's population depend on traditional medicine, there is an interest in examining whether phytochemicals may be used for the treatment of antibiotic-resistant bacteria [112]. Alpha-viniferin is a phytochemical substance obtained from the medicinal plant Carex humilis, which is found in several eastern Asian nations [113]. Additionally, it was recognized in Caragana Sinica, Caragana chamlagu, and Iris clarkei. Alpha-viniferin has a range of biological properties, including antioxidant, antitumor, anticancer, and anti-arthritis properties [113]. Additionally, cyclooxygenase, acetylcholinesterase, and prostaglandin H-2 synthase have all been documented to be inhibited by it [113]. Further studies documented the inhibitor effect of alpha-viniferin on both drug-susceptible and drug-resistant strains of *Mycobacterium* TB and *Staphylococcus* species [113]. Therefore, Rahim et al. tested whether alpha-viniferin could eradicate *S. aureus* from the nasal passages.

Specifically, Rahim et al. examined the antibacterial efficacy of alpha-viniferin against *S. aureus* in a ten-day clinical trial [93]. The study enrolled 20 Korean adult females aged between 20 and 60 years with overall good health and physical fitness and the willingness to avoid topical agents applied to the nares during the entire trial. Alpha-viniferin, the study medication, was placed in sequentially numbered containers and given to the subjects in numerical order in accordance with the randomization process. Healthcare professionals gathered nare samples on day 0 and day 10. On days 0, 4, and 8 of the study, the skin moisture content of each participant was assessed using a corneometer [93]. The corneometer measurement was carried out five times on each measurement day at the same location and in the same manner, with the same temperature and humidity, and the average result was immediately recorded. The samples were then examined to determine the moisturizing ability of alpha-viniferin since the moisturizing ability is important for maintaining the skin barrier [93]. The nasal isolates obtained from the patients were then used to assess the antibacterial activity of alpha-viniferin. In comparison to vancomycin and methicillin, alpha-viniferin demonstrated excellent efficacy against three Staphylococcus species, including methicillin-susceptible *S. aureus* (MSSA), methicillin-resistant *S. aureus* (MRSA), and methicillin-resistant *S. epidermidis* (MRSE), with no toxicity to other bacterial strains. In the culture and RT-PCR-based analysis of the collected nasal swab samples, *S. aureus* was reduced. Alpha-viniferin also inhibited *S. aureus* and MRSA while protecting the natural nasal microbiome. Additionally, the skin's moisture content was enhanced by alpha-viniferin, which is crucial for maintaining skin flexibility and barrier integrity without toxicity. Specifically, the 16S ribosomal RNA based amplicon sequencing analysis showed that *S. aureus* was reduced from 51.03% to 23.99% [93]. Given its effectiveness in reducing *S. aureus* species while preserving the microbial flora, Rahim et al. suggested further studies should be performed with larger sample sizes and comparison groups of other phytochemicals to assess the safety and efficacy of alpha-viniferin.

In addition to the nasal mucosa, other clinical studies examined the use of phytochemicals on the skin. Sebaceous follicle inflammation in the skin is the main cause of acne vulgaris [114]. Some bacterial species, such as *Propionibacterium acnes*, *S. aureus*, and *S. epidermidis*, are responsible for its onset. Due to its capacity to activate complements and metabolize sebaceous triglycerides into fatty acids, which then chemotactically attract neutrophils, *P. acnes*, an obligate anaerobic microorganism, causes inflammatory acne [114]. On the other hand, superficial infections within the skin's sebaceous unit (the hair follicle, arrector pili muscle, and sebaceous gland) are typically caused by aerobic *Staphylococcus* species [114]. Benzoyl peroxide, retinoids, and antibiotics such as erythromycin or clindamycin can all be applied topically to treat acne vulgaris [115]. Oral drugs from the tetracycline and azithromycin classes can also be used to treat acne vulgaris. Due to the development of antibiotic resistance in these bacteria and side effects from current treatment protocols, novel therapeutic medicines for acne vulgaris must be introduced [115]. Different civilizations have employed the seeds of *Nigella sativa L.* (black cumin) for centuries to cure dermatological diseases, including acne vulgaris, burns, wounds, and other inflammatory skin conditions [116]. Data that demonstrated *N. sativa* oil extract in a lotion formulation, which is the primary treatment for mild to moderate acne vulgaris, had superior efficacy and was less toxic than a 5% benzoyl peroxide lotion corroborated these conventional assertions [116]. Additionally, *N. sativa* is a key ingredient in a number of topical preparations used in traditional medicine to treat acne vulgaris and is widely used in Sri Lankan folklore medicine as a dermatological cure [116]. Given these observations, Nawarathne et al. planned to develop topical cosmeceutical formulations incorporating *N. sativa* and evaluate the antibacterial activity of those formulations against selected acne-causing bacteria [94].

The agar-well diffusion method was initially used to test the antibacterial activity of seed extracts against *S. aureus* and *P. acnes* [94]. After that, topical gels were created using three different strengths of ethyl acetate extracted from *N. sativa* seeds. These topical formulations underwent antimicrobial activity and stability tests over a 30-day period [94]. The formulation with 15% seed extract had the best antibacterial activity of the three and was able to stop the growth of *S. aureus* and *P. acnes*. This formulation's antibacterial efficacy against *S. aureus* outperformed commercial products [94]. Additionally, no changes in color, odor, homogeneity, washability, consistency, or pH were noted, and the antibacterial potency was maintained during storage. Furthermore, a small test on 50 subjects showed that only 7 (14%) developed signs of hypersensitivity, while the majority of the participants (86%) were unaffected by the application of the herbal gel formulation [94]. Overall, the results showed that the phytochemicals in the seeds had a strong antibacterial activity in topical gel formulations made from *N. sativa's* ethyl acetate, suggesting their suitability to be used in place of the currently available anti-acne drugs.

In addition, phytochemicals from *Plantago lanceolata* herbal tea were shown to be effective antimicrobial agents for controlling bacterial species in the oral cavity [117–119]. Different kinds of *Streptococcus* and *Lactobacillus* bacteria play a major part in the onset and progression of caries [117–119]. Reduced levels of these microorganisms in the oral cavity will add another justification for dental caries prevention because they are the most significant elements in the process [117–119]. Antimicrobial therapies, such as those derived from plant extracts that fight bacteria and lower the levels of cariogenic microflora in saliva, are potential alternatives [120–123]. About 275 species make up the *Plantago* genus (*Plantaginaceaeare*) found throughout the globe. Some *Plantago* species exhibit strong antiviral, anti-inflammatory, and antioxidant properties [124,125]. Additionally, the genus Plantago has a high concentration of phenolic chemicals (flavonoids and tannins). Particularly, phenolic chemicals regulate bacterial growth, which prevents tooth decay by limiting the proliferation and virulence of pathogenic oral flora [124,125]. A study by Ferrazzano et al. examined the effectiveness of a mouthwash made from an infusion of dried *P. lanceolata* leaves in lowering cariogenic microflora salivary counts [95]. The antimicrobial activity of a *P. lanceolata* tea against cariogenic bacterial strains of the species *Streptococcus* and *Lactobacillus* isolated from clinical samples was evaluated in vitro [95].

To examine the efficacy of this mouthwash, Ferrazzano et al. used clinical isolates of *L. casei, S. bovis, S. mutans, S. mitis, S. parasanguinis, S. viridans, and S. sobrinus* from specimens obtained from 44 adolescents (24 males and 20 females) at the Diagnostic Unit of Bacteriology and Mycology of the University of Naples [95]. Patients were randomly assigned to the test and control groups using blocked randomization from a computer-generated list. The placebo and treatment rinse formulations were prepared using the clinical isolates obtained from the patients. The experimental mouth rinse was prepared with an infusion of *P. lanceolata leaves* and flowers, while the placebo mouthwash was prepared with Amorosa water colored with food dye [95]. The placebo group was instructed to rinse with 10 mL of a placebo mouth wash that did not contain phenolic substances for 60 s after performing oral hygiene three times a day (after breakfast, after lunch, and before sleeping) for 7 days [95]. After seven days, Ferrazzano et al. observed a reduction in Streptococci (28.6% vs. 85.7%) species in the treatment group compared to the control groups; however, there was no difference in the Lactobacilli group (65% vs. 75%). A further analysis using mass spectroscopy showed that the flavonoids, coumarins, lipids, cinnamic acids, lignans, and phenolic compounds were likely responsible for the antimicrobial effect from the *P. lanceolata* mouthwash [95]. However, Ferrazzano et al. only examined the short-term efficacy of *P. lanceolata* against oral streptococci and lactobacilli. Further research into other bacterial species and longer time points are needed to determine whether the ability to lower mutans streptococci salivary numbers can be sustained over time and whether resistance will develop. In addition, it is important to evaluate the patients' long-term acceptability and compliance [95].

A similar study by Kerdar et al. examined the use of the Scrophularia striata plant against *Streptococcus mutans* [96]. The Iranian flowering plants in the *Scrophularia* genus, such as *Scrophularia striata*, are used in traditional medicine to alleviate inflammation throughout the body [126]. The biologically active substances iridoids, flavonoids, phenyl propanoids, and phenolic acids with anti-inflammatory and antimicrobial activities are abundant in the genus *Scrophularia* [126]. An oral inflammatory condition called chronic periodontitis damages the soft tissues as well as the alveolar bone, periodontal ligament, and cementum. The most common bacteria associated with tooth plaque, which is a sticky substance made from leftover food particles and saliva in your mouth, is associated with periodontitis secondary to *S. mutans* infection [127]. The disease is brought on by an interplay between the body's defense mechanism and the biofilm retention of the gum sulcus [127].

In this study, Kerdar et al. investigated a mouthwash using *Scrophularia striata* in vitro for chronic periodontitis disease. The study was a randomized clinical trial that incorporated 50 people between 20 and 50 years old who had chronic periodontitis. These patients were given either a Listerine (control/placebo) or an *S. striata* mouthwash. The patients were asked to gargle 15 mL of mouthwash for 30 s, followed by at least 45 min of fasting. After using the mouthwash for two and four weeks, participants were observed for three clinical criteria: the plaque index, gingival bleeding, and probing depth (the distance measured from the base of the pocket to the most apical point on the gingival margin). Saliva samples were taken to assess the mouthwash's antibacterial efficacy [96]. The analysis revealed a significant difference in bleeding on probing (bleeding induced by the gentle manipulation of the tissue) during the initial examination in the *S. striata* group following two weeks of mouthwash use. Between two and four weeks of treatment, there were no appreciable changes. In the treatment group, bleeding on probing was not significantly different between the first and second examinations after taking the mouthwash, but a comparison of the first and last evaluations showed that the mouthwashes decreased bleeding on probing [96]. In addition, a substantial change in the plaque index (PI) was seen in the treatment group after the initial evaluation following two weeks of mouthwash use. During the second examination, neither group experienced any appreciable changes. The plaque index showed a significant difference in the treatment group during the first examination, but no significant difference was seen in the second examination. The mean value of the PI in the treatment group was considerably lower than in the control group [96]. Overall, Kerdar et al. showed that the *S. striata* plant extract is effective in treating chronic periodontitis disease and is more potent in comparison to other mouthwash products. In the short term, *S. striata* may improve the plaque index, pocket depth, and bleeding on probing.

Beyond treating mucosal infections, phytochemicals were also shown to be effective against tropical parasitic infections. Human Trypanosomiasis is caused by two subspecies of *Trypanosoma* brucei: *T. brucei gambiense* and *T. brucei* rhodesiense [128]. Due to its effects on people's settlement patterns, especially land use and farming, the disease has a significant economic impact in Africa [128]. In Africa, trypanocides are used to treat the illness, but the medications are outdated, expensive, ineffective, and have a problem with drug resistance. *Khaya senegalensis, Piliostigma reticulatum, Securidaca longepedunculata, Ximenia americana, and Artemisia abyssinica* are a few examples of herbal treatments that have been utilized to treat this disease and are highly trypanocidal [129]. The biennial plant *Verbascum sinaiticum* is used to treat several conditions, including wounds, stomach aches, and viral infections [129]. Given these observations, Mergia et al. performed an in vitro randomized experiment using Swiss albino mice infected with a field isolate of *T. congolense* to assess the effectiveness of *V. sinaiticum* extracts [97].

The *V. sinaiticum* extracts were injected intraperitoneally for 7 days at doses of 100, 200, and 400 mg/kg at 12 days postinfection, when the peak parasitemia level was around 108 trypanosomes/mL. As indicators for gauging the effectiveness of the extracts, the parasitemia, packed cell volume, mean survival time, and change in body weight were used [97]. To examine the trypanocidal properties of the *V. sinaiticum* extracts, forty healthy Swiss albino mice were intraperitoneally injected with 0.2 mL of *T. congolense*-infected blood (10^4 trypanosomes/mL). Eight groups of five mice were formed by randomly dividing the mice. On the 12th day after infection, when the infected mice displayed maximal parasitemia (10^8 trypanosomes/mL), the mice in each group were treated with the extracts. *V. sinaiticum* was administered to groups I–III at doses of 100, 200, and 400 mg/kg, and to groups IV–VI at doses of 100, 200, and 400 mg/kg, respectively. Diminazine aceturate was administered to group VII, the positive control, in a single dose of 28 mg/kg [97]. The extracts had no toxicological effect on Swiss albino mice. Alkaloids, flavonoids, glycosides, saponins, steroids, phenolic compounds, and tannins were among the phytochemicals examined in *V. sinaiticum*. On day 14 of treatment, the mice treated with 400 mg/kg of *V. sinaiticum* showed considerably lower mean parasitemia than the negative control group. When compared to the negative control at the end of the observation period, animals treated with the same dose had significantly higher packed cell volume values and body weights as well as a maximum mean survival time of approximately 40 days [97]. Overall, Mergia et al. showed that *V. sinaiticum* has the potential to be used as a trypanocidal treatment, but more research is needed to pinpoint the biologically active compounds in the extract as well as to test the extracts in human studies.

Phytochemicals were also effective against urogenital infections, such as bacterial vaginosis. Bacterial vaginosis affects adult females when *Lactobacillus* spp. are replaced by *Gardnerella vaginalis, Mobiluncus curtisii, M. mulieris*, or *Mycoplasma hominis* [130,131]. Another kind of vaginosis among 10–25% of either pregnant or non-pregnant women is caused by *Trichomonas vaginalis* [130,131]. Due to the emergence of antibiotic-resistant strains, bacterial vaginosis recurs in 30% of patients within the first month and 59% within six months [130,131]. Given the adverse effects of antibiotics for bacterial vaginitis, natural products such as boric acid, douching, *Melaleuca alternifolia* essential oil, garlic, and propolis have been used for the treatment of bacterial vaginitis. As such, Askari et al. examined the effectiveness of a myrtle and oak gall suppository (MGOS) in the treating vaginosis. In the randomized control trial, 120 of the 150 patients (40 in the metronidazole group, 40 in the MOGS group, and 40 in the placebo group) finished the prescribed course of treatment. According to test results, metronidazole was superior to a placebo in treating bacterial vaginosis and was also the best therapy for achieving a negative Nugent score [98]. On the other hand, MOGS was also more effective in treating vaginal trichomoniasis. Overall, the clinical study by Askari et al. demonstrated that MGOS was more effective than metronidazole in the treatment of bacterial vaginosis, without experiencing significant side effects.

Lastly, clinical studies found that phytochemicals are effective antibacterial agents to treat wound infections. *Acinetobacter baumannii* has become a significant human pathogen, especially when it comes to infections contracted in hospitals [132,133]. *A. baumannii* has developed resistance to the majority of the currently available antibiotics over the past few years. In addition, *A. baumannii* is a significant pathogen that causes persistent wound infections in burn patients, which can result in the loss of skin grafts and slow wound healing [132,133]. As a result, systemic antibiotics are ineffective in reducing pathogen loads in granulation wounds [132,133]. Thus, alternative approaches to treat *A. baumannii*-related wound infections are now necessary due to the pathogen's multidrug resistance. In recent years, several plant-derived compounds have been investigated for their potential would healing properties [132,133]. In this an vitro study, Karumathil et al. examined whether transcinnamaldehyde (TC) and eugenol (EG), two naturally occurring plant-derived antimicrobials (PDAs), could reduce *A. baumannii* adherence to and invasion of human keratinocytes (HEK001 cells) [99]. In the study, Karumathil et al. used two clinical isolates of *A. baumannii* obtained from infected wounds (Navel-17 and OIFC-109). Compared to the control keratinocytes, TC and EG both significantly decreased *A. baumannii* adherence and invasion to HEK001 by about 2 to 3 log colony-forming units /mL. In addition, TC and EG reduced the production of *A. baumannii* biofilms. An RT-qPCR analysis showed that the two phytochemicals significantly reduced the transcription of genes linked to the development of *A. baumannii* biofilm. The findings imply that both TC and EG might be utilized to treat *A. baumannii* wound infections. However, further research in human patients is required to confirm their effectiveness.

4. Conclusions

As the threat of antibiotic resistance increases, alternative antimicrobial methods are needed. Phytochemicals remain an attractive alternative for addressing this need. As shown previously, phytochemicals show antimicrobial activities in several different clinical scenarios, which makes them versatile agents against several microbial species. Further randomized clinical trials using a greater number of subjects are needed to assess their efficacy and applicability in other infections, particularly viral infections. Despite the long history of utilizing natural products, any medications have the possibility of being dangerous to the consumer. Despite their availability, plant extracts and other natural products are neither regulated nor quality controlled. As a result, further research on the safety and effects of phytochemicals remains to be investigated [11,13,17,134]. However, given their simplicity, efficacy, and affordability, phytochemicals are a promising alternative to antibiotics.

Author Contributions: Conception and design: J.M. and J.K. Manuscript writing: J.M., J.K. and A.H. Final approval of manuscript: A.H. All authors have read and agreed to the published version of the manuscript.

Funding: This work was partially funded by a grant from the CH foundation.

Institutional Review Board Statement: Not applicable.

Informed Consent Statement: Not applicable.

Data Availability Statement: Not applicable.

Conflicts of Interest: The authors declare no conflict of interest.

References

1. Durand, G.A.; Raoult, D.; Dubourg, G. Antibiotic discovery: History, methods and perspectives. *Int. J. Antimicrob. Agents* **2019**, *53*, 371–382. [CrossRef] [PubMed]
2. Peláez, F. The historical delivery of antibiotics from microbial natural products—Can history repeat? *Biochem. Pharmacol.* **2006**, *71*, 981–990. [CrossRef] [PubMed]
3. Rhee, C.; Kadri, S.S.; Dekker, J.P.; Danner, R.L.; Chen, H.-C.; Fram, D.; Zhang, F.; Wang, R.; Klompas, M.; CDC Prevention Epicenters Program. Prevalence of Antibiotic-Resistant Pathogens in Culture-Proven Sepsis and Outcomes Associated with Inadequate and Broad-Spectrum Empiric Antibiotic Use. *JAMA Netw. Open* **2020**, *3*, e202899. [CrossRef] [PubMed]
4. Baym, M.; Stone, L.K.; Kishony, R. Multidrug evolutionary strategies to reverse antibiotic resistance. *Science* **2016**, *351*, aad3292. [CrossRef]
5. Balasubramanian, D.; Harper, L.; Shopsin, B.; Torres, V.J. *Staphylococcus aureus* pathogenesis in diverse host environments. *Pathog. Dis.* **2017**, *75*, ftx005. [CrossRef] [PubMed]
6. Sabir, N.; Ikram, A.; Zaman, G.; Satti, L.; Gardezi, A.; Ahmed, A.; Ahmed, P. Bacterial biofilm-based catheter-associated urinary tract infections: Causative pathogens and antibiotic resistance. *Am. J. Infect. Control.* **2017**, *45*, 1101–1105. [CrossRef] [PubMed]
7. Brown, D. Antibiotic resistance breakers: Can repurposed drugs fill the antibiotic discovery void? *Nat. Rev. Drug Discov.* **2015**, *14*, 821–832. [CrossRef]
8. Rana, R.; Sharma, R.; Kumar, A. Repurposing of Existing Statin Drugs for Treatment of Microbial Infections: How Much Promising? *Infect. Disord.—Drug Targets* **2019**, *19*, 224–237. [CrossRef]
9. Donlan, R.M.; Costerton, J.W. Biofilms: Survival Mechanisms of Clinically Relevant Microorganisms. *Clin. Microbiol. Rev.* **2002**, *15*, 167–193. [CrossRef]
10. Walsh, C. Molecular mechanisms that confer antibacterial drug resistance. *Nature* **2000**, *406*, 775–781. [CrossRef]
11. Cowan, M.M. Plant Products as Antimicrobial Agents. *Clin. Microbiol. Rev.* **1999**, *12*, 564–582. [CrossRef]
12. Hintz, T.; Matthews, K.K.; Di, R. The Use of Plant Antimicrobial Compounds for Food Preservation. *BioMed Res. Int.* **2015**, *2015*, 246264. [CrossRef]
13. Patra, A.K. An overview of antimicrobial properties of different classes of phytochemicals. In *Dietary Phytochemicals and Microbes*; Springer: Dordrecht, The Netherlands, 2012; pp. 1–32. ISBN 978-94-007-3926-0.
14. Dahiya, P.; Purkayastha, S. Phytochemical screening and antimicrobial activity of some medicinal plants against multi-drug resistant bacteria from clinical isolates. *Indian J. Pharm. Sci.* **2012**, *74*, 443–450. [CrossRef] [PubMed]
15. Khameneh, B.; Diab, R.; Ghazvini, K.; Bazzaz, B.S.F. Breakthroughs in bacterial resistance mechanisms and the potential ways to combat them. *Microb. Pathog.* **2016**, *95*, 32–42. [CrossRef] [PubMed]
16. Shakeri, A.; Sharifi, M.J.; Bazzaz, B.S.F.; Emami, A.; Soheili, V.; Sahebkar, A.; Asili, J. Bioautography Detection of Antimicrobial Compounds from the Essential Oil of Salvia Pachystachys. *Curr. Bioact. Compd.* **2018**, *14*, 80–85. [CrossRef]
17. Barbieri, R.; Coppo, E.; Marchese, A.; Daglia, M.; Sobarzo-Sánchez, E.; Nabavi, S.F.; Nabavi, S.M. Phytochemicals for human disease: An update on plant-derived compounds antibacterial activity. *Microbiol. Res.* **2017**, *196*, 44–68. [CrossRef]
18. Khameneh, B.; Iranshahy, M.; Soheili, V.; Bazzaz, B.S.F. Review on plant antimicrobials: A mechanistic viewpoint. *Antimicrob. Resist. Infect. Control* **2019**, *8*, 118. [CrossRef]
19. Khare, T.; Anand, U.; Dey, A.; Assaraf, Y.G.; Chen, Z.-S.; Liu, Z.; Kumar, V. Exploring Phytochemicals for Combating Antibiotic Resistance in Microbial Pathogens. *Front. Pharmacol.* **2021**, *12*, 720726. [CrossRef]
20. Cushnie, T.T.; Cushnie, B.; Lamb, A.J. Alkaloids: An overview of their antibacterial, antibiotic-enhancing and antivirulence activities. *Int. J. Antimicrob. Agents* **2014**, *44*, 377–386. [CrossRef]
21. Laines-Hidalgo, J.I.; Muñoz-Sánchez, J.A.; Loza-Müller, L.; Vázquez-Flota, F. An Update of the Sanguinarine and Benzophenanthridine Alkaloids' Biosynthesis and Their Applications. *Molecules* **2022**, *27*, 1378. [CrossRef]
22. Obiang-Obounou, B.W.; Kang, O.-H.; Choi, J.-G.; Keum, J.-H.; Kim, S.-B.; Mun, S.-H.; Shin, D.-W.; Kim, K.W.; Park, C.-B.; Kim, Y.-G.; et al. The mechanism of action of sanguinarine against methicillin-resistant *Staphylococcus aureus*. *J. Toxicol. Sci.* **2011**, *36*, 277–283. [CrossRef] [PubMed]
23. Jubair, N.; Rajagopal, M.; Chinnappan, S.; Abdullah, N.B.; Fatima, A. Review on the Antibacterial Mechanism of Plant-Derived Compounds against Multidrug-Resistant Bacteria (MDR). *Evid.-Based Complement. Altern. Med.* **2021**, *2021*, 3663315. [CrossRef] [PubMed]
24. Hubert, D.; Réglier-Poupet, H.; Sermet-Gaudelus, I.; Ferroni, A.; Le Bourgeois, M.; Burgel, P.-R.; Serreau, R.; Dusser, D.; Poyart, C.; Coste, J. Association between *Staphylococcus aureus* alone or combined with *Pseudomonas aeruginosa* and the clinical condition of patients with cystic fibrosis. *J. Cyst. Fibros.* **2013**, *12*, 497–503. [CrossRef] [PubMed]
25. Mitchell, G.; Lafrance, M.; Boulanger, S.; Séguin, D.L.; Guay, I.; Gattuso, M.; Marsault, E.; Bouarab, K.; Malouin, F. Tomatidine acts in synergy with aminoglycoside antibiotics against multiresistant *Staphylococcus aureus* and prevents virulence gene expression. *J. Antimicrob. Chemother.* **2011**, *67*, 559–568. [CrossRef] [PubMed]
26. Guay, I.; Boulanger, S.; Isabelle, C.; Brouillette, E.; Chagnon, F.; Bouarab, K.; Marsault, E.; Malouin, F. Tomatidine and analog FC04-100 possess bactericidal activities against *Listeria*, *Bacillus* and *Staphylococcus* spp. *BMC Pharmacol. Toxicol.* **2018**, *19*, 7. [CrossRef]

27. Dorsaz, S.; Snäkä, T.; Favre-Godal, Q.; Maudens, P.; Boulens, N.; Furrer, P.; Ebrahimi, S.N.; Hamburger, M.; Allémann, E.; Gindro, K.; et al. Identification and Mode of Action of a Plant Natural Product Targeting Human Fungal Pathogens. *Antimicrob. Agents Chemother.* **2017**, *61*, e00829-17. [CrossRef]
28. Medina, J.M.; Rodrigues, J.C.F.; De Souza, W.; Atella, G.C.; Barrabin, H. Tomatidine promotes the inhibition of 24-alkylated sterol biosynthesis and mitochondrial dysfunction in *Leishmania amazonensis* promastigotes. *Parasitology* **2012**, *139*, 1253–1265. [CrossRef]
29. Desmond, E.; Gribaldo, S. Phylogenomics of Sterol Synthesis: Insights into the Origin, Evolution, and Diversity of a Key Eukaryotic Feature. *Genome Biol. Evol.* **2009**, *1*, 364–381. [CrossRef]
30. Simons, V.; Morrissey, J.P.; Latijnhouwers, M.; Csukai, M.; Cleaver, A.; Yarrow, C.; Osbourn, A. Dual Effects of Plant Steroidal Alkaloids on *Saccharomyces cerevisiae*. *Antimicrob. Agents Chemother.* **2006**, *50*, 2732–2740. [CrossRef]
31. Khan, I.A.; Mirza, Z.M.; Kumar, A.; Verma, V.; Qazi, G.N. Piperine, a Phytochemical Potentiator of Ciprofloxacin against *Staphylococcus aureus*. *Antimicrob. Agents Chemother.* **2006**, *50*, 810–812. [CrossRef]
32. Khameneh, B.; Iranshahy, M.; Ghandadi, M.; Atashbeyk, D.G.; Bazzaz, B.S.F.; Iranshahi, M. Investigation of the antibacterial activity and efflux pump inhibitory effect of co-loaded piperine and gentamicin nanoliposomes in methicillin-resistant *Staphylococcus aureus*. *Drug Dev. Ind. Pharm.* **2014**, *41*, 989–994. [CrossRef] [PubMed]
33. Verma, V.C.; Lobkovsky, E.B.; Gange, A.C.; Singh, S.K.; Prakash, S. Piperine production by endophytic fungus *Periconia* sp. Isolated from *Piper longum* L. *J. Antibiot.* **2011**, *64*, 427–431. [CrossRef] [PubMed]
34. Jin, J.; Zhang, J.; Guo, N.; Feng, H.; Li, L.; Liang, J.; Sun, K.; Wu, X.; Wang, X.; Liu, M.; et al. The plant alkaloid piperine as a potential inhibitor of ethidium bromide efflux in *Mycobacterium smegmatis*. *J. Med. Microbiol.* **2011**, *60*, 223–229. [CrossRef] [PubMed]
35. Sharma, S.; Kalia, N.P.; Suden, P.; Chauhan, P.S.; Kumar, M.; Ram, A.B.; Khajuria, A.; Bani, S.; Khan, I.A. Protective efficacy of piperine against *Mycobacterium tuberculosis*. *Tuberculosis* **2014**, *94*, 389–396. [CrossRef]
36. Iwasa, K.; Moriyasu, M.; Yamori, T.; Turuo, T.; Lee, D.-U.; Wiegrebe, W. In Vitro Cytotoxicity of the Protoberberine-Type Alkaloids. *J. Nat. Prod.* **2001**, *64*, 896–898. [CrossRef]
37. Yi, Z.-B.; Yu, Y.; Liang, Y.-Z.; Zeng, B. Evaluation of the antimicrobial mode of berberine by LC/ESI-MS combined with principal component analysis. *J. Pharm. Biomed. Anal.* **2007**, *44*, 301–304. [CrossRef]
38. Domadia, P.N.; Bhunia, A.; Sivaraman, J.; Swarup, S.; Dasgupta, D. Berberine Targets Assembly of *Escherichia coli* Cell Division Protein FtsZ. *Biochemistry* **2008**, *47*, 3225–3234. [CrossRef]
39. Peng, L.; Kang, S.; Yin, Z.; Jia, R.; Song, X.; Li, L.; Li, Z.; Zou, Y.; Liang, X.; He, C.; et al. Antibacterial activity and mechanism of berberine against *Streptococcus agalactiae*. *Int. J. Clin. Exp. Pathol.* **2015**, *8*, 5217–5223.
40. Leng, B.-F.; Qiu, J.-Z.; Dai, X.-H.; Dong, J.; Wang, J.-F.; Luo, M.-J.; Li, H.-E.; Niu, X.-D.; Zhang, Y.; Ai, Y.-X.; et al. Allicin Reduces the Production of α-Toxin by *Staphylococcus aureus*. *Molecules* **2011**, *16*, 7958–7968. [CrossRef]
41. Khodavandi, A.; Alizadeh, F.; Aala, F.; Sekawi, Z.; Chong, P.P. In Vitro Investigation of Antifungal Activity of Allicin Alone and in Combination with Azoles Against *Candida* Species. *Mycopathologia* **2009**, *169*, 287–295. [CrossRef]
42. Choo, S.; Chin, V.K.; Wong, E.H.; Madhavan, P.; Tay, S.T.; Yong, P.V.C.; Chong, P.P. Review: Antimicrobial properties of allicin used alone or in combination with other medications. *Folia Microbiol.* **2020**, *65*, 451–465. [CrossRef]
43. Strehlow, B.; Bakowsky, U. A Novel Microparticulate Formulation with Allicin In Situ Synthesis. *J. Pharm. Drug Deliv. Res.* **2016**, *5*, 10-4172. [CrossRef]
44. Fahey, J.W.; Stephenson, K.K.; Wade, K.L.; Talalay, P. Urease from *Helicobacter pylori* is inactivated by sulforaphane and other isothiocyanates. *Biochem. Biophys. Res. Commun.* **2013**, *435*, 1–7. [CrossRef]
45. Dufour, V.; Stahl, M.; Baysse, C. The antibacterial properties of isothiocyanates. *Microbiology* **2015**, *161*, 229–243. [CrossRef] [PubMed]
46. Lu, Z.; Dockery, C.R.; Crosby, M.; Chavarria, K.; Patterson, B.; Giedd, M. Antibacterial Activities of Wasabi against *Escherichia coli* O157:H7 and *Staphylococcus aureus*. *Front. Microbiol.* **2016**, *7*, 1403. [CrossRef]
47. Palaniappan, K.; Holley, R.A. Use of natural antimicrobials to increase antibiotic susceptibility of drug resistant bacteria. *Int. J. Food Microbiol.* **2010**, *140*, 164–168. [CrossRef]
48. Saavedra, M.; Borges, A.; Dias, C.; Aires, A.; Bennett, R.; Rosa, E.; Simões, M. Antimicrobial Activity of Phenolics and Glucosinolate Hydrolysis Products and their Synergy with Streptomycin against Pathogenic Bacteria. *Med. Chem.* **2010**, *6*, 174–183. [CrossRef]
49. Nedorostova, L.; Kloucek, P.; Kokoska, L.; Stolcova, M.; Pulkrabek, J. Antimicrobial properties of selected essential oils in vapour phase against foodborne bacteria. *Food Control* **2009**, *20*, 157–160. [CrossRef]
50. Luciano, F.B.; Holley, R.A. Enzymatic inhibition by allyl isothiocyanate and factors affecting its antimicrobial action against *Escherichia coli* O157:H. *Int. J. Food Microbiol.* **2009**, *131*, 240–245. [CrossRef]
51. Prati, D.; Bossdorf, O. Allelopathic inhibition of germination by *Alliaria petiolata* (Brassicaceae). *Am. J. Bot.* **2004**, *91*, 285–288. [CrossRef] [PubMed]
52. Aires, A.; Mota, V.; Saavedra, M.J.; Rosa, E.; Bennett, R. The antimicrobial effects of glucosinolates and their respective enzymatic hydrolysis products on bacteria isolated from the human intestinal tract. *J. Appl. Microbiol.* **2009**, *106*, 2086–2095. [CrossRef] [PubMed]
53. Drobnica, L.; Zemanová, M.; Nemec, P.; Antoš, K.; Kristián, P.; Martvoň, A.; Závodská, E. Antifungal Activity of Isothiocyanates and Related Compounds. *Appl. Environ. Microbiol.* **1968**, *16*, 582–587. [CrossRef] [PubMed]

54. Calmes, B.; N'Guyen, G.; Dumur, J.; Brisach, C.A.; Campion, C.; Iacomi, B.; Pigné, S.; Dias, E.; Macherel, D.; Guillemette, T.; et al. Glucosinolate-derived isothiocyanates impact mitochondrial function in fungal cells and elicit an oxidative stress response necessary for growth recovery. *Front. Plant Sci.* **2015**, *6*, 414. [CrossRef] [PubMed]
55. Lima, M.; de Sousa, C.P.; Fernandez-Prada, C.; Harel, J.; Dubreuil, J.; de Souza, E. A review of the current evidence of fruit phenolic compounds as potential antimicrobials against pathogenic bacteria. *Microb. Pathog.* **2019**, *130*, 259–270. [CrossRef] [PubMed]
56. Quideau, S.; Deffieux, D.; Douat-Casassus, C.; Pouységu, L. Plant Polyphenols: Chemical Properties, Biological Activities, and Synthesis. *Angew. Chem. Int. Ed.* **2011**, *50*, 586–621. [CrossRef]
57. Cushnie, T.P.T.; Lamb, A.J. Recent advances in understanding the antibacterial properties of flavonoids. *Int. J. Antimicrob. Agents* **2011**, *38*, 99–107. [CrossRef]
58. Babu, K.S.; Babu, T.H.; Srinivas, P.; Sastry, B.; Kishore, K.H.; Murty, U.; Rao, J.M. Synthesis and in vitro study of novel 7-O-acyl derivatives of Oroxylin A as antibacterial agents. *Bioorganic Med. Chem. Lett.* **2005**, *15*, 3953–3956. [CrossRef]
59. Tsuchiya, H.; Sato, M.; Miyazaki, T.; Fujiwara, S.; Tanigaki, S.; Ohyama, M.; Tanaka, T.; Iinuma, M. Comparative study on the antibacterial activity of phytochemical flavanones against methicillin-resistant *Staphylococcus aureus*. *J. Ethnopharmacol.* **1996**, *50*, 27–34. [CrossRef]
60. Ávila, H.P.; Smânia, E.D.F.A.; Monache, F.D.; Smânia, A. Structure-activity relationship of antibacterial chalcones. *Bioorganic Med. Chem.* **2008**, *16*, 9790–9794. [CrossRef]
61. Nielsen, S.F.; Boesen, T.; Larsen, M.; Schønning, K.; Kromann, H. Antibacterial chalcones—bioisosteric replacement of the 4′-hydroxy group. *Bioorganic Med. Chem.* **2004**, *12*, 3047–3054. [CrossRef]
62. Shamsudin, N.F.; Ahmed, Q.U.; Mahmood, S.; Ali Shah, S.A.; Khatib, A.; Mukhtar, S.; Alsharif, M.A.; Parveen, H.; Zakaria, Z.A. Antibacterial Effects of Flavonoids and Their Structure-Activity Relationship Study: A Comparative Interpretation. *Molecules* **2022**, *27*, 1149. [CrossRef] [PubMed]
63. Nowakowska, Z.; Kędzia, B.; Schroeder, G. Synthesis, physicochemical properties and antimicrobial evaluation of new (E)-chalcones. *Eur. J. Med. Chem.* **2008**, *43*, 707–713. [CrossRef] [PubMed]
64. Batovska, D.; Parushev, S.; Stamboliyska, B.; Tsvetkova, I.; Ninova, M.; Najdenski, H. Examination of growth inhibitory properties of synthetic chalcones for which antibacterial activity was predicted. *Eur. J. Med. Chem.* **2009**, *44*, 2211–2218. [CrossRef]
65. Alcaráz, L.; Blanco, S.; Puig, O.; Tomás, F.; Ferretti, F. Antibacterial Activity of Flavonoids Against Methicillin-Resistant *Staphylococcus aureus* strains. *J. Theor. Biol.* **2000**, *205*, 231–240. [CrossRef]
66. Otsuka, N.; Liu, M.-H.; Shiota, S.; Ogawa, W.; Kuroda, T.; Hatano, T.; Tsuchiya, T. Anti-Methicillin Resistant *Staphylococcus aureus* (MRSA) Compounds Isolated from *Laurus nobilis*. *Biol. Pharm. Bull.* **2008**, *31*, 1794–1797. [CrossRef]
67. Mughal, E.U.; Ayaz, M.; Hussain, Z.; Hasan, A.; Sadiq, A.; Riaz, M.; Malik, A.; Hussain, S.; Choudhary, M.I. Synthesis and antibacterial activity of substituted flavones, 4-thioflavones and 4-iminoflavones. *Bioorganic Med. Chem.* **2006**, *14*, 4704–4711. [CrossRef] [PubMed]
68. Cushnie, T.; Hamilton, V.; Chapman, D.; Taylor, P.; Lamb, A. Aggregation of *Staphylococcus aureus* following treatment with the antibacterial flavonol galangin. *J. Appl. Microbiol.* **2007**, *103*, 1562–1567. [CrossRef]
69. Cushnie, T.T.; Lamb, A. Detection of galangin-induced cytoplasmic membrane damage in *Staphylococcus aureus* by measuring potassium loss. *J. Ethnopharmacol.* **2005**, *101*, 243–248. [CrossRef]
70. Lin, R.-D.; Chin, Y.-P.; Hou, W.-C.; Lee, M.-H. The Effects of Antibiotics Combined with Natural Polyphenols against Clinical Methicillin-Resistant *Staphylococcus aureus* (MRSA). *Planta Med.* **2008**, *74*, 840–846. [CrossRef]
71. Qu, S.; Dai, C.; Shen, Z.; Tang, Q.; Wang, H.; Zhai, B.; Zhao, L.; Hao, Z. Mechanism of Synergy between Tetracycline and Quercetin against Antibiotic Resistant *Escherichia coli*. *Front. Microbiol.* **2019**, *10*, 2536. [CrossRef] [PubMed]
72. Zhao, W.-H.; Hu, Z.-Q.; Okubo, S.; Hara, Y.; Shimamura, T. Mechanism of Synergy between Epigallocatechin Gallate and β-Lactams against Methicillin-Resistant *Staphylococcus aureus*. *Antimicrob. Agents Chemother.* **2001**, *45*, 1737–1742. [CrossRef]
73. Yi, S.; Wang, W.; Bai, F.; Zhu, J.; Li, J.; Li, X.; Xu, Y.; Sun, T.; He, Y. Antimicrobial effect and membrane-active mechanism of tea polyphenols against *Serratia marcescens*. *World J. Microbiol. Biotechnol.* **2013**, *30*, 451–460. [CrossRef]
74. Yoda, Y.; Hu, Z.-Q.; Shimamura, T.; Zhao, W.-H. Different susceptibilities of *Staphylococcus* and Gram-negative rods to epigallocatechin gallate. *J. Infect. Chemother.* **2004**, *10*, 55–58. [CrossRef] [PubMed]
75. Stapleton, P.D.; Shah, S.; Hara, Y.; Taylor, P.W. Potentiation of Catechin Gallate-Mediated Sensitization of *Staphylococcus aureus* to Oxacillin by Nongalloylated Catechins. *Antimicrob. Agents Chemother.* **2006**, *50*, 752–755. [CrossRef] [PubMed]
76. Caturla, N. The relationship between the antioxidant and the antibacterial properties of galloylated catechins and the structure of phospholipid model membranes. *Free Radic. Biol. Med.* **2003**, *34*, 648–662. [CrossRef] [PubMed]
77. Kajiya, K.; Kumazawa, S.; Nakayama, T. Steric Effects on Interaction of Tea Catechins with Lipid Bilayers. *Biosci. Biotechnol. Biochem.* **2001**, *65*, 2638–2643. [CrossRef]
78. Kajiya, K.; Kumazawa, S.; Nakayama, T. Effects of External Factors on the Interaction of Tea Catechins with Lipid Bilayers. *Biosci. Biotechnol. Biochem.* **2002**, *66*, 2330–2335. [CrossRef]
79. Kubo, I.; Xiao, P.; Fujita, K. Anti-MRSA activity of alkyl gallates. *Bioorganic Med. Chem. Lett.* **2001**, *12*, 113–116. [CrossRef]
80. Bernal, P.; Zloh, M.; Taylor, P.W. Disruption of d-alanyl esterification of *Staphylococcus aureus* cell wall teichoic acid by the β-lactam resistance modifier (−)-epicatechin gallate. *J. Antimicrob. Chemother.* **2009**, *63*, 1156–1162. [CrossRef]

81. Yam, T.S.; Hamilton-Miller, J.M.; Shah, S. The effect of a component of tea (*Camellia sinensis*) on methicillin resistance, PBP2′ synthesis, and beta-lactamase production in *Staphylococcus aureus*. *J. Antimicrob. Chemother.* **1998**, *42*, 211–216. [CrossRef]
82. Stapleton, P.D.; Taylor, P.W. Methicillin Resistance in *Staphylococcus aureus*: Mechanisms and Modulation. *Sci. Prog.* **2002**, *85*, 57–72. [CrossRef]
83. Bernal, P.; Lemaire, S.; Pinho, M.; Mobashery, S.; Hinds, J.; Taylor, P.W. Insertion of Epicatechin Gallate into the Cytoplasmic Membrane of Methicillin-resistant *Staphylococcus aureus* Disrupts Penicillin-binding Protein (PBP) 2a-mediated β-Lactam Resistance by Delocalizing PBP2. *J. Biol. Chem.* **2010**, *285*, 24055–24065. [CrossRef]
84. Miklasińska-Majdanik, M.; Kępa, M.; Wojtyczka, R.D.; Idzik, D.; Wąsik, T.J. Phenolic Compounds Diminish Antibiotic Resistance of *Staphylococcus aureus* Clinical Strains. *Int. J. Environ. Res. Public Health* **2018**, *15*, 2321. [CrossRef] [PubMed]
85. Zhao, Y.; Chen, M.; Zhao, Z.; Yu, S. The antibiotic activity and mechanisms of sugarcane (*Saccharum officinarum* L.) bagasse extract against food-borne pathogens. *Food Chem.* **2015**, *185*, 112–118. [CrossRef] [PubMed]
86. Ling, W.; Jones, P. Dietary phytosterols: A review of metabolism, benefits and side effects. *Life Sci.* **1995**, *57*, 195–206. [CrossRef] [PubMed]
87. Donald, P.R.; Lamprecht, J.H.; Freestone, M.; Albrecht, C.F.; Bouic, P.J.; Kotze, D.; van Jaarsveld, P.P. A randomised place-bo-controlled trial of the efficacy of beta-sitosterol and its glucoside as adjuvants in the treatment of pulmonary tuberculosis. *Int. J. Tuberc. Lung Dis.* **1997**, *1*, 518–522. [PubMed]
88. Raicht, R.F.; I Cohen, B.; Fazzini, E.P.; Sarwal, A.N.; Takahashi, M. Protective effect of plant sterols against chemically induced colon tumors in rats. *Cancer Res.* **1980**, *40*, 403–405.
89. Yamada, H.; Yoshino, M.; Matsumoto, T.; Nagai, T.; Kiyohara, H.; Cyong, J.-C.; Nakagawa, A.; Tanaka, H.; Omura, S. Effects of phytosterols on anti-complementary activity. *Chem. Pharm. Bull.* **1987**, *35*, 4851–4855. [CrossRef]
90. Berges, R.R.; Windeler, J.; Trampisch, H.J.; Senge, T. Randomised, placebo-controlled, double-blind clinical trial of beta-sitosterol in patients with benign prostatic hyperplasia. Beta-sitosterol Study Group. *Lancet* **1995**, *345*, 1529–1532. [CrossRef]
91. Bouic, P.J.; Etsebeth, S.; Liebenberg, R.W.; Albrecht, C.F.; Pegel, K.; Van Jaarsveld, P.P. Beta-Sitosterol and beta-sitosterol glu-coside stimulate human peripheral blood lymphocyte proliferation: Implications for their use as an immunomodulatory vitamin combination. *Int. J. Immunopharmacol.* **1996**, *18*, 693–700. [CrossRef]
92. Ahmed, S.; Ullah, N.; Parveen, S.; Javed, I.; Jalil, N.A.C.; Das Murtey, M.; Sheikh, I.S.; Khan, S.; Ojha, S.C.; Chen, K. Effect of Silymarin as an Adjunct Therapy in Combination with Sofosbuvir and Ribavirin in Hepatitis C Patients: A Miniature Clinical Trial. *Oxid. Med. Cell. Longev.* **2022**, *2022*, 9199190. [CrossRef]
93. Rahim, A.; Seo, H.; Kim, S.; Jeong, Y.K.; Tajdozian, H.; Kim, M.; Lee, S.; Song, H.-Y. A Clinical Trial to Evaluate the Efficacy of α-Viniferin in *Staphylococcus aureus*—Specific Decolonization without Depleting the Normal Microbiota of Nares. *Pol. J. Microbiol.* **2021**, *70*, 117–130. [CrossRef] [PubMed]
94. Nawarathne, N.W.; Wijesekera, K.; Wijayaratne, W.M.D.G.B.; Napagoda, M. Development of Novel Topical Cosmeceutical Formulations from *Nigella sativa* L. with Antimicrobial Activity against Acne-Causing Microorganisms. *Sci. World J.* **2019**, *2019*, 5985207. [CrossRef] [PubMed]
95. Ferrazzano, G.F.; Cantile, T.; Roberto, L.; Ingenito, A.; Catania, M.R.; Roscetto, E.; Palumbo, G.; Zarrelli, A.; Pollio, A. Determination of the In Vitro and In Vivo Antimicrobial Activity on Salivary Streptococci and Lactobacilli and Chemical Characterisation of the Phenolic Content of a *Plantago lanceolata* Infusion. *BioMed Res. Int.* **2015**, *2015*, 286817. [CrossRef]
96. Kerdar, T.; Rabienejad, N.; Alikhani, Y.; Moradkhani, S.; Dastan, D. Clinical, in vitro and phytochemical, studies of *Scrophularia striata* mouthwash on chronic periodontitis disease. *J. Ethnopharmacol.* **2019**, *239*, 111872. [CrossRef]
97. Mergia, E.; Shibeshi, W.; Terefe, G.; Teklehaymanot, T. Antitrypanosomal activity of *Verbascum sinaiticum* Benth. (Scrophulariaceae) against *Trypanosoma congolense* isolates. *BMC Complement. Altern. Med.* **2016**, *16*, 362. [CrossRef]
98. Askari, S.F.; Jahromi, B.N.; Dehghanian, A.; Zarei, A.; Tansaz, M.; Badr, P.; Azadi, A.; Mohagheghzadeh, A. Effect of a novel herbal vaginal suppository containing myrtle and oak gall in the treatment of vaginitis: A randomized clinical trial. *DARU J. Pharm. Sci.* **2020**, *28*, 603–614. [CrossRef] [PubMed]
99. Karumathil, D.P.; Surendran-Nair, M.; Venkitanarayanan, K. Efficacy of *Trans*-Cinnamaldehyde and Eugenol in Reducing *Acinetobacter baumannii* Adhesion to and Invasion of Human Keratinocytes and Controlling Wound Infection In Vitro. *Phytother. Res.* **2016**, *30*, 2053–2059. [CrossRef]
100. Hanafiah, K.M.; Groeger, J.; Flaxman, A.D.; Wiersma, S.T. Global epidemiology of hepatitis C virus infection: New estimates of age-specific antibody to HCV seroprevalence. *Hepatology* **2013**, *57*, 1333–1342. [CrossRef]
101. Averhoff, F.M.; Glass, N.; Holtzman, D. Global Burden of Hepatitis C: Considerations for Healthcare Providers in the United States. *Clin. Infect. Dis.* **2012**, *55*, S10–S15. [CrossRef]
102. Travasso, C. Indian government plans 10 regional laboratories to estimate hepatitis burden. *BMJ* **2014**, *349*, g5021. [CrossRef] [PubMed]
103. Li, C.-F.; Tsao, S.-M.; Liao, H.-H.; Chen, S.-C.; Lee, Y.-T. Treatment of chronic hepatitis C regiments containing with recombinant interferon in patients with sustained virological response predicts risk of hepatocellular carcinoma. *Medicine* **2020**, *99*, e22435. [CrossRef] [PubMed]
104. Khezri, H.D.; Salehifar, E.; Kosaryan, M.; Aliasgharian, A.; Jalali, H.; Amree, A.H. Potential Effects of Silymarin and Its Flavonolignan Components in Patients with *β*-Thalassemia Major: A Comprehensive Review in 2015. *Adv. Pharmacol. Sci.* **2016**, *2016*, 3046373. [CrossRef]

105. Amniattalab, A.; Malekinejad, H.; Rezabakhsh, A.; Rokhsartalab-Azar, S.; Alizade-Fanalou, S. Silymarin: A Novel Natural Agent to Restore Defective Pancreatic β Cells in Streptozotocin (STZ)-Induced Diabetic Rats. *Iran. J. Pharm. Res. IJPR* **2016**, *15*, 493–500.
106. Gharagozloo, M.; Jafari, S.; Esmaeil, N.; Javid, E.N.; Bagherpour, B.; Rezaei, A. Immunosuppressive Effect of Silymarin on Mitogen-Activated Protein Kinase Signalling Pathway: The Impact on T Cell Proliferation and Cytokine Production. *Basic Clin. Pharmacol. Toxicol.* **2013**, *113*, 209–214. [CrossRef]
107. Gharagozloo, M.; Velardi, E.; Bruscoli, S.; Agostini, M.; Di Sante, M.; Donato, V.; Amirghofran, Z.; Riccardi, C. Silymarin suppress CD4+ T cell activation and proliferation: Effects on NF-κB activity and IL-2 production. *Pharmacol. Res.* **2010**, *61*, 405–409. [CrossRef]
108. Balouchi, S.; Gharagozloo, M.; Esmaeil, N.; Mirmoghtadaei, M.; Moayedi, B. Serum levels of TGFβ IL-10, IL-17, and IL-23 cytokines in β-thalassemia major patients: The impact of silymarin therapy. *Immunopharmacol. Immunotoxicol.* **2014**, *36*, 271–274. [CrossRef]
109. Islam, Z.; Johannesen, T.B.; Lilje, B.; Urth, T.R.; Larsen, A.R.; Angen, Ø.; Larsen, J. Investigation of the human nasal microbiome in persons with long- and short-term exposure to methicillin-resistant *Staphylococcus aureus* and other bacteria from the pig farm environment. *PLoS ONE* **2020**, *15*, e0232456. [CrossRef]
110. Lowy, F.D. Antimicrobial resistance: The example of *Staphylococcus aureus*. *J. Clin. Investig.* **2003**, *111*, 1265–1273. [CrossRef]
111. Tong, S.Y.; Davis, J.S.; Eichenberger, E.; Holland, T.L.; Fowler, V.G., Jr. *Staphylococcus aureus* infections: Epidemiology, pathophysiology, clinical manifestations, and management. *Clin. Microbiol. Rev.* **2015**, *28*, 603–661. [CrossRef]
112. Chew, Y.L.; Chan, E.W.L.; Tan, P.L.; Lim, Y.Y.; Stanslas, J.; Goh, J.K. Assessment of phytochemical content, polyphenolic composition, antioxidant and antibacterial activities of Leguminosae medicinal plants in Peninsular Malaysia. *BMC Complement. Altern. Med.* **2011**, *11*, 12. [CrossRef]
113. Seo, H.; Kim, M.; Kim, S.; Al Mahmud, H.; Islam, I.; Nam, K.-W.; Cho, M.-L.; Kwon, H.-S.; Song, H.-Y. In vitro activity of alpha-viniferin isolated from the roots of *Carex humilis* against *Mycobacterium tuberculosis*. *Pulm. Pharmacol. Ther.* **2017**, *46*, 41–47. [CrossRef]
114. Budhiraja, A.; Dhingra, G. Development and characterization of a novel antiacne niosomal gel of rosmarinic acid. *Drug Deliv.* **2015**, *22*, 723–730. [CrossRef] [PubMed]
115. Nelson, K.; Lyles, J.T.; Li, T.; Saitta, A.; Addie-Noye, E.; Tyler, P.; Quave, C.L. Anti-Acne Activity of Italian Medicinal Plants Used for Skin Infection. *Front. Pharmacol.* **2016**, *7*, 425. [CrossRef] [PubMed]
116. Eid, A.M.; Elmarzugi, N.A.; Abu Ayyash, L.M.; Sawafta, M.N.; Daana, H.I. A Review on the Cosmeceutical and External Applications of *Nigella sativa*. *J. Trop. Med.* **2017**, *2017*, 7092514. [CrossRef] [PubMed]
117. Campus, G.; Condò, S.G.; Di Renzo, G.; Ferro, R.; Gatto, R.; Giuca, M.R.; Giuliana, G.; Majorana, A.; Marzo, G.; Ottolenghi, L.; et al. National Italian Guidelines for caries prevention in 0 to 12 years-old children. *Eur. J. Paediatr. Dent.* **2007**, *8*, 153–159. [PubMed]
118. Marsh, P.D. Are dental diseases examples of ecological catastrophes? *Microbiology* **2003**, *149*, 279–294. [CrossRef]
119. Seminario, A.; Broukal, Z.; Ivancakova, R.K. Mutans streptococci and the development of dental plaque. *Prague Med. Rep.* **2005**, *106*, 349–358.
120. Palombo, E.A. Traditional medicinal plant extracts and natural products with activity against oral bacteria: Potential application in the prevention and treatment of oral diseases. *Evid.-Based Complement. Altern. Med.* **2011**, *2011*, 680354. [CrossRef]
121. Ferrazzano, G.F.; Amato, I.; Ingenito, A.; Zarrelli, A.; Pinto, G.; Pollio, A. Plant Polyphenols and Their Anti-Cariogenic Properties: A Review. *Molecules* **2011**, *16*, 1486–1507. [CrossRef]
122. Ferrazzano, G.F.; Roberto, L.; Catania, M.R.; Chiaviello, A.; De Natale, A.; Roscetto, E.; Pinto, G.; Pollio, A.; Ingenito, A.; Palumbo, G. Screening and Scoring of Antimicrobial and Biological Activities of Italian Vulnerary Plants against Major Oral Pathogenic Bacteria. *Evid.-Based Complement. Altern. Med.* **2013**, *2013*, 316280. [CrossRef] [PubMed]
123. Ferrazzano, G.F.; Roberto, L.; Amato, I.; Cantile, T.; Sangianantoni, G.; Ingenito, A. Antimicrobial Properties of Green Tea Extract Against Cariogenic Microflora: An In Vivo Study. *J. Med. Food* **2011**, *14*, 907–911. [CrossRef] [PubMed]
124. Gálvez, M.; Martín-Cordero, C.; Houghton, P.J.; Ayuso, M.J. Antioxidant activity of *Plantago bellardii* All. *Phytother. Res.* **2005**, *19*, 1074–1076. [CrossRef] [PubMed]
125. Gálvez, M.; Martín-Cordero, C.; Houghton, P.J.; Ayuso, M.J. Antioxidant Activity of Methanol Extracts Obtained from *Plantago* Species. *J. Agric. Food Chem.* **2005**, *53*, 1927–1933. [CrossRef]
126. Shahbazi, Y. Chemical Composition and In Vitro Antibacterial Activity of *Mentha spicata* Essential Oil against Common Food-Borne Pathogenic Bacteria. *J. Pathog.* **2015**, *2015*, 916305. [CrossRef]
127. Savage, A.; Eaton, K.A.; Moles, D.R.; Needleman, I. A systematic review of definitions of periodontitis and methods that have been used to identify this disease. *J. Clin. Periodontol.* **2009**, *36*, 458–467. [CrossRef]
128. Feyera, T.; Terefe, G.; Shibeshi, W. Phytochemical Screening and In Vitro Antitrypanosomal Activity of the Aerial Parts of *Artemisia abyssinica* Against *Trypanosoma congolense* Field Isolate. *Ethiop. Pharm. J.* **2013**, *29*, 137–142. [CrossRef]
129. Afewerk, Y.; Clausen, P.-H.; Abebe, G.; Tilahun, G.; Mehlitz, D. Multiple-drug resistant *Trypanosoma congolense* populations in village cattle of Metekel district, north-west Ethiopia. *Acta Trop.* **2000**, *76*, 231–238. [CrossRef]
130. Mills, B.B. Vaginitis: Beyond the Basics. *Obstet. Gynecol. Clin. N. Am.* **2017**, *44*, 159–177. [CrossRef]
131. Paladine, H.L.; Desai, U.A. Vaginitis: Diagnosis and Treatment. *Am. Fam. Physician* **2018**, *97*, 321–329.

132. Esterly, J.S.; Griffith, M.; Qi, C.; Malczynski, M.; Postelnick, M.J.; Scheetz, M.H. Impact of Carbapenem Resistance and Receipt of Active Antimicrobial Therapy on Clinical Outcomes of *Acinetobacter baumannii* Bloodstream Infections. *Antimicrob. Agents Chemother.* **2011**, *55*, 4844–4849. [CrossRef] [PubMed]
133. Gaddy, J.A.; Tomaras, A.P.; Actis, L.A. The *Acinetobacter baumannii* 19606 OmpA Protein Plays a Role in Biofilm Formation on Abiotic Surfaces and in the Interaction of This Pathogen with Eukaryotic Cells. *Infect. Immun.* **2009**, *77*, 3150–3160. [CrossRef] [PubMed]
134. Radji, M.; Agustama, R.A.; Elya, B.; Tjampakasari, C.R. Antimicrobial activity of green tea extract against isolates of methicillin-resistant *Staphylococcus aureus* and multi-drug resistant *Pseudomonas aeruginosa*. *Asian Pac. J. Trop. Biomed.* **2013**, *3*, 663–666. [CrossRef] [PubMed]

antibiotics

Article

The Anti-Virulence Effect of *Vismia guianensis* against *Candida albicans* and *Candida glabrata*

Elizangela Pestana Motta [1,2], Josivan Regis Farias [1], Arthur André Castro da Costa [1], Anderson França da Silva [1], Alberto Jorge Oliveira Lopes [3,4], Maria do Socorro Sousa Cartágenes [3], Roberto Nicolete [5], Afonso Gomes Abreu [6], Elizabeth Soares Fernandes [7,8], Flavia Raquel Fernandes Nascimento [1,2], Cláudia Quintino da Rocha [9], Cristina Andrade Monteiro [10] and Rosane Nassar Meireles Guerra [1,2,*]

1 Laboratório de Imunofisiologia, Departamento de Patologia, Universidade Federal do Maranhão, Avenida dos Portugueses, 1966, Ensino Integrado, Bloco 1, São Luís 65080-805, MA, Brazil
2 Programa de Pós-Graduação em Ciências da Saúde, Centro de Ciências Biológicas e da Saúde, Universidade Federal do Maranhão, Avenida dos Portugueses, 1966, São Luís 65080-805, MA, Brazil
3 Laboratório Experimental de Estudos da Dor, Departamento de Ciências Fisiológicas, Universidade Federal do Maranhão, São Luís 65080-805, MA, Brazil
4 Instituto Federal de Ciências e Educação do Maranhão-Campus Santa Inês, Rua Castelo Branco, 1, Santa Inês 65300-000, MA, Brazil
5 Fiocruz Ceará-Rua São José, S/N-Precabura, Eusébio 61773-270, CE, Brazil
6 Laboratóio de Patogenicidade Microbiana, Programa de Pós-Graduação em Biologia Microbiana, Universidade UNICEUMA, Rua Josué Montelo, 1-Renascença, São Luís 65075-120, MA, Brazil
7 Instituto Pelé Pequeno Príncipe, Av. Silva Jardim, 1632-Água Verde, Curitiba 80250-060, PR, Brazil
8 Programa de Pós-Graduação em Biotecnologia Aplicada à Saúde da Criança e do Adolescente, Faculdades Pequeno Príncipe, Av. Iguaçú, 333-Rebouças, Curitiba 80230-020, PR, Brazil
9 Laboratório de Química de Produtos Naturais, Centro de Ciências Exatas e Tecnológicas, Universidade Federal do Maranhão, São Luís 65080-805, MA, Brazil
10 Departamento de Biologia, Instituto Federal do Maranhão, Avenida Getúlio Vargas, No 4, Monte Castelo, São Luís 65030-005, MA, Brazil
* Correspondence: rosane.guerra@ufma.br; Tel.: +55-98-3272-8548

Citation: Motta, E.P.; Farias, J.R.; Costa, A.A.C.d.; Silva, A.F.d.; Oliveira Lopes, A.J.; Cartágenes, M.d.S.S.; Nicolete, R.; Abreu, A.G.; Fernandes, E.S.; Nascimento, F.R.F.; et al. The Anti-Virulence Effect of *Vismia guianensis* against *Candida albicans* and *Candida glabrata*. *Antibiotics* **2022**, *11*, 1834. https://doi.org/10.3390/antibiotics11121834

Academic Editor: Jesus Simal-Gandara

Received: 11 November 2022
Accepted: 9 December 2022
Published: 16 December 2022

Abstract: In folk medicine, *Vismia guianensis* is used to treat skin diseases and mycoses in the Amazon region. We evaluated the anti-*Candida* activity of the hydroalcoholic extract from the leaves of *Vismia guianensis* (EHVG). HPLC-PDA and FIA-ESI-IT-MS^n were used to chemically characterize EHVG. The anti-*Candida* activity was determined in vitro by the minimum inhibitory concentrations (MIC) against *Candida glabrata* (ATCC-2001); *Candida albicans* (ATCC-90028, ATCC-14053, and ATCC-SC5314), and *C. albicans* clinical isolates. EHVG effects on adhesion, growth, and biofilm formation were also determined. Molecular docking was used to predict targets for EHVG compounds. The main compounds identified included anthraquinone, vismione D, kaempferol, quercetin, and vitexin. EHVG was fungicidal against all tested strains. *C. albicans* ATCC 14053 and *C. glabrata* ATCC 2001 were the most sensitive strains, as the extract inhibited their virulence factors. In silico analysis indicated that vismione D presented the best antifungal activity, since it was the most effective in inhibiting CaCYP51, and may act as anti-inflammatory and antioxidant agent, according to the online PASS prediction. Overall, the data demonstrate that EHVG has an anti-*Candida* effect by inhibiting virulence factors of the fungi. This activity may be related to its vismione D content, indicating this compound may represent a new perspective for treating diseases caused by *Candida* sp.

Keywords: *Vismia guianensis*; *Candida albicans*; vismione D; *Candida glabrata*; CaCYP51; antifungals; lacre

1. Introduction

Candidemia has become one of the most common invasive *Candida* bloodstream infections. The high global incidence of candidiasis can be explained by the increase in the number of susceptible hosts, such as patients submitted to immunosuppressive treatments, long-term use of broad-spectrum antibiotics, use of catheters and probes, and hematopoietic transplants, amongst others. In addition, fungal resistance to antimicrobials has also increased [1,2].

Candidiasis is the fourth most common cause of nosocomial infections in the world, and *C. albicans* is the most frequent causative agent of medical relevance, due to its prevalence in both healthy hosts and those with underlying conditions or immune impairment. *C. albicans* is one of the most prevalent fungal species in the human microbiota, since it asymptomatically colonizes healthy individuals. This microorganism is, therefore, easily detected in the oral mucosa, gastrointestinal tract, urogenital tract, and skin of humans from birth [3].

Although *C. albicans* is still the leading cause of candidemia, increasing proportions of cases in recent years have been attributed to non-*albicans* species that are often resistant to antifungal drugs, including *Candida glabrata*, *C. parapsilosis*, *C. tropicalis*, and *C. krusei* [4].

The medical impact of *C. albicans* infections usually depends on the pathogen's ability to form biofilms [5]. Adhesion is the first step of the pathogenic process. It is complex and still not completely understood [6]. Among the known virulence factors, adhesion and subsequent biofilm formation by *Candida* sp. confer to the yeast the ability to persist and grow more easily, thus promoting the persistence of infection [1,7].

Several drugs are used to treat *Candida* infections; however, there has been an increase in the number of *C. albicans* and *C. glabrata* cases that are resistant to the available therapies. The latter is an emerging pathogenic fungus that is resistant to Fluconazole [8]. In this context, plant-derived alternatives are an interesting option for the bioprospecting of new compounds which could be used alone or in combination as antifungal agents [9,10]. Thus, plant extracts and their bioactive molecules might be a promising alternative for the treatment of candidiasis, especially extracts with a high content of phenolic compounds [11].

The medicinal effects of *Vismia guianensis* (Aubl.) Chosy, family Clusiaceae, include its antimicrobial activity, which has been associated especially with orange latex exudate from the branches of this tree. In folk medicine, this species is used to treat wounds, ulcerations, skin diseases, dermatomycoses, and herpes, particularly in the Amazon region [12]. *V. guianensis* leaves contain phenolic compounds such as anthraquinones, flavonoids, xanthones, and benzophenones, with anthraquinones being the most common compounds [13–16]. The plant also contains some other pharmacologically active compounds, such as vismione D and ferruginin, which possess immunosuppressive [13], antioxidant [17], antibacterial, and antifungal activities [18].

Considering the widespread use of *V. guianensis* by the population and its reported biological effects, this study evaluated its anti-*Candida* activity, as well as its effects on early and mature biofilm formation and adhesion. It also sought to provide data from molecular docking and PASS online prediction to help to guide the identification of the most promising *V. guianensis* compounds with antifungal activity against *Candida* sp, which may also show anti-inflammatory and antioxidant activity.

2. Results

2.1. Characterization of the Plant Material and Extract

We determined the particle sizes of dried and powdered leaves to characterize the plant material and the yield after extraction.

Figure 1 shows the particle size of the leaves and the product obtained from *V. guianensis*, which was classified as a semi-fine powder, with a large portion being retained in mesh M60 (34.09%), followed by mesh M80 (20.73%).

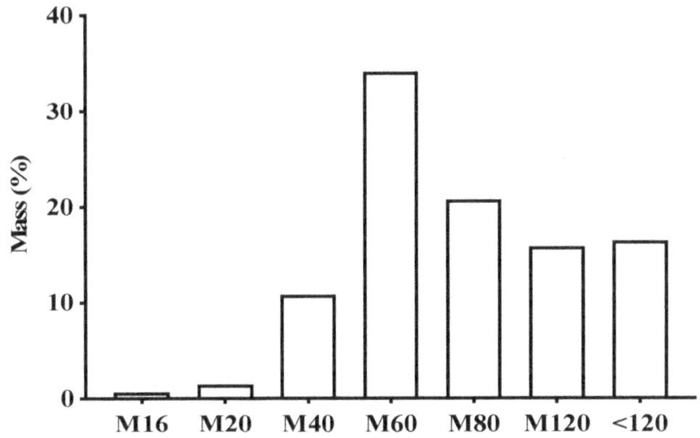

Figure 1. Percentage of different particles sizes in the powder of dry *Vismia guianensis* leaves after grinding and sifting using different mesh sizes.

Table 1 shows the results related to the ash content and the total amount of residue from *V. guianensis*. The different hydromodules used to prepare the hydroethanolic extract of the *V. guianensis* leaves (EHVG) presented similar yields, but different ash content and total residues. The hydromodule 1:10 showed the best results, with 97% organic matter, and, therefore, it was chosen for chemical analysis and experiments.

Table 1. Yield of the different hydromodules of the hydroalcoholic leaf extract of *Vismia guianensis*.

Hydromodule	Dry Residue (X ± SD) [a]	Total Residue	Extract Yield (%)
1:5	21 ± 0.0018	210	11
1:10	13 ± 0.0004	258	13
1:15	8 ± 0.0006	119	12
1:20	7 ± 0.0017	15	13

[a]: X ± SD = mean ± standard deviation.

2.2. Chemical Profile of EHVG

To determine the chemical profile of EHVG, different hydromodules were evaluated qualitatively by HPLC-PDA at 254 nm. Figure 2 shows the chromatographic profile of the hydromodule 1:10, which qualitatively exhibited the best extraction profile in terms of peak area.

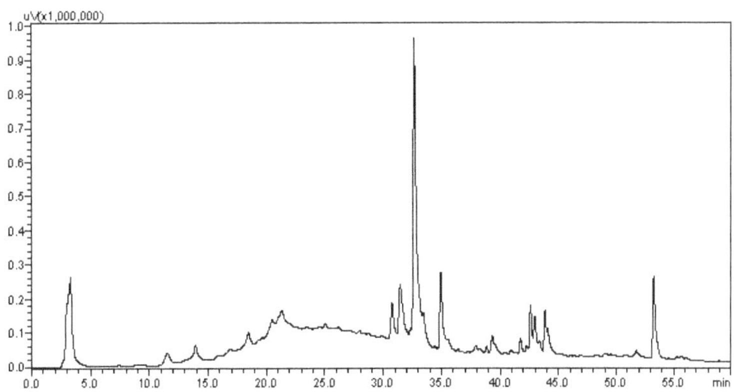

Figure 2. Chromatogram of the hydroalcoholic extracts of *Vismia guianensis* obtained by HPLC with UV detection (HPLC–PDA) for the hydromodule 1:10 with defined peaks and areas.

Figure 3 shows the identification of EHVG chemical compounds by *full-scan* spectrum (hydromodule 1:10).

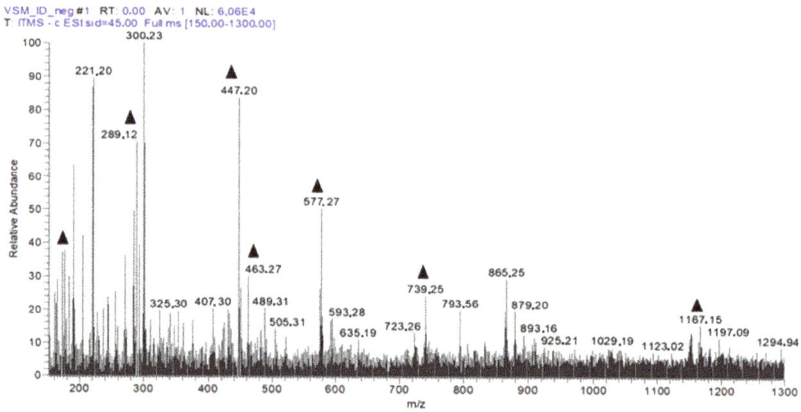

Figure 3. First-order spectrum of direct flow injection analysis (FIA-ESI-IT-MS) obtained in the negative mode for EHVG. (▲) Fragments of the chemical compounds identified.

The fourteen compounds identified by flow injection analysis (FIA-ESI-IT-MS) are depicted in Table 2. It is possible to observe a predominance of anthraquinones, flavonoids, and Vismione D.

Table 2. Compounds identified by MS^n in *Vismia guianensis* extract.

Number	[M−H]−	MS^n Ions	Proposed Compound
1	191	173, 111, 85	Quinic acid
2	431	269	Isovitexin
3	447	429, 357	Isoorientin
4	289		Catechin
5	447	429, 301, 269, 229	Orientin
6	431	285, 163	Kaempferol-*O*-rhamnoside
7	731	285, 255	Kaempferol galactoside-rhamnoside
8	1153	1001, 983, 789	Catechin tetramer
9	463	301, 283, 273, 229, 179, 121	Quercetin glycoside
10	577	425, 407, 285, 257, 213	Catechin dimer
11	1167	1015, 863, 711	A-type procyanidin trimer
12	409	273, 255	Vismione D
13	283	269, 239	Anthraquinone F
14	285		Kaempferol

Figure 4 shows the structures of the compounds identified in the leaf extract of *V. guianensis* (EHVG), according to the spectrum shown in Figure 3 and the compounds presented in Table 2. Herein, it is important to highlight, the structures of Vismione D, anthraquinone F, and kaempferol, which are important markers of the genus *Vismia*.

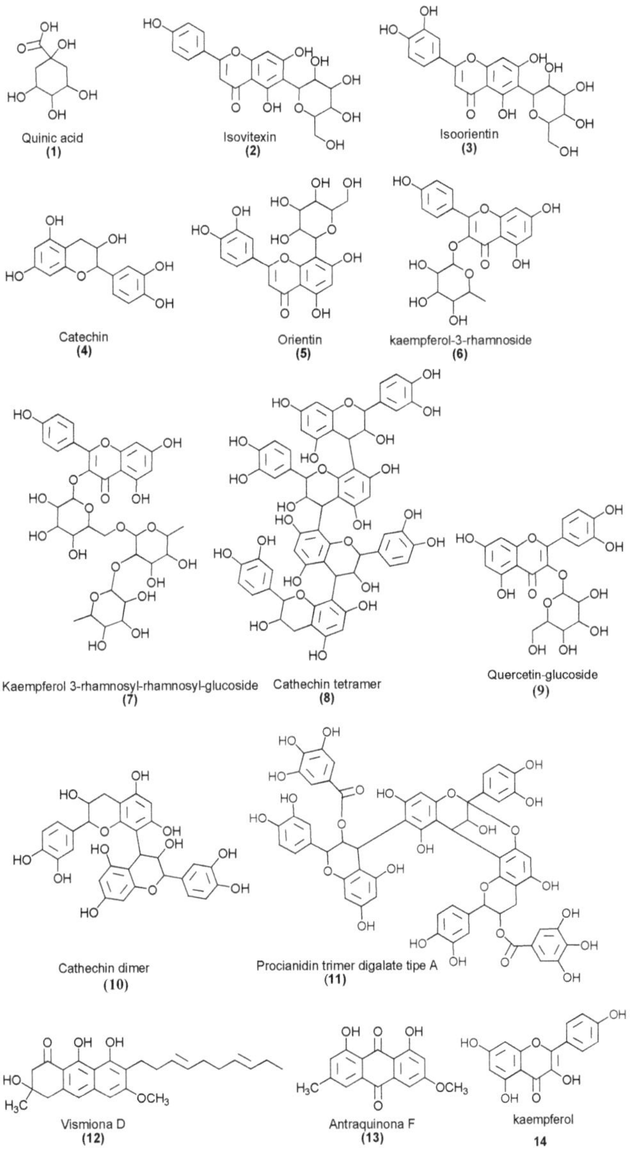

Figure 4. Chemical structures of the compounds identified in the *Vismia guianensis* leaf extract.

2.3. EHVG Showed Low Toxicity Using Different Assays

To define the best concentrations of the extract of *Vismia guianensis* (EHVG) for the antifungal assays, its hemolytic and cytotoxic activities were evaluated. The extract showed low toxicity, inducing less than 10% cell death and hemolysis (Figure 5A,B) at concentrations $\leq$ 5 mg/mL. Cell viability was greater than 80% at concentrations $\leq$ 10 mg/mL (Supplementary Materials: Figure S1).

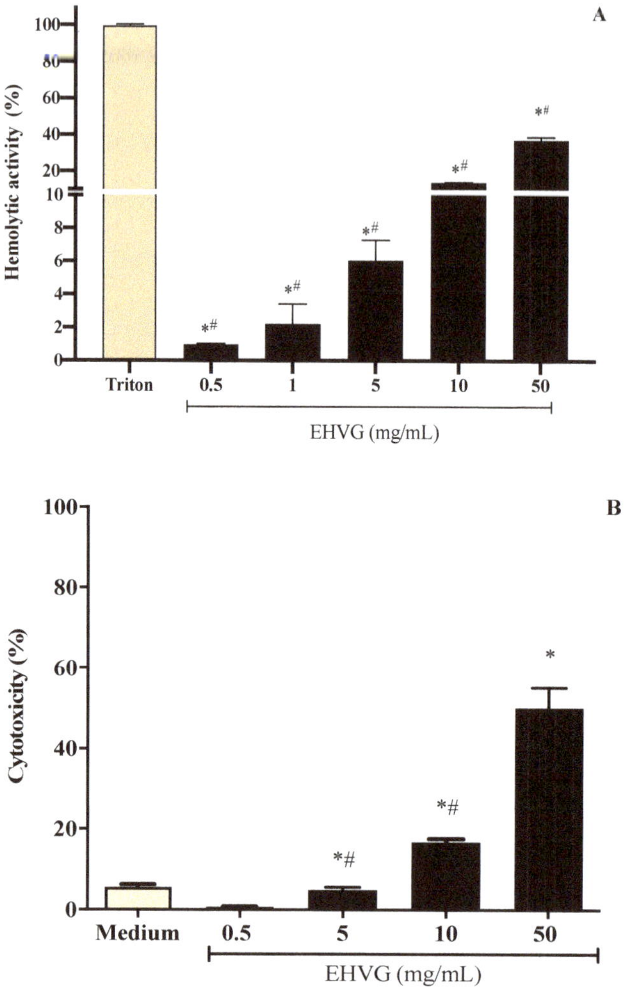

Figure 5. Different concentrations of the hydroalcoholic leaf extract of *Vismia guianensis* (EHVG) showed low toxicity when evaluated by the hemolysis (**A**) and MTT assay (**B**). Triton was the positive control for hemolysis (**A**) and medium was the negative control in the cytotoxic assay. Data represent the mean ± standard deviation of individual samples tested in quadruplicate. (*) $p < 0.05$ compared to the control; (#) $p < 0.05$ compared with other concentrations.

2.4. Antifungal Activity of EHVG against C. albicans and C. glabrata

EHVG showed antifungal potential and fungicidal activity against all tested strains (Table 3). Against *C. albicans* isolates, EHVG MIC values ranged from 3.125–6.25 mg/mL, MFC from 3.125–12.5 mg/mL, and the CFM/MIC ratio ranged from 1–2. Against *C. glabrata*, the MIC was 3.125 mg/mL, and the MFC was 6.25 mg/mL. MIC values were used as a parameter for selecting two strains for the subsequent assays: *C. albicans* (SC5314) and *C. glabrata* (ATCC 2001), both of which are of clinical interest.

Table 3. Minimum inhibitory (MIC) and minimum fungicidal (MFC) concentrations of the hydroalcoholic leaf extract of *Vismia guianensis* (EHVG) against *Candida* spp. strains.

Candida Strain	*Vismia guianensis* (EHVG)			Antifungal	
	MIC [a]	MFC [a]	MFC/MIC Ratio	Ampho B [b]	Flu [c]
C. glabrata (ATCC 2001) [d]	3.125	6.25	2	0.25	16
C. albicans (ATCC 90028)	3.125	3.125	1	1	8
C. albicans (ATCC 14053)	6.25	6.25	1	0.5	8
C. albicans (SC 5314)	6.25	6.25	1	0.5	16
A1 [e] *C. albicans*	6.25	12.5	2	1	8
A2 *C. albicans*	3.125	3.125	1	0.5	8
A3 *C. albicans*	6.25	6.25	1	0.25	4
A4 *C. albicans*	3.125	6.25	2	0.5	16
A5 *C. albicans*	3.125	3.125	1	0.5	16
A6 *C. albicans*	3.125	3.125	1	0.5	16
A7 *C. albicans*	6.25	6.25	1	1	16

(a) Values are expressed as mg/mL. (b) Ampho B = Amphotericin B concentration: 0.03 to 16 μg/mL. (c) Flu = Fluconazole concentration: 0.125 to 256 μg/mL. (d) ATCC® (American Type Culture Collection). (e) A1–A7: clinical samples.

2.4.1. Treatment with EHVG Inhibits the Growth of *C. albicans* and *C. glabrata*

Figure 6 shows the fungicidal activity of EHVG. EHVG inhibited the growth of *C. albicans* and *C. glabrata* when tested at two different concentrations. Figure 6A shows the results for *C. albicans* SC5314, after 36 (2× MIC = 12.5 mg/mL) and 48 h (1× MIC = 6.25 mg/mL). Figure 6B shows similar results against *C. glabrata* (ATCC 2001), in a shorter period, for both concentrations, since the inhibition was observed after 24 and 36 h at concentrations of 2× (6.25 mg/mL) and 1× the MIC (3.125 mg/mL), respectively.

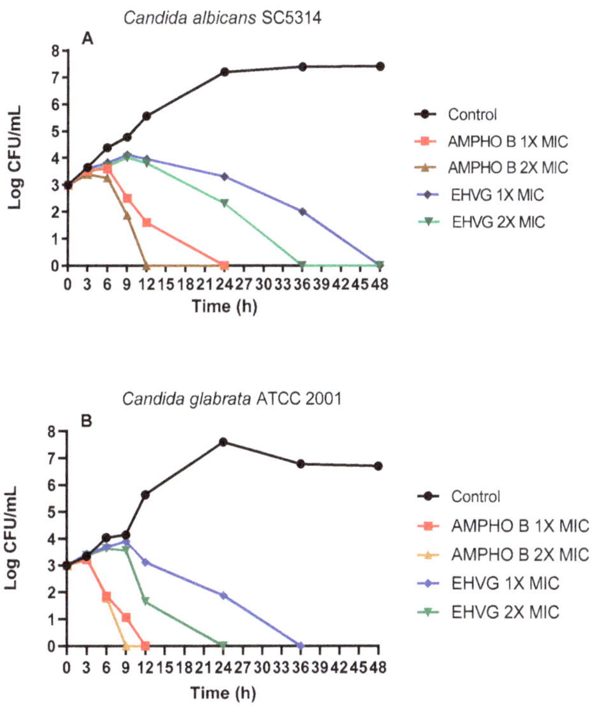

Figure 6. Effects of EHVG on time-kill curves. The effects of EHVG against *Candida albicans* SC5314 (**A**) and *Candida glabrata* ATCC 2001 (**B**) were tested at two different concentrations corresponding to the MIC or 2× MIC, then compared to untreated cultures (Control) or with cultures treated with Amphotericin B (AMPHO B; MIC—0.5 μg/mL and 2× MIC—1 μg/mL).

2.4.2. EHVG Inhibits *C. albicans* and *C. glabrata* Adhesion

C. albicans (ATCC 90028, SC5314) and *C. glabrata* ATCC 2001 were used to investigate the effects of EHVG on pathogen adhesion. Figure 7A shows that EHVG was more effective than Amphotericin B in inhibiting *C. glabrata* (ATCC 2001) adhesion at the lowest tested concentration (1/2× MIC). At the highest concentration (1× MIC), all treatments had similar inhibitory effects in comparison with the untreated control. Figure 7B,C demonstrate that EHVG effects were equivalent to those of reference drugs (FLU and Ampho B), regardless of the concentration used, against the two different *C. albicans* strains.

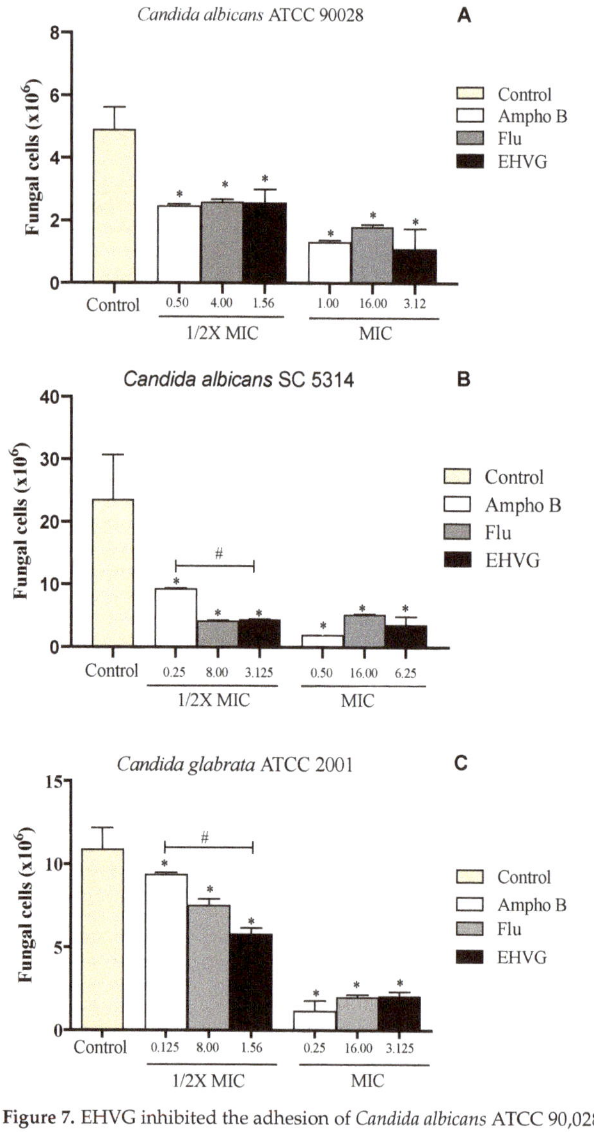

Figure 7. EHVG inhibited the adhesion of *Candida albicans* ATCC 90,028 (**A**), *C. albicans* SC5314 (**B**), and *C. glabrata* ATCC 2001 (**C**) in cultures treated with concentrations of 1/2 MIC and MIC when compared to untreated samples (control). The EHVG-treated cultures were also compared to those treated with the reference drugs Amphotericin B (Ampho B) or Fluconazole (Flu). (*) $p < 0.05$ differs from untreated control; (#) $p < 0.05$ differs from Amphotericin B.

2.4.3. EHVG Disrupts Early and Mature Biofilm Formation

Figure 8 shows that treatment with EHVG impaired early *Candida* biofilms formed within 48 h (Figure 8A,C,E) and mature biofilms formed within 72 h (Figure 8B,D,F). This effect was similar to that of Amphotericin B or Fluconazole compared to untreated controls.

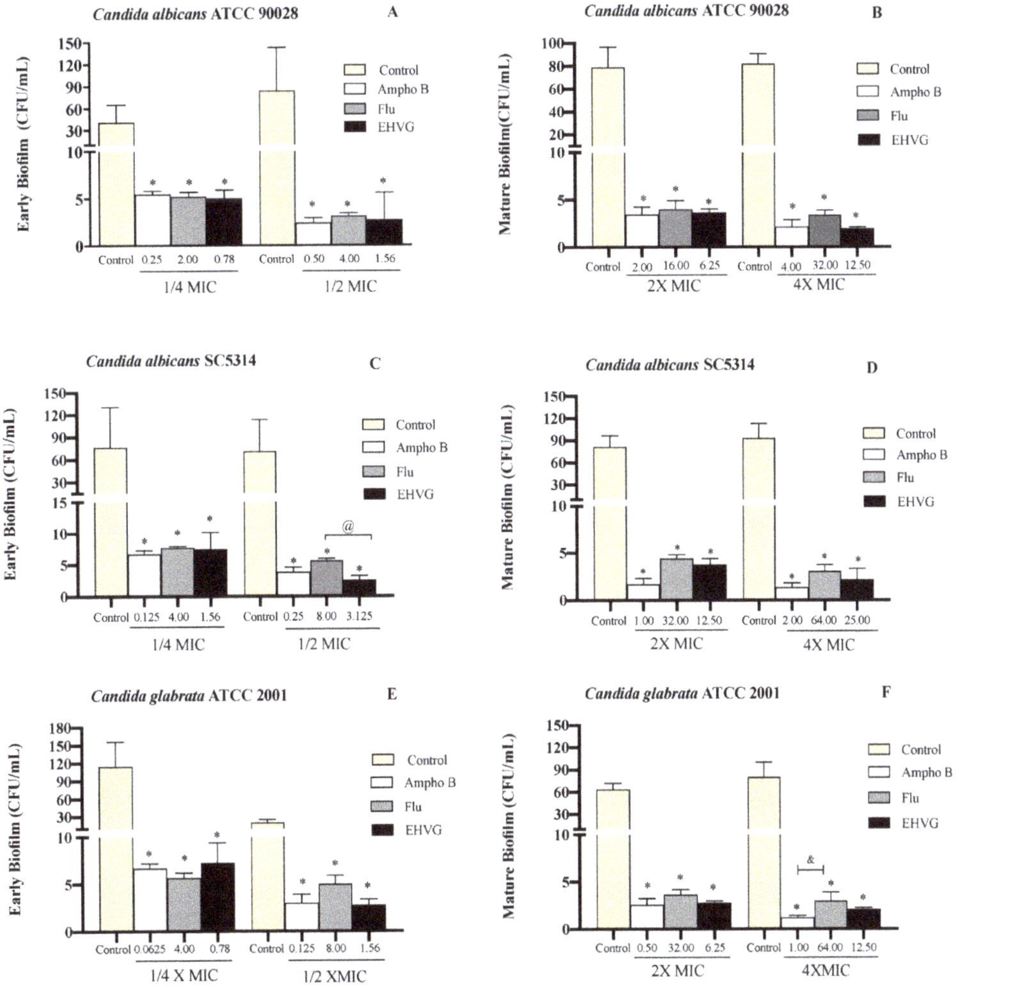

Figure 8. EHVG inhibited the formation of early (**A**,**C**,**E**) and mature (**B**,**D**,**F**) biofilms. The CFU/mL values were established in cultures of *Candida albicans* (ATCC 90028, **A**,**B**; SC5314, **C**,**D**) and *C. glabrata* (ATCC 2001, **E**,**F**) after 48 h and 72 h. The samples treated with different concentrations of EHVG (1/4 and 1/2× MIC for early biofilms, and 2× and 4× MIC for mature biofilms) were compared to cultures treated with Amphotericin B (Ampho B) or fluconazole (Flu), and with the untreated control. (*) $p < 0.05$ in comparison to control; (@) in comparison to Fluconazole, and (&) $p < 0.05$ Fluconazole compared to Amphotericin B.

Treatment with EHVG inhibited biofilm formation with an efficacy similar to that of Fluconazole or Amphotericin B, even at sub-inhibitory concentrations (1/4 MIC and 1/2 MIC). For *Candida albicans* SC5314, the reduction was more significant than that observed with Amphotericin B (Figure 9).

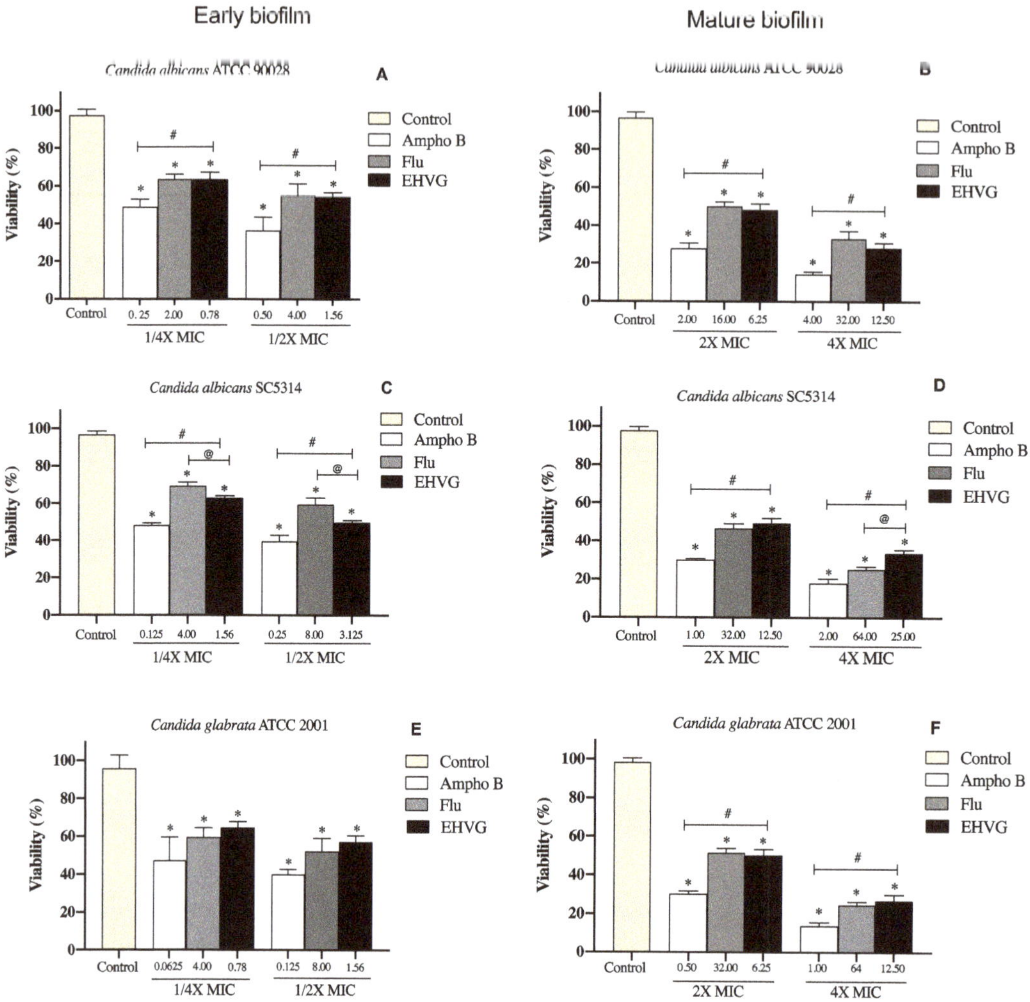

Figure 9. Percentage of inhibition on EHVG-treated early (**A**,**C**,**E**) and mature biofilms (**B**,**D**,**F**) of *Candida albicans* ATCC 90,028 (**A**,**B**), *C. albicans* SC5314 (**C**,**D**) and *C. glabrata* ATCC 2001 (**E**,**F**) after 48 and 72 h, evaluated by MTT assay. Samples were treated with EHVG (1/4 and 1/2 for early biofilms, and 2× and 4× MIC for mature biofilms) and compared to untreated controls or the reference drugs Amphotericin B (Ampho B) and Fluconazole (Flu). (*) $p < 0.05$ compared to control. (#) $p < 0.05$ compared to Amphotericin B and (@) $p < 0.05$ compared to fluconazole.

Figure 10 shows the reduction in biofilm biomass in the groups treated with EHVG for 48 and 72 h, with the same intensity as that observed for Amphotericin B and sometimes higher than that for Fluconazole. This confirms the data obtained from the MTT assay.

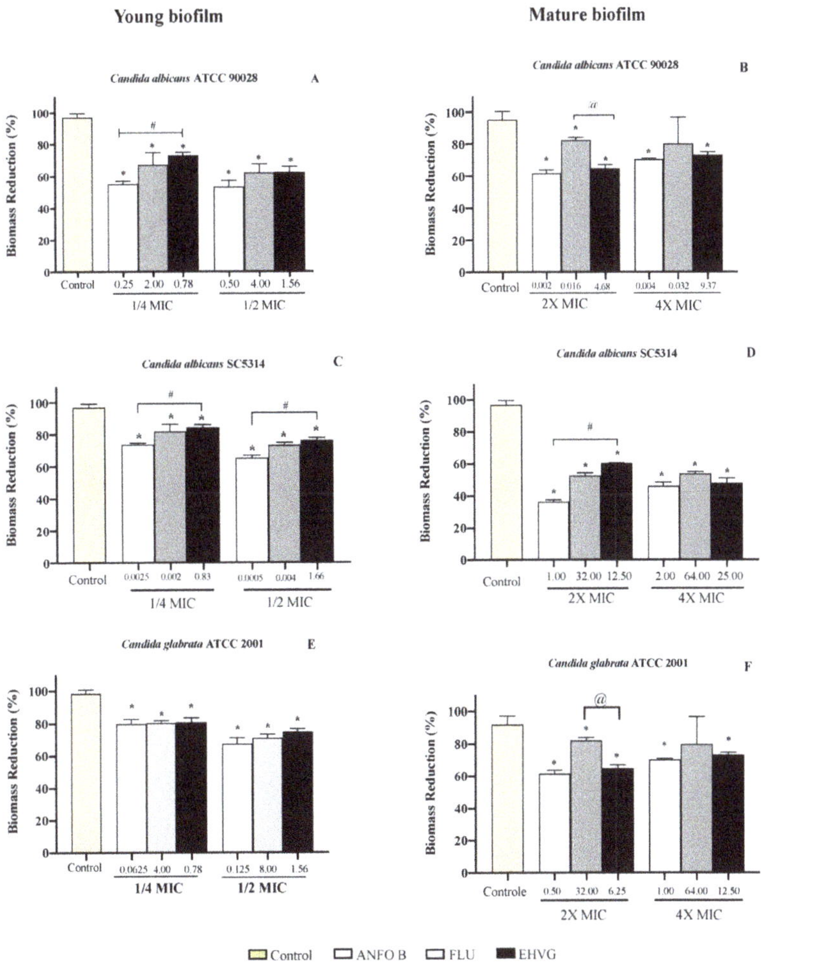

Figure 10. Reductions in biofilm biomass stained with crystal violet (CV) of *Candida* spp. (ATCC 90028, **A**,**B**), (ATCC 2001, **C**,**D**) and (SC5314, **E**,**F**) after 48 and 72 h. The samples were treated with EHVG and compared to untreated controls or to the reference drugs Amphotericin B (AMHO B) and Fluconazole (FLU). (*) $p < 0.05$ differs from control; (#) $p < 0.05$ differs from Amphotericin B and (@) $p < 0.05$ compared to fluconazole.

2.5. In Silico Biological Activity and Toxicity for the Compounds Identified in EHVG

The PASS online tool evaluated the antifungal, anti-inflammatory, and antioxidant activities of the five selected compounds detected in EHVG, including catechin, kaempferol, anthraquinone, and Vismione D, in comparison to those of Fluconazole and Amphotericin. The compounds showed greater Pa than Pi as antifungal, considering Pa values higher than Pi (>0.3) due to the strong molecular potency. They also exhibited a high probability of gut permeability, and no violation to Lipinski's rules.

Vismione D showed the highest predictive antifungal and also showed high values as anti-inflammatory agent when compared to other compounds found in EHVG (Table 4). Kaempferol and quercetin showed the highest predictive value as anti-inflammatory. Quercetin and catechin showed the highest values as antioxidant compounds, and anthraquinone, fluconazole, and amphotericin B had no antioxidant predictive activities with Pa values lower than 0.3.

Table 4. PASS prediction of commercial drugs and three selected compounds detected in the *Vismia guianensis* extract, considering the Potential activity (Pa) higher than potential inactivity (Pi), as well as the antifungal, antioxidant, and anti-inflammatory activities.

Activity/ Compounds	Antifungal		Anti-Inflammatory		Antioxidant	
	Pa [a]	Pi [b]	Pa	Pi	Pa	Pi
Vismione D	0.684	0.011	0.606	0.030	0.478	0.008
Catechin	0.552	0.023	0.548	0.044	0.810	0.003
Kaempferol	0.495	0.031	0.676	0.019	0.856	0.003
Quercetin	0.490	0.032	0.689	0.017	0.872	0.003
Anthraquinone	0.351	0.063	0.410	0.090	-	-
Fluconazole	0.726	0.008	-	-	-	-
Amphotericin	0.977	0.000	0.330	0.136	-	-

[a]: Pa (probability "to be active") estimates the chance that the studied compound belongs to the sub-class of active compounds. [b]: Pi (probability "to be inactive") estimates the chance that the studied compound belongs to the sub-class of inactive compounds.

2.6. Compounds Present in EHVG, Especially Vismione D, Interact with CaCYP51

The following compounds identified in EHVG were used for molecular docking: anthraquinone F, catechin, kaempferol, and vismione D. Figure 11 illustrates the molecular interaction between vismione D and the enzyme CaCYP51 of *C. albicans*. Analysis of the vismione D + CaCYP51 complex is characterized by the formation of hydrogen bonds between the ligand and the enzyme residues Tyr118, Ser378, and Met508, as well as hydrophobic contact with residues Leu121, Phe228, Pro230, Phe233, Gly303, Ile304, Gly307, Gly308, Thr311, Leu376, Phe380, and Ser508, including interactions with the heme group.

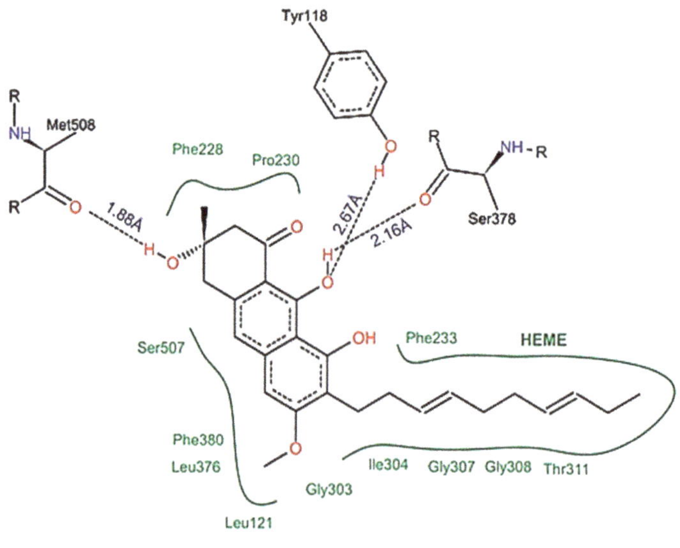

Figure 11. Schematic representation of the interactions between vismione D and CaCYP51, identified by molecular docking. It is possible to observe the formation of hydrogen bonds and hydrophobic contacts between the ligand and the enzyme residues, including interactions with the heme group. The figure was obtained with PoseView.

Molecular docking analysis indicated that vismione D presents the highest affinity for the fungal enzyme CaCYP51, with a binding free energy (ΔGbind) of 10.96 kcal/mol and an inhibition constant (Ki) of 0.009 μM, when compared to the reference drugs posaconazole and fluconazole. Anthraquinone provided results similar to those of posaconazole, and the results for catechin and kaempferol were similar to those of fluconazole (Table 5).

Table 5. Free binding energies and inhibition constants between fourteen compounds identified in *Vismia guianensis* extract and CaCYP51, obtained through molecular docking.

Ligand	ΔGbind (kcal/mol) *	Ki (μM) **
Vismione D	−10.96	0.009
Anthraquinone F	−7.92	1.56
Catechin	−6.97	7.79
Kaempferol	−6.70	12.35
Quercetin	−5.60	20.98
Posaconazole	−8.43	0.57
Fluconazole	−6.89	11.61

(*) Bind free energy; (**) inhibition constant.

Figure 12 shows the spatial conformation obtained for the molecular docking analysis of vismione D and CaCYP51 structures. Several binding sites could be observed, which favor a possible action of the compound on the enzyme.

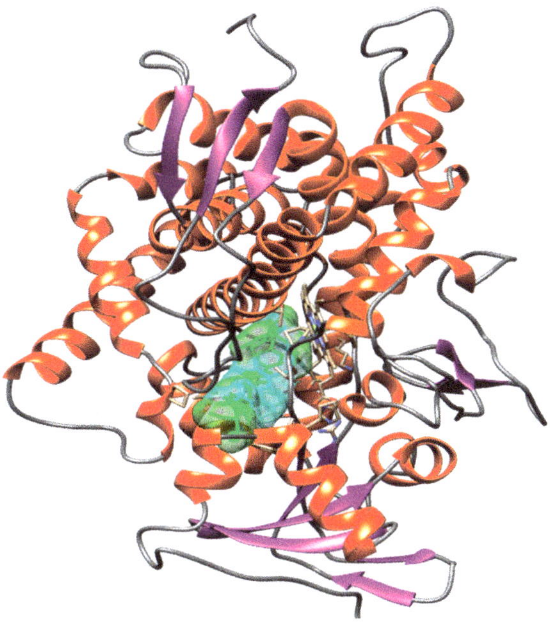

Figure 12. Spatial conformation showing several common binding sites between the CaCYP51 enzyme (PDB: 5FSA) and Vismione D (blue) or the antifungal posaconazole (green) through molecular docking. The image was obtained with USCF Chimera.

3. Discussion

V. guianensis is a native plant found in different regions of Brazil, especially in the Amazon region, where its leaf extracts are used to treat skin inflammatory and infectious skin diseases [13]. Little is known about its mechanisms of action. Thus, our aim was to investigate its anti *Candida* activity from the hydroethanolic extract prepared with a vegetal part (EHVG), considering that plants are still the main source of bioactive compounds and that there is a growing resistance to commercial antifungal drugs.

For the extract preparation, the particle size of the leaves and the product obtained were evaluated and classified as a semi-fine powder, with a large part retained in mesh M60, followed by mesh M80. It is important to highlight that the particle size of the dry plant material is important for the extract preparation, since the more homogenous the extraction is, the better, resulting in an increased yield and collection of chemical components, maximizing biological activities [19].

The preparation of EHVG with different hydromodules resulted in a similar yield, but different dry and total residues after evaluation of the ash content. The hydromodule 1:10 showed the best result for the chemical analysis, as it exhibited the best qualitative profile due to the more intense peaks and larger and better-defined areas, to confirm the presence of secondary metabolites in the extract. For this reason, it was used for antifungal assays.

Direct flow injection analysis of EHVG with ionization in negative mode confirmed the presence of phenolic compounds, including anthraquinones, catechins, epicatechins, kaempferol, vismione D, and flavonoids. The presence of these compounds was also previously reported by Seo et al. [20], who isolated five benzophenones, vismiaguianones A–E, and two benzocoumarins, vismiaguianins from chloroformic root extract during the investigation of the antitumoral effects of *V. guianensis*. In 2004, Politi et al. [14] described the presence of anthraquinones, vismiones, flavonoids, xanthones, and benzophenones in *n*-hexane, as well as dichloromethane, when investigating leaf methanol extracts in comparison with those from roots. Later, Hussain et al. [21], in a review related to the genus *Vismia*, described the presence of the same compounds which we found, and discussed anthraquinones and vismione as important chemical markers of this group.

According to Costa et al. [22], antifungal activities may be related to the presence of mono- and sesquiterpene metabolites, components also identified in the present study. This previous study demonstrated that such metabolites could promote cell membrane rupture, causing fungal death as well as impairments of membrane synthesis, cellular respiration, and spore germination. Flavonoids may also cause rupture of the microbial membrane due to their lipophilic nature [23]. Other *V. guianensis* chemical components, such as humulene epoxide II (not reported in this study) isolated from the oil of leaves, exhibited antifungal activity against several *Candida* species, including *C. albicans* and *C. glabrata* [22].

EHVG exhibited low toxicity using different techniques and different concentrations. The decision to investigate this by different assays was based on previous observations on the MTT assay; neutral red and crystal violet staining assays provided complementary data for cell viability [24].

The antifungal activity of EHVG was evaluated by time-kill assays, using the *C. albicans* (SC5314) and *C. glabrata* (ATCC 2001) strains. These strains were chosen due to their clinical interest [22,25,26]. EHVG showed a fungicidal effect against both strains, inhibiting all *Candida* strains over 48 h, especially for *C. glabrata* after the first 12 h.

As EHVG was antimicrobial against *C. albicans* and *C. glabrata*, we investigated its effects on *Candida* virulence (adhesion and biofilm formation). The extract impaired the adhesion of all tested strains with similar efficacy to that observed for reference drugs such as fluconazole. This is an important action, considering that adhesion is the first step of a successful fungal colonization and pathogenesis [1,7].

EHVG exerted an inhibitory concentration-dependent effect on early biofilm formation by *C. albicans* (ATCC90028) and *C. glabrata* (ATCC2001). The observed inhibitory effects were equivalent to Amphotericin B, but no differences were observed after incubation of sub-inhibitory concentrations of EHVG. The extract also exerted inhibitory activity on mature biofilms produced by *C. albicans* (ATCC 90028) and *C. glabrata* (ATCC 2001), with similar performance to those of amphotericin B and fluconazole when tested at 2× MIC. For *C. albicans* (SC5314), EHVG showed better results than fluconazole.

The biofilm acts as a protective barrier for the microorganism, rendering their cells more resistant to the immune system, as well as to the effects of antifungal therapies. Biofilms arising from *Candida* infections are difficult to treat and, depending on the biofilm phase, higher concentrations of antifungal compounds are necessary in comparison to planktonic cells [27]. In addition, biofilm formation may vary among *Candida* strains in response to different antifungals, contributing to increased virulence [5,7]. EHVG exhibited anti-*Candida* activity, reducing both adhesion and biofilm formation by all tested strains.

In order to better understanding the underlying mechanisms of EHVG anti-*Candida* activities, molecular modeling was used to identify the probable targets of EHVG compounds. We found that all tested compounds exhibited favorable interactions with CaCYP51 (cytochrome P450 lanosterol 14a-demethylase). However, vismione D was the most potent inhibitor of the enzyme, with a docking score of −10.96kcal/mol. Vismione D also exhibited the highest affinity for CaCYP51, whilst presenting the lowest inhibition constant and high negative binding free energy, with values even greater than those of posaconazole and fluconazole, used herein as reference drugs. These findings indicate vismione D as the most promising anti-*Candida* compound of EHVG. It is important to note that the other tested compounds were also predicted to interact with CaCYP51, allowing us to conclude that EHVG activity may be due to a phytocomplex present in the EHVG, in which vismione D may be the most effective.

Binding free energy is a parameter used to indicate how spontaneously the interaction between a molecule and a biological target occurs. Negative binding free energies indicate that these interactions favor the formation of ligand-receptor complexes [28].

Intermolecular interactions, such as hydrogen bonds with amino acid residues (Tyr-118, Ser-378 and Met-508), hydrophobic interactions, and interactions with the heme group, confer the ability to stabilize a given ligand at the binding site, forming ligand-receptor complexes. In the present study, vismione D was predicted to interact with practically the same amino acids as posaconazole, a commercially available fungicide, but more spontaneously and with a greater interaction potential. In addition, the spatial conformation obtained indicated a more linear structure of vismione D contributing to the molecular interactions.

Values lower than 2Å indicate that the docking protocol is valid, i.e., shows similarity with the experimental structure [28]. The CaCYP51 enzyme plays a key role in the synthesis of ergosterol from its precursor, lanosterol [29,30]. Blockade of this enzyme thus inhibits the process necessary for *Candida* survival [31]. Based on this, it is reasonable to propose that EHVG's anti-*Candida* activities are due to its ability to affect ergosterol synthesis.

It is important to emphasize the relevance of phytocomplexes present in EHVG for the development of new antifungals with, possibly, a broader spectrum of action than traditional agents [22], since the extract compounds may act additively as antifungals. The in silico absorption, distribution, metabolism, excretion (ADME), and toxicity modeling is an important tool for rational drug design for studying interactions of ligands with biological targets at the atomic scale. As shown by the in silico results, the evaluated compounds, including vismione D, presented good predictive values with high gut absorption, as well as accordance to the Lipinski's rule, an important characteristic for drug-likeness [32,33].

The in silico absorption, distribution, metabolism, excretion (ADME), and toxicity (T) modeling is an important tool for rational drug design and to study the interactions of ligands with the biological targets in atomic scale, especially if we consider the several restrictions for animal use in the world.

The molecular docking findings suggest that the compounds present in EHVG meet Lipinski's rules, which indicates that they might act as effective drug candidates with a low risk level for oral use. Furthermore, the ADMET predictor confirmed the in vitro results related to the low toxicity as an important characteristic to show the plant compounds as a good medicinal product [32,33].

Additionally, based on PASS prediction simulation, it is possible to propose other biological activities for vismione D, since it showed a higher value of Pa (potential activity) [34] as an antifungal compound, and good values as an anti-inflammatory and antioxidant compound.

To the best of our knowledge, there are no available data evaluating the in silico activities of compounds from *V. guianensis* extract (EHVG) or any other species of the genus *Vismia*. This highlights the importance of this analysis for elucidating the possible mechanisms underlying the antifungal action of this species.

4. Materials and Methods

4.1. Collection and Identification of Vismia guianensis

The leaves were collected between October and November 2016, in the morning, in São Luís, MA, Brazil (2°28′47.1″ S and 44°13′17.4″ W). The collected material was cleaned and left to dry for 7 days, protected from sunlight, under natural ventilation, and at room temperature (±24 °C). The plant material was identified and deposited in the Herbarium of Maranhão, Federal University of Maranhão (No. 11.078), and is registered with the National System for the Management of Genetic Heritage and Associated Traditional Knowledge (SisGen) (Registration No. AFE7A08) as regulated by Law No. 13.123/15.

4.2. Preparation of the Extract

The dried leaves were ground in an electrical grinder. The powder obtained was used to prepare the extract and the different hydromodules The extract yield was calculated as % of the total biomass of the plant material. The dependence of the extract yield on the hydromodule was studied to determine the best extraction process, and the choice of 1:10 was based on the highest extract yield and best chemical profile.

For the hydromodules, 10 g of the powdered leaves was transferred into a Falcon polyethylene test tube, and added to different Extract/Ethanol ratios (hydromodule) as follows: hydomodule 1:5 (50 mL); hydromodule 1:10 (100 mL); hydromodule 1:15 (150 mL); and 1:20 (200 mL). After this, the extraction process was carried out.

The extract was obtained by maceration for 7 days, protected from light. After this period, the solvent was evaporated in a rotary evaporator, and then the material was lyophilized and stored in a refrigerator for subsequent assessment.

4.3. Characterization of the Plant Material

4.3.1. Particle Size Analysis of the Leaves

For characterization of the plant material, two samples of 100 g of dried leaves were placed in a mesh strainer, with mesh sizes No. 16, 20, 40, 60, 80, and 120, and openings of 1190, 850, 425, 250, 180, and 125 μm, respectively, according to Brazilian normalization rules. This lasted for 20 min, with the rheostat being adjusted to intensity 7. The samples were removed and weighed for yield calculation [35].

4.3.2. Ash Content and Extract Yield

Total ash content was quantified as previously described [35]. Samples of the *V. guianensis* leaf powder (3 g), in duplicate, were placed in crucibles, stabilized for 30 min in a desiccator, and heated separately in a muffle furnace (Magnus Ltd.a.®) at an initial temperature of 200 °C, with increments of 200 °C every 2 h, until a temperature of 600 °C was reached. The samples were then removed, transferred to a desiccator, maintained at room temperature, and weighed for calculation of the ash volume. At the end of this process, the yield was calculated by placing 1 mL of the extract on Petri dishes and incubated at 37 °C until completely dried. Individual samples were tested in triplicate, and the yield was calculated considering the dry weight of the extract.

4.4. Chemical Characterization

4.4.1. Analysis of EHVG by HPLC-PDA

The extract was analyzed by HPLC-PDA. A cleaning step was performed to remove any contaminants. Sample solution (30 mg/mL, 1 mL) was run through solid phase extraction (SPE) using Phenomenex Strata C18 cartridges (500.0 mg stationary phase), which had previously been activated with 5.0 mL MeOH and then equilibrated with 5.0 mL MeOH:H_2O (1:1, v/v). The sample was then filtered through a 0.22 μm PTFE filter, and subsequently dried. The dried sample was dissolved at a concentration of 10.0 mg/mL in HPLC-grade methanol solvent. An aliquot of 10.0 μL was injected directly into the HPLC-PDA. The system used was a Shimadzu model HPLC system (Shimadzu Corp., Kyoto, Japan), consisting of a solvent injection module with a binary pump and PDA detector (SPA-20A. The column used was Luna 5.0 μm C18 100 ^{0}A (150.0 μm × 4.6 μm). The elution solvents were A (2% acetic acid in water) and B (2% acetic acid in methanol). The sample was eluted according to the following gradient: 5% to 60% B in 60 min. The flow rate was 1 mL/min. The data were processed using the LC Solution software (Shimadzu) [36].

4.4.2. Analysis of EHVG by FIA-ESI-IT-MS^n

For direct flow injection analysis (FIA-ESI-IT-MS^n), 10 mg of the crude extract was dissolved in 1 mL of MeOH: H_2O (1:1, v/v). After incubation in an ultrasound bath for 5 min, the extract was filtered (0.22 μm), and 20-μL aliquots at a concentration of 5 ppm were directly injected into the FIA-ESI-IT-MS^n system.

Full-scan mass spectra were recorded in the range of 100–1000 m/z. Multistage fragmentations (ESI-MS^n) were obtained by the collision-induced dissociation (CID) method, using helium for ion activation. The different compounds of EHVG were identified, based on the comparison of UV spectra and characteristic fragmentations, to those from the data in the literature [30].

4.5. *In Vitro Tests for Anti-Candida Activity*

4.5.1. Isolation of Microorganisms

Four *Candida* spp. reference strains (ATCC), including three *C. albicans* strains and one *C. glabrata* strain (ATCC 2001), as well as eight clinical isolates, were used after approval by the Brazilian Ethics Committee (No. 813.402/2014) (Table 6). The strains were obtained from the collection of the Laboratory of Applied Microbiology, Universidade CEUMA, São Luís, MA, Brazil. All *Candida* isolates were identified by multiplex PCR. The identification was performed as previously described [37], using the combination of eight different species-specific primers in a single PCR tube, by combining two yeast-specific fragments from the ITS1 and ITS2 regions and species-specific primers for *C. albicans* and *C. glabrata*. This method was chosen as it allows for the identification of clinical isolates with high specificity, and has the potential to discriminate individual *Candida* species in polyfungal infections. For the experiments, the isolates were subcultured on Sabouraud-dextrose agar (SDA) with chloramphenicol for 24 h at 37 °C.

Table 6. Reference strains tested.

Identification	Type of Strain
ATCC 2001 [a]—*C. glabrata*	Reference
ATCC 90,028—*C. albicans*	Reference
ATCC 14,053—*C. albicans*	Reference
ATCC MYA 2876 (SC 5314)—*C. albicans*	Reference–Wild type
A1 *C. albicans*	Clinical isolate–vagina
A2 *C. albicans*	Clinical isolate–vagina
A3 *C. albicans*	Clinical isolate–vagina
A4 *C. albicans*	Clinical isolate–vagina
A5 *C. albicans*	Clinical isolate–vagina
A6 *C. albicans*	Clinical isolate–vagina
A7 *C. albicans*	Clinical isolate–oral
A8 *C. albicans*	Clinical isolate–oral

[a]: ATCC—American Type Culture Collection.

4.5.2. Inoculum Preparation

The different strains were activated on SDA with chloramphenicol by seeding and incubation for 24 h at 35 °C. The colonies were resuspended in 5 mL of 0.85% sterile saline (0.145 mol/L; 8.5 g/L NaCl). The resulting suspension was homogenized in a Vortex® for 15 s. The cell density was adjusted in a spectrophotometer (Global Trade Technology), at a wavelength of 530 nm, by the addition of saline to obtain an absorbance corresponding to 0.5 McFarland standards.

4.5.3. Determination of Minimum Inhibitory Concentration (MIC)

Antifungal susceptibility tests were standardized following previously established guidelines (M27-A3, [38]), and according to Ostrosky et al. [39]. For the test, 100 µL RPMI 1640 medium (with glutamine and phenol red, no bicarbonate) buffered with MOPS [3-(N-morpholine) propanesulfonic acid] was added to the wells of a 96-well sterile microtiter plate, which contained a cell suspension of *Candida* spp. (1×10^6 CFU/mL). Next, 100 µL EHVG at a concentration of 50 mg/mL was added to the first column, followed by serial dilutions up to 0.095 mg/mL. Amphotericin B (0.0313–16 µg/mL) and Fluconazole (0.125–256 µg/mL) (Sigma-Aldrich, São Paulo, Brazil) were used as positive controls. RPMI 1640 medium, without extract or an antifungal agent, served as the growth control. The *Candida* spp. inoculum plus EHVG or antifungal agent was incubated for 24 h at 37 °C.

The result was analyzed visually and read in a microplate reader at 540 nm. The MIC was defined as the lowest concentration of the extract or antifungal agent at which no growth was visible or detected. The test was carried out in triplicate in two different experiments.

4.5.4. Determination of Minimum Fungicidal Concentration (MFC)

The MFC was determined based on the MIC results. A 10 µL aliquot of the wells corresponding to up to 4× the MIC was seeded onto SDA plates with chloramphenicol. The plates were incubated for 24–48 h at 37 °C, and the MFC was defined as the lowest concentration of the EHVG and antifungal agents that inhibited fungal growth. All tests were carried out in triplicate. The MFC/MIC ratio was calculated in order to determine whether the extract had fungistatic (MFC/MIC $\geq$ 4) or fungicidal activity (MFC/MIC $\leq$ 4) [40].

4.5.5. Time-Kill Assay

C. albicans ATCC 2001 and *C. albicans* SC 5314 were used to establish the time-kill curves according to [26], with modifications. The inoculum was diluted in RPMI 1640 medium to a final concentration of 5×10^3 CFU/mL, and added to 96-well plates. EHVG was used at concentrations corresponding to the MIC and 2× MIC, using 1% DMSO as vehicle. Amphotericin B plus inoculum was used as a positive control, and RPMI 1640 plus inoculum as a negative control. The plates were incubated for 24 h at 37 °C, and the number of CFUs was determined.

Growth kinetics were evaluated after different periods of incubation: 0, 3, 6, 12, 24, 36, and 48 h. Serial dilutions were prepared for each sample (10^{-1} to 10^{-4}), and 10 µL was transferred to Petri dishes containing SDA with chloramphenicol. After colony counting, the following formula was applied: CFU/mL = number of counted colonies $\times$ $10^n/q$, where n corresponds to the absolute value of the chosen dilution and q to the volume (mL) of each dilution seeded on the plates [41]. The assay was carried out in triplicate in two different experiments.

4.5.6. Adhesion Assay

Candida spp. was activated as described in item 4.5.2, washed with PBS (3 mL), and centrifuged twice at 2060× *g* for 5 min. After the last wash, the inoculum was resuspended in 5 mL PBS, read at 530 nm, and adjusted to 1×10^7 CFU/mL [42].

For adhesion assessment, 100 µL PBS and 100 µL EHVG, at concentrations corresponding to the MIC and 1/2 MIC, were added to each *Candida* strain. Serial dilutions were prepared, and the samples were incubated for 90 min at 37 °C. The supernatant was then discarded, and the plates were washed twice with PBS. In the last wash, PBS was discarded completely, and the cells were resuspended in 100 μL PBS. The microdroplet technique was used to seed dilutions of 10^{-2} to 10^{-4} on SDA plates with chloramphenicol, which were incubated for 24–48 h at 37 °C. The assay was carried out in triplicate in two different experiments.

4.5.7. Effect of EHVG on Biofilm Formation

The experiments were carried out as described previously [24], with modifications. Sub-inhibitory concentrations of EHVG (200 μL), corresponding to 1/4 MIC and 1/2 MIC, were used to assess the extract's interference with biofilm formation. The plates were incubated for 24 h at 37 °C. After this period, the supernatant was removed, and the pellet was washed twice with PBS. The number of CFUs was determined, and the metabolic activity was evaluated by the MTT assay, or by biomass formation using crystal violet staining. For the MTT (3-(4,5-dimethyl-2-thiazyl)-2,5-diphenyl-2H-tetrazolium bromide) assay, a concentration of 5 mg/mL was added, and the samples were read in a microplate reader at 570 nm. For crystal violet staining, 0.2 mg/mL of the dye was added, and the samples were read at 595 nm, as described by Zago et al. [41].

For the quantification of CFU, dilutions of 10^{-2} to 10^{-4} were seeded on SDA plates with chloramphenicol by the microdroplet technique, and the plates were incubated for 24–48 h at 37 °C. For CFU quantification, the formula described in item 2.5.6 was applied. The assay was carried out in triplicate in two different experiments.

The effect of the extract on the mature biofilm was analyzed following the same procedure as described above, and the microplates were incubated for 48 h at 37 °C. The culture medium was changed after 24 h. The supernatant was aspirated, and the biofilm was washed twice with PBS. Concentrations of EHVG (200 μL) corresponding to 2×- and 4× the MIC were used. The plates were incubated at 37 °C for an additional 24 h. After this period, the plates were washed twice with 100 μL PBS, and the number of CFUs, the metabolic activity (MTT assay), and the biomass (crystal violet staining) were analyzed [43] (Seneviratne et al., 2016). For CFU quantification, dilutions of 10^{-2} to 10^{-4} were seeded on SDA plates with chloramphenicol by the microdroplet technique, and the plates were incubated for 24–48 h at 37 °C. The number of CFUs was determined by applying the formula proposed previously [41]. In all experiments, biofilms without extract were used as negative controls, and biofilms with Amphotericin B and Fluconazole were used as positive controls.

4.6. Determination of EHVG Cytotoxicity by the MTT Assay

The cytotoxicity of the extracts against murine macrophages (RAW 264.7; APABCAM, Rio de Janeiro, Brazil) was evaluated. The cells were maintained in cell culture flasks (TPP) in RPMI medium supplemented with 10% fetal bovine serum (Invitrogen, New York, NY, USA) and 1% penicillin-streptomycin (Gibco, Grand Island, NE, USA), and then incubated for 1 h at 37 °C in the presence of 5% CO_2. After reaching sub-confluence, the cells were detached with a cell scraper (TPP) and centrifuged at 2000 rpm for 5 min.

The supernatant was discarded, and the cells were, again, resuspended in RPMI medium and transferred to microtiter plates (100 μL, 1 × 10^6/well). The plates were incubated for 4 h in the presence of 5% CO_2 for adherence of the cells. After this period, increasing concentrations of EHVG (0.5 to 50 mg/mL) were added, and the plates were incubated for 24 h at 37 °C in the presence of 5% CO_2. As growth controls, the cells were incubated with RPMI without extract and 1% DMSO as vehicle.

After incubation, the supernatant was removed, and adherent cells were evaluated by the addition of 10 μL MTT solution (5 mg/mL) (Sigma, St. Louis, MO, USA) to the wells in the dark. The plates were incubated for 4 h at 37 °C. The supernatant was removed, and formazan was extracted by the addition of 100 μL sodium dodecyl sulfate (SDS) and incubation overnight. Absorbance was read at 570 nm after 24 h of incubation.

4.7. The Hemolytic Assay

Defibrinated sheep blood (EB FARMA, Rio de Janeiro, Brazil) was used in this study. Red blood cells were isolated by centrifugation at 290× *g* (5810R Centrifuge, Eppendorf) for 10 min at 4 °C. After the removal of plasma, the red blood cells were washed three times with PBS (pH 7.4) and immediately resuspended in 2% (*v*/*v*) of the same buffer. To evaluate the hemolytic activity of the extracts and fractions, 100-μL aliquots of the red blood cell solution were added to flat-bottom 96-well microplates (Kasvi, Italy), together with serial dilutions of the extract and fractions (0.05 to 50 mg/mL). Total hemolysis was achieved with 1% Triton X-100 (Sigma-Aldrich), and PBS was used as a negative control. DMSO (1%, vehicle) was also included as a control. After incubation for 60 min at room temperature, the cells were centrifuged at 300× *g* for 10 min, and absorbance was measured in the supernatant at 540 nm [11]. Relative hemolytic activity was expressed in relation to Triton X-100, and was calculated using the following formula: relative hemolytic activity (%) = [(As − Ab) × 100]/(Ac − Ab), where Ac is the absorbance of the control (blank, without extract), As is the absorbance in the presence of the extract, and Ab is the absorbance in the presence of Triton X-100. The assays were carried out in quadruplicate on three different experiments.

4.8. Evaluation of Cell Viability by the Neutral Red Assay

Neutral red (Vetec/Sigma Aldrich, São Paulo, Brazil) was prepared in PBS (20 μg/mL), and 100 μL/well of this solution was placed in the wells that contained RAW cells (100 μL/1 × 10^6) treated with different concentrations of EHVG, starting at 50 mg/mL (25, 12.5, 6.25, 3.125, 1.76, 0.78, and 0.39 mg/mL). After 2 h of incubation while protected from light, the plates were centrifuged, and the supernatant was discarded. Next, ethanol P.A. (100 μL) was added to the wells. The plates were shaken on an orbital shaker for 15 min. Absorbance was then determined at 570 nm, and the values obtained were converted into percentage of cell viability [44].

4.9. In Silico Analysis

4.9.1. Ligands and Target Preparations

The GaussView 5.0.8 program was used to illustrate the three-dimensional (3D) structure of the compounds Vismione D, anthraquinone, kaempferol, and catechin, identified by HPLC-UV/Vis and FIA-ESI-IT-MS in the crude extract of *V. guianensis* leaves. The geometrical and vibrational properties were calculated in vacuum by the density functional theory combined with the 6-311++G** (d, p) basis set, using Gaussian 09. The 3D structure of 14-alpha-demethylase from *C. albicans* (CaCYP51) was obtained from the Protein Data Bank (PDB) (# 5FSA), then resolved by X-ray crystallography with a resolution of 2.86 Å.

4.9.2. Molecular Docking

The structures of CaCPY51 and of the ligands were prepared for the molecular docking calculations using the AutoDock Tools (ADT), version 1.5.6. The structure of CaCYP51 was considered rigid, while each ligand was considered flexible. Gasteiger partial charges were calculated after the addition of all hydrogens. Next, the apolar hydrogens of CaCYP51 and of the ligands were subsequently fused. A cubic box of 120 × 120 × 120 points, with spacing of 0.35 Å, was generated throughout the portion that corresponded to the active site of the macromolecule.

The grid box was centered on the heme group of CaCPY51. For molecular docking, a global search was performed using Lamarckian Genetic Algorithms (AGL), and a local search (LS) using pseudo-Solis and Wets. Each ligand was submitted to 100 independent simulations of ligand binding, and the remaining fit parameters were set to default values. The initial coordinates of the interactions between CaCYP51 and the compounds present in EHVG were chosen using the criterion of clustering low-energy conformations, combined with visual inspection [37]. Molecular docking was performed using AutoDock 4.2.

4.9.3. Pharmacokinetics and Toxicity Measurement

The SwissADME online method was used to assess the pharmacokinetic properties (ADME) of the compounds. Lipinski's five rule was determined based on the following parameters: molecular weight not more than 500 Dalton; H-bond donors ≤ 5; H-bond acceptors ≤ 10; molar refractivity ranging from 40 to 130; and lipophilicity positive drug-like properties of any compound [45]. In addition, the online tool admetSAR (http://lmmd.ecust.edu.cn/admetsar2, accessed on 28 September 2022) was used to calculate the toxicological properties of all the compounds.

4.9.4. Pass Prediction

The PASS online tool (http://www.pharmaexpert.ru/passonline/predict.php, accessed on 28 September 2022) was used to determine the other potential biological activities of the selected compounds. The potential biological activity considered the Pa and Pi values to range from 0.000 to 1.000 [46,47], and Pa values higher than the Pi values were considered to have the following characteristics: Pa < 0.7 suggests high drug activities; 0.5 < Pa < 0.7 shows moderate therapeutic potentials; and Pa < 0.5 shows poor pharmaceutical activity [48,49].

4.10. Statistical Analysis

The results were expressed as the mean ± standard deviation. One-way analysis of variance (ANOVA) was used, followed by Dunn's post-test for multiple comparisons and the Student *t*-test for comparison between two groups. All analyses were performed using the GraphPad Prism 8.0 software. The significance adopted was 5% ($p < 0.05$).

5. Conclusions

EHVG has important anti-*Candida* activity, being able to attenuate the growth and virulence factors of *Candida* ATCC strains and clinical isolates of interest. This anti-*Candida* effect may be due to the presence of vismione D, which may act synergistically to other compounds. This inhibits the Candida CaCYP51 enzyme, and, therefore, impairs the biosynthesis of ergosterol.

Overall, the results suggest that EHVG is a promising anti-*Candida* alternative that can be used as a reference to generate a lead compound for the purpose of antifungal drug discovery.

Supplementary Materials: The following supporting information can be downloaded at: https://www.mdpi.com/article/10.3390/antibiotics11121834/s1, Figure S1: Effect of different concentrations of the hydroalcoholic leaf extract of *Vismia guianensis* (EHVG) on RAW 264.7 cells viability tested by neutral red assay.

Author Contributions: E.P.M.: anti-*Candida* assays, writing—original draft; J.R.F. and A.A.C.d.C.: anti-*Candida* assays; A.J.O.L. and M.d.S.S.C.: molecular docking evaluation, writing—original draft, review and editing; C.Q.d.R.: methodology, chemical investigation; C.A.M.: anti-*Candida* assays, review and editing; F.R.F.N., A.G.A., E.S.F., A.F.d.S. and R.N.: writing —original draft, review and editing; E.S.F.: writing—original draft, review and editing; R.N.M.G.: project administration and coordination; funding acquisition, data interpretation, writing—original draft, review and editing. All authors have read and agreed to the published version of the manuscript.

Funding: This research was funded by FAPEMA (Fundação de Amparo à Pesquisa do Maranhão)—Project NATUMED proc. No IECT-02885/17 as part of State Institute of Research in Biotechnology; CNPq (Conselho Nacional de Desenvolvimento e Tecnológico) and CAPES (Financial code 01).

Institutional Review Board Statement: The study is registered with the National System for the Management of Genetic Heritage and Associated Traditional Knowledge (SisGen) (Registration No. AFE7A08) as regulated by Brazilian Law No. 13.123/15.

Informed Consent Statement: Not applicable.

Data Availability Statement: On demand.

Acknowledgments: We thank FAPEMA (Fundação de Amparo à Pesquisa do Maranhão); CNPq (Conselho Nacional de Desenvolvimento e Tecnológico); Laboratório de Imunofisiologia (LIF), Departamento de Ciências Biológicas e da Saúde, Universidade CEUMA, Laboratório de Produtos Naturais, Departamento de Química, UFMA, Laboratório de Química de Produtos Naturais (LQPN).

Conflicts of Interest: The authors declare no conflict of interest.

Abbreviations

UFMA: Universidade Federal do Maranhão; EHVG: hydroethanolic extract of leaves from *Vismia guianensis*; HPLC-PDA: High Performance Liquid Chromatography—photo diode array; FIA-ESI-IT-MS^n: direct flow injection analysis and mass spectroscopy; ATCC: American-type colony collection; CaCYP51: Candida albicans cytochrome P51; PBS: phosphate-buffered saline; Ampho B: Amphotericin B; Flu, Fluconazole; Pa: potential activity; Pi: potential inactivity; AGL: Lamarckian Genetic Algorithms; LS: local search.

References

1. Santos, G.C.O.; Vasconcelos, C.C.; Lopes, A.J.O.; Cartagenes, M.S.S.; Dualibe Filho, A.K.; Nascimento, F.R.F.; Ramos, R.M.; Pires, E.R.R.B.; Andrade, M.S.; Rocha, F.M.G.; et al. *Candida* infections and therapeutic strategies: Mechanisms of action for traditional and alternative agents. *Front Microbiol.* **2018**, *9*, 1351. [CrossRef] [PubMed]
2. Seagle, E.E.; Willinam, S.L.; Chiller, T.M. Recent Trends in the Epidemiology of Fungal Infections. *Infect. Dis. Clin. N. Am.* **2021**, *35*, 237–260. [CrossRef] [PubMed]
3. Santana, D.P.; Ribeiro, E.L.; Menezes, A.C.S.; Naves, P.L.F. Novas abordagens sobre os fatores de virulência de *Candida albicans*. *J. Medl. Biol. Sci.* **2013**, *12*, 229–233.
4. Lockhart, S. Current epidemiology of *Candida* infection. *Clin. Microbiol. News* **2014**, *36*, 131–136. [CrossRef]
5. Lohse, M.B.; Gulati, M.; Johnson, A.D.; Nobile, C.J. Development and regulation of single- and multi-species *Candida albicans* biofilms. *Nat. Rev. Microbiol.* **2018**, *16*, 19–31. [CrossRef]
6. Sardi, J.C.O.; Scorzoni, L.; Bernardi, A.M.; Fusco-Almeida, A.M.; Mendes Giannini, M.J.S. *Candida* species: Current epidemiology, pathogenicity, biofilm formation, natural antifungal products and new therapeutic options. *J. Med. Microbiol.* **2013**, *62*, 10–24. [CrossRef]
7. Roscetto, E.; Contursi, P.; Vollaro, A.; Fusco, S.; Notomista, E.; Catania, M.R. Antifungical and anti-biofilm activity of the first cryptic antimicrobial peptide from an archaeal protein against *Candida* spp. clinical isolates. *Sci. Rep.* **2018**, *8*, 17570. [CrossRef]
8. Colombo, A.L.; Garnica, M.; Camargo, L.F.A.; Cunha, C.A.; Bandeira, A.C.; Borghi, D.; Campos, T.; Senna, A.L.; Didier, M.E.V.; Dias, V.C.; et al. *Candida glabrata*: An emerging pathogen in Brazilian tertiary care hospitals. *Med. Mycol.* **2013**, *51*, 38–44. [CrossRef]
9. Balunas, M.J.; Kinghorn, A.D. Drug discovery from medicinal plants. *Life Sci.* **2005**, *78*, 431. [CrossRef]
10. Brandão, H.N.; David, J.P.; Couto, R.D.; Nascimento, J.A.P.; David, J.M. Química e farmacologia de quimioterápicos antineoplásicos derivados de plantas. *Quim. Nova* **2010**, *33*, 1359–1369. [CrossRef]
11. Silva, A.R.; Andrade Neto, J.B.; Silva, C.R.; Campos, R.D.S.; Costa Silva, R.A.; Freitas, D.D.; Nascimento, F.B.; Andrade, L.N.D.; Sampaio, L.S.; Grangeiro, T.B.; et al. Berberine antifungal activity in Fluconazole-resistant pathogenic yeasts: Action mechanism evaluated by flow cytometry and biofilm growth inhibition in *Candida* spp. *Antimicrob. Agents Chemother.* **2016**, *60*, 3551–3557. [CrossRef] [PubMed]
12. Camelo, S.R.P.; Costa, R.S.; Ribeiro-Costa, R.M.; Barbosa, W.L.R.; Vasconcelos, F.; Vieira, J.M.S.; Silva Junior, J.O.C. Phytochemical evaluation and antimicrobial activity of ethanolic extract of *Vismia guianensis* (Aubl.) Choisy. *Int. J. Pharm. Sci. Res.* **2011**, *2*, 3224–3229.
13. Gonzales, J.G.; delle Monache, F.; delle Monache, G.; Marini-Bettolo, G.B. Chemistry of the genus Vismia. Part VII.Vismione A from the leaves of *Vismia guianensis*. *Planta Medica* **1980**, *40*, 347–350.
14. Guerra, R.N.M. Immunosuppressive Activity of *Vismia reichardtiana* Fruits. Ph.D. Thesis, Universidade de São Paulo, São Paulo, Brazil, 1997.
15. Politi, M.; Sanogo, R.; Ndjoko, K.; Guilet, D.; Wolfender, J.L.; Hostettmann, K.; Morelli, I. HPLC-UV/PAD; HPLC-MS Analyses of leaf and root extracts of *Vismia guianensis* and isolation and identification of two new bianthrones. *Phytochem. Anal.* **2004**, *15*, 355–364. [CrossRef] [PubMed]
16. Tala, M.F.; Jeanne, E.; Lantovololona, R.; Talontsi, F.M.; Wabo, H.K.; Tane, P.; Laatsch, H. Anthraquinones and triterpenoids from seeds of *Vismia guianensis*. *Biochem. Syst. Ecol.* **2013**, *50*, 310–312. [CrossRef]
17. Álvarez, E.R.; Gil, J.H.G.; Jímenez, O.J.G.; Posada, C.M.A.; Rojano, B.A.; García, C.M.P.; Durango, D.L.R. Actividad antioxidante y contenido fenólico de los extractos provenientes de las bayas de especies del género *Vismia* (Guttiferae). *Vitae* **2008**, *15*, 165–172.
18. Lins, A.; Agra, M.; Conceição, D.; Pinto, F.; Camara, C.; Silva, T. Chemical Constituents and Antioxidant Activity from Aerial Parts of *Clusia paralicola* and *Vismia guianensis*. *Rev. Virtual Quím.* **2016**, *8*, 157–168. [CrossRef]

19. Zaiter, A.; Becker, L.; Karam, M.C.; Dicko, A. Effect of particle size on antioxidant activity and catechin content of green tea powders. *J. Food Sci. Technol.* **2016**, *53*, 2025–2032. [CrossRef]
20. Seo, E.K.; Mukherjee, R.; Wani, M.C.; Wall, M.E.; Navarro, H.; Farnsworth, N.R.; Kinghorn, A.D. New bioactive aromatic compounds from *Vismia guianensis*. *Phytochemistry* **2000**, *55*, 35–42. [CrossRef]
21. Hussain, H.; Hussain, J.; Al-Harrasi, A.; Saleem, M.; Green, I.R.; van Ree, T.; Ghulam, A. Chemistry and biology of genus *Vismia*. *Pharm. Biol.* **2012**, *11*, 1448–1462. [CrossRef]
22. Costa, M.D.C.; Silva, A.G.D.; Silva, A.P.S.D.; Lima, V.L.M.; Bezerra Silva, P.C.; Rocha, S.K.L.D.; Navarro, D.M.D.; Correia, M.T.D.S.; Napoleão, T.H.; Silva, M.V.D.; et al. Essential oils from leaves of medicinal plants of Brazilian flora: Chemical composition and activity against *Candida* Species. *Medicines* **2017**, *4*, 27. [CrossRef] [PubMed]
23. Sachikonye, M.; Mukanganyama, S. Antifungal and drug efflux inhibitory activity of selected flavonoids against *Candida albicans* and *Candida krusei*. *J. Biol. Act. Prod. Nat.* **2016**, *6*, 223–236.
24. Oliveira, A.H.; Oliveira, G.G.; Carnevale Neto, D.F.; Portuondo, A.; Batista-Duharte, A.; Carlos, I.Z. Anti-inflammatory activity of *Vismia guianensis* (Aubl.) Pers. extracts and antifungal activity against *Sporothrix schenckii*. *J. Ethnopharmacol.* **2017**, *195*, 266–274. [CrossRef] [PubMed]
25. Reis, A.J.; Mata-Santos, T.; Carrion, L.L.; Rodrigues, K.; Fenalti, J.M.; Mesquita, D.W.O.; Scaini, C.J.; Martins, D.; Mesquita, A.S.S.; Silva, P.E.A.; et al. Evaluation of antifungal, antimycobacterial and larvicide activity of the *Duroia macrophylla* and *D. saccifera*. *J. Epidemiol. Infect. Control* **2016**, *6*, 108–124. [CrossRef]
26. Seleem, D.; Benso, B.; Noguti, J.; Pardi, V.; Murata, R.M. In vitro and in vivo antifungical activity of Lichochalcone-A against *Candida albicans* biofilms. *PLoS ONE* **2016**, *11*, e0157188. [CrossRef] [PubMed]
27. Hargrove, T.Y.; Friggeri, L.; Wawrzak, Z.; QI, A.; Hoekstra, W.J.; Schotzinger, R.J.; York, J.D.; Guengerich, F.; Lepesheva, G.I. Structural analyses of *Candida albicans* sterol 14α-demethylase complexed with azole drugs address the molecular basis of azole-mediated inhibition of fungal sterol biosynthesis. *J. Biol. Chem.* **2017**, *292*, 6728–6743. [CrossRef] [PubMed]
28. Guimarães, A.G.; Scotti, L.; Scotti, M.T.; Mendonça, F.J.B.; Melo, N.S.R.; Alves, R.S.; De Gomez-García, O.; Andrade-Pavón, D.; Campos-Aldrete, E.; Ballinas-Indilí, R.; et al. Synthesis, molecular docking, and antimycotic evaluation of some 3-acyl imidazo[1,2-a] pyrimidines. *Molecules* **2018**, *23*, 599.
29. Oliveira-Santos, G.C. Efeito de *Terminalia catappa* L. em leveduras de *Candida*: Avaliação *In Sílico*, In Vitro e *In Vivo*. Ph.D. Thesis, Rede Nordeste de Biotecnologia (Renorbio) Universidade Federal do Maranhão, Sao Luís, Brazil, 2018.
30. Ahmad, A.; Khan, A.; Manzoor, N.; Khan, L.A. Evolution of ergosterol biosynthesis inhibitors as fungicidal against *Candida*. *Microb. Pathoge* **2010**, *48*, 35–41. [CrossRef]
31. Cheng, F.X.; Li, W.H.; Liu, G.X.; Tang, Y. In silico ADMET prediction: Recent advances, current challenges and future trends. *Curr. Top. Med. Chem.* **2013**, *13*, 1273–1289. [CrossRef]
32. Cheng, Y.; Li, H.L.; Zhou, Z.W.; Long, H.Z.; Luo, H.Y.; Wen, D.D.; Cheng, L.; Gao, L.C. Isoliensinine: A Natural Compound with "Drug-Like" Potential. *Front. Pharmacol.* **2021**, *12*, 630385. [CrossRef]
33. Poroikov, V.; Filimonov, D. PASS: Prediction of biological activity spectra for substances predictive toxicology. In *Predictive Toxicology*, 1st ed.; Helma, C., Ed.; CRC Press: Boca Raton, FL, USA, 2005. [CrossRef]
34. ANVISA—Farmacopeia Brasileira, Vol 2—Agência Nacional de Vigilância Sanitária. Brasília: 1. Substâncias Farmacêuticas Químicas, Vegetais e Biológicas. 2. Medicamentos e Correlatos. 3. Especificações e Métodos de Análise. 2010; 546p, 1v/il. Available online: https://www.gov.br/anvisa/pt-br/assuntos/farmacopeia/farmacopeia-brasileira/6a-edicao-volume-2 (accessed on 8 August 2022).
35. Medeiros, A.C.; Almeida, E.; Quintino-da-Rocha, C.; Tangerina, M.; Lima-Neto, J.S.; Silva, A.; Rocha, C.; Martins, L. Antiparasitic activities of hydroethanolic extracts of *Ipomoea imperati* (Vahl) Griseb. (Convolvulaceae). *PLoS ONE* **2019**, *14*, e0211372.
36. Silva, A.F.; Rocha, C.Q.; Silva, L.C.N.; Carvalho-Júnior, A.R.; Mendes, I.N.F.V.; Araruna, A.B.; Motta, E.P.; Silva, R.S.; Campos, C.D.L.; Farias, J.R.; et al. Antifungal and anti-virulence activities of hydroalcoholic extract and fractions of *Platonia insignis* leaves against vaginal isolates of *Candida* species. *Pathogens* **2020**, *9*, 84. [CrossRef] [PubMed]
37. *CLSI—Clinical and Laboratory Standards Institute—Method for Broth Dilution Antifungal Susceptibility Testing of Yeasts: Approved M27-A3*; CLSI: Wayne, PA, USA, 2012.
38. Ostrosky, E.A.; Mizumoto, M.K.; Lima, M.E.L.; Kaneko, T.M.; Nishikawa, S.O.; Freitas, B.R. Methods for evaluation of the antimicrobial activity and determination of Minimum Inhibitory Concentration (MIC) of plant extracts. *Braz. J. Pharmacogn.* **2008**, *18*, 301–307. [CrossRef]
39. Sidiqui, Z.N.; Farooq, F.; Musthafa, T.N.M.; Ahmad, A.; Khan, A.U. Synthesis, characterization and antimicrobial evaluation of novel halopyrazole derivatives. *J. Saudi. Chem. Soc.* **2013**, *17*, 237–243. [CrossRef]
40. Zago, C.E.; Silva, S.; Sanitá, P.V.; Barbugli, P.A.; Dias, C.M.I.; Lordello, V.B.; Vergani, C.E. Dynamics of biofilm formation and the interaction between *Candida albicans* and methicillin-susceptible (MSSA) and -resistant *Staphylococcus aureus* (MRSA). *PLoS ONE* **2015**, *13*, e0123206. [CrossRef]
41. Gulati, M.; Lohse, M.B.; Ennis, C.L.; Gonzalez, R.E.; Perry, A.M.; Bapat, P.; Valle Arevalo, A.; Rodriguez, D.L.; Nobile, C.J. In vitro culturing and screening of *Candida albicans* biofilms. *Curr. Protoc. Microbiol.* **2018**, *50*, e60. [CrossRef]
42. Seneviratne, C.J.; Rajan, S.; Wong, S.W.; Tsang, D.N.C.; Lai, C.K.C.; Samaranayake, L.P.; Jin, L. Antinfungal susceptibility in serum and virulence determinants of *Candida* isolates from Hong Kong. *Front. Microbiol.* **2016**, *7*, 216. [CrossRef]

43. Seneviratne, C.J.; Silva, W.J.; Jin, L.J.; Samaranayake, Y.H.; Samaranayake, L.P. Architectural analysis, viability assessment and growth kinetics of *Candida albicans* and *Candida glabrata* biofilms. *Arch. Oral Biol.* **2009**, *54*, 1052–1060. [CrossRef]
44. Lipinski, C.A.; Lomabardo, F.; Dominy, B.W.; Feeney, P.J. Experimental and computational approaches to estimate solubility and permeability in drug discovery and development settings. *Adv. Drug Deliv. Rev.* **2012**, *64*, 4–17. [CrossRef]
45. Sliwoski, G.; Kothiwale, S.; Meiler, J.; Lowe, E.W. Computational methods in drug discovery. *Pharmocol. Rev.* **2014**, *66*, 334–395. [CrossRef]
46. Mojumdar, M.; Paul, A.; Kabir, M.S.H.; Rahman, G.; Zohora, F.T.; Hasan, M.S.; Ahmed, T.; Rahman, M.R.; Akter, Y.; Rahman, M. Molecular docking and PASS prediction for analgesic activity of some isolated compounds from acalypha indica l and ADME/T property analysis of the compounds. *World J. Pharm.* **2016**, *5*, 1761–1770.
47. Goel, R.K.; Singh, D.; Lagunin, A.; Poroikov, V. PASS-assisted exploration of new therapeutic potential of natural products. *Med. Chem. Res.* **2011**, *20*, 1509–1514. [CrossRef]
48. Khurana, N.; Ishar, M.P.S.; Gajbhiye, A.; Goel, R.K. PASS assisted prediction and pharmacological evaluation of novel nicotinic analogs for nootropic activity in mice. *Eur. J. Pharmacol.* **2011**, *662*, 22–30. [CrossRef] [PubMed]
49. Rudik, A.V.; Dmitriev, A.V.; Lagunin, A.A.; Filimonov, D.A.; Poroikov, V.V. PASS-based prediction of metabolites detection in biological systems. *SAR QSAR Environ. Res.* **2019**, *10*, 751–758. [CrossRef] [PubMed]

antibiotics

Article

Effects of *Coleus amboinicus* L. Essential Oil and Ethanolic Extracts on Planktonic Cells and Biofilm Formation of *Microsporum canis* Isolated from Feline Dermatophytosis

Arpron Leesombun [1], Karnchanarat Thanapakdeechaikul [1], Jiraporn Suwannawiang [1], Pipada Mukto [1], Sivapong Sungpradit [1], Norasuthi Bangphoomi [1], Tanasak Changbunjong [1,2], Orathai Thongjuy [3], Thekhawet Weluwanarak [2] and Sookruetai Boonmasawai [1,*]

1 Department of Pre-Clinic and Applied Animal Science, Faculty of Veterinary Science, Mahidol University, Nakhon Pathom 73170, Thailand
2 The Monitoring and Surveillance Center for Zoonotic Diseases in Wildlife and Exotic Animals (MoZWE), Faculty of Veterinary Science, Mahidol University, Nakhon Pathom 73170, Thailand
3 The Center of Veterinary Diagnosis, Faculty of Veterinary Science, Mahidol University, Nakhon Pathom 73170, Thailand
* Correspondence: sookruetai.boo@mahidol.edu

Citation: Leesombun, A.; Thanapakdeechaikul, K.; Suwannawiang, J.; Mukto, P.; Sungpradit, S.; Bangphoomi, N.; Changbunjong, T.; Thongjuy, O.; Weluwanarak, T.; Boonmasawai, S. Effects of *Coleus amboinicus* L. Essential Oil and Ethanolic Extracts on Planktonic Cells and Biofilm Formation of *Microsporum canis* Isolated from Feline Dermatophytosis. *Antibiotics* **2022**, *11*, 1734. https://doi.org/10.3390/antibiotics11121734

Academic Editors: Valério Monteiro-Neto and Elizabeth S. Fernandes

Received: 27 October 2022
Accepted: 28 November 2022
Published: 1 December 2022

Abstract: *Microsporum canis* is an important zoonotic fungus that causes dermatophytosis in domestic animals and their owners. Domestic cats are the primary reservoir for *M. canis*. Antifungal drugs frequently produce adverse effects on the host animal, increasing the demand for novel alternative treatments derived from nature. We evaluated the antifungal activity of *Coleus amboinicus* essential oil (CEO) and ethanolic extracts (CEE) against *M. canis* in planktonic and biofilm growth. Twelve clinical isolates of *M. canis* were identified in feline dermatophyte samples. Using GC-MS, 18 compounds were identified in CEO, with carvacrol being the major constituent. HPLC analysis of CEE revealed that it contained rosmarinic acid, apigenin, and caffeic acid. The planktonic growth of all *M. canis* isolates was inhibited by *C. amboinicus* extracts. The minimum inhibitory concentration at which $\geq$50% of the isolates were inhibited (MIC_{50}) was 128 μg/mL (32–256 μg/mL) for both CEO and CEE. The MIC_{90} values of CEO and CEE were 128 and 256 μg/mL, respectively. CEO at MIC (128 μg/mL) and 2× MIC (256 μg/mL) significantly inhibited the biofilm formation of weak, moderate, and strong biofilm-producing *M. canis*. CEE at 2× MIC (256 μg/mL) significantly inhibited the biofilm formation of all isolates. Overall, *C. amboinicus* extracts inhibited planktonic growth and exhibited a significant antibiofilm effect against *M. canis*. Thus, *C. amboinicus* is a potential source of natural antifungal compounds.

Keywords: *Microsporum canis*; *Coleus amboinicus*; essential oil; ethanolic extracts; antifungal; biofilm formation

1. Introduction

Microsporum canis is an important zoophilic dermatophyte in domestic cats and dogs; however, it also causes dermatophytosis in humans. Human infections are acquired from domestic animals through direct contact with clinically or subclinically infected animals [1]. *M. canis* is a major species of keratinophilic and keratinolytic filamentous fungi that cause superficial fungal infections worldwide [2], particularly in Europe, the eastern Mediterranean, and South America [3–5]. *M. canis* infections in cats, dogs, and other domestic animals such as rabbits generally manifest as multifocal alopecia, scaling, and circular lesions [6,7]. Both stray and domesticated cats are considered common reservoirs for *M. canis* [8–10] and anyone in direct contact with these animals, including owners and veterinarians, is at risk of becoming infected [9]. Treatments include oral and topical preparations of antifungal agents such as amphotericin B, griseofulvin, terbinafine, itraconazole, flucytosine, and fluconazole, administered alone or in combination. These agents cause adverse reactions, such as nephrotoxicity, hepatotoxicity, and neurotoxicity and have been associated with treatment failure secondary to drug resistance [11,12]. One antifungal-resistant strain of *M. canis* is resistant to terbinafine and has been reported in some feline patients [13]. Factors mitigating drug resistance include a lack of patient compliance, poor tissue penetration, and variable medication bioavailability [11,14]. Biofilms produced by *M. canis* are an additional concern [15,16]. Biofilm refers to the matrix that surrounds the microbial population and acts as a physical barrier, protecting the microorganisms in it. Biofilms are typically associated with increased resistance to antifungal agents [15].

Plants, including their extracts, contain phytochemical compounds that are used as antimicrobials and antibiofilm agents [17]. Identifying novel antifungal targets and compounds derived from natural plants will help develop novel antifungal strategies and improve existing ones [15]. Several previous (and ongoing) studies have focused on the use of plant extracts as alternative treatments for fungal infections [12]. Plant extracts that have antifungal activity include lemon grass (*Cymbopogon citrates*), lantana (*Lantana camara*), nerium (*Nerium oleander*), basil (*Ocimum basilicum*), and olive leaves (*Olea europaea*) [18]. The ethanolic extracts of *Echinophora platyloba* inhibit *Candida albicans* ATCC 10231 [19]. The aqueous and ethanolic extracts and the essential oil of *Thymus capitatus* exhibit antifungal activity against *C. albicans* and *M. canis* [20]. Extracts of *Ocimum gratissimum* leaves display antifungal activity against *M. canis, Microsporum gypseum, Trichophyton rubrum,* and *Trichophyton mentagrophytes* [21]. Methanolic leaf extracts of *Eucalyptus camaldulensis* were investigated for suspected *in vitro* antifungal activities against *M. canis, M. gypseum, T. rubrum, Trichophyton schoenleinii, T. mentagrophytes,* and *Epedermophyton floccosum* [22].

Coleus amboinicus Lour. (synonym: *Plectranthus amboinicus* [Lour.] Spreng), whose common name is Indian borage, is widely cultivated in tropical Africa, Asia, and Australia. This perennial herb belonging to the family Lamiaceae contains several phytochemicals, including monoterpenoids, diterpenoids, triterpenoids, sesquiterpenoids, phenolics, flavonoids, and esters [23,24]. The geographic location of *C. amboinicus* cultivation influences the phytochemical composition of plant extracts [25]. The stem, leaf, and root extracts of *C. amboinicus* contain high concentrations of polyphenolics such as caffeic acid, rosmarinic acid, apigenin, chrysoeriol, 5-*O*-methyl-luteolin, and 5,8-dihydroxy-7,2′,3′,5′-tetramethoxyflavone, which is a novel flavonoid [26–28]. Several studies have reported the wide range of pharmacological properties of *C. amboinicus* extracts and essential oil, including antioxidant activity [25], antidandruff action [29], antiproliferative effects on cancer cells [30], analgesic and anti-inflammatory activities [31], antirheumatoid arthritis [32], anti-inflammatory effect following bone injury [33], antibacterial activity against methicillin-resistant *Staphylococcus aureus* [34], mosquitocidal and water sedimentation properties [35], and insect repellent [36] and insecticidal effects [37].

Although *C. amboinicus* exhibits antifungal activity against several fungal species [38], little is known of its antifungal effects on dermatophytes and the biofilms they produce. Therefore, we evaluated the antifungal effects of *C. amboinicus* essential oil (CEO) and *C. amboinicus* ethanolic extracts (CEE) against planktonic cells and biofilm formation of *M. canis* clinically isolated from a feline dermatophyte.

2. Results

2.1. Chemical Composition of CEO and CEE

The yield of essential oil obtained from fresh *C. amboinicus* leaves was 0.08% (v/w), and the oil was clear and light yellow. The refractive index, density (g/cm^3), and specific gravity of the oil at 20 °C were 1.505, 0.935, and 0.937, respectively. Figure 1 presents the chromatogram of the main components of CEO. Gas chromatography–mass spectrometry (GC-MS) analysis revealed 18 compounds, representing 99.84% of the total composition of CEO (Table 1). The mass spectrum of each compound is presented in Supplementary Figure S1. The concentrations of carvacrol, β-caryophyllene, and thymol in CEO were determined and found to be 3.4 ± 0.2, 0.35 ± 0.16, and 0.013 ± 0.08 mg/mL, respectively.

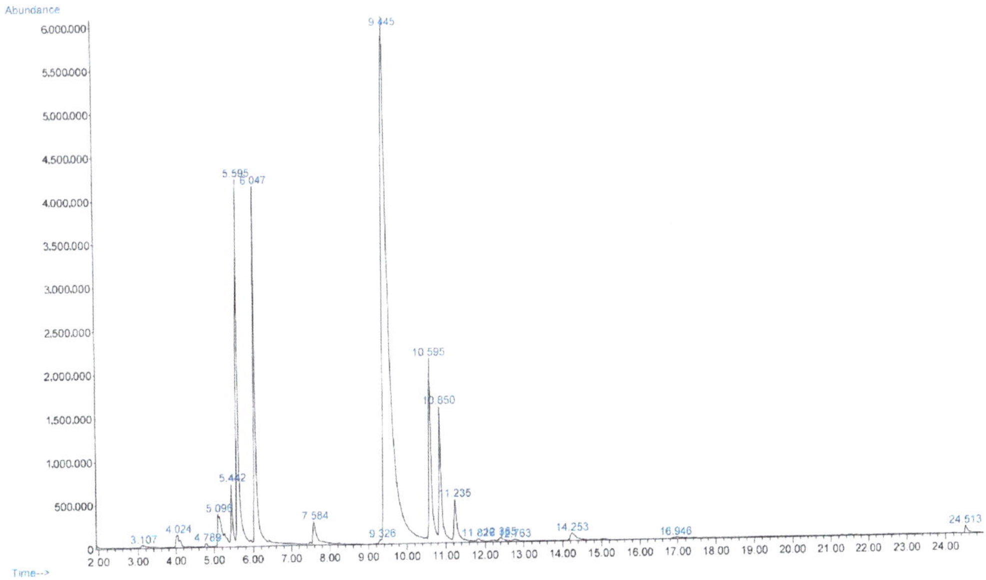

Figure 1. Chromatogram of the main components of *Coleus amboinicus* essential oil, as determined via gas chromatography–mass spectrometry.

Table 1. Chemical composition of *Coleus amboinicus* essential oil.

No.	Retention Time (min)	Classes	Compounds	Formula	Chemical Structure	Peak Area (%)	Similarity Index (%)
1	3.10	Aldehydes	Hexenal	$C_6H_{10}O$		0.17	94
2	4.02	Monoterpene	α-Thujene	$C_{10}H_{16}$		0.71	94
3	4.78	Monoterpene	Sabinene	$C_{10}H_{16}$		0.12	91
4	5.09	Monoterpene	β-Myrcene	$C_{10}H_{16}$	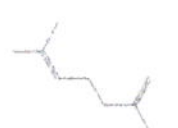	2.57	97
5	5.44	Monoterpene	α-Terpinene	$C_{10}H_{16}$		1.62	98
6	5.59	Monoterpene	p-Cymene	$C_{10}H_{14}$		10.89	97
7	6.04	Monoterpene	γ-Terpinene	$C_{10}H_{16}$		9.33	97

Table 1. *Cont.*

No.	Retention Time (min)	Classes	Compounds	Formula	Chemical Structure	Peak Area (%)	Similarity Index (%)
8	7.58	Monoterpene	4-Terpineol	$C_{10}H_{18}O$	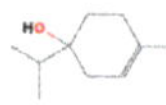	1.26	96
9	9.32	Monoterpene	Thymol	$C_{10}H_{14}O$	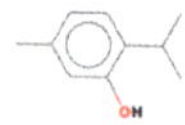	0.17	90
10	9.44	Monoterpene	Carvacrol	$C_{10}H_{14}O$	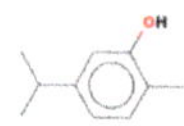	56.65	97
11	10.59	Sesquiterpene	β-Caryophyllene	$C_{15}H_{24}$	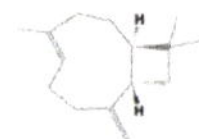	7.12	99
12	10.85	Sesquiterpene	α-Farnesene	$C_{15}H_{24}$		5.41	91
13	11.23	Sesquiterpene	α-Humulene	$C_{15}H_{24}$	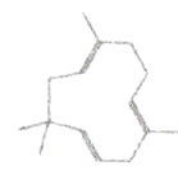	2.28	97
14	11.82	Sesquiterpene	β-Farnesene	$C_{15}H_{24}$		0.19	92
15	12.38	Sesquiterpene	β-Bisabolene	$C_{15}H_{24}$		0.26	95
16	12.76	Sesquiterpene	β-Sesquiphellandrene	$C_{15}H_{24}$	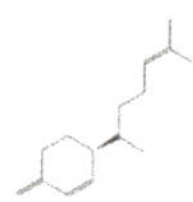	0.11	95
17	14.25	Sesquiterpene	(-)-Caryophyllene oxide	$C_{15}H_{24}O$	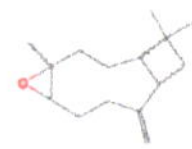	0.69	74
18	24.51	Phthalates	Bis(2-ethylhexyl) phthalate	$C_{24}H_{38}O_4$		0.29	91
Total						99.84	

The yield of CEE was 2.2% *w/w*, appearing as a dark green and highly viscous solid. The total phenolic content was 666.0 ± 9.1 mg gallic acid equivalent (GAE)/g sample and the total flavonoid content was 462.3 ± 3.1 mg quercetin equivalent (QE)/g sample. The three compound standards, namely rosmarinic acid, apigenin, and caffeic acid, were qualified via high-performance liquid chromatography ([HPLC]; Figure 2). The concentrations of rosmarinic acid, apigenin, and caffeic acid used for quantification were 1.251, 1.175, and 0.732 mg/g samples, respectively (Table 2).

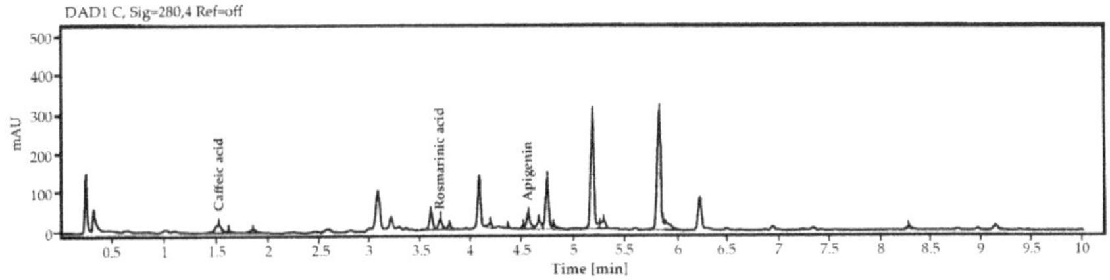

Figure 2. High-performance liquid chromatography of *Coleus amboinicus* ethanolic extracts.

Table 2. Retention time, peak area, and concentration of compounds identified in *Coleus amboinicus* ethanolic extracts using high-performance liquid chromatography.

No.	Retention Time (min)	Compounds	Peak Area	Concentration (mg/g Sample)
1	1.52	Caffeic acid	96.78	0.73
2	3.70	Rosmarinic acid	67.43	1.25
3	4.55	Apigenin	105.93	1.17

2.2. Fungal Isolation and Biofilm Formation

We identified 12 *M. canis* isolates based on morphology, polymerase chain reaction (PCR), and gene sequencing. A total of 720 base pair (bp) amplicons were obtained from positive *M. canis* samples using targeted ITS1-5.8S-ITS2 PCR. Bidirectional DNA sequencing was performed to demonstrate that all tested feline specimens were *M. canis* ITS1-5.8S-ITS2 sequences, which shared 100% sequence identity with those isolated from Thai and Belgium cats (sampled from hair and skin) and are deposited in relevant databases (MT487850 and OW988573). Additionally, the sequences identified in this study were perfectly matched with *M. canis* ITS1-5.8S-ITS2 sequences isolated from dog (ON527777 and KT155637), horse (LC623726 and OW984765), rabbit (OW987260 and OW987262), and human (OW988577) specimens (Supplementary Figure S2).

All 12 *M. canis* isolates were classified as weak (25%), moderate (50%), or strong biofilm producers (25%) based on their biofilm production. *C. albicans* and *T. rubrum* were classified as strong and moderate biofilm producers, respectively (Table 3). The biofilm production of *M. canis* is shown in Supplementary Figure S3.

Table 3. Biofilm classification and minimum inhibitory concentration (MIC) of CEO, CEE, and fluconazole against planktonic cells of fungal isolates.

No.	Fungi	Biofilm Classification	MIC (μg/mL)		
			Fluconazole	CEO	CEE
1	*Candida albicans*	strong	8	128	64
2	*Trichophyton rubrum*	moderate	4	32	256
3	W01	weak	8	32	128
4	W02	weak	8	128	128
5	W03	weak	16	128	128
6	M01	moderate	4	32	128
7	M02	moderate	4	256	128
8	M03	moderate	8	64	128
9	M04	moderate	8	128	128
10	M05	moderate	8	128	256
11	M06	moderate	16	128	32
12	S01	strong	4	64	256
13	S02	strong	4	64	256
14	S03	strong	8	128	64

CEO, *Coleus amboinicus* essential oil; CEE, *C. amboinicus* ethanolic extracts.

2.3. Effect of CEO and CEE on Planktonic Cells

CEO and CEE inhibited the planktonic cell growth of all *M. canis* isolates. The results of the minimum inhibitory concentration (MIC) analysis of CEO and CEE are presented in Tables 3 and 4. The minimal fungicidal concentration (MFC) and MIC values of each treatment were comparable. CEO exhibited antifungal activity comparable to that of CEE, with an MIC_{50} of 128 μg/mL (32–256 μg/mL). The MIC_{90} of CEE was twice that of CEO. Fluconazole had an MIC_{50} and MIC_{90} of 8 μg/mL (4–16 μg/mL) and 16 μg/mL, respectively (Table 4).

Table 4. MIC_{50}, MIC_{90}, GM of MIC, and MIC range of fluconazole, *Coleus amboinicus* essential oil (CEO), and *C. amboinicus* ethanolic extracts (CEE) against planktonic cells of *Microsporum canis*.

Treatment	Minimum Inhibitory Concentration (MIC) (μg/mL)			
	MIC_{50}	MIC_{90}	GM of MIC	MIC Range
Fluconazole	8	16	8	4–16
CEO	128	128	90.51	32–256
CEE	128	256	128	32–256

MIC_{50}, the minimum inhibitory concentration at which ≥50% of the isolates were inhibited; MIC_{90}, the minimum inhibitory concentration at which ≥90% of the isolates were inhibited; GM: geometric mean.

2.4. Effect of CEO and CEE on Biofilm Formation

CEO at MIC (128 μg/mL) and 2× MIC (256 μg/mL) significantly inhibited the biofilm formation of *C. albicans*, *T. rubrum*, and all weak, moderate, and strong biofilm-producing *M. canis* isolates (Figure 3). CEE at MIC (128 μg/mL) significantly inhibited the biofilm formation of *C. albicans*, *T. rubrum*, and all weak (100%), five moderate (83.3%), and two strong (66.7%) biofilm-producing *M. canis* isolates. CEE at 2× MIC (256 μg/mL) significantly inhibited the biofilm formation of all *M. canis* isolates (Figure 4). Fluconazole at MIC (8 μg/mL) and 2× MIC (16 μg/mL) did not affect the biofilm formation of *C. albicans* and *T. rubrum*.

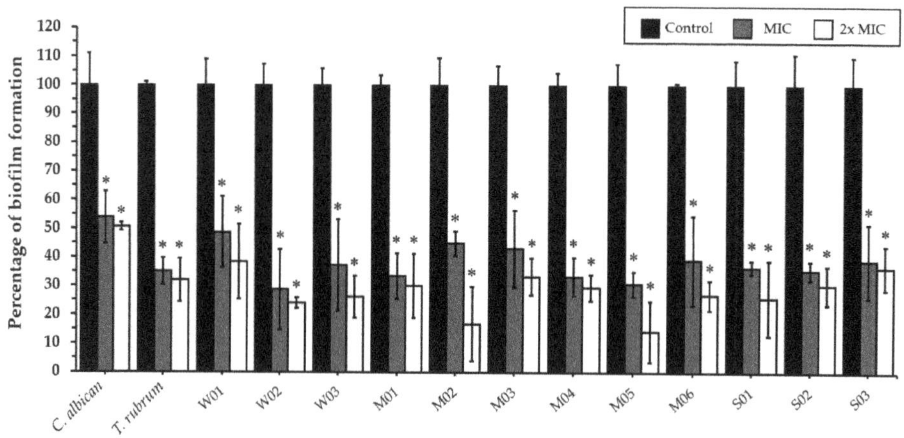

Figure 3. Effects of *Coleus amboinicus* essential oil on the biofilm formation of *Candida albicans* ATCC 90028, *Trichophyton rubrum*, and *Microsporum canis* after 96 h of treatment. Each bar represents the mean ± SD of three experiments per group. * differences between the control and treatments were statistically significant ($p < 0.05$).

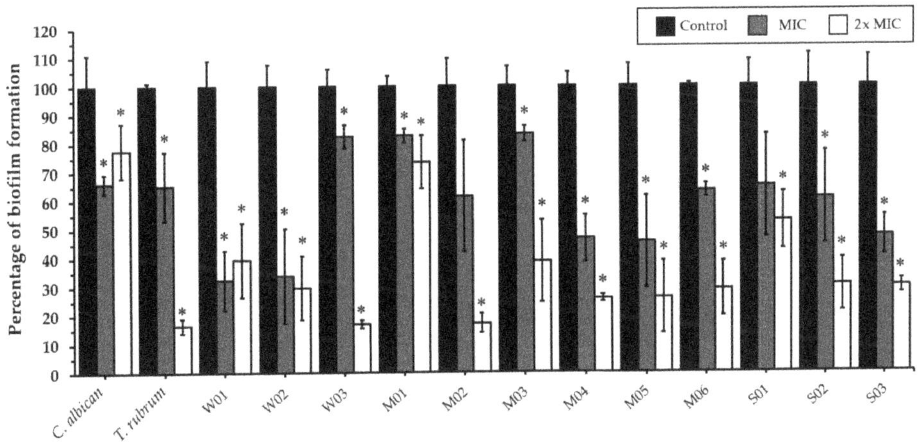

Figure 4. Effects of *Coleus amboinicus* ethanolic extracts on the biofilm formation of *Candida albicans* ATCC 90028, *Trichophyton rubrum*, and *Microsporum canis* after 96 h of treatment. Each bar represents the mean ± SD of three experiments per group. * differences between the control and treatments were statistically significant ($p < 0.05$).

3. Discussion

To the best of our knowledge, this is the first report of the antifungal activity of *C. amboinicus* extracts against planktonic cells and the biofilm formation of *M. canis*. *M. canis* biofilms are composed of a multidirectionally expanded network of hyphae linked together by a polysaccharide extracellular matrix [16]. Biofilm reduces the penetrability of antifungal agents, thus contributing to treatment failure and recurrent infection [39,40]. The inhibitory effect of antifungal agents on biofilm formation was observed at concentrations higher than those required to inhibit the growth of planktonic cells [41]. Fungal biofilm formation is a key factor in fungal virulence, persistence, and invasion as well as recurrent fungal infections and conventional antifungal resistance [40,42]. The time-dependent adherence of arthroconidia was observed, starting at 2 h and up to 6 h after inoculation. *M. canis* produced keratinolytic enzymes and secreted endo and exoproteases during adhesion; this process was likely inhibited by chymostatin, a serine protease inhibitor [43]. After biofilm formation for 72 h, a polysaccharide extracellular matrix that links fungal hyphae was observed [16,44]. The extracellular matrices of poor, moderate, and strong biofilm-producing *M. canis* appear to be related to mechanisms of antifungal resistance; however, further investigations are needed to confirm this. Flucytosine or fluconazole treatment at every 6–24 h could not completely destroy the biofilms of *Candida* spp. Poor drug penetration might not be a major mechanism of antifungal resistance for *Candida* biofilms [45]. During the early stage of *C. albicans* biofilm formation, genes encoding efflux pumps are upregulated, thereby mediating antifungal resistance [46]. Developing new compounds or alternative inhibitors to treat biofilm-related drug resistant fungal infections is essential to veterinary and human medicine [40,42,47]. In this study, fluconazole (4–16 μg/mL) had no effect on the mature biofilms of *M. canis* isolates. This result was similar to those reported by Bila et al., who found that fluconazole only inhibited the metabolic activity of early-stage biofilms of *T. mentagrophytes* at 32 mg/L but did not exhibit antibiofilm activity on mature biofilms, even at the highest concentration (512 mg/mL) [48].

Our study also demonstrated that *C. amboinicus* can inhibit planktonic cell growth and biofilm formation of feline zoonotic *M. canis*. CEO and CEE significantly inhibited the planktonic cell growth of *M. canis* at 128 μg/mL (32–256 μg/mL). Considering the MIC_{90} values, CEO was found to have a higher potency more potent than CEE against all *M. canis* isolates (MIC_{90} of 128 μg/mL vs. 256 μg/mL). *C. amboinicus* has been reported to exhibit antifungal activity against several fungi, including *Aspergillus clavatus*, *Aspergillus niger*, *Cladosporium cladosporioides*, *Chaetomium globosum*, *Myrothecium verrucaria*, *Penicillium citrinum*, *Trichoderma viride*, and *Mucor* sp. [38,49]. It also inhibits the biofilm formation of other pathogenic microorganisms, such as *Streptococcus mutans*, *Streptococcus pyogenes*, and *S. aureus* [50–52]. *C. amboinicus* is rich in monoterpenes, including carvacrol, thymol, eugenol, chavicol, and ethyl salicylate [37,49,53]. The 18 compounds identified in CEO in this study represent 99.84% of the total essential oil and include carvacrol (56.65%), p-cymene (10.89%), and γ-terpinene (9.33%); these three compounds alone comprise 76.87 % of the total essential oil. The concentrations of carvacrol, β-caryophyllene, and thymol were found to be 3.4 ± 0.2, 0.35 ± 0.16, and 0.013 ± 0.08 mg/mL, respectively. This result differs from that reported by da Costa et al., who found thymol to be the major constituent (64.3%), followed by p-cymene (10.3%), γ-terpinene (9.9%), and β-caryophyllene (2.8%) [54]. Previous studies have reported that the phytochemical composition of CEO is significantly influenced by the cultivation location, processes, and methods of essential oil extraction [25,37]; for example, steam distillation produced higher levels of carvacrol in *C. amboinicus* essential oil than those produced via the hydrodistillation method [37]. CEO at MIC had excellent effects against all clinical isolates. The high potency of CEO may be attributed to the hydrophobic property of essential oil, which adversely affects every step of biofilm formation, including adhesion, growth, maturation, and dissemination. The antibiofilm mechanisms of essential oil include reducing bacterial adhesions, preventing fresh biofilm formation, and destroying existing biofilm [55,56].

Carvacrol appears to be a major contributor to the antifungal properties of CEO. Carvacrol, p-cymene, and γ-terpinene are monoterpenes that exhibit various biological activities, including antioxidative, anti-inflammatory, anxiolytic, antineoplastic, and antimicrobial effects [57]. The antimicrobial effects of carvacrol are effective against various microorganisms, including bacteria such as *S. aureus* and *Pseudomonas aeruginosa* and fungi such as *C. albicans*, *Candida glabrata*, and *Candida parapsilosis* [58]. In a previous study, *Lavandula multifida* L. essential oil containing carvacrol as the main constituent was effective against dermatophytes (MIC: 0.16 μL/mL) and *Cryptococcus neoformans* (MIC: 0.32 μL/mL) [59]. Carvacrol also exhibits antifungal activity against *Aspergillus* spp. (MIC: 100 μg/mL) and *Cladosporium* spp. (MIC: 100 μg/mL) [60,61], possibly by binding to sterols in the fungi. The sterols residing on planktonic cell membranes are essential for their survival [62], and their hydrophilic properties allow carvacrol to penetrate the polysaccharide layer of their biofilm matrix, thereby destabilizing the biofilm [63]. A recent study reported that p-cymene exhibited no antifungal activity against *A. niger* (MIC: >300 μL/mL) and *Rhizopus oryzae* (MIC: >1024 μg/mL) [64], whereas γ-terpinene has been shown to exhibit antifungal activity against *Sporothrix schenckii* (MIC = 62.5–500 μg/mL) and *Sporothrix brasiliensis* (125–500 μg/mL) [65].

In the present study, CEE effectively eradicated the biofilm formation of weak, moderate, and strong biofilm producers at 2× MIC. Total phenolic and flavonoid contents were positively correlated with the antimicrobial activity of the plant extracts [66]. We found higher total phenolic and flavonoid levels in CEE than those reported in previous studies. For example, the ethanolic extract of *C. amboinicus* leaves obtained from Vietnam had a total phenolic and total flavonoid content of 26.84 ± 0.91 μg GAE/mg sample and 12.14 ± 0.42 μg QE/mg sample, respectively [67]. A methanolic extract of the *C. amboinicus* stem obtained from India had a total phenolic content of 49.91 mg GAE/g sample and total flavonoid content of 26.6 mg rutin equivalent/g sample [68]. Flavonoids inhibit nucleic acid biosynthesis and spore germination in plant pathogens [69,70]. High phenolic and flavonoid levels may thus be related to the significant antifungal effects of CEE. Importantly, CEE contained remarkable levels of rosmarinic acid (1.251 mg/g sample), apigenin (1.175 mg/g sample), and caffeic acid (0.732 mg/g sample) in this study. Rosmarinic and caffeic acid compounds have significant antifungal effects against *Fusarium oxysporum* [71]. The antifungal mechanism of rosmarinic acid is poorly understood but is believed to be related to the RTPase enzyme [72]. Apigenin at a concentration of 5 μg/mL exhibited antifungal activity against *C. albicans*, *C. parapsilosis*, *Malassezia furfur*, *T. rubrum*, and *T. beigelii* by inhibiting biofilm formation and efflux-mediated pumps of fungi. It also induced cell death by interfering with membrane function and increasing cell permeability [73,74]. Mice infected with *T. mentagrophytes* recovered after treatment with apigenin ointment administered at concentrations of 2.5 and 5 mg/g on the 12th and 16th days, respectively [75]. Caffeic acid phenethyl ester, a major active component of propolis (*Apis trigona*), has been shown to exert concentration-dependent effects on planktonic cells and biofilm formation of different *Candida* species [76] and synergistically enhance the antifungal activity of fluconazole against resistant clinical isolates of *C. albicans* [77]. Another study reported that the fungicidal activity of caffeic acid against *T. rubrum* was observed at 86.59 µM; this activity was mediated via plasma membrane damage and reduced ergosterol production, where caffeic acid reduced isocitrate lyase activity and downregulated critical genes (*ERG1*, *ERG6*, and *ERG11*) required for ergosterol synthesis [78].

Although CEO and CEE had different chemical constituents, both exhibited excellent and comparable inhibitory activities against all fungal isolates obtained from feline dermatophyte samples. Our findings suggest that both CEO and CEE act as natural antifungal agents against planktonic cells and biofilm-producing *M. canis*. Future investigations of the relationship between plant-based compounds, such as carvacrol in CEO and apigenin in CEE, their mechanisms of action, and classification based on biofilm production may contribute to a better understanding and guide the development of safe and effective antifungal agents derived from natural sources.

4. Materials and Methods

4.1. Plant Preparation and Extractions

C. amboinicus Lour. was harvested from a pesticide-free garden in the Nonthaburi Province, Thailand (13.862162, 100.409385). The plants were identified and housed at the herbarium within the Faculty of Pharmacy, Mahidol University, Thailand. The voucher specimen was PBM-005507-08. The hydrodistillation method was used to process the fresh leaves of *C. amboinicus* for essential oil extraction. The extraction was performed using a Clevenger-type apparatus operating at atmospheric pressure. The collected CEO was dried with anhydrous sodium sulfate, transferred to amber glass bottles, and stored at 40 °C. The physical properties of the CEO, including color, density, refractive index, and specific gravity, were evaluated and recorded. The yield of CEO was determined based on the weight of the fresh plant material before processing and was expressed in % (*v*/*w*) [37].

During the ethanolic extraction process, *C. amboinicus* leaves were dried in a hot air oven at 60 °C for 72 h and ground into small pieces. The leaf fragments were macerated in 95% ethanol at room temperature (RT) for 5 days. The extract solution was filtered through sterile gauze and a vaporized solvent using a rotary evaporator at 40 °C (BÜCHI, Flawil, Switzerland). The CEE was lyophilized in Labconco FreeZone 4.5 L Freeze Dryer equipped with Lyo-Works™ Operating System (Labconco, Kansas City, MO, USA) and stored at −20 °C. The yield of CEE was determined based on dry weight, weight after lyophilization, and weight of the leaf fragments before processing, and expressed in % (*w*/*w*).

4.2. GC-MS

The chemical composition of CEO was analyzed via GC-MS. The samples were diluted with methanol and injected in the split mode (1:10 split ratio) into the GC-MS model Agilent 7890A/5977B GC/MSD system equipped with a DB-5HT capillary column (0.1 µm film thickness × 0.25 mm diameter × 30 m length; Agilent Tech., Santa Clara, CA, USA) at a flow rate of 1 mL·min^{-1} in helium (carrier gas) and an injector temperature of 250 °C. The initial oven temperature was 40 °C (5 min), which was then increased to 250 °C at a rate of 10 °C/min and maintained there for 5 min. The following MS settings were used: ion source temperature, 230 °C; ionization energy, 70 eV; and mass scan range, 35–550 *m*/*z*. Compounds were identified by matching their mass spectra against those specified in Wiley Registry 7th Edition MS libraries. The concentrations of the major components were calculated by comparing the peak area of samples with the peak area of standard compounds.

4.3. Determination of Total Phenolic Content

The total phenolic content of CEE was determined using Folin–Ciocalteu's colorimetric assay, with slight modifications. The stock extract solution (1000 μg/mL) was mixed with 125 μL of Folin–Ciocalteu reagent (Merck, Darmstadt, Germany) in a 1:1 ratio for 5 min. Subsequently, 400 μL of 7.5% sodium carbonate was added to the mixture, followed by incubation at RT for 30 min. The absorbance of the final mixture was measured at 760 nm using the Synergy H1 Hybrid Multi-Mode Microplate Reader (BIOTEK, Winooski, VT, USA). Gallic acid was used to prepare the standard curve (with a 40–240 μg/mL calibration range). The gallic acid solutions and the results are expressed in GAE/g of the crude extracts [79].

4.4. Determination of Flavonoid Content

The modified aluminum chloride colorimetric method was used to determine the flavonoid content of the plant extracts [80]. A 250-μL aliquot of the extract solution (1000 μg/mL) was mixed with 1.25 mL of deionized water, after which a 5% sodium nitrite solution (75 μL) (Sigma Aldrich, St. Louis, MO, USA) was added and the mixture was allowed to stand for 5 min. Subsequently, 150 μL of 10% aluminum chloride (Sigma Aldrich, St. Louis, MO, USA) was added to the extract solution, followed by 500 μL of 1 M sodium hydroxide. The solution was further diluted with 275 μL of deionized water and allowed to stand for 6 min. Finally, the absorbance was measured at a wavelength of 510 nm using the Synergy H1 Hybrid Multi-Mode Microplate Reader (BIOTEK, Winooski, VT, USA). A quercetin solution (30–300 μg/mL) was used to prepare the standard calibration curves.

4.5. HPLC

CEE was analyzed using HPLC. The HPLC 1290 Infinity II system: Zorbrax Eclipse Plus C18 column (2.1 × 50 mm, 1.8-Micron; Becton, Dickinson and Company, Franklin Lakes, NJ, USA) with an ultraviolet (UV) detector (280 nm) was used at a gradient flow of 0.5 mL/min. The mobile phase composition was 3% acetic acid in water: 1% acetic acid in water acetonitrile. The injection volume was 2 μL. The column temperature was maintained at 30 °C. The stock extract solution (10 mg/mL) was dissolved in methanol and filtered through 0.45-μm nylon membrane filters before performing HPLC. The compounds present in the extracts were characterized according to their UV–vis spectra and identified in terms of their retention time relative to that of known standards: rosmarinic acid, apigenin, and caffeic acid (Sigma Aldrich, St. Louis, MO, USA) [25]. A standard graph was generated using standard solutions of 5–500 μg/mL. The regression equation correlating to the area under the peak (Y) and standard (X) was as follows: Y = 5.04 (X: rosmarinic acid) + 4.37, Y = 8.98(X: apigenin) + 0.31 and Y = 13.04 (X: caffeic acid) + 1.31.

4.6. Sample Collection and Fungal Identification

The study protocol was approved by the Faculty of Veterinary Science, Animal Care and Use Committee (MUVS-2019-09-45) and the Faculty of Veterinary Science-Institutional Biosafety Committee (IBC/MUVS-B-005-2562). Skin, nail, and hair specimens were randomly collected from cat patients with feline dermatophytosis during 2019–2020 at Prasu-Arthorn Animal Hospital, Faculty of Veterinary Science, Mahidol University, Thailand. The samples were placed on Difco™ Potato Dextrose Agar (PDA) (Becton, Dickinson and Company, NJ, USA) plates supplemented with 0.1% chloramphenicol. The *M. canis* isolates were screened based on the morphology of the colonies, including their size, texture, and color. The characteristics, size, and arrangement of microconidia and macroconidia were evaluated by lactophenol cotton blue staining and observed under a light microscope at 10× and 40× magnification [81].

4.7. Confirmation of M. canis Using Molecular Techniques

PCR was used to confirm the species of *M. canis*. Twelve fungal samples obtained from feline patients were grown on PDA at 27 °C for 7 days before DNA extraction using the

QIAamp DNA Micro Kit (Qiagen, Hilden, Germany). Samples stored at −80 °C were used to determine DNA concentration and purity using the Beckman Coulter DU 730 Uv/Vis Spectrophotometer (Beckman Coulter, Brea, CA, USA).

PCR was performed using the universal fungal primers of the ITS1-5.8S-ITS2 gene, i.e., ITS1: 5′-TCCGTAGGTGAACCTGCGG-3′ and ITS4: 5′-TCCTCCGCTTATTGATATG-3′ [82,83]. The reaction mixture (25 µL) contained 12.5 µL of GoTaq® Green Master Mix (Promega, Madison, Wisconsin, USA), 10 pmol of ITS1 and ITS4 primers, and 100 ng of DNA template and ultrapure water (Milli-Q; pH 6.5). PCR was performed under the following conditions: 94 °C for 2 min; 35 cycles of 94 °C for 30 s, 55 °C for 60 s, and 72 °C for 90 s; the final extension was performed at 72 °C for 10 min. Subsequently, 3 µL of PCR products was mixed with 3 µL of loading dye and 3 µL of GelStar™ Nucleic Acid Gel Stain (Lonza, Basel, Switzerland) and examined via 1.5% agarose gel electrophoresis and visualized via UV irradiation. Six (50% of each biofilm classification) of the amplified PCR products (720 bp) were cleaned using the ExoSAP-IT™ PCR Product Cleanup Reagent (Thermo Fisher, Waltham, MA, USA) according to the manufacturer's protocol. The PCR products were amplified by mixing them with a 5× sequencing buffer (Thermo Fisher, Waltham, MA, USA). The amplification was performed at 96 °C for 1 min, 96 °C for 10 s for 25 cycles, 50 °C for 5 s, 60 °C for 4 s, and finally cooling at 15 °C. The nucleotide sequences were determined using the NextSeq 550 sequencing system (Illumina Inc., San Diego, CA, USA) and compared to reference strains from the GenBank® database (National Center for Biotechnology Information, Bethesda, MD, USA).

4.8. Bioinformatic and Phylogenetic Analysis

All *M. canis* ITS1-5.8S-ITS2 sequences were compared with sequences available in the GenBank database using The Basic Local Alignment Search Tool (http://blast.ncbi.nlm.nih.gov/Blast.cgi: accessed on 18 August 2022). Multiple alignments of all nucleotide sequences were conducted using the ClustalW web-based tool (https://www.genome.jp/tools-bin/clustalw: accessed on 18 August 2022) [84]. The final dataset comprised 387 positions. Phylogenetic trees were reconstructed using maximum likelihood analysis with bootstrapping (100 replications) in the advanced mode of the phylogeny.fr web server (http://www.phylogene.fr/: accessed on 18 August 2022) [85]. Published sequences in the GenBank database originating from other global locations were used to compare all sequences. *S. schenckii* (MG976612), *T. rubrum* (MK027017), *C. albicans* (OW988269), and *C. neoformans* var. *neoformans* (KP068909) were included as the outgroup for the ITS1-5.8S-ITS2 phylogenetic tree. The *M. canis* ITS1-5.8S-ITS2 nucleotide sequence data obtained in this study are available in GenBank using the accession numbers OP227140–OP227145.

4.9. Biofilm Formation Classification

All feline dermatophyte samples were grown in Sabouraud dextrose agar (SDA) plates (Becton, Dickinson and Company, Franklin Lakes, NJ, USA) until conidia formation was identified. The fungal inoculum was prepared by covering the colonies with 5 mL of 0.85% sterile saline and gently scraping the colony's surfaces with a sterile cotton swab. The conidial suspension was transferred into sterile tubes, which were allowed to stand for 5–10 min to allow for sedimentation of the hyphae. The supernatant, which contained microconidia and macroconidia, was transferred into new sterile tubes. Turbidity was adjusted to 0.5 McFarland (approximately 2×10^6 CFU/mL), and the microconidia and macroconidia were counted using a hemocytometer [86].

The biofilm formation assay was performed as previously described [41]. Briefly, 200 µL of each fungal inocula was added to sterile 96-well polystyrene plates (Kartell S.p.A., Noviglio, MI, Italy) and incubated at 37 °C for 3 h to allow for adhesion. The wells were then washed with 200 µL of phosphate-buffered saline (PBS) at a pH of 7.0; then, 200 µL of RPMI 1640 (Sigma Aldrich, St. Louis, MO, USA) was added. The plates were incubated at 37 °C for 72 h without shaking. Each fungal isolate preparation was performed in triplicate. *C. albicans* ATCC 90028 and *T. rubrum* (reference strain from the Center of

Veterinary Diagnosis, Faculty of Veterinary Science, Mahidol University, Thailand) were used as the experimental controls.

Each biofilm biomass was quantified using crystal violet assay. The planktonic cells were discarded, and the attached cells were gently washed twice with PBS. After drying the plates at RT for 10 min, the cells were fixed with 200 µL of absolute methanol for 10 min and subsequently dried for 10 min. After 10 min of drying, 100 µL of aqueous 0.3% crystal violet solution was added to each well and the plates were incubated at RT for 20 min. The remaining dye was removed, and the biofilms were washed with PBS to remove any excess dye. After drying for another 10 min, the crystal violet that had accumulated in the biofilm cells was decolorized using 150 µL of 33% acetic acid for 30 s. Finally, each solution was transferred to a new plate, and the optical density was measured immediately at 590 nm using the BIOTEK ELx808 spectrophotometer (BIOTEK, Winooski, VT, USA) [87].

The biofilm formation cutoff was established according to optical density "ODc," defined as the mean absorbance at 590 nm of control RPMI 1640. "OD" refers to the optical density of the treated biofilm solution. Biofilm formation ability was classified as none (OD $\leq$ ODc), weak (ODc $<$ OD $\leq$ 2$\times$ ODc), moderate (2$\times$ ODc $<$ OD $\leq$ 4$\times$ ODc), or strong (4$\times$ ODc $<$ OD) [87].

4.10. *Effects of CEO and CEE on Planktonic Cells*

The microdilution method (based on standard Clinical and Laboratory Standards Institute Guidelines, 2008) [88], was used to determine the MIC. CEO and CEE were mixed with 100 mg/mL dimethyl sulfoxide in a RPMI 1640 medium supplemented with L-glutamine, which was buffered to pH 7.0 with 3-[N-morpholino] propane sulfonic acid (Sigma Aldrich, MO, USA) [41,48], and then subjected to two-fold serial dilution to obtain concentrations from 1024 µg/mL to 2 µg/mL. Fluconazole (Sigma Aldrich, St. Louis, MO, USA), an antifungal agent used as the positive control, was prepared at the same concentration as CEO and CEE. Subsequently, 100 µL of each concentration was added into the wells, followed by inoculation with 100 µL of each fungal suspension to obtain a final concentration of 2×10^3 CFU/mL. All procedures were performed in triplicate. Culture media was used as the negative control. *C. albicans* and *T. rubrum* were used for internal quality control. The microplates were incubated at 35 °C for 96 h. The MIC was defined as the lowest concentration of extract at which no visible growth was observed under an inverted microscope.

MFCs were established by streaking the subcultures taken from the MIC wells without visible growth on an SDA plate. After incubation at 37 °C for 96 h in aerobic conditions, viable fungal growth was evaluated. The MFC was defined as the lowest concentration of extract at which no fungal growth was observed under an inverted microscope.

4.11. *Effects of CEO and CEE on the Biofilm Formation of M. canis*

The susceptibility of mature *M. canis* biofilms to CEO and CEE was evaluated by exposing the biofilms to extracts at MIC and 2$\times$ MIC. The extracts were diluted in RPMI 1640 and added to each well of a 96-well plate containing a mature biofilm, following which the plates were incubated at 37 °C for 96 h. Each treatment was performed in triplicate. After incubation, the biofilm metabolic activity of the fungi was determined via XTT (2,3-bis(2-methoxy-4-nitro-5-sulfophenyl)-2H-tetrazolium-5-carboxanilide) reduction assay [89]. XTT (Sigma Aldrich, St. Louis, MO, USA) was prepared according to the manufacturer's instructions, and 100 µL of XTT-menadione solution was added to each well. The plates were then incubated in the dark for 1–2 h at 37 °C, after which absorbance at 490 nm was measured using the Synergy H1 Hybrid Multi-Mode Microplate Reader (BIOTEK, Winooski, VT, USA). The results are expressed as the average of absorbance values.

4.12. Statistical Analysis

Data were evaluated for normal distribution using the Shapiro–Wilk test prior to one-way analysis of variance. All statistical analyses were performed using IBM SPSS Statistics version 21.0. A p value of < 0.05 was considered statistically significant.

5. Conclusions

The increasing resistance in zoonotic fungi and adverse reactions to antifungal agents are major challenges for the development of natural-based agents. We found CEO and CEE to be extremely effective against the planktonic cell growth of clinical *M. canis* isolates. *C. amboinicus* also had significant antibiofilm effects on weak, moderate, and strong fungal biofilm producers. Thus, *C. amboinicus* may emerge as a novel source of natural antifungals. The antifungal mechanisms of *C. amboinicus* and drug formulations warrant further investigation for developing safe and effective treatments for zoonotic *M. canis* infections.

Supplementary Materials: The following supporting information can be downloaded at https://www.mdpi.com/article/10.3390/antibiotics11121734/s1, Figure S1: Mass spectrum of the main components of *Coleus amboinicus* essential oil determined via gas chromatography–mass spectrometry; Figure S2: Phylogenetic tree analysis of *Microsporum canis* based on nucleotide sequences from a 720 bp fragment of ITS1-5.8S-ITS2 using the neighbor-joining method. Sequences from this study are marked with stars; Figure S3: Microscopic appearance of *Microsporum canis* biofilm classified as strong (a), moderate (b), and weak (c) biofilm producers after 72 h incubation (40× magnification).

Author Contributions: Conceptualization, A.L., N.B., S.S. and S.B.; methodology, A.L., N.B., S.S. and S.B.; validation, A.L., S.S. and S.B.; investigation A.L., K.T., J.S., P.M., N.B., S.S., T.C., O.T., T.W. and S.B.; resources, A.L., N.B., S.S. and S.B.; data curation, A.L., S.S. and S.B.; writing—original draft preparation, A.L., S.S. and S.B.; writing—review and editing, A.L., S.S. and S.B.; project administration, S.B.; funding acquisition, S.B. All authors have read and agreed to the published version of the manuscript.

Funding: This study was supported by the Faculty of Veterinary Science, Mahidol University, Thailand.

Institutional Review Board Statement: The study protocol was approved by the Faculty of Veterinary Science-Animal Care and Use Committee (MUVS-2019-09-45) and the Faculty of Veterinary Science-Institutional Biosafety Committee (IBC/MUVS-B-005-2562).

Informed Consent Statement: Not applicable.

Data Availability Statement: The data presented in this study are available within the article.

Acknowledgments: We would like to thank the Center of Veterinary Diagnosis, Faculty of Veterinary Science, Mahidol University for kindly providing the fungal samples. We would also like to thank Phirom Prompiram and Kanaporn Poltep for assisting with the use of laboratory equipment.

Conflicts of Interest: The authors declare no conflict of interest.

References

1. Moriello, K.A.; Coyner, K.; Paterson, S.; Mignon, B. Diagnosis and treatment of dermatophytosis in dogs and cats.: Clinical Consensus Guidelines of the World Association for Veterinary Dermatology. *Vet. Dermatol.* **2017**, *28*, 266-e68.
2. Gnat, S.; Nowakiewicz, A.; Łagowski, D.; Zięba, P. Host- and pathogen-dependent susceptibility and predisposition to dermatophytosis. *J. Med. Microbiol.* **2019**, *68*, 823–836.
3. Ginter-Hanselmayer, G.; Weger, W.; Ilkit, M.; Smolle, J. Epidemiology of tinea capitis in Europe: Current state and changing patterns. *Mycoses* **2007**, *50*, 6–13.
4. Uhrlaß, S.; Krüger, C.; Nenoff, P. *Microsporum canis*: Current data on the prevalence of the zoophilic dermatophyte in central Germany. *Hautarzt* **2015**, *66*, 855–862.
5. Pasquetti, M.; Min, A.R.M.; Scacchetti, S.; Dogliero, A.; Peano, A. Infection by *Microsporum canis* in paediatric patients: A veterinary perspective. *Vet. Sci.* **2017**, *4*, 46.
6. Bond, R. Superficial veterinary mycoses. *Clin. Dermatol.* **2010**, *28*, 226–236.
7. Chupia, V.; Ninsuwon, J.; Piyarungsri, K.; Sodarat, C.; Prachasilchai, W.; Suriyasathaporn, W.; Pikulkaew, S. Prevalence of *Microsporum canis* from pet cats in Small Animal Hospitals, Chiang Mai, Thailand. *Vet. Sci.* **2022**, *9*, 21.

8. Cafarchia, C.; Romito, D.; Capelli, G.; Guillot, J.; Otranto, D. Isolation of *Microsporum canis* from the hair coat of pet dogs and cats belonging to owners diagnosed with *M. canis* tinea corporis. *Vet. Dermatol.* **2006**, *17*, 327–331.
9. Brosh-Nissimov, T.; Ben-Ami, R.; Astman, N.; Malin, A.; Baruch, Y.; Galor, I. An outbreak of *Microsporum canis* infection at a military base associated with stray cat exposure and person-to-person transmission. *Mycoses* **2018**, *61*, 472–476.
10. Yamada, S.; Aizawa, K.; Mochizuki, T. An epidemiological study of feline and canine dermatophytoses in Japan. *Med. Mycol. J.* **2019**, *60*, 39–44.
11. Uma, K.; Huang, X.; Kumar, B.A. Antifungal effect of plant extract and essential oil. *Chin. J. Integr. Med.* **2017**, *23*, 233–239.
12. Aneke, C.I.; Otranto, D.; Cafarchia, C. Therapy and antifungal susceptibility profile of *Microsporum canis*. *J. Fungi* **2018**, *4*, 107.
13. Hsiao, Y.H.; Chen, C.; Han, H.S.; Kano, R. The first report of terbinafine resistance *Microsporum canis* from a cat. *J. Vet. Med. Sci.* **2018**, *80*, 898–900.
14. Cowen, L.E.; Sanglard, D.; Howard, S.J.; Rogers, P.D.; Perlin, D.S. Mechanisms of antifungal drug resistance. *Cold Spring Harb. Perspect. Med.* **2014**, *5*, a019752.
15. Martinez-Rossi, N.M.; Bitencourt, T.A.; Peres, N.T.A.; Lang, E.A.S.; Gomes, E.V.; Quaresemin, N.R.; Martins, M.P.; Lopes, L.; Rossi, A. Dermatophyte resistance to antifungal drugs: Mechanisms and prospectus. *Front. Microbiol.* **2018**, *9*, 1108.
16. Danielli, L.J.; Lopes, W.; Vainstein, M.H.; Fuentefria, A.M.; Apel, M.A. Biofilm formation by *Microsporum canis*. *Clin. Microbiol. Infect.* **2017**, *23*, 941–942.
17. Merghni, A.; Haddaji, N.; Bouali, N.; Alabbosh, K.F.; Adnan, M.; Snoussi, M.; Noumi, E. Comparative study of antibacterial, antibiofilm, antiswarming and antiquorum sensing activities of *Origanum vulgare* essential oil and terpinene-4-ol against pathogenic bacteria. *Life* **2022**, *12*, 1616.
18. Saadabi, A. Antifungal activity of some Saudi plants used in traditional medicine. *Asian J. Plant Sci.* **2006**, *5*, 907–909.
19. Avijgan, M.; Hafizi, M.; Mehdi, S.; Nilforoushzadeh, M.A. Antifungal effect of *Echinophora Platyloba*'s extract against *Candida albicans*. *IJPR* **2006**, *4*, 285–289.
20. Benoutman, A.; Erbiai, E.H.; Edderdaki, F.Z.; Cherif, E.K.; Saidi, R.; Lamrani, Z.; Pintado, M.; Pinto, E.; da Silva, J.C.G.E.; Maouni, A. Phytochemical composition, antioxidant and antifungal activity of *Thymus capitatus*, a medicinal plant collected from Northern Morocco. *Antibiotics* **2022**, *11*, 681.
21. Silva, M.R.R.; Oliveira, J.G., Jr.; Fernandes, O.F.L.; Passos, X.S.; Costa, C.R.; Souza, L.K.H.; Lemos, J.A.; Paula, J.R. Antifungal activity of *Ocimum gratissimum* towards dermatophytes. *Mycoses* **2005**, *48*, 172–175.
22. Falahati, M.; Tabrizib, N.O.; Jahaniani, F. Antidermatophyte activities of *Eucalyptus camadulensis* in comparison with griseofulvin. *IJPT* **2005**, *4*, 80–83.
23. Arumugam, G.; Swamy, M.K.; Sinniah, U.R. *Plectranthus amboinicus* (Lour.) Spreng: Botanical, phytochemical, pharmacological, and nutritional significance. *Molecules* **2016**, *21*, 369.
24. Lukhoba, C.W.; Simmonds, M.S.; Paton, A.J. *Plectranthus*: A review of ethnobotanical uses. *J. Ethnopharmacol.* **2006**, *103*, 1–24.
25. Ślusarczyk, S.; Cieślak, A.; Yanza, Y.R.; Szumacher-Strabel, M.; Varadyova, Z.; Stafiniak, M.; Wojnicz, D.; Matkowski, A. Phytochemical profile and antioxidant activities of *Coleus amboinicus* Lour. cultivated in Indonesia and Poland. *Molecules* **2021**, *26*, 2915.
26. Cheng, C.Y.; Kao, C.L.; Li, H.T.; Yeh, H.C.; Fang, Z.Y.; Li, W.J.; Wu, H.M. A new flavonoid from *Plectranthus amboinicus*. *Chem. Nat. Compd.* **2021**, *57*, 28–29.
27. Ruan, T.Z.; Kao, C.L.; Hsieh, Y.L.; Li, H.T.; Chen, C.Y. Chemical constituents of the leaves of *Plectranthus amboinicus*. *Chem. Nat. Compd.* **2019**, *55*, 124–126.
28. El-hawary, S.S.; El-sofany, R.H.; Abdel-Monem, A.R.; Ashour, R.S.; Sleem, A.A. Polyphenolics content and biological activity of *Plectranthus amboinicus* (Lour.) spreng growing in Egypt (Lamiaceae). *Pharmacogn. J.* **2012**, *32*, 45–54.
29. Selvakumar, P.; Naveena, B.E.; Prakash, S.D. Studies on the antidandruff activity of the essential oil of *Coleus amboinicus* and *Eucalyptus globulus*. *Asian Pacific. J. Trop. Dis.* **2012**, *2*, S715–S719.
30. Laila, F.; Fardiaz, D.; Yuliana, N.D.; Damanik, M.R.M.; Dewi, F.N.A. Methanol extract of *Coleus amboinicus* (Lour) exhibited antiproliferative activity and induced programmed cell death in Colon Cancer Cell WiDr. *Int. J. Food Sci.* **2020**, *24*, 9068326.
31. Chiu, Y.J.; Huang, T.H.; Chiu, C.S.; Lu, T.C.; Chen, Y.W.; Peng, W.H.; Chen, C.Y. Analgesic and antiinflammatory activities of the aqueous extract from *Plectranthus amboinicus* (Lour.) Spreng. both *in vitro* and *in vivo*. *Evid. Based Complement. Altern. Med.* **2012**, *2012*, 508137.
32. Chang, J.M.; Cheng, C.M.; Hung, L.M.; Chung, Y.S.; Wu, R.Y. Potential use of *Plectranthus amboinicus* in the treatment of rheumatoid arthritis. *Evid. Based Complement. Altern. Med.* **2010**, *7*, 115–120.
33. Hsu, Y.C.; Cheng, C.P.; Chang, D.M. *Plectranthus amboinicus* attenuates inflammatory bone erosion in mice with collagen-induced arthritis by downregulation of RANKL-induced NFATc1 expression. *J. Rheumatol.* **2011**, *38*, 1844–1857.
34. de Oliveira, F.F.; Torres, A.F.; Gonçalves, T.B.; Santiago, G.M.; de Carvalho, C.B.; Aguiar, M.B.; Camara, L.M.; Rabenhorst, S.H.; Martins, A.M.; Valença Junior, J.T.; et al. Efficacy of *Plectranthus amboinicus* (Lour.) Spreng in a murine model of methicillin-resistant *Staphylococcus aureus* skin abscesses. *Evid. Based Complementary Altern. Med.* **2013**, *2013*, 291592.
35. Arjunan, N.; Murugan, K.; Madhiyazhagan, P.; Kovendan, K.; Prasannakumar, K.; Thangamani, S.; Barnard, D.R. Mosquitocidal and water purification properties of *Cynodon dactylon*, *Aloe vera*, *Hemidesmus indicus* and *Coleus amboinicus* leaf extracts against the mosquito vectors. *Parasitol. Res.* **2012**, *110*, 1435–1443.
36. Pouët, C.; Deletre, E.; Rhino, B. Repellency of wild oregano plant volatiles, *Plectranthus amboinicus*, and their essential oils to the silverleaf whitefly, *Bemisia Tabaci*, on tomato. *Neotrop. Entomol.* **2022**, *51*, 133–142.

37. Leesombun, A.; Sungpradit, S.; Boonmasawai, S.; Weluwanarak, T.; Klinsrithong, S.; Ruangsittichai, J.; Ampawong, S.; Masmeatathip, R.; Changbunjong, T. Insecticidal activity of *Plectranthus amboinicus* essential oil against the stable fly *Stomoxys calcitrans* (Diptera: Muscidae) and the horse fly *Tabanus megalops* (Diptera: Tabanidae). *Insects* **2022**, *13*, 255
38. Hsu, K.-P.; Ho, C.-L. Antimildew effects of *Plectranthus amboinicus* leaf essential oil on paper. *Nat. Prod. Commun.* **2019**, *14*, 1934578X19862903.
39. Ramage, G.; Rajendran, R.; Sherry, L.; Williams, C. Fungal biofilm resistance. *Int. J. Microbiol.* **2012**, *2012*, 528521.
40. Roudbary, M.; Vahedi-Shahandashti, R.; Santos, A.L.S.D.; Roudbar, M.S.; Aslani, P.; Lass-Flörl, C.; Rodrigues, C.F. Biofilm formation in clinically relevant filamentous fungi: A therapeutic challenge. *Crit. Rev. Microbiol.* **2022**, *48*, 197–221.
41. Brilhante, R.S.N.; Correia, E.E.M.; Guedes, G.M.M.; de Oliveira, J.S.; Castelo-Branco, D.; Cordeiro, R.A.; Pinheiro, A.Q.; Chaves, L.J.Q.; Pereira Neto, W.A.; Sidrim, J.J.C.; et al. In vitro activity of azole derivatives and griseofulvin against planktonic and biofilm growth of clinical isolates of dermatophytes. *Mycoses* **2018**, *61*, 449–454.
42. Wu, S.; Wang, Y.; Liu, N.; Dong, G.; Sheng, C. Tackling fungal resistance by biofilm inhibitors. *J. Med. Chem.* **2017**, *23*, 2193–2211.
43. Baldo, A.; Monod, M.; Mathy, A.; Cambier, L.; Bagut, E.T.; Defaweux, V.; Symoen, S.F.; Antoine, N.; Mignon, B. Mechanisms of skin adherence and invasion by dermatophytes. *Mycoses* **2012**, *55*, 218–223.
44. Brilhante, R.S.N.; Aguiar, L.; Sales, J.A.; Araújo, G.D.S.; Pereira, V.S.; Pereira-Neto, W.A.; Pinheiro, A.Q.; Paixão, G.C.; Cordeiro, R.A.; Sidrim, J.J.C.; et al. Ex Vivo biofilm-forming ability of dermatophytes using dog and cat hair: An ethically viable approach for an infection model. *Biofouling* **2019**, *35*, 392–400.
45. Al-Fattani, M.A.; Douglas, L.J. Penetration of *Candida* biofilms by antifungal agents. *Antimicrob. Agents Chemother.* **2004**, *48*, 3291–3297.
46. Shi, C.; Liu, J.; Li, W.; Zhao, Y.; Meng, L.; Xiang, M. Expression of fluconazole resistance-associated genes in biofilm from 23 clinical isolates of *Candida albicans*. *Braz. J. Microbiol.* **2019**, *50*, 157–163.
47. Jabra-Rizk, M.A.; Falkler, W.A.; Meiller, T.F. Fungal biofilms and drug resistance. *Emerg. Infect. Dis.* **2004**, *10*, 14–19.
48. Bila, N.M.; Costa-Orlandi, C.B.; Vaso, C.O.; Bonatti, J.L.C.; de Assis, L.R.; Regasini, L.O.; Fontana, C.R.; Fusco-Almeida, A.M.; Mendes-Giannini, M.J.S. 2-Hydroxychalcone as a potent compound and photosensitizer against dermatophyte biofilms. *Front. Cell Infect. Microbiol.* **2021**, *11*, 679470.
49. Girish, K. Antimicrobial activities of *Coleus aromaticus* Benth. *J. Pharm. Res.* **2016**, *10*, 635–646.
50. Umayal, S.; Geetha, R.V. Comparative evaluation of antibiofilm formation activity of *Plectranthus amboinicus* extract against *Streptococcus mutans*. *Drug Invent. Today* **2019**, *12*, 2253–2255.
51. Vasconcelos, S.E.C.B.; Melo, H.M.; Cavalcante, T.T.A.; Júnior, F.E.A.C.; de Carvalho, M.G.; Menezes, F.G.R.; de Sousa, O.V.; Costa, R.A. *Plectranthus amboinicus* essential oil and carvacrol bioactive against planktonic and biofilm of oxacillin- and vancomycin-resistant *Staphylococcus aureus*. *BMC Complement. Altern. Med.* **2017**, *17*, 462.
52. Manimekalai, K.; Srinivasan, P.; Dineshbabu, J.; Guna, G.; Darsini, D. Anti-biofilm efficacy of *Plectranthus amboinicus* against *Streptococcus pyogenes* isolated from pharyngitis patients. *Asian J. Pharm. Clin. Res.* **2016**, *9*, 348–354.
53. Wadikar, D.D.; Patki, P.E. *Coleus aromaticus*: A therapeutic herb with multiple potentials. *J. Food Sci. Technol.* **2016**, *53*, 2895–2901.
54. da Costa, J.G.M.; Pereira, C.K.B.; Rodrigues, F.F.G.; de Lima, S.G. Chemical composition, antibacterial and fungicidal activities of leaf oil of *Plectranthus amboinicus* (Lour.) Spreng. *JEOR* **2010**, *22*, 183–185.
55. Dhifi, W.; Bellili, S.; Jazi, S.; Bahloul, N.; Mnif, W. Essential oils' chemical characterization and investigation of some biological activities: A critical review. *Medicines* **2016**, *3*, 25.
56. Wang, F.; Wei, F.; Song, C.; Jiang, B.; Tian, S.; Yi, J.; Yu, C.; Song, Z.; Sun, L.; Bao, Y.; et al. *Dodartia orientalis* L. essential oil exerts antibacterial activity by mechanisms of disrupting cell structure and resisting biofilm. *Ind. Crops Prod.* **2017**, *109*, 358–366.
57. Kordali, S.; Cakir, A.; Ozer, H.; Cakmakci, R.; Kesdek, M.; Mete, E. Antifungal, phytotoxic and insecticidal properties of essential oil isolated from Turkish *Origanum acutidens* and its three components, carvacrol, thymol and p-cymene. *Bioresour. Technol.* **2008**, *99*, 8788–8795.
58. Rao, A.; Zhang, Y.; Muend, S.; Rao, R. Mechanism of antifungal activity of terpenoid phenols resembles calcium stress and inhibition of the TOR pathway. *Antimicrob. Agents Chemother.* **2010**, *54*, 5062–5069.
59. Zuzarte, M.; Vale-Silva, L.; Gonçalves, M.J.; Cavaleiro, C.; Vaz, S.; Canhoto, J.; Pinto, E.; Salgueiro, L. Antifungal activity of phenolic-rich *Lavandula multifida* L. essential oil. *Eur. J. Clin. Microbiol. Infect. Dis.* **2012**, *31*, 1359–1366.
60. Abbaszadeh, S.; Sharifzadeh, A.; Shokri, H.; Khosravi, A.R.; Abbaszadeh, A. Antifungal efficacy of thymol, carvacrol, eugenol and menthol as alternative agents to control the growth of food-relevant fungi. *J. Mycol. Med.* **2014**, *24*, e51–e56.
61. Tian, F.; Woo, S.Y.; Lee, S.Y.; Chun, H.S. p-Cymene and its derivatives exhibit antiaflatoxigenic activities against *Aspergillus flavus* through multiple modes of action. *Appl. Biol. Chem.* **2018**, *61*, 489–497.
62. Niu, C.; Wang, C.; Yang, Y.; Chen, R.; Zhang, J.; Chen, H.; Zhuge, Y.; Li, J.; Cheng, J.; Xu, K.; et al. Carvacrol induces *Candida albicans* apoptosis associated with Ca^{2+}/calcineurin pathway. *Front. Cell. Infect. Microbiol.* **2020**, *10*, 192.
63. Alves Resende, J.; Soares Toneto, D.; Cruz Albuquerque, M.C.; Areas Bastos, K.; Fontes Pinheiro, P.; Drummond Costa Ignacchiti, M. Antibacterial and anti-biofilm potential of *Plectranthus amboinicus* (Lour.) Spreng essential oil and carvacrol against *Staphylococcus aureus* and *Escherichia coli*. *Rev. Ciênc. Méd. Biol.* **2022**, *21*, 11–17.
64. Marchese, A.; Arciola, C.R.; Barbieri, R.; Silva, A.S.; Nabavi, S.F.; Tsetegho Sokeng, A.J.; Morteza Izadi, M.; Jonaidi Jafari, N.; Suntar, I.; Daglia, M.; et al. Update on monoterpenes as antimicrobial agents: A particular focus on p-Cymene. *Materials* **2017**, *10*, 947.

65. Couto, C.; Raposo, N.; Rozental, S.; Borba-Santos, L.; Bezerra, L.; Almeida, P.; Brandão, M. Chemical composition and antifungal properties of essential oil of *Origanum vulgare* Linnaeus (Lamiaceae) against *Sporothrix schenckii* and *Sporothrix brasiliensis*. *Trop. J. Pharm. Res.* **2015**, *14*, 1207–1212.
66. Baba, S.A.; Malik, S.A. Determination of total phenolic and flavonoid content, antimicrobial and antioxidant activity of a root extract of *Arisaema jacquemontii* Blume. *J. Taibah Univ. Med. Sci.* **2015**, *9*, 449–454.
67. Nguyen, N.Q.; Minh, L.V.; Trieu, L.H.; Bui, L.M.; Lam, T.D.; Hieu, V.Q.; Khang, T.V.; Trung, L.N.Y. Evaluation of total polyphenol content, total flavonoid content, and antioxidant activity of *Plectranthus amboinicus* leaves. *IOP Conf. Ser. Mater. Sci. Eng.* **2020**, *736*, 062017.
68. Bhatt, P.; Joseph, G.S.; Negi, P.S.; Varadaraj, M.C. Chemical composition and nutraceutical potential of Indian Borage (*Plectranthus amboinicus*) stem extract. *J. Chem.* **2013**, *2013*, 320329.
69. Harborne, J.B.; Williams, C.A. Advances in flavonoid research since 1992. *Phytochemistry* **2000**, *55*, 481–504.
70. Cushnie, T.P.T.; Lamb, A.J. Antimicrobial activity of flavonoids. *Int. J. Antimicrob. Agents* **2005**, *26*, 343–356.
71. Ahmad, H.; Matsubara, Y.I. Effect of lemon balm water extract on *Fusarium* Wilt control in strawberry and antifungal properties of secondary metabolites. *Hort. J.* **2020**, *89*, 175–181.
72. Swari, D.A.M.A.; Santika, I.W.M.; Aman, I.G.M. Antifungal activities of ethanolic extract of rosemary leaf (*Rosemarinus officinalis* L.) against *Candida albicans*. *JPSA* **2020**, *2*, 28–35.
73. Aboody, M.S.A.; Mickymaray, S. Anti-fungal efficacy and mechanisms of flavonoids. *Antibiotics* **2020**, *9*, 45.
74. Lee, H.; Woo, E.R.; Lee, D.G. Apigenin induces cell shrinkage in *Candida albicans* by membrane perturbation. *FEMS Yeast Res.* **2018**, *18*, foy003.
75. Singh, G.; Kumar, P.; Joshi, S.C. Treatment of dermatophytosis by a new antifungal agent 'apigenin'. *Mycoses* **2014**, *57*, 497–506.
76. Alfarrayeh, I.; Pollák, E.; Czéh, Á.; Vida, A.; Das, S.; Papp, G. Antifungal and anti-biofilm effects of caffeic acid phenethyl ester on different *Candida* Species. *Antibiotics* **2021**, *10*, 1359.
77. Sun, L.; Liao, K.; Hang, C. Caffeic acid phenethyl ester synergistically enhances the antifungal activity of fluconazole against resistant *Candida albicans*. *Phytomedicine* **2018**, *40*, 55–58.
78. Cantelli, B.A.M.; Bitencourt, T.A.; Komoto, T.T.; Beleboni, R.O.; Marins, M.; Fachin, A.L. Caffeic acid and licochalcone A interfere with the glyoxylate cycle of *Trichophyton rubrum*. *Biomed. Pharmacother.* **2017**, *96*, 1389–1394.
79. Wolfe, K.; Wu, X.; Liu, R.H. Antioxidant activity of apple peels. *J. Agric. Food Chem.* **2003**, *51*, 609–614.
80. Matić, P.; Sabljić, M.; Jakobek, L. Validation of spectrophotometric methods for the determination of total polyphenol and total flavonoid content. *J. AOAC Int.* **2019**, *100*, 1795–1803.
81. Petrucelli, M.F.; de Abreu, M.H.; Cantelli, B.A.M.; Segura, G.G.; Nishimura, F.G.; Bitencourt, T.A.; Marins, M.; Fachin, A.L. Epidemiology and diagnostic perspectives of dermatophytoses. *J. Fungi* **2020**, *6*, 310.
82. Sugita, T.; Nishikawa, A.; Shinoda, T. Identification of *Trichosporon asahii* by PCR based on sequences of the internal transcribed spacer regions. *J. Clin. Microbiol.* **1998**, *36*, 2742–2744.
83. Turin, L.; Riva, F.; Galbiati, G.; Cainelli, T. Fast, simple and highly sensitive double-rounded polymerase chain reaction assay to detect medically relevant fungi in dermatological specimens. *Eur. J. Clin. Investig.* **2000**, *30*, 511–518.
84. Larkin, M.A.; Blackshields, G.; Brown, N.P.; Chenna, R.; McGettigan, P.A.; McWilliam, H.; Valentin, F.; Wallace, I.M.; Wilm, A.; Lopez, R.; et al. Clustal W and Clustal X version 2.0. *Bioinformatics* **2007**, *23*, 2947–2948.
85. Dereeper, A.; Guignon, V.; Blanc, G.; Audic, S.; Buffet, S.; Chevenet, F.; Dufayard, J.F.; Guindon, S.; Lefort, V.; Lescot, M.; et al. Phylogeny.fr: Robust phylogenetic analysis for the non-specialist. *Nucleic Acids Res.* **2008**, *1*, 465–469.
86. Sen, S.; Borah, S.N.; Bora, A.; Deka, S. Rhamnolipid exhibits anti-biofilm activity against the dermatophytic fungi *Trichophyton rubrum* and *Trichophyton mentagrophytes*. *Biotechnol. Rep.* **2020**, *27*, e00516.
87. Sales, G.; Medeiros, S.; Soares, I.; Sampaio, T.; Bandeira, M.; Nogueira, N.; Queiroz, M. Antifungal and modulatory activity of *Lemon Balm* (Lippia alba (MILL.) N. E. BROWN) essential oil. *Sci. Pharm.* **2022**, *90*, 31.
88. *M38-A2*; Reference Method for Broth Dilution Antifungal Susceptibility Testing of Filamentous Fungi. Clinical and Laboratory Standards Institute (CLSI): Wayne, PA, USA, 2008.
89. Kim, D.J.; Lee, M.W.; Choi, J.S.; Lee, S.G.; Park, J.Y.; Kim, S.W. Inhibitory activity of hinokitiol against biofilm formation in fluconazole-resistant *Candida* species. *PLoS ONE* **2017**, *12*, e0171244.

antibiotics

MDPI

Article

Phytochemical Profile of the Ethanol Extract of *Malvaviscus arboreus* Red Flower and Investigation of the Antioxidant, Antimicrobial, and Cytotoxic Activities

Hanaa S. S. Gazwi [1,*], Nagwa A. Shoeib [2], Magda E. Mahmoud [1], Osama I. A. Soltan [3], Moaz M. Hamed [4] and Amany E. Ragab [2,*]

1 Department of Agricultural Chemistry, Faculty of Agriculture, Minia University, El-Minya 61519, Egypt
2 Department of Pharmacognosy, Faculty of Pharmacy, Tanta University, Tanta 31257, Egypt
3 Department of Food Science, Faculty of Agriculture, Minia University, El-Minia 61519, Egypt
4 Marine Microbiology Laboratory, National Institute of Oceanography and Fisheries, Cairo 11562, Egypt
* Correspondence: hanaa.saleh@mu.edu.eg (H.S.S.G.); amany.ragab@pharm.tanta.edu.eg (A.E.R.)

Abstract: Flowers are rich sources of bioactive antimicrobial, antioxidant, and anticancer components. This study aimed to determine the constituents of the ethanol extract of *Malvaviscus arboreus* red flower (ERF) by GC-MS analysis and HPLC identification of phenolic compounds and flavonoids, in addition to the ^{1}HNMR fingerprint. The antimicrobial, antioxidant, and cytotoxic activities of the ERF were investigated. The GC-MS analysis revealed twenty-one components, while HPLC analysis revealed the presence of phenolic and flavonoid compounds. The ERF showed antifungal and antibacterial activity. The highest antibacterial activity was found against *Vibrio damsela* where a time-kill assay revealed a decline in the amount of viable *V. damsela*. For fungi, the highest activity was observed against *Aspergillus terreus*. Using the SRB test on HepG2, the anti-proliferative efficacy of the ERF was evaluated. Cell cycle analysis was utilized to determine autophagic cell death. The ERF prevented the proliferation of the HepG2 cell line with an IC_{50} of 67.182 µg/µL. The extract primarily promoted apoptosis in HepG2 cells by accumulating hypodiploid cells in the sub-G0/G1 phase, increased caspase 3/7 activity, and caused considerable autophagic cell death in apoptosis-deficient cells. Finally, the observed elevation of cancer cell death indicated that ERF had substantial anticancer potential against HepG2 cells.

Keywords: *Malvaviscus arboreus*; GC-MS; HPLC; antimicrobial; antioxidant; HepG2

Citation: Gazwi, H.S.S.; Shoeib, N.A.; Mahmoud, M.E.; Soltan, O.I.A.; Hamed, M.M.; Ragab, A.E. Phytochemical Profile of the Ethanol Extract of *Malvaviscus arboreus* Red Flower and Investigation of the Antioxidant, Antimicrobial, and Cytotoxic Activities. *Antibiotics* **2022**, *11*, 1652. https://doi.org/10.3390/antibiotics11111652

Academic Editors: Valério Monteiro-Neto and Elizabeth S. Fernandes

Received: 2 November 2022
Accepted: 16 November 2022
Published: 18 November 2022

1. Introduction

Cancer is one of the biggest causes of mortality in the world, accounting for an estimated 9.9 million lives lost in 2020 [1]. A recent study found that hepatocellular carcinoma (HCC) was the fourth greatest cause of cancer-related fatalities worldwide [2]. The prognosis for this form of cancer is dismal [3], as it is typically diagnosed late. Unlike several other cancers, these strike more frequently in developing nations. HCC usually happens alongside cirrhosis, which can be caused by the hepatitis C virus, hepatitis B virus, alcoholism, Wilson's disease, type 2 diabetes, hemochromatosis, and hemophilia. Still, the hepatitis B virus and hepatitis C virus are the main causes of liver cancer [4]. Previous studies indicated that oxidative stress plays a role in liver cancer [5], but its mechanisms and impacts remain unclear. Reactive oxygen species (ROS) such as superoxide anion (O_2^-), hydrogen peroxide (H_2O_2), and hydroxyl radical (HO), which are mostly made by breathing, inflammation, or metabolism, can cause mutations or lesions in larger genomic sites. In addition, H_2O_2 is a signaling molecule that balances inflammation, separation, growth, protection, metastasis, autophagy, division, and metabolic pathways. In cancer, the activity of these pathways is a key determinant of malignancy [6]. Antioxidants and peroxidants are kept in balance in check within a healthy cell. Oncogenesis and tumor growth in HCC are triggered by an imbalance of peroxidants and antioxidants [7]. Chemotherapeutic drugs are currently limited in treating cancer due to side effects and tumor resistance. New and safe anticancer drugs can be found in natural sources [8].

Antimicrobial resistance is a worldwide problem urging the research for new pipelines of natural or synthetic sources. Therefore, it is a significant challenge to find innovative and safe therapeutic choices [9].

Bioactive compounds are abundant in plant extracts. This is due to a range of chemical ingredients, for example, alkaloids, polyphenols, and flavonoids, all of which play an essential part in the drug development process [10]. In Africa, many medicinal plants are utilized to cure different illnesses since they are a viable option, especially in developing countries. *Malvaviscus arboreus* Cav. is a tropical and subtropical perennial deciduous shrub endemic to Central and South America. This plant has multiple common names, encompassing Wax mallow, Drummond Wax Mallow, Turk's cap, and Sleeping Hibiscus. The leaves of *M. arboreus* contain compounds such as protocatechuic acid, chlorogenic acid, gallic acid, *p*-coumaric acid, ferulic acid, and hydroxybenzoic acid [11]. *Malvaviscus arboreus* has been used in traditional medicine. The leaf decoction is used for cystitis, diarrhea, fever, and gastritis [12]. The flower decoction is used as a gargle for sore throat, nursing infants with cold, bronchitis, diarrhea, thrush, and tonsillitis.

There is no information on the chemical components and the biological effects of the ethanol extract of red flowers (ERF) of *M. arboreus* in the literature. The current study investigated chemical constituents of ERF of *M. arboreus*, and its antimicrobial and antioxidant effects, in addition to the cytotoxic action on the HepG2 cell cline.

2. Materials and Methods

2.1. Preparation of the Extract

The *Malvaviscus arboreus* red flowers were collected from the campus of Minia University in May 2021 and authenticated by Professor Raga A. Taha, Horticulture Department, Faculty of Agriculture, Minia University. The flowers were washed with distilled water and kept at room temperature to dry. The dry flowers were ground and soaked in ethanol (100 mL ethanol for a 10 g dry sample) at room temperature for 24 h, filtered through Whatman No.4 filter paper (Whatman®Prepleated Qualitative Filter Paper, Grade 4 V, Sigma-Aldrich Company Ltd. (St. Louis, MO, USA)), and the extract was evaporated using a rotary evaporator (Büchi Rotavapor R-114 a Waterbath. B-480, Buchi, Switzerland) at 40 °C to obtain the crude extract. Then the extract was kept at 4 °C until used in the analysis.

2.2. Phytochemical Examination

The presence of coumarins, saponins, tannins, flavonoids, glycosides, phenols, steroids, terpenoids, emodins, anthocyanins, and alkaloids in the ERF of *M. arboreus* was investigated using qualitative assays as previously reported [13].

2.3. Determination of Total Phenolic Content (TPC) and Total Flavonoid Content (TFC)

As previously reported [14], total flavonoid content (TFC) in the ERF of *M. arboreus* was examined using aluminum chloride ($AlCl_3$) colorimetric test. The flavonoid concentration was calculated as mg of quercetin equivalent/g extract. The total phenolic content (TPC) in the ERF of *M. arboreus* was recorded by the Folin–Ciocalteu assay [14]. The amount of phenolics was calculated as mg of gallic acid equivalent/g extract.

2.4. Antioxidant Activities (ABTS+, FRAP, DPPH, Metal Chelating Property, and ORAC)

Different assays were utilized to evaluate the antioxidant potential of the ERF of *M. arboreus*. The radical scavenging activity of 2,2-azinobis 3-ethylbenzothiazoline-6-sulfonic acid (ABTS+) and 1,1-diphenyl-2-picrylhydrazyl (DPPH), in addition to the ferric reducing antioxidant power (FRAP) were measured following the procedures published by Adedapo et al. [15]. The capacity of the extract to chelate iron (II) was evaluated using the procedure described by Gülc et al. [16]. For the metal chelating activity test, the ORAC assay was performed as published by Ou et al. [17].

2.5. GC-MS Analysis

The GC-MS analysis of ERF of *M. arboreus* was performed following a published procedure [18] using a Trace GC1310-ISQ mass spectrometer (Thermo Scientific, Austin, TX, USA) with a direct capillary column TG–5MS (30 mm × 0.25 mm × 0.25 μm film thickness, Thermo Scientific, Austin, TX, USA). The column's oven temperature was kept at 50 °C; following that, it was set to reach 200 °C at 7 °C/min, held for 2 min, and then set to reach 290 °C, increased at 15 °C /min and maintained for 2 min. The temperature in the injector was kept at 260 °C. At a steady flow rate of 1 mL/min, helium was utilized as the carrier gas. After a 4-min solvent delay, an AS3000 autosampler and GC in the split mode were employed to automatically inject 1 μL of the diluted sample. At 70 eV ionization voltages spanning the m/z 50–650 range, EI mass spectra were acquired in full scan mode. The ion source and transfer line were adjusted to 270 and 250 degrees, respectively. By contrasting the components' mass spectra retention times to the NIST 11 and WILEY 09 mass spectral databases, the components were identified.

2.6. HPLC Determination of Phenolics and Flavonoids

The phenolic and flavonoid components of the ERF were determined using an injection volume of 25 μL of the extract as previously reported [19] in an HPLC system (Agilent 1100; Santa Clara, CA, USA).

The extract phenolic components were identified using an HPLC system (Agilent 1100; Santa Clara, CA, USA) with a UV/Vis detector at a wavelength of 250 nm using a C18 column (125 × 4.60 mm, particle size 5 μm). The Agilent Chem Station was used to acquire and analyze chromatograms. To completely separate the components of phenolic acids, a mobile gradient phase of two solvents methanol [A] and acetic acid in water (1:25) [B] was used. The gradient program started at 100% B and stayed for the first three min. This was followed by 5 min of 50% eluent A, 2 min of 80% A, 5 min of 50% A, and the detection wavelength was at 250 nm.

The same HPLC system was used to identify the flavonoid components in the extract using a C18 column (Agilent; Santa Clara, CA, USA) (250 × 4.6 mm, 5 μm) and a UV/Vis detector at a wavelength of 360 nm. Acetonitrile (A) and 0.2% (v/v) aqueous formic acid (B) were used as the mobile phase with an isocratic elution (70:30) procedure.

2.7. ^{1}HNMR Fingerprint Analysis

The ^{1}HNMR fingerprint was analyzed at 400 MHz using a Bruker Avance 400 spectrophotometer (Karlsruh, Germany), using DMSO as a solvent and tetramethylsilane (TMS) as an internal standard.

2.8. Antibacterial Activity

2.8.1. Test Microorganism

The bacterial strains *Enterococcus faecalis*, *Bacillus subtillus*, *Bacillus cereus*, *Staphylococcus aureus*, *Vibrio fluvialis*, *Vibrio damsela*, *Pseudomonas aeruginosa*, and *Salmonella typhimurium*; and the *fungal strains Aspergillus fumigatus*, *Aspergillus terreus*, *Aspergillus niger*, *Aspergillus flavus*, *Aspergillus parasiticus*, and *Penicillium oxalicum* used in this work were provided by the Department of Microbiology, National Institute of Oceanography and Fishers, Red Sea branch, Egypt. These strains were isolated from marine sources and identified by Dr. Moaz M. Hamed. The strains were kept at 2 °C on nutrient agar slants for bacteria and Potato Dextrose Agar (PDA) (Neogen Corporation, Lansing, MI, USA) for fungi slants. The slants were folded with 25% glycerol to ensure long-term preservation.

2.8.2. Bacterial Inactivation by ERF

The agar well diffusion assay technique was used to measure antibacterial activity. Antibacterial susceptibility assay of ERF was performed against the selected pathogens. In a petri-dish containing 20 mL of Muller Hinton agar media (composed of g/L is: beef ex-tract 2.0; acid hydrolysate of casein 17.5; Starch 1.5 and agar 17.5), the agar plate surface was inoculated by spreading a volume of the microbial inoculum (0.1 mL of bacterial suspension containing 10^6 CFU/mL) over the entire agar surface. Then, a hole with a diameter of 8 mm was punched aseptically with a sterile cork borer, and a volume of ERF (100 μL) was introduced into the well. In agar wells of control plates, we applied DMSO (0.5%) (Was purchased from R&M Marketing, Essex, UK) as a negative control and amoxicil-lin/clavulanic acid (20/10 mcg) as a positive control, and then incubated the plates at 37 °C for 24 h [20].

2.8.3. Minimum Inhibitory Concentration (MIC)

A tetrazolium microplate assay was used to determine the minimum inhibitory concentrations (MICs) of the test organisms [21]. A 96-well clear microtiter plate was used for the experiment. Each well of the 96-well plate was inoculated with a suspension of freshly isolated bacteria (0.1 mL) at a concentration of 5×10^5 CFU/mL. Different concentrations, 15 to 0.25 mg/mL, of the test extract were diluted in series with Muller–Hinton broth (Becton Dickinson, Sparks, MD, USA). A volume of 200 μL of each concentration was added in triplicate to the wells and the plates were then incubated for 18–24 h at 37 °C ± 0.5. After incubation, in each well, 50 μL of 3-(4, 5-dimethylthiazol-2-yl)-2, 5-diphenyltetrazolium bromide (MTT), with a concentration of 0.2 mg/mL, was added and the plate was incubated at 37 °C for 30 min. The bacterial suspension without extract served as the positive control, while the corresponding solvent blank (DMSO) served as the negative control. The percentage reduction of the dye (representing the inhibition of bacterial growth) was determined by measuring the absorbance at 570 nm relative to a reference wavelength of 650 nm, which was accomplished by introducing DMSO to the spectrophotometer [22].

2.8.4. Fungal Inactivation by ERF

The minimum inhibitory concentration (MIC) technique employing diffusion discs was used to evaluate antifungal activity. ERF was diluted to 25% in DMSO, followed by different concentrations of ERF (0.5–2.0 mg/mL). The strains (0.2 mL spore suspension (10^6 spores per mL) of the tested fungal isolate) of *A. fumigatus*, *A. terreus*, *A. niger*, *A. flavus*, *A. parasiticus*, and *P. oxalicum* were activated for 24 h in a liquid culture medium, Czapek Dox broth (composition (g/L): Sucrose: 30; $NaNO_3$: 3; KH_2PO_4: 1; $MgSO_4{\cdot}7H_2O$: 0.5; KCl: 0.5 and $FeSO_4{\cdot}7H_2O$: 0.01), at a temperature of 25 °C, and then brought to a concentration of 0.5 McFarland by spectrophotometric reading. Czapek Dox agar was used to inoculate petri dishes with the fungal strains that had already been produced. Six-millimeter sterile discs were set atop the culture medium, and 10 μL of the diluted extract was pipetted onto each one. The cultures were incubated at 25 °C for 72 h. The MIC of ERF was defined as the lowest concentration that effectively suppressed fungal growth. DMSO was used as a negative control [23].

2.8.5. Time-Kill Assay

According to the preliminary findings, the ERF of *M. arboreus* had the highest level of antimicrobial activity on marine *V. damsela*. An investigation on the bactericidal effects of the ERF on *V. damsela* was carried out utilizing a time-kill test. A bacterial culture (5×10^6 CFU/mL) was added to Mueller Hinton broth (MHB) containing the extract at 4 × MIC, 2 × MIC, MIC ½ × MIC, and ¼ × MIC, and untreated cultures were incubated at 37 °C. Tryptic soy agar (TSA) plates were used to culture samples for 0, 2, 4, 6, 8, 10, 12, and 24 h. A control incubation was performed with 1% DMSO. Surviving colony bacteria were counted, and $\log_{10}$ CFU/mL was calculated. A time-kill curve was analyzed by plotting log CFU/mL against time (min) [24].

2.8.6. Synergistic Activity

The ERF was tested in conjunction with amoxicillin/clavulanic acid using the standard disc diffusion method against selected marine *V. damsela*. The antibacterial activity was evaluated on an agar plate using discs made by combining amoxicillin/clavulanic acid (20/10 mcg) with different doses of ERF (250, 500, 750, and 1000 µg/mL). The antibacterial effectiveness of the ERF and amoxicillin/clavulanic combination was evaluated by measuring the size of the zone of inhibition after 24 h of incubation at 37 °C [25].

2.9. Cytotoxic Study

2.9.1. Cell Lines

Nawah Scientific Inc. provided HepG2: Hepatocellular carcinoma (Mokatam, Cairo, Egypt). In a humidified 5% (*v*/*v*) CO_2 atmosphere, cells were kept at 37 °C in Dulbecco's Minimum Essential Medium (DMEM, Lonza, Basel, Switzerland) media enriched with 100 units/mL penicillin, 100 mg/mL streptomycin, and 10% heat-inactivated fetal bovine serum (FBS).

2.9.2. Cytotoxicity Assay

The cell viability was measured utilizing the Sulforhodamine B (SRB) test. One hundred microliter cell suspension (5×10^3 cells) aliquots were incubated in complete media for 24 h in 96-well plates. A further aliquot of 100 µL media comprising ERF in varying amounts was administered to the cells. The cells were fixed after 72 h of ERF treatment by changing the medium with 150 µL of 10% Trichloroacetic acid (TCA) and incubated for 1 h at 4 °C. The cells were rinsed five times with distilled water after the TCA solution was removed. Seventy microliters of SRB solution (0.4% *w*/*v*) was administrated in aliquots and incubated for 10 min in the dark at room temperature. Before being air-dried overnight, plates were washed thrice in 1% acetic acid, then 150 µL of TRIS (10 mM) was administrated to disperse the protein-bound SRB dye, the absorbance was observed at 540 nm utilizing the FlUOstar Optima Microplate Reader (BMG LABTECH, Ortenberg, Germany).

2.9.3. Analysis of Cell Cycle Distribution

A previously published procedure was followed for the analysis of the cell cycle distribution [26]. One hundred and five cells were trypsinized and rinsed twice with ice-cold phosphate buffered saline (PBS) after being treated with test drugs for 24 or 48 h and paclitaxel (1 µM) for 24 h as a positive control (pH 7.4). The cells were fixed by resuspending them in 2 mL of 60% ice-cold ethanol and incubating them for one h at 4 °C. After being rinsed twice with PBS, the fixed pellet was resuspended in 1 mL of PBS (pH 7.4) with 50 µg/mL RNAase A and 10 µg/mL propidium iodide (PI). An FL2 (λex/em 535/617 nm) signal detector was used to determine the DNA content of cells after 20 min of incubation in the dark at 37 °C. (ACEA Novocyte™ flow cytometer, ACEA Biosciences Inc., San Diego, CA, USA). Each specimen was made up of 12,000 events in total. The ACEA NovoExpress application was used to calculate the cell cycle dispersion (ACEA Biosciences Inc., San Diego, CA, USA).

2.9.4. Apoptosis Assay

Flow cytometry with two fluorescent channels and an annexin V-FITC apoptosis detection kit were employed to identify apoptosis and necrosis in cell populations (Abcam Inc., Cambridge Science Park, Cambridge, UK) using a published procedure [26]. After 24/48 or 72 h of treatment with ERF and doxorubicin (10 µM) as a positive control, cells (10^5) were trypsinized and rinsed twice with ice-cold PBS (pH 7.4). The cells were then maintained at room temperature in the dark for 30 min with Annexin V-FITC/PI solution 0.5 mL, as directed by the manufacturer. After labeling, cells were added to an ACEA Novocyte™ flow cytometer (ACEA Biosciences Inc., San Diego, CA, USA) and measured for PI and FTIC fluorescent signals with FL1 and FL2 signal detectors ($\lambda_{ex}/_{em}$ 488/530 nm for FITC and $\lambda_{ex}/_{em}$ 535/617 nm for PI, respectively). ACEA NovoExpress™ software was used to assess the positive FITC or PI cells for each sample, utilizing quadrant analysis (ACEA Biosciences Inc., San Diego, CA, USA).

2.9.5. Autophagy Assay

Autophagic cell death was measured by flow cytometry and acridine orange lysosomal staining. A total of 10^5 cells were trypsinized and rinsed twice with ice-cold PBS after treatment with ERF for 24/48 or 72 h and chloroquine (10 μM) as a positive control for 24/48 or 72 h (pH 7.4). The cells were stained with acridine orange (10 μM) and incubated for 30 min at 37 °C in the dark. The acridine orange fluorescence signals using an FL1 signal detector ($\lambda_{ex}/_{em}$ 488/530 nm) in an ACEA Novocyte™ flow cytometer (ACEA Biosciences Inc., San Diego, CA, USA). The ACEA NovoExpress™ software was used to calculate net fluorescence intensity (NFI) from 12,000 incidences per specimen (ACEA Biosciences Inc., San Diego, CA, USA).

2.9.6. Caspase-Glo 3/7 Activity

The impact of the IC_{50} of ERF on caspase 3/7 activity in HepG2 cells was evaluated using the Caspase-Glo 3/7 Assay kit (Promega, Walldorf, Germany), according to the manufacturer's instructions. Caspase activity was expressed as a proportion of the untreated control [27].

2.10. Statistical Analysis

The Graphpad Prism 6 software was used to conduct all statistical analyses. A one-way analysis of variance was utilized to compare the results (ANOVA). The statistical significance was determined as a *p*-value < 0.05.

3. Results and Discussion

Plant extracts have substantial therapeutic potential with few negative adverse effects for treating infectious diseases, making medicinal herbs an appealing source of new medicinal components. The therapeutic potentials are related to the phytochemical components. The phytochemical profile ERF of *M. arboreus* was investigated.

3.1. Phytochemical Evaluation of ERF of M. arboreus

3.1.1. Phytochemical Screening

The preliminary screening of ERF of *M. arboreus* showed the presence of many phytoconstituents, for example flavonoids, tannins, coumarins, saponins, glycosides, phenols, terpenoids, steroids, emodins, alkaloids, and anthocyanins, which might account for their medicinal effects (Table 1).

Table 1. Phytochemical screening of ERF of *M. arboreus*.

Tests	Result
Flavonoids	+
Tannins	+
Coumarins	+
Saponins	±
Steroids	±
Glycosides	+
Phenols	+
Terpenoids	+
Emodins	+
Anthocyanins	+
Alkaloids	±

(+) positive test; (±) faint.

3.1.2. Total Flavonoid and Phenolic Contents

The ERF of *M. arboreus* showed TFC and TPC of 23.83 ± 2.9 mg quercetin equivalent/g extract and 46.25 ± 2.1 mg gallic acid equivalent/g extract, respectively, as well as high antioxidant activity (Table 2). These phytoconstituents were shown to have a variety of therapeutic activities and were known to be biologically active compounds [28].

Table 2. Total flavonoid and phenolics contents of ERF of *M. arboreus*.

Parameters	Result
TPC (mg GAE/g extract)	46.25 ± 2.1
TFC (mg QE/g extract)	23.83 ± 2.9

Variables are shown as mean ± SD (standard deviation, n = 3). GAE: gallic acid equivalent; QE: quercetin equivalent.

3.1.3. GC/MS Analysis

Figure 1 shows a total scan gas chromatogram of the ERF of *M. arboreus*. It demonstrated the presence of several bioactive chemicals with varying retention times (RT). Table 3 shows the molecular weight, RT, and percent peak area, as well as chemical formulae of the identified compounds. Additionally, the biological functions of the identified compounds, as anticipated by Dr. Duke's phytochemical and ethnobotanical databases (USDA, Agricultural Research Service, 1992–2016), are also summarized in Table 3.

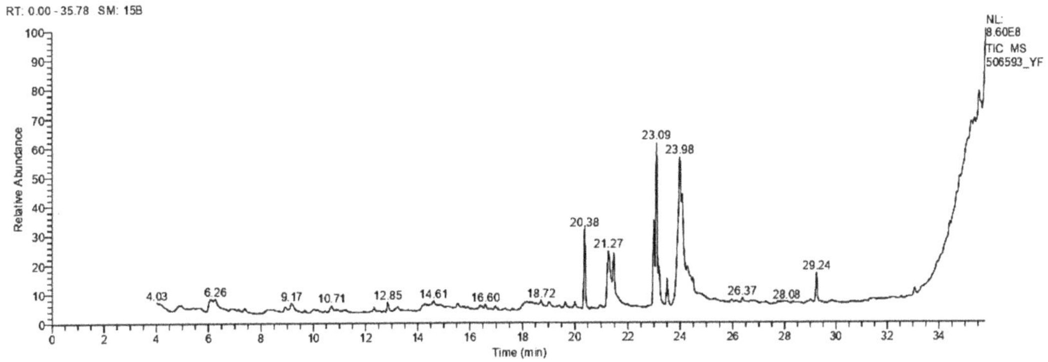

Figure 1. GC-MS chromatogram of the ERF of *M. arboreus*.

Table 3. The identified compounds in the ERF of *M. arboreus* RF by GC/MS analysis.

	RT	Name of the Compound	MF	MW	Peak Area (%)	Biological Activity **
1	6.07	4H-Pyran-4-one, 2,3-dihydro-3,5-dihydroxy-6-methyl	$C_6H_8O_4$	144	2.85	Antimicrobial, anti-inflammatory
2	6.26	Octadecanoic acid, ethyl ester	$C_{20}H_{40}O_2$	312	1.46	Anti-microbial
3	9.17	3,5-Heptadienal, 2-ethylidene-6-methyl-	$C_{10}H_{14}O$	150	1.14	Anti-inflammatory, antitumor, antiviral
4	10.71	4-(3,3-dimethyl-1-butynyl)-4-hydroxy-2,6,6-trimethyl-2-cyclohexen-1-one	$C_{15}H_{22}O_2$	234	1.09	Antioxidant, anti-inflammatory
5	12.33	10,13-Octadecadiynoic acid, methyl ester	$C_{19}H_{30}O_2$	290	0.60	No activity reported
6	12.84	1-(3-Methoxy-5-methylphenyl)-*N*-methylpropan-2-amine	$C_{12}H_{19}NO$	193	1.20	No activity reported

Table 3. *Cont.*

	RT	Name of the Compound	MF	MW	Peak Area (%)	Biological Activity **
7	18.72	9,12,15-Octadecatrienoic acid, 2,3-bis [(trimethylsilyl)oxy] propyl ester, (z, z, z)-	$C_{27}H_{52}O_4Si_2$	496	0.66	Anticancer, hepatoprotective
8	19.64	Cis-13-Eicosenoic acid	$C_{20}H_{38}O_2$	310	0.81	Anti-inflammatory activity
9	20.01	9-octadecenoic acid, (2-phenyl-1,3-dioxolan-4-yl) methyl ester, cis	$C_{28}H_{44}O_4$	444	0.72	Antimicrobial, anti-inflammatory
10	20.38	Hexadecanoic acid, methyl ester	$C_{17}H34O_2$	270	8.88	Antioxidant, antimicrobial, antihypercholesterolemic property
11	21.27	Hexadecanoic acid	$C_{16}H_{32}O_2$	256	8.52	Anti-inflammatory, antioxidant, antihypercholesterolemic
12	21.47	Hexadecanoic acid, ethyl ester	$C_{18}H_{36}O_2$	284	3.66	Antioxidant, antihypercholesterolemic antiandrogenic
13	22.99	9,12-Octadecadienoic acid (Z, Z)-, methyl ester	$C_{19}H_{34}O_2$	294	9.46	Hepatoprotective, antihistamine, hypocholesterolemia, anti-eczema
14	23.09	11-Octadecenoic acid, methyl ester	$C_{19}H_{36}O_2$		19.49	Antioxidant, antimicrobial
15	23.19	16-Octadecenoic acid, methyl ester	$C_{19}H_{36}O_2$	296	3.86	Selectively inhibit eukaryotic DNA polymerase activities in vitro
16	23.50	Octadecanoic acid, methyl ester	$C_{19}H_{38}O_2$	298	3.03	Antimicrobial
17	23.98	9,12-Octadecadienoic acid (Z, Z)-, 2-hydroxy-1-(hydroxymethyl)ethyl ester	$C_{21}H_{38}O_4$	354	18.41	Antiarthritic, hepatoprotective, antiandrogenic, anticoronary, antieczemic, anticancer
18	24.08	Ethyl oleate	$C_{18}H_{34}O_2$	282	7.73	Antibacterial, anticancer
19	24.28	Oleic acid	$C_{20}H_{38}O_2$	310	1.25	It is used as a vehicle for intramuscular drug delivery, progesterone
20	24.41	Linoleic acid ethyl ester	$C_{20}H_{36}O_2$	308	0.71	Anti-arthritic, anti-acne, hepatoprotective, anti-histaminic, anti-coronary
21	24.49	Octadecanoic acid, 2,3-dihydroxypropyl ester	$C_{21}H_{42}O_4$	358	1.00	Anticancer, antimicrobial

** Dr. Duke's Phytochemical and Ethnobotanical Databases. RT: retention time; MF: molecular formula; MW: molecular weight.

In the ERF of *M. arboreus*, 21 components were found. The predominant compounds were 11-octadecenoic acid methyl ester (19.49%), 9,12-octadecadienoic acid (Z, Z)-2-hydroxy-1-(hydroxymethyl)ethyl ester (18.41%), 9,12-Octadecadienoic acid (Z, Z)-methyl ester (9.46%), hexadecanoic acid methyl ester (8.88%), hexadecanoic acid (8.52%), and oleic acid (7.73%).

3.1.4. Identification and Quantification of Phenolics and Flavonoids

HPLC examination of the ERF of *M. arboreus* revealed the identification and quantification of 13 polyphenolic compounds (6 flavonoids and 7 phenolic acids), as shown in Table 4. The compounds were identified by comparison to authentic samples analyzed using the same procedures. Hesperidin and luteolin were the major flavonoids identified at concentrations of 8.78 and 7.55 µg/mg of the ERF. Gallic acid was the predominant phenolic component identified in the extract (7.39 µg/mg), followed by syringic acid (7.16 µg/mg) and cinnamic acid (6.44 µg/mg) (Figure 2, Table 4).

Table 4. HPLC analysis of phenolics and flavonoids in the ERF of *M. arboreus*.

Components	RT (min)	Conc. (µg/mg)
	Flavonoid compounds	
Naringin	4.6	2.63
Hesperidin	10.0	8.78
Kaempferol	8.1	1.02
Luteolin	9.0	7.55
Apigenin	15.0	5.14
Catechin	12.0	4.27
	Phenolic compounds	
Caffeic acid	8.1	4.78
Cinnamic acid	13.0	6.44
Gallic acid	10.0	7.39
Syringic acid	5.2	7.16
Benzoic acid	7.0	3.55
Ellagic acid	15. 6	1.23

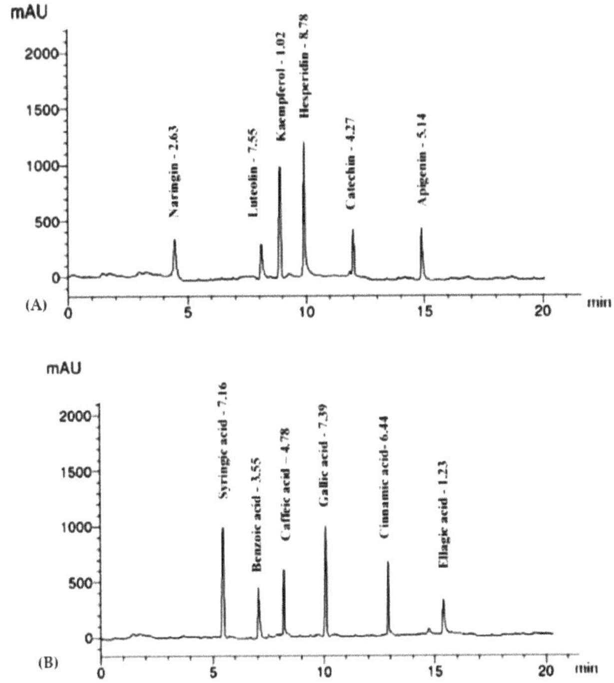

Figure 2. HPLC chromatogram showing identified flavonoids (**A**) and phenolic acids (**B**) in the ERF.

3.1.5. ^{1}HNMR Fingerprint of the ERF

The ^{1}HNMR spectrum of the ERF at 400 MHz (Figure 3) revealed that the extract is rich in oxygenated saturated and unsaturated hydrocarbon compounds. Signals in the range 0.0–4.0 ppm are predominant in the spectrum while signals in the aromatic range 6.0–8.0 ppm are weak. In correlation to the GC and HPLC analysis, the extract is rich in fatty acids (saturated and unsaturated), which explains why the fingerprint pattern as the characteristic signals for fatty acids are 2.0–2.5 ppm for (CH_2), 3.0–4.0 ppm for (-CHOH-, CH_3-CO-, -CH_2-CO-), 5.0–6.0 ppm (-CH=CH-) [29].

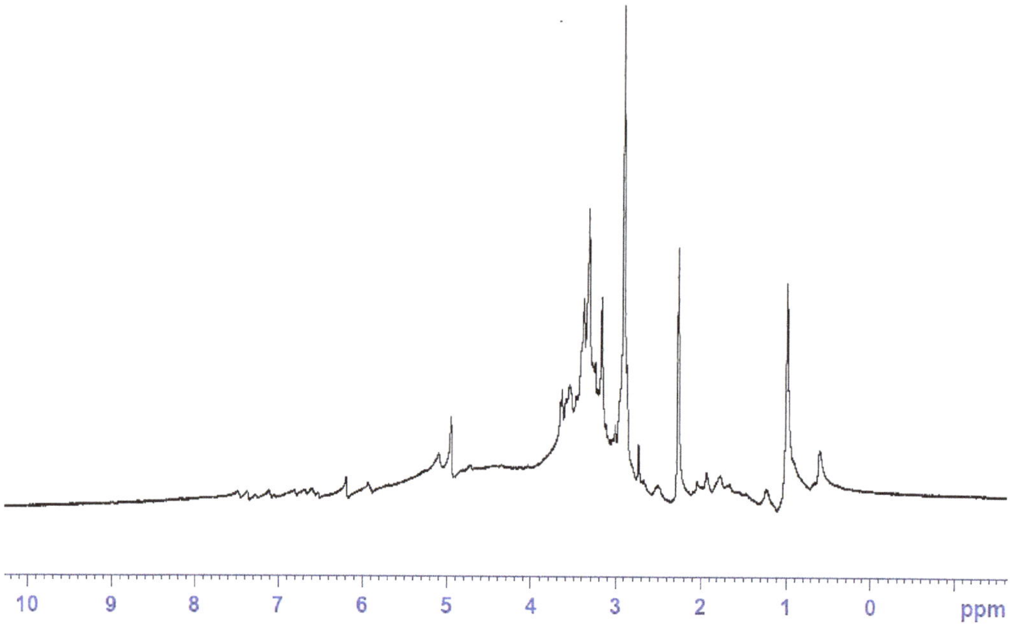

Figure 3. ^{1}HNMR spectrum of the ERF of *M. arboreus* in DMSO (400 MHz).

3.2. Antioxidant Capacities of ERF of M. arboreus

The equivalent antioxidant capacities of trolox (TE) as compared to the ERF were 716.45 ± 16.12, 99.15 ± 4.96, 1138.11 ± 79.65 μM TE/mg extract in ABTS, FRAP, and ORAC assays, respectively. ERF exerted high free radical scavenging activity against DPPH radical (IC_{50} = 115.6 ± 16.9 μg /mL) and high ability of its metal chelating property (57.58 ± 3.5 μM EDTA eq/mg extract) (Table 5).

Table 5. Antioxidant capacities of the ERF of *M. arboreus*.

Parameters	Result
DPPH (IC_{50} μg/mL)	115.6 ± 16.9
ABTS (μM TE/mg extract)	716.45 ± 16.12
FRAP (μM TE/mg extract)	99.15 ± 4.96
ORAC (μM TE/mg extract)	1138.11 ± 79.65
Metal chelating property (μM EDTA eq/mg extract)	57.58 ± 3.5

Variables are shown as mean ± SD (standard deviation, n = 3).

Antioxidant properties are well known in phenolic compounds by acting as reducing agents, free radical scavengers, or metal chelators [30]. The most abundant plant phenolics are flavonoids and phenolic acids, which have a substantial antioxidant activity both in vitro as well as in vivo [31].

3.3. Antibacterial and Antifungal Activities of Extract

In many regions of the world, there is a great deal of interest in medicinal plants as therapeutic medications because of the rise in drug-resistant bacteria and the emergence of more pathogenic bacterial species. Many medicinal plants have been studied in vitro against bacterial strains, with extracts and pure components of several medicinal plants being particularly beneficial [32].

Eight different strains of marine pathogenic bacteria were selected in this study, including *B. subtillus*, *E. faecalis*, *B. cereus*, *S. typhimurium*, *P. aeruginosa*, *V. fluvialis*, *S. aureus*, and *V. damsela*. The ERF of *M. arboreus* showed an antibacterial effect against most of the tested strains with average inhibition zones ranging between 10 and 20 mm compared to the positive control amoxicillin/clavulanic acid (Table 6). The ERF of *M. arboreus* exhibited a strong antibacterial activity against *V. damsela* with an inhibition zone of 20 ± 0.2 mm, moderate antibacterial activity against *V. fluvialis* and *S. typhimurium* with the inhibition zones being 16 mm, and showed lower effects against *E. faecalis*, *S. aureus*, and *P. aeruginosa* with inhibition zones of 10, 12, and 14 mm, respectively. On the other hand, the ERF of *M. arboreus* was ineffective against *B. subtilus* and *B. cereus*. The negative control (DMSO) showed no zone of inhibition.

Table 6. Antimicrobial activity of the crude extract of red flower of *M. arboreus* using well-cut diffusion method.

Pathogens	Inhibition Zone (mm)	Amoxicillin/Clavulanic (Positive Control)	DMSO (Negative Control)
B. subtillus	0.0	14.0 ± 0.5	0.0
S. aureus	12.0 ± 0.6	12.0. ± 0.2	0.0
E. faecalis	10.0 ± 0.1	18.0 ± 0.5	0.0
P. aeruginosa	14.0 ± 0.1	14.0 ± 0.2	0.0
V. fluvialis	16.0 ± 0.4	22.0 ± 0.6	0.0
V. damsela	20.0 ± 0.2	24.0 ± 0.2	0.0
B. cereus	0.0	10.0 ± 0.3	0.0
S. typhimurium	16.0 ± 0.2	20.0 ± 0.3	0.0

The data are represented as mean ± SD in mm of inhibition zone demonstrated, contrasted utilizing ANOVA, with a significance level (p-value) ≤ 0.05.

In our study, the ERF presented activity against *A. terreus*, *A. fumigatus*, and *A. flavus* respectively, with no effect on the other strains. The ERF of *M. arboreus* did not show any effect on either *A. parasiticus* and *P. oxalicum*. To establish the susceptibilities of ERF against the tested strains, the minimum inhibitory concentration (MIC) values were determined (Table 7). The ERF of *M. arboreus* exhibited the lowest MIC for *V. damsela* (1.5 ± 0.02 mg/mL). The MIC values for *E. faecalis*, *S. aureus*, *P. aeruginosa*, *V. fluvialis*, and *S. typhimurium* were 12.5 ± 0.02, 10.0 ± 0.06, 10.0 ± 0.01, 2.5 ± 0.05, and 5.0 ± 0.01 mg/mL, respectively. On the other hand, the MIC values against *A. fumigatus*, *A. flavus*, *A. niger*, and *A. terreus* were 1.0 ± 0.02, 1.25 ± 0.01, 1.75 ± 0.06, and 0.75 ± 0.01 mg/mL, respectively.

Table 7. MIC values of the ERF against selected pathogens.

Pathogens	MIC (mg/mL)
S. aureus	10.0 ± 0.06
E. faecalis	12.5 ± 0.02
P. aeruginosa	10.0 ± 0.01
V. fluvialis	2.5 ± 0.05
V. damsela	1.5 ± 0.02
S. typhimurium	5.0 ± 0.01
A. fumigatus	1.0 ± 0.02
A. niger	1.75 ± 0.06
A. flavus	1.25 ± 0.01
A. terreus	0.75 ± 0.01

3.3.1. Bacterial Killing Kinetics Assay of ERF against Marine *V. damsela*

A time-kill kinetic assay of the ERF against marine *V. damsela* was investigated, with the results demonstrated in Figure 4. As a result, time-kill curve was plotted between the logarithmic number of CFU/mL and incubation time. At 4 × MIC concentration, the ERF showed a decrease in the amount of viable *V. damsela* at 8–24 h. The extent by which bacteria was inhibited by the plant extract by time varied greatly, as shown by killing analyses [33]. Therefore, the capacity of plant secondary metabolites to possess antibacterial characteristics may be taken into consideration, as well as their response to microbial infection [34].

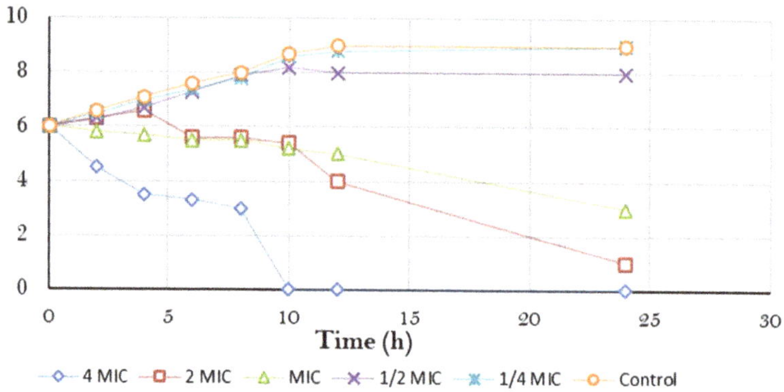

Figure 4. Time-kill curve of *V. damsela* by *M. arboreus* extract.

3.3.2. Analysis of the Synergistic Impact of ERF

Figure 5 and Table 8 display the findings of an evaluation of the synergistic effect of ERF and amoxicillin/clavulanic acid against the selected pathogen marine *V. damsela*. Amoxicillin/clavulanic acid at a concentration of (20/10 mcg) demonstrated moderate effective action against the *V. damsela* that were examined. When compared to 250 and 750 μg/mL, the antibacterial activity displayed by the combined effect of antibiotics and ERF was significantly stronger against the selected pathogen at a concentration of 1000 μg/mL, with a zone of inhibition ranging in diameter from 26 ± 0.2 to 28 ± 0.1 mm (Table 6). Due to the synergistic action of the ERF and amoxicillin/clavulanic, it was hypothesized that this combination therapy would be successful against the *V. damsela* that were tested.

The chemical composition of the ERF of M. arboreus revealed the existence of noteworthy chemicals such as octadecenoic acid methyl ester, hexadecanoic acid, oleic acid, 11-octadecenoic acid, and octadecanoic acid (Table 3). These compounds have proven antimicrobial activity. Cinnamic acid and its hydroxylated derivatives demonstrated antifungal properties, reducing antityrosinase enzyme activity and fungal spore germination [35]. Cinnamic acids suppressed fungal expansion via interacting with the enzyme benzoate 4-hydroxylase, which is involved in the detoxification of aromatic compounds [36]. Hexadecanoic acid reacted with the lipopolysaccharides' hydroxyl group, an element of the bacterial cell wall, causing the lipopolysaccharide membrane structure's asymmetric conversion, as per Johannes et al. [37]. Therefore, the lipid structure of the membrane was disrupted. The cell swelled, the cytoplasm membrane was damaged, and the cell was distended and lysed due to the alteration in the cell membrane. The hydroxyl group of hexadecanoic acid has been noticed to be toxic to the cell protoplasm, as the compound permeates the cell wall and causes damage [38].

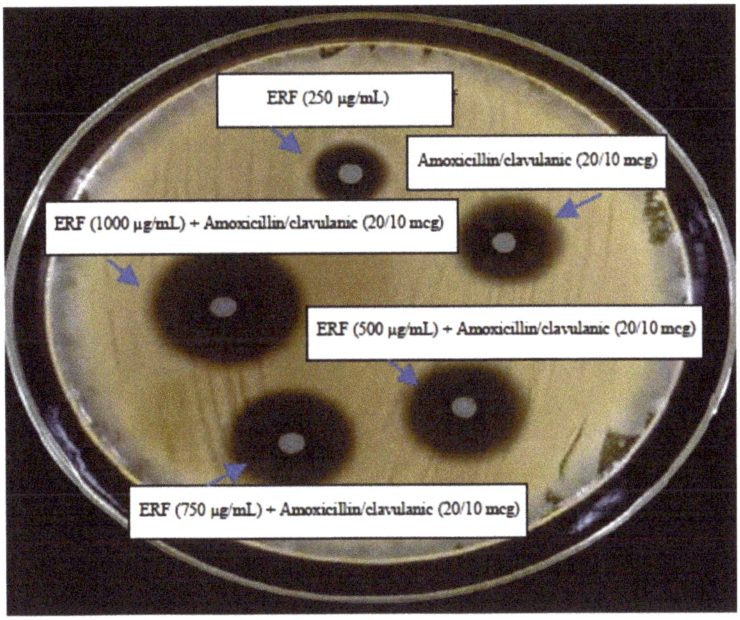

Figure 5. Synergistic effect of deferent concentrations of ERF and amoxicillin/clavulanic (20/10 mcg) acid against selected pathogen marine *V. damsela*.

Table 8. Combined activity (Inhibition zone (mm)) of ERF with Amoxicillin/clavulanic against different *V. damsela* (10^{-6} CFU/mL).

Amoxicillin/Clavulanic & ERF	Inhibition Zone (mm) of *V. damsela* (10^{-6} CFU/mL)
Amoxicillin/clavulanic (20/10 mcg)	24.0 ± 0.2
ERF (250 µg/mL)	20.0 ± 0.2
ERF (500 µg/mL) + Amoxicillin/clavulanic (20/10 mcg)	26.0 ± 0.2
ERF (750 µg/mL) + Amoxicillin/clavulanic (20/10 mcg)	26.9 ± 0.2
ERF (1000 µg/mL) + Amoxicillin/clavulanic (20/10 mcg)	28.0 ± 0.1

V. damsela is one of the pathogens associated with infections caused by seafood; thus, the ERF may be an option for treating this infection.

3.4. Cytotoxic Activity

This study aims to examine the impacts of the ERF on liver cancer in vitro, utilizing the most common cell line for hepatotoxicity and drug metabolism studies, hepatocellular carcinoma HepG2. HepG2 cells are nontumorigenic, increase rapidly, have an epithelial-like shape, and are capable of performing a wide variety of differentiated liver activities [39].

After 72 h of incubation, the SRB test was utilized to analyze the cytotoxicity of the ERF on the hepatocellular carcinoma (HepG2) cell line. The results showed that ERF significantly reduced HepG2 cell proliferation in a dose-dependent manner, with an IC_{50} value of 67.182 µg/µL (Figures 6 and 7).

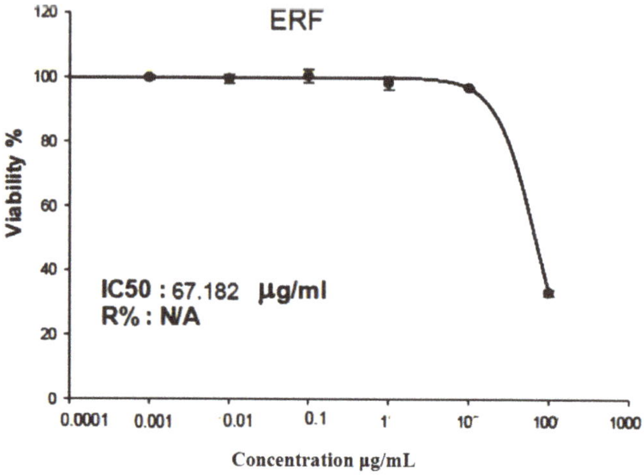

Figure 6. SRB assay on HepG2 cells to validate the ERF impact on the cell viability after 72 h.

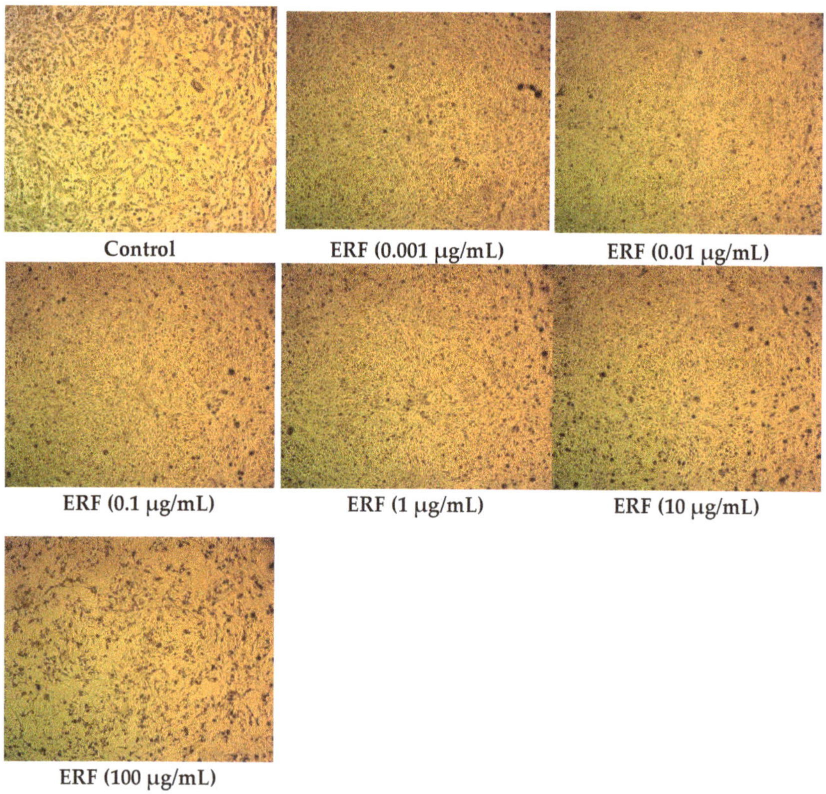

Figure 7. Photomicrographs of HepG2 cells treated with ERF (100× magnification).

Phenolics, which come in various types, are known for causing apoptosis and cytotoxicity in cancer cell lines. The capability of phenolic compounds to scavenge radicals and their antioxidant capabilities are primarily responsible for their anticancer effects.

Hesperidin, a primary flavonoid in the extract under investigation, protected the rat liver against CCl_4-induced oxidative stress and dysfunction linked to its antioxidant properties [40]. Hesperidin's impact on the MCF-7 human breast cancer cells and prostate cancer cell proliferation was studied [41]. Abd El-Azim et al. [42] found that 4-hydroxybenzoic acid, a phenolic acid in excessive levels in the extract, had substantial cytotoxic action on both colon (HCT116) and liver (HepG2) cancer cell lines. Polyphenolic substances reduce mutagenesis and carcinogenesis in humans when consumed in up to 1.0 g per day from a diet rich in fruits, vegetables, and other plants [43].

3.4.1. Cell Cycle Analysis

To explore the impact of the ERF on cell cycle distribution, HepG2 cells were treated for 48 h with the pre-determined IC_{50} of the ERF, and DNA content was measured utilizing flow cytometry. The results in Figure 8 revealed an apparent change in the distribution of different phases. In G0/G1-phase cells, ERF did not further increase antiproliferative effects (38.1 ± 1.19) compared to untreated cells (41.39 ± 0.46%).

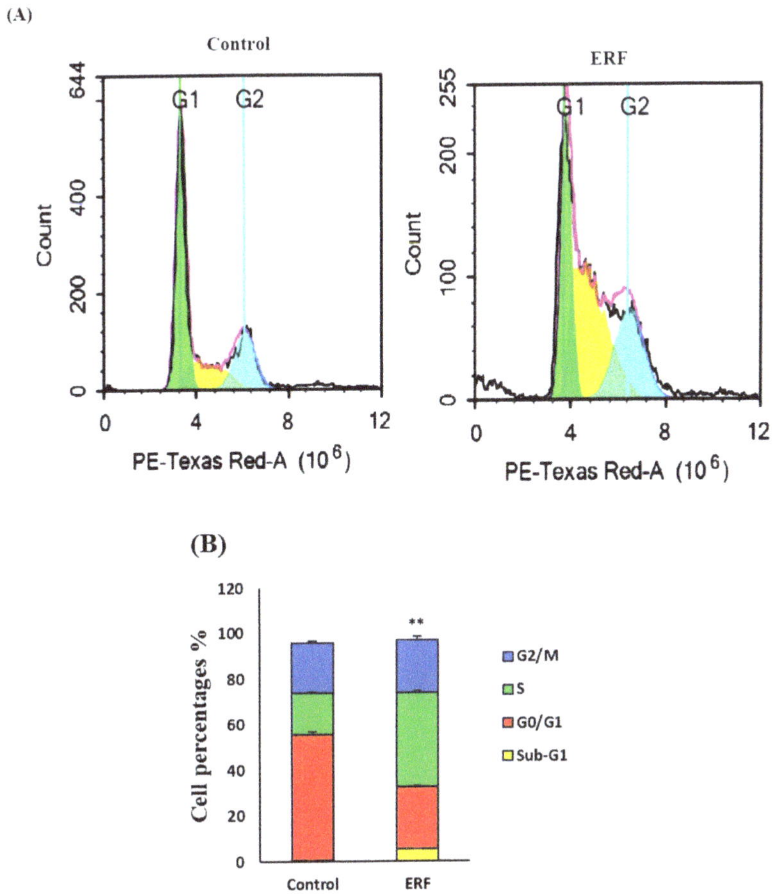

Figure 8. Representative flow cytometry analysis of HepG2 cells treated for 48 h with the IC_{50} of the ERF. DNA cytometry analysis was performed to examine cell cycle distribution, and diverse cell phases (**A**) were plotted as percentages of total occurrences (**B**). ** Significantly different from the control group; results are provided as mean ± SD; n = 3.

Compared to the untreated cells (50.4 ± 3.2%), the ERF caused S-phase arrest, and thus increased the cell population (38.0 ± 1.9%). Compared to the untreated cells (0.59 ± 0.03%), ERF dramatically accelerated cell mortality as observed by an elevation in the sub-G1 phase cell population (5.31 ± 0.34%). The findings implied HepG2 cell death by exposure to the study extract.

The ERF significantly induced more cell death manifested by an increased pre-G phase cell population (5.31 ± 0.34%) compared to untreated cells (0.59 ± 0.03%). The results suggest that HepG2 cells underwent apoptosis upon treatment with the study extract.

A range of processes, including apoptosis and cell cycle arrest, were involved in the cytotoxic effects of the ERF extract. The ability of anticancer drugs to induce cell cycle arrest in cancer cells was measured [44]. A significant hypodiploid sub-G0/G1 peak was visible in the production of apoptotic cells, which was easily observed with substantial damage to cellular DNA and might be distinguished by flow cytometry [45].

According to these data, the ERF extract could produce substantial DNA loss to cause apoptosis in the present investigation, as the concentration of hypodiploid cells in the sub-G0/G1 phase was a sign of apoptotic cell death. Apoptosis would be confirmed by several intracellular pathways, such as caspase activation and MMP disruption.

3.4.2. Assessing Cell Apoptosis with Annexin V-FITC

The ERF impact on the growth suppression of HepG2 cells was associated with apoptosis, as determined by apoptotic and necrotic cells' Annexin V analysis. The cells were double-labeled with PI, which produced red fluorescence in necrotic cells, and Annexin V-FITC, which caused cytoplasmic green labeling in apoptotic cells after 24 h of treatment with the ERF extract's IC_{50}. In fluorescence microscopy images, viable cells were negative for Annexin V and PI (Figure 9). A considerable amount of green and red labeling was observed in ERF, indicating apoptotic and necrotic cells. When cells were treated with ERF, many apoptotic cells were found, indicating that this extract was primarily responsible for apoptosis.

Figure 9 demonstrates the HepG2 cells' distribution in four quadrants (Q1 = necrosis phase, Q2 = late apoptosis, Q3 = normal intact cells, Q4 = early apoptosis phase) and represents one of three independent tests carried out. Cells that experienced apoptosis would shift from the viable quadrant (Q3) to the early apoptosis quadrant (Q4) and finally end up in the late apoptosis quadrant (Q2). Necrosis, in contrast, caused cells to move from the viable quadrant (Q3) to the late necrosis quadrant (Q2). Untreated cells had a proportion of viable cells of 98.07 ± 0.08%, dead cells of 1.93 ± 0.08%, late apoptosis of 0.37 ± 0.07%, and early apoptosis of 0.32 ± 0.09%. ERF increased the late apoptotic population to 4.67 ± 0.31%. A noticeable decrease was indicated in necrotic cells, with proportions of 0.37 ± 0.07% upon treatment, compared to untreated cells at 0.52 ± 0.12. Lastly, early apoptotic cells, expressed by Q4, demonstrated only a slight elevation in cell distribution due to treatment with the ERF to 5.05 ± 0.17%.

Phosphatidyl-serine (PS) on the outer layer of the plasma membrane served as a recognition site for phagocytes during the early stages of apoptosis [46]. Annexin V, a calcium-dependent protein, could bind to the exposed phosphatidyl-serine on the membrane's exterior layer (PS) [47]. The percentage of cells going through late apoptosis rose exponentially in this study, indicating that apoptosis was one of the primary mechanisms in which the plant extract induced cell death in the four studies.

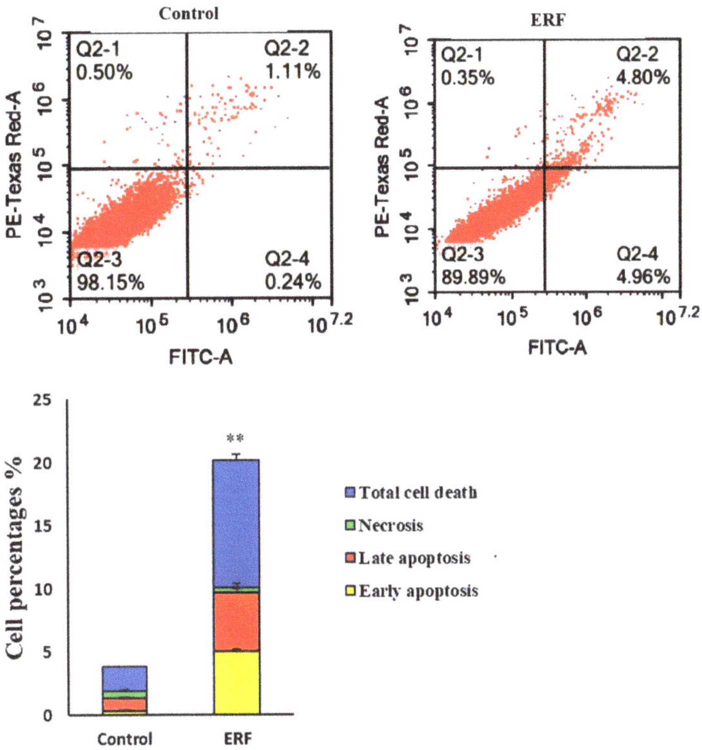

Figure 9. In contrast to untreated cells, apoptosis and necrosis were detected utilizing Annexin V-FITC and PI dual staining after 24 h of treating HepG2 cells with IC_{50} of the study crude extract RF. ** Significantly different from the control group; data are presented as mean ± SD; $n = 3$.

3.4.3. Assessment of Autophagy

Autophagy-mediated programmed cell death, such as apoptosis, is a significant issue in science. Using the Cyto-ID autophagy detection dye and flow cytometry, we studied the influence of ERF on the autophagy process in HepG2 cells. In comparison to the untreated cells, ERF treatment significantly boosted autophagic cell death (Figure 10).

Autophagy was another hypothesized cell death route, but its significance in cancer cell death was convoluted and controversial [48]. ERF caused considerable autophagic cell death in HepG2 cells, which could be a pro-death mechanism due to poor apoptosis in this cell type [49].

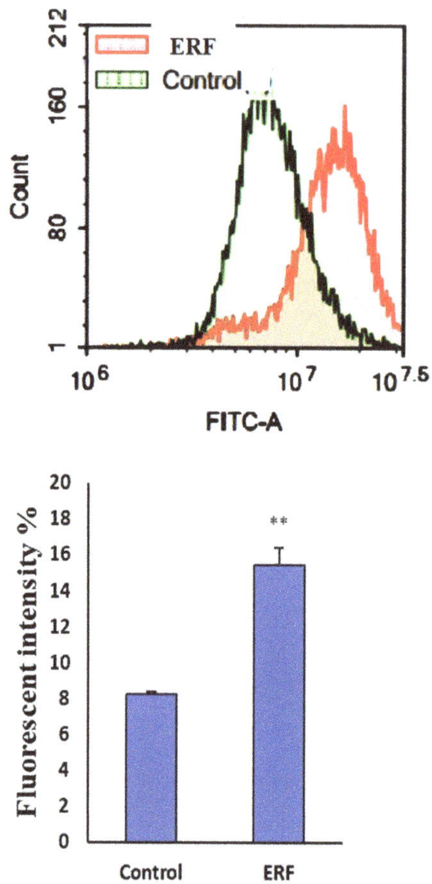

Figure 10. After being exposed to ERF, autophagic cell death was assessed in HepG2 cells. Cells were stained with a Cyto-ID autophagosome tracker after exposure to ERF for 24 h. ** Significantly different from the control group; data are presented as mean SD; $n = 3$.

3.4.4. Effect of ERF on the Activity of Caspase 3/7

The activation of caspases is required for the last step of apoptosis [50]. Understanding the stimulation route by cytotoxic substances could help model improved therapeutic options [51].

The caspase 3/7 activity was evaluated on HepG2 cells treated with the concentration of IC_{50} values of the extract for 24 h to determine if the apoptotic effect generated by the ERF was dependent on caspase activation (Figure 11). The activation of caspase 3/7 was increased by 8.71 $\pm$ 0.99-fold in ERF-treated cells compared to 1.28 $\pm$ 0.17-fold in the untreated cells, confirming the effect of this extract on apoptotic cell death formerly demonstrated in cell cycle studies and Annexin V. The cleavage of several caspases triggered apoptosis. Understanding the effects of caspase cleavage could help us understand cell death, as well as other biological processes [52]. The increased caspase-3 activity in the treated HepG2 suggested that extrinsic and intrinsic caspase-3 activation pathways were utilized at this dose (IC_{50}). Apoptosis triggered by caspases could activate either the death receptor (extrinsic) or mitochondrial (intrinsic) pathways or both [53].

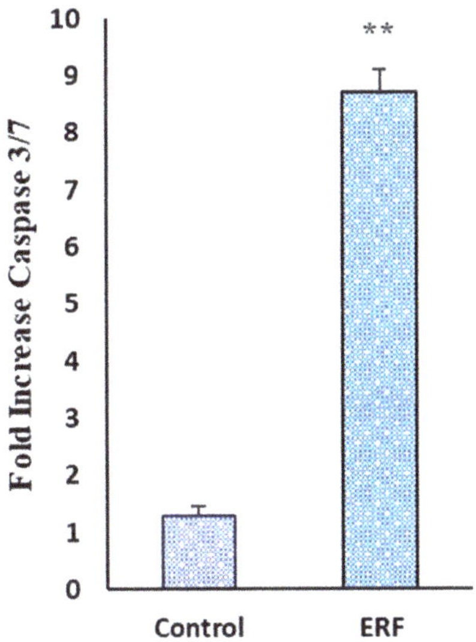

Figure 11. Effect of ERF on the Activity of Caspase 3/7 on HepG2 cells. All data were expressed as mean ± standard deviation (SD) at a significance level of $p < 0.05$ and indicated by **.

4. Conclusions

This is the first antimicrobial, anticancer evaluation of the ERF of *M. arboreus* against hepatocarcinoma cell line HepG2. The ERF of *M. arboreus* demonstrated anticancer effectiveness against HepG2 and an in vitro growth inhibitory impact against microbiological growth. Our GC-MS and HLPC analyses showed the existence of many phytochemical compounds that might influence the antibacterial and anticancer properties of *M. arboreus* red flower ethanolic extract. As a result, it is suggested that the antibacterial and anticancer efficacy of the GC-MS and HLPC found compounds to be evaluated to develop a novel perspective on antimicrobial and anticancer medicine and assess the mode of action used to combat anticancer recovery.

Author Contributions: H.S.S.G. conceived the project. H.S.S.G., A.E.R. and M.M.H. designed and performed the experiments. H.S.S.G., A.E.R. and O.I.A.S. analyzed the data. N.A.S. supervised the project. H.S.S.G., A.E.R., M.M.H., O.I.A.S. and M.E.M. wrote the manuscript. H.S.S.G., A.E.R., N.A.S., M.M.H., O.I.A.S. and M.E.M. provided critical discussion, editing, and final approval of the manuscript. All authors have read and agreed to the published version of the manuscript.

Funding: This research received no external funding.

Institutional Review Board Statement: Not applicable.

Informed Consent Statement: Not applicable.

Data Availability Statement: Not applicable.

Conflicts of Interest: The authors declare no conflict of interest.

References

1. Garcia-Oliveira, P.; Otero, P.; Pereira, A.; Chamorro, F.; Carpena, M.; Echave, J.; Fraga-Corral, M.; Simal-Gandara, J.; Prieto, M. Status and Challenges of Plant-Anticancer Compounds in Cancer Treatment. *Pharmaceuticals* **2021**, *14*, 157. [CrossRef] [PubMed]
2. Villanueva, A. Hepatocellular carcinoma. *N. Engl. J. Med.* **2019**, *380*, 1450–1462. [CrossRef] [PubMed]
3. Di Bisceglie, A.M.; Rustgi, V.K.; Hoofnagle, J.H.; Dusheiko, G.M.; Lotze, M.T. NIH conference. Hepatocellular carcinoma. *Ann. Intern. Med.* **1988**, *108*, 390–401. [CrossRef] [PubMed]

4. Alhaqbani, A.H.; Almusalam, A.; Alnadhari, A.; AlSalem, A.; Alrasheed, M.; Alalwan, A.; Aldamkh, B. The prevalence and associated factors of viral hepatitis and cryptogenic related hepatocellular carcinoma at King Abdulaziz Medical city-Riyadh. *Int. J. Community Med. Public Health* **2019**, *6*, 561. [CrossRef]
5. Wang, Z.; Li, Z.; Ye, Y.; Xie, L.; Li, W. Oxidative Stress and Liver Cancer: Etiology and Therapeutic Targets. *Oxidative Med. Cell. Longev.* **2016**, *2016*, 7891574. [CrossRef]
6. McLoughlin, M.R.; Orlicky, D.J.; Prigge, J.R.; Krishna, P.; Talago, E.A.; Cavigli, I.R.; Eriksson, S.; Miller, C.G.; Kundert, J.A.; Sayin, V.I.; et al. TrxR1, Gsr, and oxidative stress determine hepatocellular carcinoma malignancy. *Proc. Natl. Acad. Sci. USA* **2019**, *116*, 11408–11417. [CrossRef]
7. Arteel, G.E. Oxidants and antioxidants in alcohol-induced liver disease. *Gastroenterology* **2003**, *124*, 778–790. [CrossRef]
8. Gordaliza, M. Natural products as leads to anticancer drugs. *Clin. Transl. Oncol.* **2007**, *9*, 767–776. [CrossRef]
9. Abdulaev, F.I. Plant-derived agents against cancer. In *Pharmacology and Therapeutics in the New Millennium*; Gupta, S.K., Ed.; Narosa Publishing House: New Delhi, India, 2001; p. 345.
10. Newman, D.J.; Cragg, G.M. Natural Products as Sources of New Drugs over the 30 Years from 1981 to 2010. *J. Nat. Prod.* **2012**, *75*, 311–335. [CrossRef]
11. Puckhaber, L.S.; Stipanovic, R.D.; Bost, G.A. Analyses for flavonoid aglycones in fresh and preserved Hibiscus flowers. In *Trends in New Crops and New Uses*; ASHS Press: Alexandria, Egypt, 2002; pp. 556–563.
12. Manian, R.; Anusuya, N.; Siddhuraju, P.; Manian, S. The antioxidant activity and free radical scavenging potential of two different solvent extracts of *Camellia sinensis* (L.) O. Kuntz, *Ficus bengalensis* L. and *Ficus racemosa* L. *Food Chem.* **2008**, *107*, 1000–1007. [CrossRef]
13. Laulloo, S.J.; Bhowon, M.G.; Soyfoo, S.; Chua, L.S. Nutritional and Biological Evaluation of Leaves of *Mangifera indica* from Mauritius. *J. Chem.* **2018**, *2018*, 6869294. [CrossRef]
14. Zhishen, J.; Mengcheng, T.; Jianming, W. The determination of flavonoid contents in mulberry and their scavenging effects on superoxide radicals. *Food Chem.* **1999**, *64*, 555–559. [CrossRef]
15. Adedapo, A.A.; Jimoh, F.A.; Koduru, S.; Afolayan, A.J.; Masika, P.J. Antibacterial and antioxidant properties of the methanol extracts of the leaves and stems of *Calpurnia aurea*. *BMC Complement. Altern. Med.* **2008**, *8*, 53. [CrossRef]
16. Gülçin, I.; Küfrevioğlu, Ö.İ.; Oktay, M.; Büyükokuroğlu, M.E. Antioxidant, antimicrobial, antiulcer and analgesic activities of nettle (*Urtica dioica* L.). *J. Ethnopharmacol.* **2004**, *90*, 205–215. [CrossRef]
17. Ou, B.; Hampsch-Woodill, M.; Prior, R.L. Development and Validation of an Improved Oxygen Radical Absorbance Capacity Assay Using Fluorescein as the Fluorescent Probe. *J. Agric. Food Chem.* **2001**, *49*, 4619–4626. [CrossRef]
18. Abdel-Shafi, S.; Al-Mohammadi, A.-R.; Sitohy, M.; Mosa, B.; Ismaiel, A.; Enan, G.; Osman, A. Antimicrobial Activity and Chemical Constitution of the Crude, Phenolic-Rich Extracts of *Hibiscus sabdariffa*, *Brassica oleracea* and *Beta vulgaris*. *Molecules* **2019**, *24*, 4280. [CrossRef]
19. Mizzi, L.; Chatzitzika, C.; Gatt, R.; Valdramidis, V. HPLC Analysis of Phenolic Compounds and Flavonoids with Overlapping Peaks. *Food Technol. Biotechnol.* **2020**, *58*, 12–19. [CrossRef]
20. Abdelhakim, H.; El-Sayed, E.; Rashidi, F. Biosynthesis of zinc oxide nanoparticles with antimicrobial, anticancer, antioxidant and photocatalytic activities by the endophytic *Alternaria tenuissima*. *J. Appl. Microbiol.* **2020**, *128*, 1634–1646. [CrossRef]
21. Pia, S.P.; Mahmud, R.; Perumal, S. Determination of Antibacterial Activity of Essential Oil of Myristica fragrans Houtt. using Tetrazolium Microplate Assay and its Cytotoxic Activity Against Vero Cell Line. *Int. J. Pharmacol.* **2012**, *8*, 572–576.e6. [CrossRef]
22. Pourali, P.; Badiee, S.H.; Manafi, S.; Noorani, T.; Rezaei, A.; Yahyaei, B. Biosynthesis of gold nanoparticles by two bacterial and fungal strains, *Bacillus cereus* and *Fusarium oxysporum*, and assessment and comparison of their nanotoxicity in vitro by direct and indirect assays. *Electron. J. Biotechnol.* **2017**, *29*, 86–93. [CrossRef]
23. Shabih, S.; Hajdari, A.; Mustafa, B.; Quave, C.L. Medicinal plants in the Balkans with antimicrobial properties. *Med. Plants Anti-Infect.* **2022**, *2022*, 103–138. [CrossRef]
24. Saeloh, D.; Visutthi, M. Efficacy of Thai Plant Extracts for Antibacterial and Anti-Biofilm Activities against Pathogenic Bacteria. *Antibiotics* **2021**, *10*, 1470. [CrossRef]
25. Gupta, M.; Tomar, R.S.; Kaushik, S.; Mishra, R.K.; Sharma, D. Effective Antimicrobial Activity of Green ZnO Nano Particles of *Catharanthus roseus*. *Front. Microbiol.* **2018**, *9*, 2030. [CrossRef]
26. Fekry, M.; Ezzat, S.M.; Salama, M.M.; AlShehri, O.Y.; Al-Abd, A.M. Bioactive glycoalkaloides isolated from *Solanum melongena* fruit peels with potential anticancer properties against hepatocellular carcinoma cells. *Sci. Rep.* **2019**, *9*, 1746. [CrossRef]
27. Sreevidya, N.; Mehrotra, S. Spectrometric method for estimation of precipitable with Dragendor's reagent in plant material. *J. Aoac. Int.* **2003**, *36*, 1124–1127. [CrossRef]
28. Yassien, E.E.; Hamed, M.M.; Abdelmohsen, U.R.; Hassan, H.M.; Gazwi, H.S.S. In vitro antioxidant, antibacterial, and antihyperlipidemic potential of ethanolic *Avicennia marina* leaves extract supported by metabolic profiling. *Environ. Sci. Pollut. Res.* **2021**, *28*, 27207–27217. [CrossRef]
29. Knothe, G.; Kenar, J.A. Determination of the fatty acid profile by 1H-NMR spectroscopy. *Eur. J. Lipid Sci. Technol.* **2004**, *106*, 88–96. [CrossRef]
30. Chua, L.S.; Hidayathulla, S. Phytochemical profile of fresh and senescent leaves due to storage for *Ficus deltoidea*. *Plant Biosyst. Int. J. Deal. All Asp. Plant Biol.* **2015**, *151*, 74–83. [CrossRef]

31. Kasote, D.M.; Katyare, S.S.; Hegde, M.V.; Bae, H. Significance of Antioxidant Potential of Plants and its Relevance to Therapeutic Applications. *Int. J. Biol. Sci.* **2015**, *11*, 982–991. [CrossRef]
32. Mahady, G.B.; Huang, Y.; Doyle, B.J.; Locklear, T. Natural Products as Antibacterial Agents. *Stud. Nat. Prod. Chem.* **2008**, *35*, 423–444. [CrossRef]
33. Chen, B.-J.; Jamaludin, N.S.; Khoo, C.-H.; See, T.-H.; Sim, J.-H.; Cheah, Y.-K.; Halim, S.N.A.; Seng, H.-L.; Tiekink, E.R. In vitro antibacterial and time-kill evaluation of phosphanegold (I) dithiocarbamates, R3PAu [S2CN (iPr) CH2CH2OH] for R = Ph, Cy and Et, against a broad range of Gram-positive and Gram-negative bacteria. *Gold Bull.* **2014**, *47*, 225–236.
34. Schinor, E.C.; Salvador, M.J.; Ito, I.Y.; Dias, D.A. Evaluation of the antimicrobial activity of crude extracts and isolated constituents from Chresta scapigera. *Braz. J. Microbiol.* **2007**, *38*, 145–149. [CrossRef]
35. Wu, H.-S.; Raza, W.; Fan, J.-Q.; Sun, Y.-G.; Bao, W.; Shen, Q.-R. Cinnamic Acid Inhibits Growth but Stimulates Production of Pathogenesis Factors by *in Vitro* Cultures of *Fusarium oxysporum f.sp. niveum*. *J. Agric. Food Chem.* **2008**, *56*, 1316–1321. [CrossRef] [PubMed]
36. Korošec, B.; Sova, M.; Turk, S.; Kraševec, N.; Novak, M.; Lah, L.; Stojan, J.; Podobnik, B.; Berne, S.; Zupanec, N.; et al. Antifungal activity of cinnamic acid derivatives involves inhibition of benzoate 4-hydroxylase (CYP53). *J. Appl. Microbiol.* **2014**, *116*, 955–966. [CrossRef] [PubMed]
37. Johannes, E.; Litaay, M.; Syahribulan. The bioactivity of hexadecanoic acid compound isolated from hydroid *Aglaophenia cupressina lamoureoux* as antibacterial agent against Salmonella typhi. *Int. J. Biol. Med. Res.* **2016**, *7*, 5469–5472.
38. Radiati, L.E. Mechanism of Virulence Inhibition of Enterophatogenic Bacteria by Ginger (*Zingiber officinale Roscoe*) Rhizome Extract. Ph.D. Dissertation, Postgraduate Program, Bogor Agricultural University, Bogor, Indonesia, 2002.
39. Donato, M.; Tolosa, L.; Gómez-Lechón, M.J. Culture and Functional Characterization of Human Hepatoma HepG2 Cells. *Methods Mol. Biol.* **2015**, *1250*, 77–93.
40. Tirkey, N.; Pilkhwal, S.; Kuhad, A.; Chopra, K. Hesperidin, a citrus bioflavonoid, decreases the oxidative stress produced by carbon tetrachloride in rat liver and kidney. *BMC Pharmacol.* **2005**, *5*, 2. [CrossRef]
41. Lee, C.J.; Wilson, L.; Jordan, M.A.; Nguyen, V.; Tang, J.; Smiyun, G. Hesperidin suppressed proliferations of both Human breast cancer and androgen-dependent prostate cancer cells. *Phytother. Res.* **2009**, *24*, S15–S19. [CrossRef]
42. Abd El-Azim, M.H.M.; Abdelgawad, A.A.M.; El-Gerby, M.; Ali, S.; El-Mesallamy, A.M.D. Phenolic compounds and cytotoxic activities of methanol extract of basil (*Ocimum basilicumL* L.). *J. Microb. Biochem. Technol.* **2015**, *7*, 182–185. [CrossRef]
43. Tanaka, M.; Kuei, C.W.; Nagashima, Y.; Taguchi, T. Application of antioxidative millard reaction products from histidine and glucose to sardine products. *Nippon. Suisan Gakkaishil.* **1998**, *54*, 1409–1414. [CrossRef]
44. Dickson, M.A.; Schwartz, G.K. Development of Cell-Cycle Inhibitors for Cancer Therapy. *Curr. Oncol.* **2009**, *16*, 36–43. [CrossRef]
45. Riccardi, C.; Nicoletti, I. Analysis of apoptosis by propidium iodide staining and flow cytometry. *Nat. Protoc.* **2006**, *1*, 1458–1461. [CrossRef]
46. Fadok, V.A.; Voelker, D.R.; Campbell, P.A.; Cohen, J.J.; Bratton, D.L.; Henson, P.M. Exposure of phosphatidylserine on the surface of apoptotic lymphocytes triggers specific recognition and removal by macrophages. *J. Immunol.* **1992**, *148*, 2207–2216.
47. Demchenko, A.P. Beyond annexin V: Fluorescence response of cellular membranes to apoptosis. *Cytotechnology* **2012**, *65*, 157–172. [CrossRef]
48. Towers, C.G.; Torburn, A. Terapeutic targeting of autophagy. *EBioMedicine* **2016**, *14*, 15–23. [CrossRef]
49. Hippert, M.M.; O'Toole, P.S.; Thorburn, A. Autophagy in Cancer: Good, Bad, or Both? *Cancer Res.* **2006**, *66*, 9349–9351. [CrossRef]
50. Soung, Y.H.; Jeong, E.G.; Ahn, C.H.; Kim, S.S.; Song, S.Y.; Yoo, N.J.; Lee, S.H. Mutational analysis of caspase 1, 4, and 5 genes in common human cancers. *Hum. Pathol.* **2008**, *39*, 895–900. [CrossRef]
51. Tee, T.T.; Cheah, Y.H.; Hawariah, L.P.A. F16, a fraction from *Eurycoma longifolia* jack extract, induces apoptosis via a caspase-9-independent manner in MCF-7 cells. *Anticancer Res.* **2007**, *27*, 3425–3430.
52. Shao, W.; Yeretssian, G.; Doiron, K.; Hussain, S.N.; Saleh, M. The caspase-1 digestome identifies the glycolysis pathway as a target during infection and septic shock. *J. Biol. Chem.* **2007**, *282*, 36321–36329. [CrossRef]
53. Lim, S.-W.; Loh, H.-S.; Ting, K.N.; Bradshaw, T.D.; Zeenathul, N.A. Antiproliferation and induction of caspase-8-dependent mitochondria-mediated apoptosis by β-tocotrienol in human lung and brain cancer cell lines. *Biomed. Pharmacother.* **2014**, *68*, 1105–1115. [CrossRef]

antibiotics

Article

In-Vitro and *In-Vivo* Antibacterial Effects of Frankincense Oil and Its Interaction with Some Antibiotics against Multidrug-Resistant Pathogens

Megren Bin Faisal Almutairi [1,2], Mohammed Alrouji [1], Yasir Almuhanna [1], Mohammed Asad [1] and Babu Joseph [1,*]

[1] Department of Clinical Laboratory Science, College of Applied Medical Sciences, Shaqra University, Shaqra 11961, Saudi Arabia
[2] Department of Clinical Laboratory, King Saud Hospital, Unaizah 56215, Saudi Arabia
* Correspondence: bjoseph@su.edu.sa; Tel.: +966-594837620

Abstract: Frankincense (*Boswellia sacra* oleo gum resin) is reported to possess antimicrobial activity against several pathogens in-vitro. The antimicrobial effects of frankincense oil and its interaction with imipenem and gentamicin against methicillin-resistant *Staphylococcus aureus* (MRSA) and multidrug-resistant *P. aeruginosa* were determined through in-vitro methods and an in-vivo study using a rat pneumonia model. Frankincense oil was subjected to GC-MS analysis to determine the different volatile components. Antibacterial effects against MRSA and MDR-*P. aeruginosa* was evaluated and its MIC and MBC were determined. For the rat pneumonia model (in-vivo), oil was administered at a dose of 500 mg/kg and 1000 mg/kg followed by determination of CFU in lung tissue and histological studies. Frankincense oil did not show a very potent inhibitory effect against MRSA or MDR-*P. aeruginosa*; the oil did not affect the zone of inhibition or FIC when combined with imipenem or gentamicin indicating a lack of interaction between the oil and the antibiotics. Furthermore, there was no interaction between the antibiotics and the frankincense oil in the in-vivo model. The result of the study revealed that frankincense oil has a weak inhibitory effect against MRSA and MDR-*P. aeruginosa*, and it did not show any interaction with imipenem or gentamicin.

Keywords: pneumonia; volatile oils; *Boswellia*; MIC; MBC

Citation: Almutairi, M.B.F.; Alrouji, M.; Almuhanna, Y.; Asad, M.; Joseph, B. *In-Vitro* and *In-Vivo* Antibacterial Effects of Frankincense Oil and Its Interaction with Some Antibiotics against Multidrug-Resistant Pathogens. *Antibiotics* **2022**, *11*, 1591. https://doi.org/10.3390/antibiotics11111591

Academic Editors: Valério Monteiro-Neto and Elizabeth S. Fernandes

Received: 10 October 2022
Accepted: 7 November 2022
Published: 10 November 2022

1. Introduction

Antimicrobial resistance (AMR) is a threat to public health that leads to the development of serious infections and interferes with strategies for the prevention of infections [1]. The AMR is continuously increasing and is showing no signs of slowing down. Factors such as spontaneous evolution, mutation, and transfer of resistant genes contribute significantly to AMR [2,3].

Herbal medicines and their constituents have been increasingly evaluated to combat antimicrobial resistance and to explore new and more potent antimicrobial agents with lesser adverse effects. Several plant-derived components have exhibited remarkable inhibitory properties on microbial growth [4–6]. Frankincense (*Boswellia sacra* oleo gum resin) is reported to possess antimicrobial activity in earlier studies. Oil extracted from frankincense grown in different regions of Oman was reported to inhibit *S. aureus* and *P. aeruginosa* in-vitro [7]. Frankincense oil from 20 different countries showed varying effects on 5 different micro-organisms; *S. aureus*, *B. cereus*, *E. coli*, *P. vulgaris*, and *C. albicans* [8]. An earlier study about the interaction of frankincense oil with myrrh oil showed that frankincense oil produces a synergistic or additive effect with myrrh oil depending on the type of micro-organism [9]. All of these studies were carried out using in-vitro methods and no attempt has been made so far to evaluate the antibacterial effect of frankincense oil in-vivo after oral administration.

Methicillin-resistant *Staphylococcus aureus* (MRSA) causes both nosocomial and community-acquired infections. Several clones of MRSA have emerged recently and its resistance to previously sensitive antibiotics is a matter of concern for healthcare community workers [10,11]. Multidrug-resistant *Pseudomonas aeruginosa* (MDR-*P. aeruginosa*) causes severe infections and has an outstanding ability for being selected and for spreading antimicrobial resistance in-vivo [12,13]. Furthermore, the spread of "high-risk" clones of MDR-*P. aeruginosa* is a threat to global public health [14].

The present study determined the antimicrobial effect of frankincense oil and its interaction with imipenem and gentamicin against MRSA and multidrug-resistant *P. aeruginosa* using in-vitro methods and an in-vivo study in a rat pneumonia model. Further, the study also evaluated its interaction with commonly used antibacterial agents. Lastly, the pathogens that were resistant to many commonly used antibacterial agents were employed in this study.

2. Results

2.1. GC-MS Analysis of Frankincense Oil

The analysis revealed the presence of 40 constituents (Table 1). Compounds known to be present such as α-pinene, camphene, and limonene were present in the oil. The maximum area under the curve was observed for α-pinene and the minimum area under the curve was seen with β-myrcene indicating the most prominent and least amount of volatile components in the oil, respectively (Figure 1).

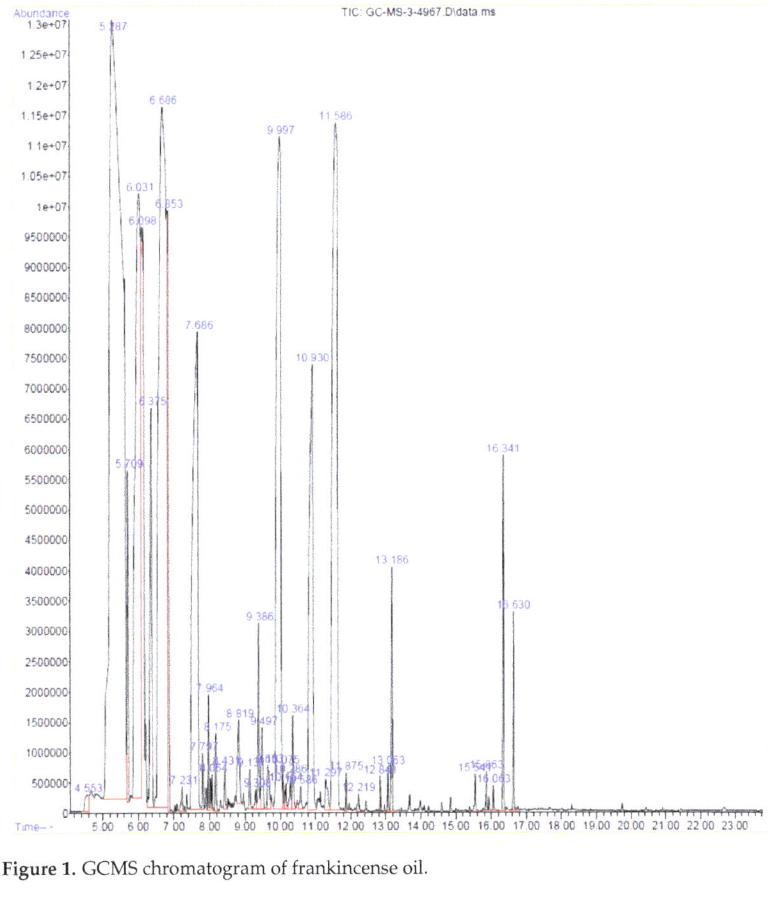

Figure 1. GCMS chromatogram of frankincense oil.

Table 1. List of constituents detected by GCMS.

Number	Name of the Constituent	RT	Area %
1.	4-Carene	4.553	0.19
2.	α-Pinene	5.287	29.31
3.	Camphene	5.709	1.09
4.	β-Pinene	6.031	9.75
5.	4-Carene	6.098	0.10
6.	α-Phellandrene	6.375	2.29
7.	1,3,8-p-Menthatriene	6.686	15.87
8.	Limonene	6.853	2.61
9.	Cycloheptene, 5-ethylidene-1-methyl	7.231	0.11
10.	1,6-Octadien-3-ol, 3,7-dimethyl-	7.686	7.42
11.	Phenylethyl Alcohol	7.797	0.17
12.	1-Propanone, 1-(5-methyl-2-furanyl)-	7.964	0.25
13.	cis-p-Mentha-2,8-dien-1-ol	8.064	0.08
14.	Bicyclo[3.1.1]hept-3-en-2-ol, 4,6,6-trimethyl	8.175	0.26
15.	Isoborneol	8.431	0.16
16.	Cyclohexane, 1-butenylidene-	8.819	0.56
17.	2-Isopropenyl-5-methylhex-4-enal	9.131	0.17
18.	2-Cyclohexen-1-ol, 2-methyl-5-(1-methylethenyl)-, trans	9.308	0.08
19.	2-Methylbicyclo[4.3.0]non-1(6)-ene	9.386	0.58
20.	1H-Pyrrole-2-carboxaldehyde, 1-methyl-	9.497	0.44
21.	1,2,3,4,4a,5,6,8a-Octahydro-naphthalene	9.653	0.26
22.	Isobornyl acetate	9.997	8.97
23.	Camphene	10.075	0.10
24.	1,6,10,14-Hexadecatetraen-3-ol, 3,7,11,15-tetramethyl-, (E,E)-	10.164	0.11
25.	Bicyclo[3.1.0]hexan-3-ol, 4-methyl-1-(1-methylethyl)-, (1.alpha., 3.beta., 4.beta., 5.alpha.)-	10.286	0.14
26.	Bicyclo[4.1.0]heptan-3-ol, 4,7,7-trimethyl-, (1.alpha., 3.alpha., 4.beta., 6.alpha.)-	10.364	0.34
27.	2,6-Octadien-1-ol, 3,7-dimethyl-acetate, (Z)-	10.586	0.11
28.	2,6-Octadien-1-ol, 3,7-dimethyl-,acetate, (E)-	10.930	4.70
29.	Bicyclo[7.2.0]undec-4-ene, 4,11,11-trimethyl-8-methylene-	11.297	0.24
30.	Caryophyllene	11.586	10.54
31.	alpha.-Caryophyllene	11.875	0.09
32.	Naphthalene, 1,2,3,5,6,7,8,8a-octahydro-1,8a-dimethyl-7-(1-methylethenyl)-, [1R-(1.alpha., 7.beta., 8a.alpha.)]-	12.219	0.07
33.	Caryophyllene oxide	12.841	0.08
34.	Diethyl Phthalate	13.063	0.10
35.	Caryophyllene oxide	13.186	0.76
36.	E,E,E)-3,7,11,15-Tetramethylhexadeca-1,3,6,10,14-pentaene	15.541	0.12
37.	2,6,11,15-Tetramethyl-hexadeca-2,6,8,10,14-pentaene	15.863	0.10
38.	β-Myrcene	16.063	0.07
39.	1,6,10-Dodecatriene, 7,11-dimethyl-3-methylene-, (Z)-	16.341	1.11
40.	2-Propenamide, N-(3,4-dichlorophenyl)-2-methyl-	16.630	0.50

2.2. Antibacterial Effect of Frankincense Oil

Frankincense oil did not show a very potent inhibitory effect against MRSA or MDR-*P. aeruginosa* and a noticeable zone of inhibition could be seen starting from 20 μL. Furthermore, the oil did not affect the zone of inhibition induced by imipenem or gentamicin, indicating a lack of interaction between the oil and the antibiotics (Figure 2A).

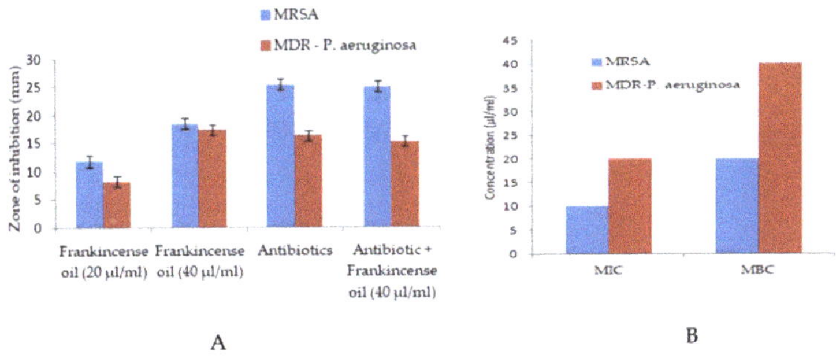

Figure 2. Antibacterial activity of frankincense oil against bacterial pathogens. (**A**) shows zone of inhibition and (**B**) indicates MIC and MBC.

Values are mean ± SEM for three independent trials. Antibiotic refers to imipenen (4 µg/mL) and gentamicin (10 µg/mL) against MRSA and MDR-*P. aeruginosa*, respectively.

The MIC of frankincense oil was 10 µL and 20 µL against MRSA MDR-*P. aeruginosa*, respectively. The MBC was found to be 20 µL for MRSA and 40 µL for MDR-*P. aeruginosa* (Figure 2B).

2.3. Interaction Study of Frankincense Oil with Antibiotics against MDR Strains

The FIC determined to evaluate the interaction between the frankincense oil and the antibiotics showed no interaction between the oil and imipenem or gentamicin (Table 2).

Table 2. The fractional inhibitory concentration of frankincense oil with different concentrations of antibiotics.

Bacterial Pathogens	Fractional Inhibitory Concentration				Outcome
	MIC Oil (µL/mL)	MIC of Ab (mg/mL)	MIC of Combination	FIC Index	
MRSA ATCC 43300	10	0.002	0.002	1	Indifference
P. aeruginosa ATCC 27853	20	0.004	0.004	1	Indifference

FIC index = $(MIC_{frankincense\ oil+antibiotic}/MIC_{frankincense\ oil}) + (MIC_{frankincense\ oil+antibiotic}/MIC_{antibiotic})$. The synergistic potential was assessed as follows; if the FIC index is ≤0.5 the combination is synergistic; at more than 0.5 and ≤2, the combination is indifferent and if the FIC index is >2, it is considered antagonistic.

2.4. Rat Pneumonia Model (In-Vivo)

The frankincense oil at both doses significantly reduced the bacterial load of MRSA in the rats' lungs after treatment for 4 days. However, it was less effective than the standard antibiotics; gentamicin and imipenem. In rats infected with MDR-*P. aeruginosa*, the lower dose of the oil (500 mg/kg) was ineffective and the effect of the higher dose of frankincense oil (1000 mg/kg) was relatively less effective compared to its effect against MRSA infection. Furthermore, there was no interaction between the antibiotics and the frankincense oil in the in-vivo model and this was similar to the results obtained in-vitro (Figure 3). Since the duration of the experiment was only 4 days, all the animals survived during the experimental period.

Histological examination of tissues from rats of different groups showed that both MRSA and MDR-*P. aeruginosa* infection had caused lung inflammation and the lung tissue was infiltrated with fluids and inflammatory cells. This inflammatory response was suppressed by frankincense and standard antibiotics. The histological changes are shown in Figure 4a–g. There was no noticeable change in the histology of lung tissue after treatment with antibiotics and frankincense oil when compared to individual treatments. Hence, photomicrographs showing the effect of combination treatment are not shown.

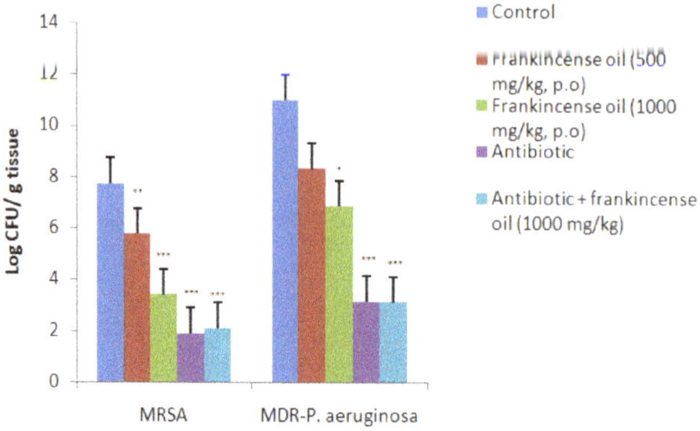

Figure 3. The bacterial load (CFU/g) in the lung tissue after treatment with Frankincense oil. Data shown are mean ± SEM, n = 6, * $p < 0.05$, ** $p < 0.01$, *** $p < 0.001$ when compared to respective controls. There was no significant difference between frankincense oil (500 mg/kg) and MDR-*P. aeruginosa* control. Antibiotic—imipenem (120 mg/kg, i.p) against MRSA, and gentamicin (10 mg/kg, i.p) against MDR-*P. aeruginosa*.

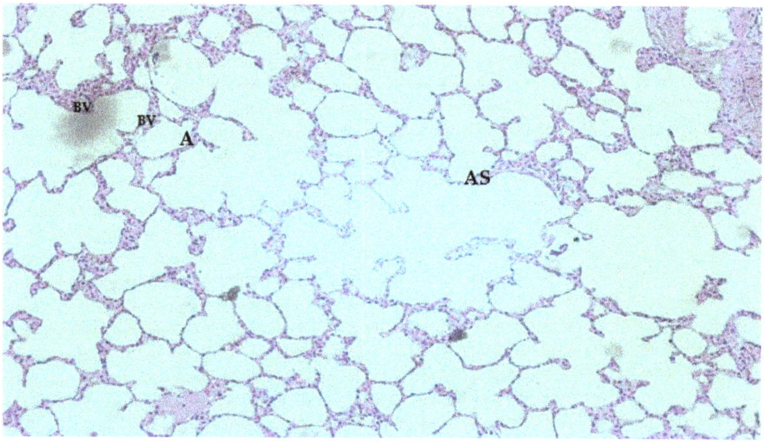

(**a**)

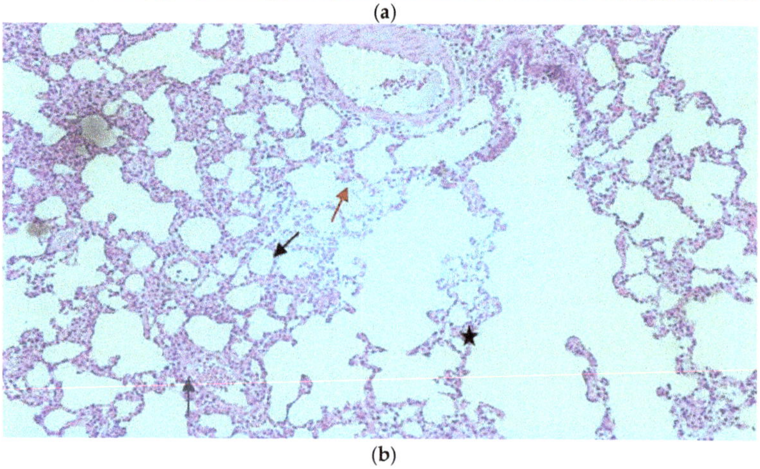

(**b**)

Figure 4. *Cont.*

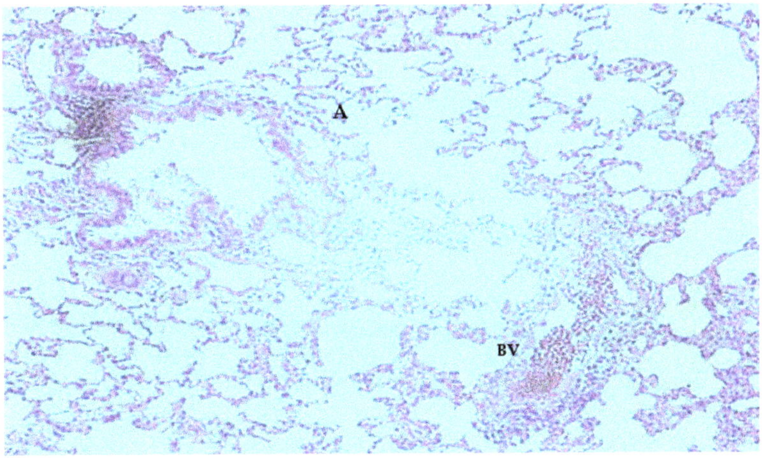

(**c**)

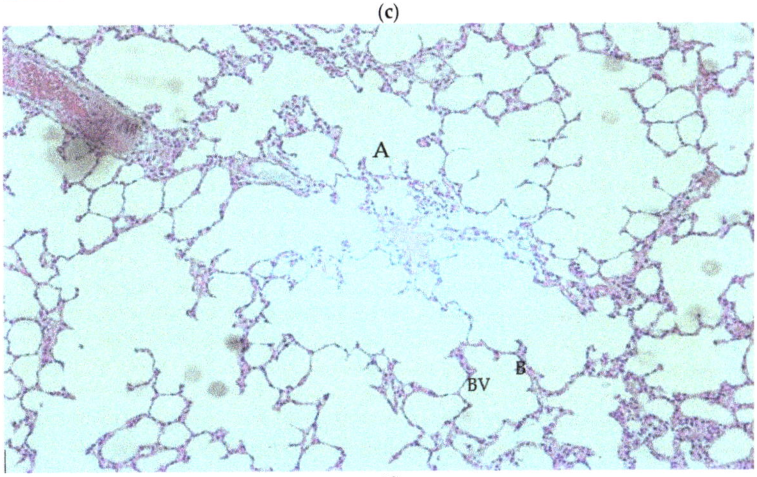

(**d**)

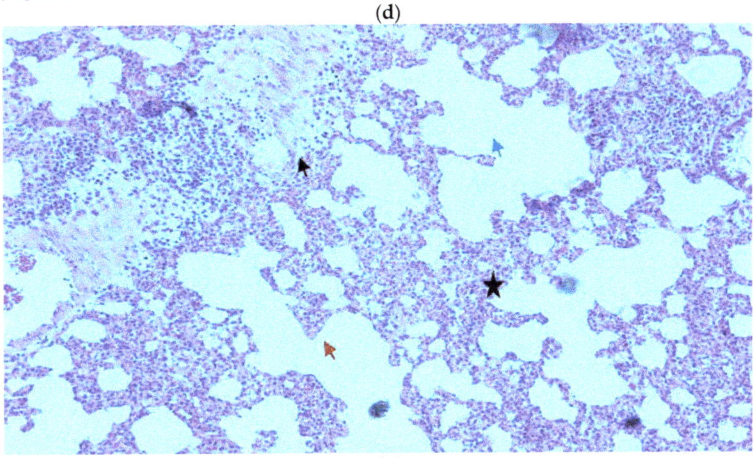

(**e**)

Figure 4. *Cont.*

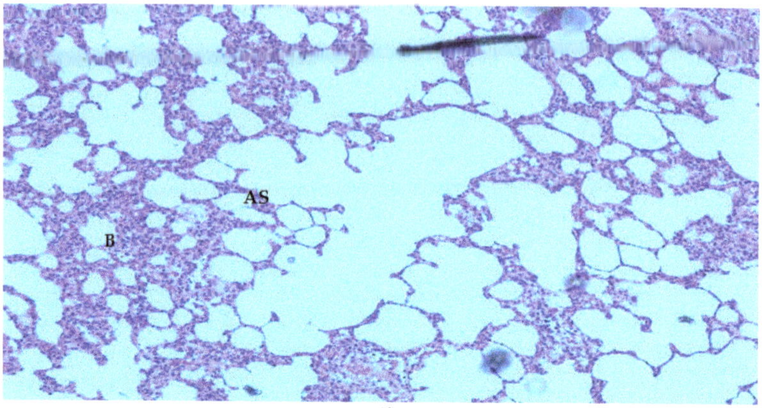

(f)

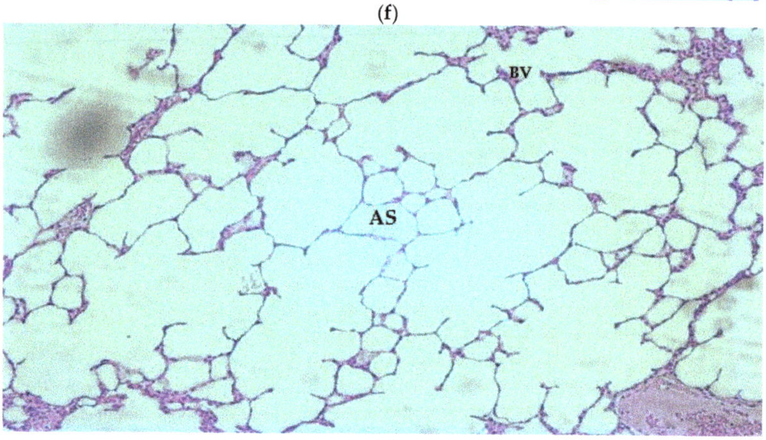

(g)

Figure 4. (**a**) Histological examination of lung tissue from normal animals showing alveolar sac (AS), alveolus (A), and blood vessel (BV) (×100). (**b**) Lung tissue histology from *P. aeruginosa* control. A distortion in the pulmonary architecture can be observed with thickened interalveolar septa (black arrows) and mononuclear cellular infiltration (red arrow), interstitial exudates (blue arrow), ruptured interalveolar septa with large irregular emphysematous air spaces (star) are seen (×100). (**c**) Histological examination of lung tissue from frankincense oil-treated group after infection with *P. aeruginosa*. A relatively protected pulmonary tissue showing alveolar sac (AS), blood vessel (BV) with less mononuclear infiltration, reduced thickness of alveolar sacs, and less interalveolar rupture (×100). (**d**) Histological examination of lung tissue from the gentamicin-treated group after infection with *P. aeruginosa*. A relatively protected pulmonary tissue showing alveolar sac (AS), bronchiole (B), and blood vessel (BV) with less mononuclear infiltration, reduced thickness of alveolar sacs, and less interalveolar rupture (×100). The effect was similar to that observed with frankincense oil. (**e**) Histological examination of lung tissue from MRSA control. A distortion in the pulmonary architecture can be observed with thickened interalveolar septa (black arrows) and mononuclear cellular infiltration (red arrow), interstitial exudates (blue arrow), ruptured interalveolar septa with large irregular emphysematous air spaces (star) are seen (×100). (**f**) Histological examination of lung tissue from frankincense oil-treated group after infection with MRSA. A relatively protected pulmonary tissue showing alveolar sac (AS), and bronchiole (B), with less mononuclear infiltration, reduced thickness of alveolar sacs, and less interalveolar rupture (×100). (**g**) Histological examination of lung tissue from imipenem treated group after infection with MRSA. A relatively protected pulmonary tissue showing alveolar sac (AS), and blood vessel (BV), with less mononuclear infiltration, reduced thickness of alveolar sacs, and less interalveolar rupture (×100).

3. Discussion

The antibacterial effects of frankincense oil have been reported earlier against a variety of pathogens [7,8]. However, this study is different from the earlier reports in many aspects. Frankincense oil has been used for an antibacterial effect by the conventional in-vitro methods, the present study determined its antibacterial effect after oral administration (in-vivo) along with in-vitro studies.

Frankincense oil showed modest antibacterial activity against MRSA and MDR-*P. aeruginosa* but there was no interaction between the oil and imipenem or gentamicin in both in-vitro and in-vivo studies. The pathogens causing common infections and resistant to many of the commonly used antibacterial agents were selected for the study.

The frankincense is given different names, such as *Boswellia serrata* for Asian and African frankincense and *Boswellia sacra* for frankincense obtained from Oman, a Middle Eastern country. In our case, we selected oil that was prepared from *Boswellia sacra* for our study. We analyzed the chemical constituents present in the commercially obtained oil by GC-MS so that if a study is repeated with frankincense having the same constituents, a similar effect may be obtained. Furthermore, analysis of the oil revealed the probable constituent(s) that might have contributed to the effect.

The in-vivo evaluation of herbal drugs is important because herbs are administered only by oral route due to several different chemical constituents present in them [15]. The oil was administered as such in the study to mimic its traditional use [16], and no derivatization was done to make it more polar for oral administration. It is well known that many antibacterial agents are not effective orally either because they are not absorbed or some of them undergo extensive first-pass metabolism or get destroyed by acid/enzymes in the digestive tract [17]. In the current study, the antibacterial effect of frankincense oil observed in-vivo after oral administration was similar to that observed in-vitro, though the effect was not very potent in both studies. This suggests that active constituent(s) of frankincense oil responsible for the antibacterial effect are well absorbed orally and reach the blood circulation in sufficient amounts to exert their effect. Frankincense oil is used traditionally in the treatment of respiratory diseases through oral administration and is considered safe for oral consumption [16]. We selected the doses of frankincense oil based on earlier reports [18]. No behavioral difference or toxicity was observed between the different treated groups.

As mentioned above, the essential oils of frankincense have been reported for antibacterial effects against many pathogens earlier. However, many of the earlier reports determined the antibacterial effects with common bacteria, and no attempt was made to evaluate the effect of essential oils from frankincense on MRSA and MDR-*P. aeruginosa*. It was shown to be effective against pathogens causing urinary tract infections [8]. It is also reported for antibacterial effects on *S. aureus*, *Escherichia coli*, and *Proteus vulgaris*, and antifungal activity against *C. albicans* and C. tropicalis [19–21]. Furthermore, an in-vitro study on the frankincense oils grown in different regions in Oman showed varying effects against *S. aureus* and *P. aeruginosa*, and a dermatological strain *P. acnes* and a good antifungal effect against *C. albicans* and *M. furfur* [7]. Another study reported antimicrobial activity against five organisms; *S. aureus*, B. cereus, *E. coli*, P. vulgaris, and *C. albicans* [8]. Frankincense is not used alone for the treatment of infections, it is usually combined with myrrh or other antimicrobial agents for antimicrobial effect [22,23]. This study determined the effect of frankincense oil against MRSA and MDR-P. aeruginosa to demonstrate its effectiveness in inhibiting the growth of resistant pathogens. The antibacterial effect of frankincense oil was relatively more against MRSA (Gram-positive) compared to P. aeruginosa (Gram-negative). This is due to the cell wall structure of Gram-negative bacteria that makes it inherently tolerant to the effect of antibacterial agents[24].

The interaction of herbs with conventional antimicrobial drugs is an area of interest and it is being thoroughly investigated to determine both beneficial and adverse reactions that may arise due to the concomitant administration of herbs with drugs [25]. Frankincense oil is a popular herb for the treatment of infections in several countries and it is common for patients to consume it along with antibiotics to obtain 'extra' benefits. Despite this, the pharmacological interaction of frankincense oil with antibiotics is not known. In the current study, antibiotics and frankincense oils showed antibacterial effects individually and it is usually expected that a combination of two such active agents produces an additive/synergistic effect [26,27]. Apart from this, some combinations of antibacterial agents such as a combination of bacteriostatic agents with bactericidal drugs induce antagonism of the bacteriostatic effect by bactericidal agents [28]. In the current study, no interaction was observed between the frankincense oil and antibiotics both in-vitro and in-vivo. The reason(s) for this cannot be explained with the present data. However, it can be ruled out that frankincense may alter the pharmacokinetics of concurrently administered antibiotics, as the effect observed was similar in both in-vitro and in-vivo.

The selection of bacterial isolates was based on animal models and pathogens causing opportunistic infections in the lungs [29,30]. MRSA is known to cause pneumonia in hospitalized patients. It is recommended that treatment for MRSA be started if 10–20% of isolates showing antibiotic resistance are MRSA [31]. Imipenem is effective against MRSA and it is one of the drugs used in the treatment of MDR infections in the lungs [32,33]. MDR-*P. aeruginosa* infection is a concern for many physicians as it causes lung function to decline and leads to the emergence of antibiotic resistance in *P. aeruginosa* strains [34]. Though different antibiotics including imipenem/cilastatin have been reported to be effective against MDR-*P. aeruginosa* infections, of late, resistance to imipenem has been reported [35,36]. The MDR strain of *P. aeruginosa* was sensitive to gentamicin, hence, gentamicin was used as a standard drug to study interaction with frankincense oil. The micro-organism selection was also based on two wide groups; the Gram-positive MRSA and Gram-negative *P. aeruginosa* to determine the spectrum of the effect of frankincense oil and its interaction with standard antibiotics in treating infection by two different groups of pathogens. As the organisms were inoculated through the intratracheal route, lung infection will be more severe than infection in other parts of the body. Hence, only the lungs were excised.

The chemical constituents present in frankincense oils differ based on geographical distribution, climatic conditions, and harvesting methods [37]. The chemical composition will vary among the different brands of frankincense oils available in the market due to the above-mentioned factors. Since chemical composition differences will affect the pharmacological activities, the antimicrobial activities of all the brands of frankincense oils cannot be predicted [8]. Hence, chromatographic analysis of essential oils is important. In the current study, 40 different constituents were identified. Different authors have identified different compounds in frankincense oil. Octyl acetate followed by 1-octanol were identified as the main compounds by Baser [38]. Limonene and (*E*)-β-ocimene were found to be the main compounds in *B. sacra* [39]. Camarda et al. [19] also found limonene to be the main component followed by α-pinene as the second component while α-pinene was identified as the main volatile component followed by octyl acetate in Saudi Arabia [40]. In the current study, the oil used contained the maximum amount of α-pinene followed by 1,3,8-p-menthatriene with a total of 40 volatile constituents. Earlier studies have reported the antibacterial activity of some of the constituents revealed by the GC-MS in the current study. The α-pinene and β-pinene have been reported for antibacterial and antifungal effects [41]. Limonene is reported to have an antibacterial effect against *S. aureus* and *P. aeruginosa* and also potentiated the effect of gentamicin in-vitro [42]. Camphene is also reported for antibacterial effect against *S. aureus* and *Enterococcus* species [43], while myrcene is reported for activity against *S. aureus* and food-borne pathogens [44]. Though individual components are known to be effective, it is believed that microbial resistance will not occur against the essential oils because the oil contains several constituents with antimicrobial effects [45]. However, it is also known that the essential oils obtained from different sources may have different effects due to variations in the physicochemical properties of essential oils and the antimicrobial effect obtained in one study cannot be compared with effects obtained with essential oils in another study. This is probably due to the loss of antimicrobial constituents or potentiating compounds during the process of distillation [46].

Our results are different from those reported by several other authors, who showed that frankincense oil has a good antibacterial effect, and a combination of frankincense oil with other antibacterial agents potentiates the effect of the latter. The exact reason(s) for this difference in the effect observed in the current study with earlier reports cannot be explained with the present data. However, the difference in the composition of the oil, the difference in the antibiotic used, and the strain of the micro-organism may have contributed to this difference in the effect. More studies with other micro-organisms and with other antibiotics may provide information to determine the variation in the effect.

For the interaction study, antibiotics were given by the parenteral route (intraperitoneal), while frankincense oil was administered orally. This is a limitation of the study as only the pharmacodynamic interaction between the oil and the standard antibiotics and pharmacokinetic interaction concerning to distribution, metabolism, and excretion of antibiotics was determined without studying the interaction in the intestinal metabolism and absorption. However, we would like to stress here that the pathogens used were resistant to several antibiotics and the best available antibiotic effective against the pathogens was used.

4. Materials and Methods

4.1. Micro-Organisms

MRSA (ATCC 43300) and MDR-*P. aeruginosa* (ATCC 27853) available at the Department of Clinical Laboratory Sciences was used. The ATCC cultures were sub-cultured in nutrient broth containing glycerin followed by storage at −80 °C. For evaluation of antibacterial effects, the microbes were inoculated in nutrient agar followed by incubation for 24 h at 36 ± 1 °C. The antibiotic susceptibility of the bacterial strains is shown in Tables 1 and 2. Frankincense oil (Losolin natural oil, Jamal Natural Factory, Medinah, Saudi Arabia) extracted from *Boswellia sacra* oleo gum resin was purchased online through a marketing website.

4.2. Animals

Adult Wistar rats weighing between 235 to 245 g and aged between 4.2 to 4.7 months were used. The animals were divided into five groups of six animals each for MRSA and MDR-*P. aeruginosa*. The first group served as control (inoculated with the pathogen), second and third groups of animals received two different doses of frankincense oil at 500 mg/kg and 1000 mg/kg, respectively. The fourth group was treated with antibiotic and the last group was administered a combination of frankincense oil (1000 mg/kg) with the antibiotic. Animals were maintained in the animal house in a separate room to prevent the spread of infection. The Ethical Research Committee of Shaqra University reviewed and approved the experimental procedures on rats (approval number—53/18909).

4.3. GC-MS Analysis of Frankincense Oil

A GC-MS 7890A GC system with 5975C VL MSD was used (Agilent Technologies, Santa Clara, CA, USA). Frankincense oil (100 µL) was mixed with 250 µL of water and 750 µL of ethyl acetate. Following this, the upper layer was separated and concentrated. To this, a 50 µL mixture of N,O-Bis (trimethylsilyl) trifluoroacetamide (49.5 µL), and trimethylchlorosilane (0.5 µL) was added followed by the addition of pyridine (10 µL). This was heated for 30 min at 60 °C and the contents were dried using liquid nitrogen before finally dissolving the dried sample (20 mg) in methanol (5 mL) for analysis. After filtration through a 0.22 µm membrane filter, 3 µL was injected through a capillary column (30 m, 0.25 mm, 0.25-micron) with an injector temperature of 270 °C and pressure at 80 kPa. The carrier gas was hydrogen and the total time for analysis was 25 min. The mass spectra obtained were used to identify different compounds by referring to the NIST mass spectral library.

4.4. Antibacterial Effect and Determination of MIC and MBC

The culture of MRSA and MDR-*P. aeruginosa* were inoculated into Muller Hinton agar (100 µL; 1.5 × 10^8 CFU/mL). Wells were made using cork borer and these were loaded with frankincense oil diluted in 10% dimethyl sulfoxide (DMSO) to different concentrations. DMSO helps in the easy diffusion of the oil through the media. Similarly, gentamicin and imipenem were loaded. The inoculated plates were subjected to incubation at a temperature of 36 ± 1 °C for a period of 24 h to measure the zone of inhibition. The antibiotics were selected after an automated antibiotic susceptibility test of MRSA and MRD-*P. aeruginosa* using a Microscan system (McHenry, IL, USA).

For the determination of MIC, Muller Hinton broth was inoculated with 50 µL of liquid cultures of pathogens (0.5 McFarland standard turbidity). Frankincense oil diluted with DMSA to different concentrations was added followed by incubation at a temperature of 36 ± 1 °C for a period of 24 h to determine the MIC. For MBC determination, nutrient agar was inoculated with a loop of culture and incubated at a temperature of 36 ± 1 °C for 24 h period. MBC was the concentration at which no growth was observed.

4.5. Interaction Study of Frankincense Oil with Antibiotics against MDR Strains

The synergistic assay of the antibiotics and *B. sacra* oil was determined by the broth dilution method using a checkerboard assay [47]. After serial dilution of the antibiotic, it was added to each well at different concentrations that included sub-inhibitory, inhibitory, and supra-inhibitory concentrations, and the MIC was calculated. Wells with different concentrations of antibiotics and without oil was considered as the MIC for antibiotics. The fractional inhibitory concentration (FIC) index was calculated as follows:

$$\text{FIC index} = \text{FIC}_{\text{frankincense oil}} + \text{FIC}_{\text{antibiotic}}$$

$$\text{FIC}_{\text{frankincense oil}} = \text{MIC}_{\text{frankincense oil+antibiotic}} / \text{MIC}_{\text{frankincense oil}}$$

$$\text{FIC}_{\text{antibiotic}} = \text{MIC}_{\text{frankincense oil+antibiotic}} / \text{MIC}_{\text{antibiotic}}.$$

The synergistic potential was assessed as if the FIC index is ≤0.5 the combination is synergistic; at more than 0.5 and ≤2 the combination is indifferent and if the FIC index is >2, it is considered antagonistic [42].

4.6. Rat Pneumonia Model (In-Vivo)

Albino Wistar rats were anesthetized by intraperitoneal administration of a mixture of ketamine and xylazine (1:10) at a dose of 1 mL/kg [48]. The trachea was exposed surgically and the animals were kept in an inclined position at 60°. The bacterial suspension (MRSA or multi-resistant *P. aeruginosa*) prepared in 1 × phosphate-buffered saline (pH7.4) was administered through the trachea at a dose of 1.2 mL/kg of body weight and the incision was closed. A set of six animals were used for each treatment. Animals were treated with two different doses of frankincense oil at 500 mg/kg/day and 1000 mg/kg/day orally [18], gentamicin (10 mg/kg, i.p) [49], imipenem (120 mg/kg, i.p) [50] alone or in combination, while one group of animals served as control each for MRSA and MDR-*P. aeruginosa*. All the animals were sacrificed after 4 days post-inoculation for assessments of bacterial growth/clearance [51,52]. The lungs were removed and the tissue samples (1 gm) were homogenized for 5 min using phosphate buffer saline (1 mL) under an aseptic technique and the bacterial count was determined after suitable dilutions. Homogenates were serially diluted (up to 10^9) and plated on nutrient agar. Plates were incubated at 37 °C and the colonies were counted and log colony-forming units (log 10 CFU) were calculated. Lungs were also subjected to histological examination after staining with H & E stain.

4.7. Statistical Analysis

Data are presented as mean ± SEM wherever indicated in footnotes. Statistical difference was done using one-way ANOVA followed by Tukey's test. $p < 0.05$ was considered significant.

5. Conclusions

This study aimed to determine the antimicrobial effect of frankincense oil and its interaction with imipenem and gentamicin against MRSA and multidrug-resistant *Pseudomonas aeruginosa*. We demonstrated that frankincense oil showed a modest inhibitory effect against MRSA and MDR-*P. aeruginosa*, the oil did not show a noticeable change in the zone of inhibition when combined with imipenem or gentamicin, indicating a lack of interaction between the oil and the antibiotics. In addition, the FIC determined to evaluate the interaction between the frankincense oil and the antibiotics showed no interaction between the oil and imipenem or gentamicin. Furthermore, there was no interaction between the antibiotics and the frankincense oil in the in-vivo model, and the antibacterial effect was similar to the results obtained in-vitro.

Author Contributions: Conceptualization, Y.A., M.A. (Mohammed Alrouji), M.A. (Mohammed Asad) and B.J.; methodology, Y.A., M.A. (Mohammed Alrouji), M.A. (Mohammed Asad), M.B.F.A. and B.J.; software, M.A. (Mohammed Asad) and B.J.; validation, Y.A., M.A. (Mohammed Alrouji), M.A. (Mohammed Asad) and B.J.; formal analysis, Y.A., M.A. (Mohammed Alrouji), M.A. (Mohammed Asad) and B.J.; investigation, Y.A., M.A. (Mohammed Alrouji), M.A. (Mohammed Asad), M.B.F.A. and B.J.; resources, Y.A. and M.A. (Mohammed Alrouji); data curation, Y.A., M.A. (Mohammed Alrouji), M.A. (Mohammed Asad), M.B.F.A. and B.J.; writing—original draft preparation, Y.A., M.A. (Mohammed Alrouji), M.A. (Mohammed Asad), M.B.F.A. and B.J.; writing—review and editing, Y.A., M.A. (Mohammed Alrouji), M.A. (Mohammed Asad), M.B.F.A. and B.J.; visualization, Y.A., M.A. (Mohammed Alrouji), M.A. (Mohammed Asad), M.B.F.A. and B.J.; supervision, M.A. (Mohammed Asad) and B.J.; project administration, Y.A. and M.A. (Mohammed Alrouji); funding acquisition, Y.A. and M.A. (Mohammed Alrouji). All authors have read and agreed to the published version of the manuscript.

Funding: This research received no external funding.

Institutional Review Board Statement: The Ethical Research Committee of Shaqra University reviewed and approved the experimental procedures on rats (approval number—53/18909).

Informed Consent Statement: Not Applicable.

Data Availability Statement: Data will be provided on request by writing to the corresponding author.

Acknowledgments: The authors would like to thank the Deanship of Scientific Research at Shaqra University for supporting this work.

Conflicts of Interest: The authors declare no conflict of interest.

References

1. Founou, R.C.; Founou, L.L.; Essack, S.Y. Clinical and economic impact of antibiotic resistance in developing countries: A systematic review and meta-analysis. *PLoS ONE* **2017**, *12*, e0189621. [CrossRef]
2. Dadgostar, P. Antimicrobial Resistance: Implications and Costs. *Infect. Drug Resist.* **2019**, *12*, 3903–3910. [CrossRef] [PubMed]
3. Shafiq, M.; Zeng, M.; Permana, B.; Bilal, H.; Anderson, J.; Yao, F.; Algammal, A.M.; Li, X.; Yuan, Y.; Jiao, X. Co-existence of bla NDM-5 and tet(X4) in international high-risk E. coli clone ST648 of Human origin in China. *Front. Microbiol.* **2022**, *10*, 4357. [CrossRef]
4. Liu, Q.; Meng, X.; Li, Y.; Zhao, C.N.; Tang, G.Y.; Li, H. Bin Antibacterial and Antifungal Activities of Spices. *Int. J. Mol. Sci.* **2017**, *18*, 1283. [CrossRef] [PubMed]
5. Shafiq, M.; Yao, F.; Bilal, H.; Rahman, S.U.; Zeng, M.; Ali, I.; Zeng, Y.; Li, X.; Yuan, Y.; Jiao, X. Synergistic Activity of Tetrandrine and Colistin against mcr-1-Harboring *Escherichia coli*. *Antibiotics* **2022**, *11*, 1346. [CrossRef] [PubMed]
6. Wang, Y.M.; Kong, L.C.; Liu, J.; Ma, H.X. Synergistic effect of eugenol with Colistin against clinical isolated Colistin-resistant Escherichia coli strains. *Antimicrob. Resist. Infect. Control* **2018**, *7*, 17. [CrossRef] [PubMed]
7. Di Stefano, V.; Schillaci, D.; Cusimano, M.G.; Rishan, M.; Rashan, L. In Vitro Antimicrobial Activity of Frankincense Oils from *Boswellia sacra* Grown in Different Locations of the Dhofar Region (Oman). *Antibiotics* **2020**, *9*, 195. [CrossRef]
8. Van Vuuren, S.F.; Kamatou, G.P.P.; Viljoen, A.M. Volatile composition and antimicrobial activity of twenty commercial frankincense essential oil samples. *S. Afri. J. Bot.* **2010**, *76*, 686–691. [CrossRef]
9. De Rapper, S.; Van Vuuren, S.F.; Kamatou, G.P.P.; Viljoen, A.M.; Dagne, E. The additive and synergistic antimicrobial effects of select frankincense and myrrh oils—A combination from the pharaonic pharmacopoeia. *Lett. Appl. Microbiol.* **2012**, *54*, 352–358. [CrossRef]

10. Lakhundi, S.; Zhang, K. Methicillin Resistant *Staphylococcus aureus*: Molecular Characterization, Evolution, and Epidemiology. *Clin. Microbiol. Rev.* **2018**, *31*, e00020-18. [CrossRef]
11. Kale, P.; Dhawan, B. The changing face of community acquired methicillin resistant *Staphylococcus aureus*. *Indian J. Med. Microbiol.* **2016**, *34*, 275–285. [CrossRef] [PubMed]
12. Poole, K. Pseudomonas Aeruginosa: Resistance to the Max. *Front. Microbiol.* **2011**, *2*, 65. [CrossRef] [PubMed]
13. Breidenstein, E.B.M.; De la Fuente-Núñez, C.; Hancock, R.E.W. Pseudomonas aeruginosa: All roads lead to resistance. *Trends Microbiol.* **2011**, *19*, 419–426. [CrossRef] [PubMed]
14. Oliver, A.; Mulet, X.; López-Causapé, C.; Juan, C. The increasing threat of *Pseudomonas aeruginosa* high-risk clones. *Drug Resist. Updat.* **2015**, *21*, 41–59. [CrossRef]
15. Fasinu, P.S.; Bouic, P.J.; Rosenkranz, B. An overview of the evidence and mechanisms of herb-drug interactions. *Front. Pharmacol.* **2012**, *3*, 69. [CrossRef]
16. Wakefield, M.E. Special Treatments: Constitutional Psychospiritual Points. *Const. Facial Acupunct.* **2014**, 267–276. [CrossRef]
17. Levison, M.E.; Levison, J.H. Pharmacokinetics and Pharma codynamics of Antibacterial Agents. *Infect. Dis. Clin. N. Am.* **2009**, *23*, 791. [CrossRef]
18. Hosny, E.N.; Elhadidy, M.E.; Sawie, H.G.; Kilany, A.; Khadrawy, Y.A. Effect of frankincense oil on the neurochemical changes induced in rat model of status epilepticus. *Clin. Phytoscience* **2020**, *61*, 11. [CrossRef]
19. Camarda, L.; Dayton, T.; Di Stefano, V.; Pitonzo, R.; Schillaci, D. Chemical composition and antimicrobial activity of some oleogum resin essential oils from *Boswellia* spp. (*Burseraceae*). *Ann. Chim.* **2007**, *97*, 837–844. [CrossRef]
20. Ljaljević Grbić, M.; Unković, N.; Dimkić, I.; Janaćković, P.; Gavrilović, M.; Stanojević, O.; Stupar, M.; Vujisić, L.; Jelikić, A.; Stanković, S.; et al. Frankincense and myrrh essential oils and burn incense fume against micro-inhabitants of sacral ambients. Wisdom of the ancients? *J. Ethnopharmacol.* **2018**, *219*, 1–14. [CrossRef]
21. Schillaci, D.; Arizza, V.; Dayton, T.; Camarda, L.; Di Stefano, V. In vitro anti-biofilm activity of *Boswellia* spp. oleogum resin essential oils. *Lett. Appl. Microbiol.* **2008**, *47*, 433–438. [CrossRef]
22. Zhang, J.; Biggs, I.; Sirdaarta, J.; White, A.; Edwin Cock, I. Antibacterial and Anticancer Properties of Boswellia carteri Birdw. and Commiphora molmol Engl. Oleo-Resin Solvent Extractions. *Pharmacogn. Commun.* **2016**, *6*, 120–136. [CrossRef]
23. Cheesman, M.J.; Ilanko, A.; Blonk, B.; Cock, I.E. Developing New Antimicrobial Therapies: Are Synergistic Combinations of Plant Extracts/Compounds with Conventional Antibiotics the Solution? *Pharmacogn. Rev.* **2017**, *11*, 57–72. [CrossRef] [PubMed]
24. Koohsari, H.; Ghaemi, E.A.; Sadegh Sheshpoli, M.; Jahedi, M.; Zahiri, M. The investigation of antibacterial activity of selected native plants from North of Iran. *J. Med. Life* **2015**, *8*, 38. [PubMed]
25. Shaikh, A.S.; Thomas, A.B.; Chitlange, S.S. Herb–drug interaction studies of herbs used in treatment of cardiovascular disorders—A narrative review of preclinical and clinical studies. *Phyther. Res.* **2020**, *34*, 1008–1026. [CrossRef] [PubMed]
26. Yang, S.K.; Yusoff, K.; Mai, C.W.; Lim, W.M.; Yap, W.S.; Lim, S.H.E.; Lai, K.S. Additivity vs. Synergism: Investigation of the Additive Interaction of Cinnamon Bark Oil and Meropenem in Combinatory Therapy. *Mol. A J. Synth. Chem. Nat. Prod. Chem.* **2017**, *22*, 1733. [CrossRef]
27. Yang, D.D.; Paterna, N.J.; Senetra, A.S.; Casey, K.R.; Trieu, P.D.; Caputo, G.A.; Vaden, T.D.; Carone, B.R. Synergistic interactions of ionic liquids and antimicrobials improve drug efficacy. *iScience* **2021**, *24*, 101853. [CrossRef]
28. Ocampo, P.S.; Lázár, V.; Papp, B.; Arnoldini, M.; Zur Wiesch, P.A.; Busa-Fekete, R.; Fekete, G.; Pál, C.; Ackermann, M.; Bonhoeffer, S. Antagonism between Bacteriostatic and Bactericidal Antibiotics Is Prevalent. *Antimicrob. Agents Chemother.* **2014**, *58*, 4573. [CrossRef]
29. Cash, H.A.; Woods, D.E.; McCullough, B.; Johanson, W.G.; Bass, J.A. A rat model of chronic respiratory infection with *Pseudomonas aeruginosa*. *Am. Rev. Respir. Dis.* **1979**, *119*, 453–459. [CrossRef]
30. Niu, H.; Yang, T.; Wang, J.; Wang, R.; Cai, Y. Immunomodulatory Effect of Colistin and its Protective Role in Rats with Methicillin-Resistant *Staphylococcus aureus*-induced *Pneumonia*. *Front. Pharmacol.* **2020**, *11*. [CrossRef]
31. Chalmers, S.J.; Wylam, M.E. Methicillin-Resistant *Staphylococcus aureus* Infection and Treatment Options. *Methods Mol. Biol.* **2020**, *2069*, 229–251. [CrossRef] [PubMed]
32. Vindhya, V.V. Sohanlal Invitro susceptibility of Imipenem by E-test in different conditions of Methicillin resistant *Staphylococcus aureus*. *IP Int. J. Med. Microbiol. Trop. Dis.* **2022**, *8*, 83–89. [CrossRef]
33. Bassetti, M.; Vena, A.; Castaldo, N.; Righi, E.; Peghin, M. New antibiotics for ventilator-associated pneumonia. *Curr. Opin. Infect. Dis.* **2018**, *31*, 177–186. [CrossRef] [PubMed]
34. Stefani, S.; Campana, S.; Cariani, L.; Carnovale, V.; Colombo, C.; Lleo, M.M.; Iula, V.D.; Minicucci, L.; Morelli, P.; Pizzamiglio, G.; et al. Relevance of multidrug-resistant Pseudomonas aeruginosa infections in cystic fibrosis. *Int. J. Med. Microbiol.* **2017**, *307*, 353–362. [CrossRef] [PubMed]
35. Nguyen, L.; Garcia, J.; Gruenberg, K.; MacDougall, C. Multidrug-Resistant Pseudomonas Infections: Hard to Treat, But Hope on the Horizon? *Curr. Infect. Dis. Rep.* **2018**, *20*, 23. [CrossRef] [PubMed]
36. Yoon, E.J.; Jeong, S.H. Mobile Carbapenemase Genes in *Pseudomonas aeruginosa*. *Front. Microbiol.* **2021**, *12*, 30. [CrossRef]
37. Mikhaeil, B.R.; Maatooq, G.T.; Badria, F.A.; Amer, M.M.A. Chemistry and immunomodulatory activity of frankincense oil. *Zeitschrift Naturforsch.-Sect. C J. Biosci.* **2003**, *58*, 230–238. [CrossRef]
38. Başer, K.H.C.; Demirci, B.; Dekebo, A.; Dagne, E. Essential oils of some *Boswellia* spp., myrrh and opopanax. *Flavour. Fragr. J.* **2003**, *18*, 153–156. [CrossRef]

39. Al-Harrasi, A.; Al-Saidi, S. Phytochemical analysis of the essential oil from botanically certified oleogum resin of *Boswellia sacra* (Omani Luban). *Molecules* **2008**, *13*, 2181–2189. [CrossRef]
40. Al-Saidi, S.; Rameshkumar, K.B.; Hisham, A.; Sivakumar, N.; Al-Kindy, S. Composition and antibacterial activity of the essential oils of four commercial grades of Omani luban, the oleo-gum resin of *Boswellia sacra* FLUECK. *Chem. Biodivers.* **2012**, *9*, 615–624. [CrossRef]
41. Da Silva, A.C.R.; Lopes, P.M.; De Azevedo, M.M.B.; Costa, D.C.M.; Alviano, C.S.; Alviano, D.S. Biological activities of α-pinene and β-pinene enantiomers. *Molecules* **2012**, *17*, 6290–6304. [CrossRef] [PubMed]
42. Nidhi, P.; Rolta, R.; Kumar, V.; Dev, K.; Sourirajan, A. Synergistic potential of *Citrus aurantium* L. essential oil with antibiotics against Candida albicans. *J. Ethnopharmacol.* **2020**, *262*, 113135. [CrossRef] [PubMed]
43. De Freitas, B.C.; Queiroz, P.A.; Baldin, V.P.; Do Amaral, P.H.R.; Rodrigues, L.L.F.; Vandresen, F.; R Caleffi-Ferracioli, K.; De, L.; Scodro, R.B.; Cardoso, R.F.; et al. (-)-Camphene-based derivatives as potential antibacterial agents against *Staphylococcus aureus* and *Enterococcus* spp. *Future Microbiol.* **2020**, *15*, 1527–1534. [CrossRef] [PubMed]
44. Surendran, S.; Qassadi, F.; Surendran, G.; Lilley, D.; Heinrich, M. Myrcene—What Are the Potential Health Benefits of This Flavouring and Aroma Agent? *Front. Nutr.* **2021**, *8*, 400. [CrossRef] [PubMed]
45. Yap, P.S.X.; Yiap, B.C.; Ping, H.C.; Lim, S.H.E. Essential oils, a new horizon in combating bacterial antibiotic resistance. *Open Microbiol. J.* **2014**, *8*, 6–14. [CrossRef] [PubMed]
46. Rashan, L.; White, A.; Haulet, M.; Favelin, N.; Das, P.; Cock, I.E. Chemical Composition, Antibacterial Activity, and Antibiotic Potentiation of *Boswellia sacra* Flueck. Oleoresin Extracts from the Dhofar Region of Oman. *Evidence-Based Complement. Altern. Med.* **2021**, *2021*, 9918935. [CrossRef] [PubMed]
47. Fankam, A.G.; Kuete, V.; Voukeng, I.K.; Kuiate, J.R.; Pages, J.M. Antibacterial activities of selected Cameroonian spices and their synergistic effects with antibiotics against multidrug-resistant phenotypes. *BMC Complement. Altern. Med.* **2011**, *11*, 104. [CrossRef]
48. Anesthesia (Guideline) | Vertebrate Animal Research. Available online: https://animal.research.uiowa.edu/iacuc-guidelines-anesthesia (accessed on 13 February 2022).
49. Kandeel, M.; Abdelaziz, I.; Elhabashy, N.; Hegazy, H.; Tolba, Y. Nephrotoxicity oxidative stress of single large dose or two divided doses of gentamicin in rats. *Pakistan J. Biol. Sci.* **2011**, *14*, 627–633. [CrossRef]
50. Guo, P.; Zhang, S.W.; Zhang, J.; Dong, J.T.; Wu, J.D.; Tang, S.T.; Yang, J.T.; Zhang, W.J.; Wu, F. Effects of imipenem combined with low-dose cyclophosphamide on the intestinal barrier in septic rats. *Exp. Ther. Med.* **2018**, *16*, 1919. [CrossRef]
51. Russo, T.A.; Davidson, B.A.; Carlino-MacDonald, U.B.; Helinski, J.D.; Priore, R.L.; Knight, P.R. The effects of Escherichia coli capsule, O-antigen, host neutrophils, and complement in a rat model of Gram-negative pneumonia. *FEMS Microbiol. Lett.* **2003**, *226*, 355–361. [CrossRef]
52. Russo, T.A.; Bartholomew, L.A.; Davidson, B.A.; Helinski, J.D.; Carlino, U.B.; Knight, P.R.; Beers, M.F.; Atochina, E.N.; Notter, R.H.; Holm, B.A. Total extracellular surfactant is increased but abnormal in a rat model of gram-negative bacterial pneumonia. *Am. J. Physiol. Lung Cell. Mol. Physiol.* **2002**, *283*, L655–L663. [CrossRef] [PubMed]

antibiotics

Article

Isolation and Characterization of a Novel *Actinomycete* Isolated from Marine Sediments and Its Antibacterial Activity against Fish Pathogens

Haimanti Mondal and John Thomas *

Centre for Nanobiotechnology, Vellore Institute of Technology (VIT), Vellore 632014, India
* Correspondence: john.thomas@vit.ac.in or th_john28@yahoo.co.in; Tel.: +91-416-220-4213; Fax: +91-416-224-3092

Abstract: Marine habitats are especially complex, with a varied diversity of living organisms. Marine organisms, while living in such intense conditions, have developed great physiological and metabolic potential to survive. This has led them to produce several potent metabolites, which their terrestrial counterparts are unable to produce. Over the past few years, marine *Actinomycetes* have been considered one of the most abundant sources of diverse and novel metabolites. In this work, an attempt was made to isolate *Actinomycetes* from marine sediments in terms of their ability to produce several novel bioactive compounds. A total of 16 different *Actinomycete* colonies were obtained from marine sediment samples. Among the 16 *Actinomycete* isolates, 2 isolates demonstrated in vitro antibacterial activity against *Aeromonas hydrophila* and *Vibrio parahemolyticus*. However, among them, only one isolate was found to have potent antibacterial activity, and hence, was taken for further analysis. This isolate was designated as *Beijerinickia fluminensis* VIT01. The bioactive components obtained were extracted and later subjected to Fourier transform infrared spectroscopy (FTIR) and gas chromatography–mass spectroscopy (GC-MS) analyses for identification. Several novel bioactive compounds were reported from the data obtained and were found to have potent antibacterial activity. Hence, they could be used as an alternative to antibiotics for treating several fish pathogens in the aquaculture industry.

Keywords: marine *Actinomycetes*; *Beijerinickia fluminensis*; antibacterial activity; FTIR; GC-MS

Citation: Mondal, H.; Thomas, J. Isolation and Characterization of a Novel *Actinomycete* Isolated from Marine Sediments and Its Antibacterial Activity against Fish Pathogens. *Antibiotics* **2022**, *11*, 1546. https://doi.org/10.3390/antibiotics11111546

Academic Editors: Valério Monteiro-Neto and Elizabeth S. Fernandes

Received: 29 September 2022
Accepted: 1 November 2022
Published: 3 November 2022

1. Introduction

Actinomycetes, belonging to the order Actinomycetales, are members of a heterogeneous group of Gram-positive bacteria, consisting of up to >55% GC content in their DNA [1]. They are anaerobic, with filamentous fungal morphology and branched growth patterns, resulting in either extensive mycelium or colony. Later, the mycelium may fall apart to form coccoid or rod-shaped forms. Most genera of *Actinomycetes* form a spore called sporangia or spore cases, found on the colony surface, aerial hyphae or maybe within the free environment. They exhibit diverse genetic, biological and functional activities and are a good source of many new secondary metabolites [2–4]. The sea has a vast treasure of resources. Due to this fact, marine *Actinomycetes* play a significant role by being a major part of it [5]. Apart from this reason, they differ from their terrestrial counterparts considerably because of distinct environmental changes prevailing in marine habitats [6,7].

The first marine *Actinomycete* was discovered in the year 1984 [4,8]. Since that period, many novel species of marine *Actinomycetes* have been found in aquatic environments worldwide [9–13]. In addition, a majority of the strains of *Actinomycetes* from the marine environment have been isolated from sediments [14]. The population size of *Actinomycetes* in the case of oceanic sediments has been reported to vary with various physicochemical parameters, such as pH, temperature, salinity, total organic carbon, pressure, etc. [4].

In general, *Actinomycetes*, and in particular, marine *Actinomycetes* have a vast range of applications in various fields. Be it in drug synthesis to cure certain diseases or biologically active products, synthesis of antibacterial, antiviral, antifungal as well as anticancer products, synthesis of enzymes, or their major role in biofouling, they have contributed a lot in various areas of the aquaculture industry [5]. Sharma et al. [15] reported the applications of *Actinomycetes* in bioremediation, cancer treatment and production of some valuable antibiotics, such as amphotericin, neomycin, vancomycin, chloramphenicol, novobiocin, gentamycin, nystatin, erythromycin, tetracycline, etc. They are also used as biocontrol tools, antifungal compounds, bio corrosion and as a source of agroactive compounds. The bioactive compounds extracted from these marine *Actinomycetes* possess different chemical structures and conformations, which hold the key to novel drug synthesis that, in future, may be potent enough to combat various resistant pathogens [16].

Actinomycetes are widely distributed in the marine environment, in marine species such as *Streptomyces*, *Nocardia* and *Micromonospora*. Their occurrence on the contact slides and dead marine algae suspended in the sea has been reported [17].

A study reported by Ref [18] revealed that some new marine *Actinomycetes* have produced a variety of bioactive metabolites. The antibacterial activity of fermentation products from *Actinomycetes* isolated from the marine environment also showed that several bioactive compounds had activities against vaccinia virus replication multidrug-resistant Gram-positive bacteria and cancer cells.

Actinomycetes contribute 70% of antibiotic sources and numerous non-antibiotic bioactive metabolites, including enzymes, enzyme inhibitors, anti-oxidation reagents, immunological regulators, etc. [19]. In other words, currently, almost two-thirds of the antibiotics developed are derived from *Actinomycetes*. Marine *Actinomycetes* have been reportedly studied for their efficacy as a potent novel antibiotic producer [20–23]. Although the use of several antibiotics has been banned in the aquaculture industry [24], they are still being used and are causing resistance. In this context, the aim of the current work was to isolate a novel bioactive compound from marine *Actinomycetes*, which is effective against pathogens in aquaculture.

2. Results

2.1. Isolation of Actinomycetes

A total of 16 different *Actinomycete* colonies were obtained from marine sediment samples. To isolate the *Actinomycetes*, specific agar medium Actinomycetes Isolation Agar (AIA) was used, since it was found to be good for the growth of actinomycetes. Some morphologically different strains of *Actinomycetes* were obtained from the sediment samples collected off the coast of Digha, West Bengal, India.

2.2. Characterization of Isolates

Among the 16 *Actinomycete* isolates, 2 isolates demonstrated in vitro antibacterial activity against *Aeromonas hydrophila* and *Vibrio parahemolyticus*. However, among them, only one isolate was found to have potent antibacterial activity, and hence, was taken for further characterization. This isolate was designated as *Beijerinckia fluminensis* VIT01.

2.2.1. Screening of Potential Actinomycete Isolates

The antibacterial activity of the isolate was confirmed by primary screening using the cross-streak method. This was followed by secondary screening using the agar well diffusion method. The crude extract of *Actinomycete* isolates showed a zone of inhibition against the two bacterial pathogens tested. Among the two pathogens, *Vibrio parahemolyticus* showed better results when compared to *Aeromonas hydrophila*. In the case of primary screening using the cross-streak method, the zones of inhibition observed in the isolate against pathogens *A. hydrophila* and *V. parahemolyticus* were 11 mm and 13 mm, respectively. In addition, the zone of inhibition observed in the secondary screening of the isolate was also found to be better in *V. parahemolyticus* compared to *A. hydrophila*.

2.2.2. Screening of the Potential Isolate in Comparison to the Antibiotic

Gentamicin was found to be sensitive to both pathogens *A. hydrophila* and *V. parahemolyticus*, showing a zone of inhibition of 11 mm and 12 mm in the case of both *V. parahemolyticus* and *A. hydrophila*, respectively, using the disc diffusion method. On the other hand, the *Actinomycete* isolate gave a better zone of inhibition compared to the antibiotic.

The zone of inhibition revealed in vitro antibacterial activity of *Beijernickia fluminensis* against both pathogens *Aeromonas hydrophila* and *Vibrio parahemolyticus*, as shown in Figure 1A. Figure 1B reveals the zone of inhibition of Gentamicin against *A. hydrophila* and *V. parahemolyticus*.

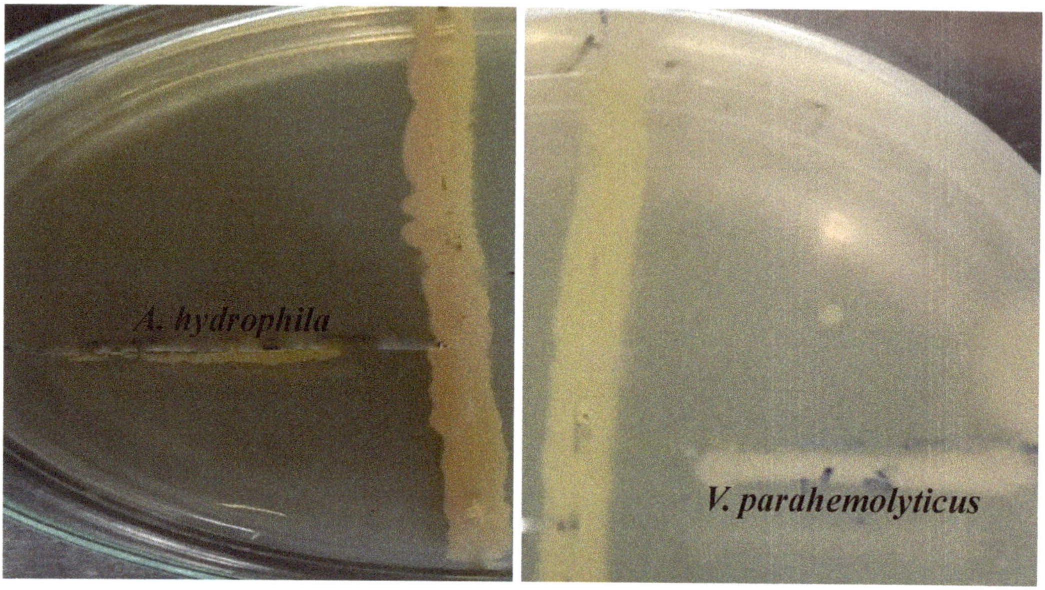

(A) (B)

Figure 1. Primary screening of the antimicrobial activity of isolate 1 (*B. fluminensis* VIT01) against selected pathogens (**A**) *A. hydrophila* (**B**) *V. parahemolyticus*.

Table 1 summarizes the antibacterial activity of some *Actinomycete* isolates against *Aeromonas hydrophila* and *Vibrio parahemolyticus* compared to Gentamicin. The data table shows that compared to Gentamicin, *Beijernickia fluminensis* gave a better zone of inhibition against both pathogens *A. hydrophila* and *V. parahemolyticus*.

Table 1. Antibacterial activity of *Actinomycete* isolates compared to Gentamicin.

S.No.	Isolates	Antibacterial Activity/Zone of Inhibition (in mm)		Zone of Inhibition in Gentamicin (in mm)	
		Aeromonas hydrophila	*Vibrio parahemolyticus*	*Aeromonas hydrophila*	*Vibrio parahemolyticus*
1	Isolate 1 (*Beijernickia fluminensis*)	11	13	12	11
2	Isolate 2	9	12	12	11

2.2.3. Morphological Characterization

Actinomycetes were isolated, and their morphological appearance was observed. The staining results showed that the bacterium was Gram-positive, cocci-shaped, non-motile, capsulated and spore-forming.

FESEM Analysis

The FESEM micrograph revealed the shape and arrangement of the isolate under different magnifications, as shown in Figure 2.

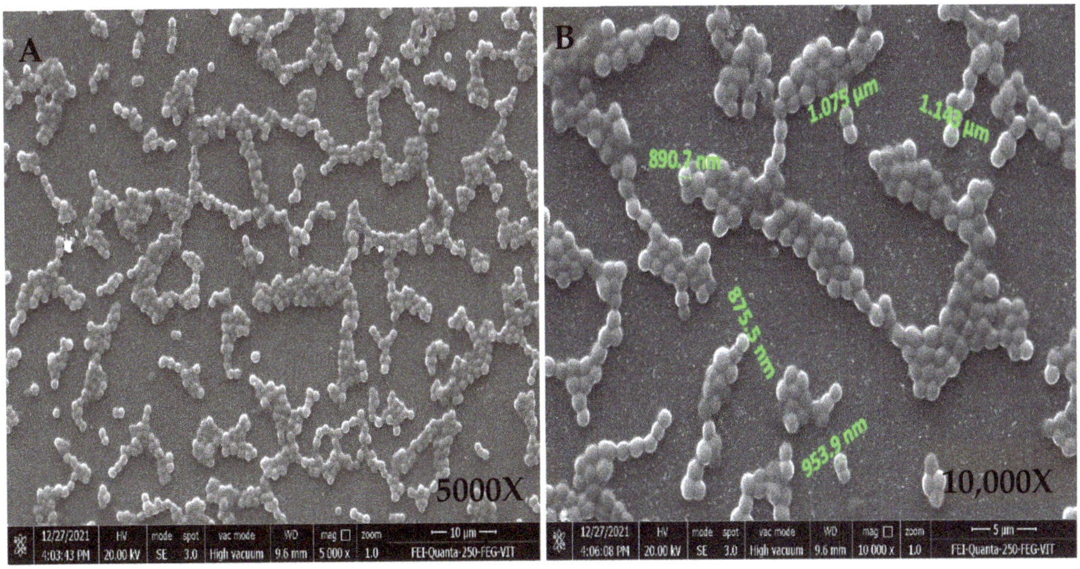

Figure 2. Scanning electron microscope images under 5000X (**A**) and 10,000X (**B**). Cocci in clusters.

Biochemical Characterization

The biochemical tests performed showed a positive result for Gram-positive bacteria. The results indicate that the isolate was capable of starch, gelatin and casein hydrolysis. The results of biochemical tests are presented in Table 2.

Physiological Characterization

The tested *Actinomycete* isolate was able to grow in 1%, 3.5%, 5% and 7% NaCl concentration but showed resistance at 10% and 20% NaCl. The ideal temperature for the growth of *Actinomycetes* was found to be between 28 and 30 °C, although they could even grow in temperatures up to 40 °C.

Molecular Characterization

The 16S rRNA sequencing revealed that the strain belongs to *Beijerinckia fluminensis* under *Actinomycetes*. This was designated as *Beijerinckia fluminensis* VIT01.

2.3. FTIR Analysis of Crude Extracts from Actinomycetes

Figure 3 summarizes the FTIR spectra of the sample ranging from 400 to 4000 cm^{-1}. The characteristic bands observed from Sample 1 are 2652.15, 2041.57, 1292.27, 1042.70, 751.34 and 556.90. The data shown in Table 2 reveal the vibrational type of the functional groups, their band strength and the compounds present in the respective peaks of the ethyl acetate extract of *B. fluminensis*. The peak at 2652.15 corresponds to the stretching vibrational type of weak thiol groups. The peak at 2041.57 was assigned to the stretching vibration of strong isothiocyanate groups. Similarly, the peaks at 1292.27 and 1042.70 correspond to the stretching vibration type of strong fluoro compound and medium amine groups, respectively. In addition, the peak at 751.34 corresponds to the bending vibration type of strong 1,2-disubstituted groups, and the peak at 556.90 corresponds to the stretching type of vibration of strong halo compound groups.

Table 2. Biochemical tests of *Beijerinickia fluminensis* VIT01 strain.

Biochemical Characteristics	*Beijerinickia fluminensis*
ONPG	Positive
Lysine Utilization	Positive
Ornithine Utilization	Positive
Urease	Positive
Phenylalanine Deamination	Negative
Nitrate Reduction	Positive
H_2S Production	Negative
Citrate Utilization	Positive
Voges–Proskauer	Negative
Methyl Red	Negative
Indole	Negative
Malonate Utilization	Negative
Esculin Hydrolysis	Positive
L-Arabinose	Negative
Xylose	Negative
Adonitol	Negative
Rhamnose	Negative
Cellobiose	Negative
Melibiose	Negative
Saccharose	Positive
Raffinose	Negative
Trehalose	Positive
Glucose	Positive
Lactose	Positive
Oxidase	Negative
Casein Utilization	Positive
Melezitose	Negative
α–Methyl–D-Mannoside	Negative
Xylitol	Positive
D-Arabinose	Negative
Sorbose	Positive

The peak values corresponding to their respective groups in the sample are shown in Table 3.

Table 3. FTIR analysis of compounds exhibiting their functional groups.

Wavelength	Functional Group/Bonds	Band Strength	Compound
Ethyl Acetate Extract of Actinomycetes (Sample)			
2652.15	S-H stretch	Weak	Thiol
2573.22	O-H stretch	Strong, Broad	Carboxylic acid
2041.57	N=C=S stretch	Strong	Isothiocyanate
2013.71	N=C=S stretch	Strong	Isothiocyanate
1292.27	C-F stretch	Strong	Fluoro compound
1167.49	C-O stretch	Strong	Tertiary alcohol
1097.84	C-N stretch	Medium	Amine
1042.705	C-N stretch	Medium	Amine
991.63	C=C bending	Strong	Alkene

Table 3. *Cont.*

Wavelength	Functional Group/Bonds	Band Strength	Compound
Ethyl Acetate Extract of Actinomycetes (Sample)			
885.41	C=C bending	Strong	Alkene
751.34	C-H bending	Strong	1,2-disubstituted
709.55	C=C bending	Strong	Alkene
556.909	C-Cl stretching	Strong	Halo compound
464.62	C-l stretching	Strong	Halo compound

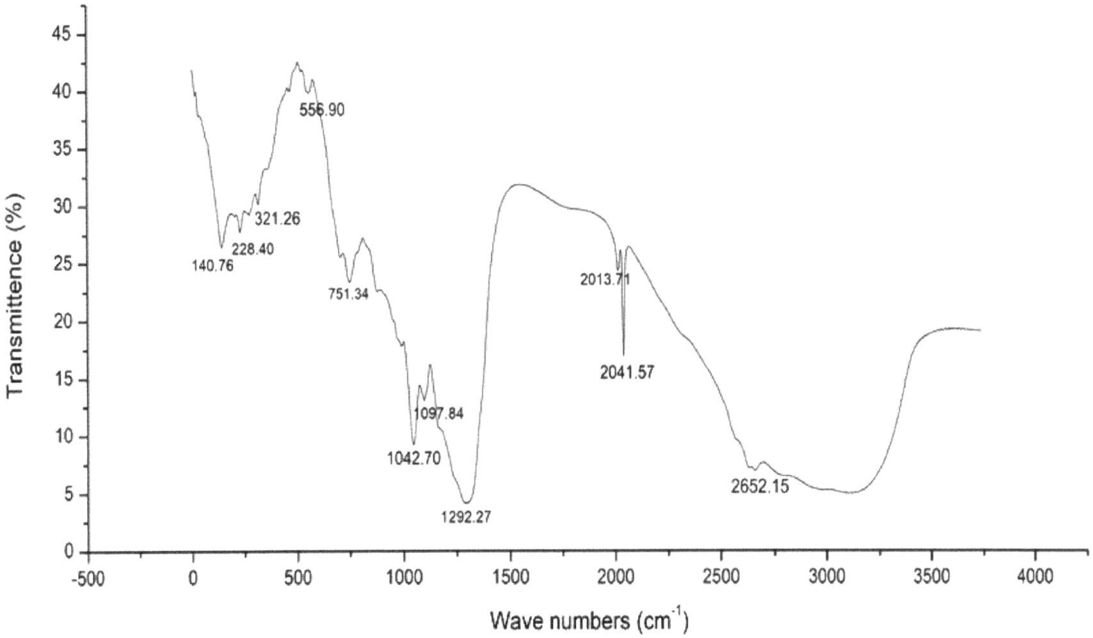

Figure 3. FTIR analysis of ethyl acetate extract of *Beijerinickia fluminensis*.

2.4. GC-MS Analysis of Crude Extracts from Actinomycetes

The GC-MS analysis of the identified active compounds isolated from the ethanolic extract of *Actinomycetes* displayed their peak percentage expressed with their retention indices in the chromatogram.

The bioactive compounds identified from the ethanolic extract of *Beijerinckia fluminensis* VIT01 showed that they contain many active components. They are responsible for several properties, including antimicrobial and antibacterial, as per Dr. Duke's Phytochemical and Ethnobotanical Database. Figure 4 represents the chromatogram of the bioactive compounds extracted from the ethanolic extract of the *Actinomycete* strain sample. The compounds are 1,6;3,4-Dianhydro-2-Deoxy-Beta-D-Lyxo-Hexopyrano; N,N-Dimethylheptanamide, Butanamide,3,N-Dihydroxy; (S)-Isopropyl Lactate; 3-O-Acetyl-exo-1,2-O-Ethylidene-Alpha-D-Erythrof; Glycine,N-Octyl-,Ethyl Ester; D-Glucitol,1-Deoxy-1-(Octylamino); Glycyl-L-Proline; 7-Tetradecene, (E); 2,3-Anhydro-D-Galactosan, etc. The data from Table 4 show the major peak, the retention time of the compounds and the activities present in the ethyl acetate extract of *B. fluminensis*. The results revealed that compounds such as N, N-Dimethylheptanamide, Glycine,N-Octyl-,Ethyl Ester, Glycyl-L-Proline, Actinomycin C2, (S)-3,4-Dimethylpentanol and 7-Tetradecene, (E), have antibacterial properties. Therefore, these compounds can be used as potent antibacterial agents.

Table 4 summarizes the main compounds, retention time, major peak area and major activities present in the sample. These bioactive compounds were later used for in vivo and in vitro studies.

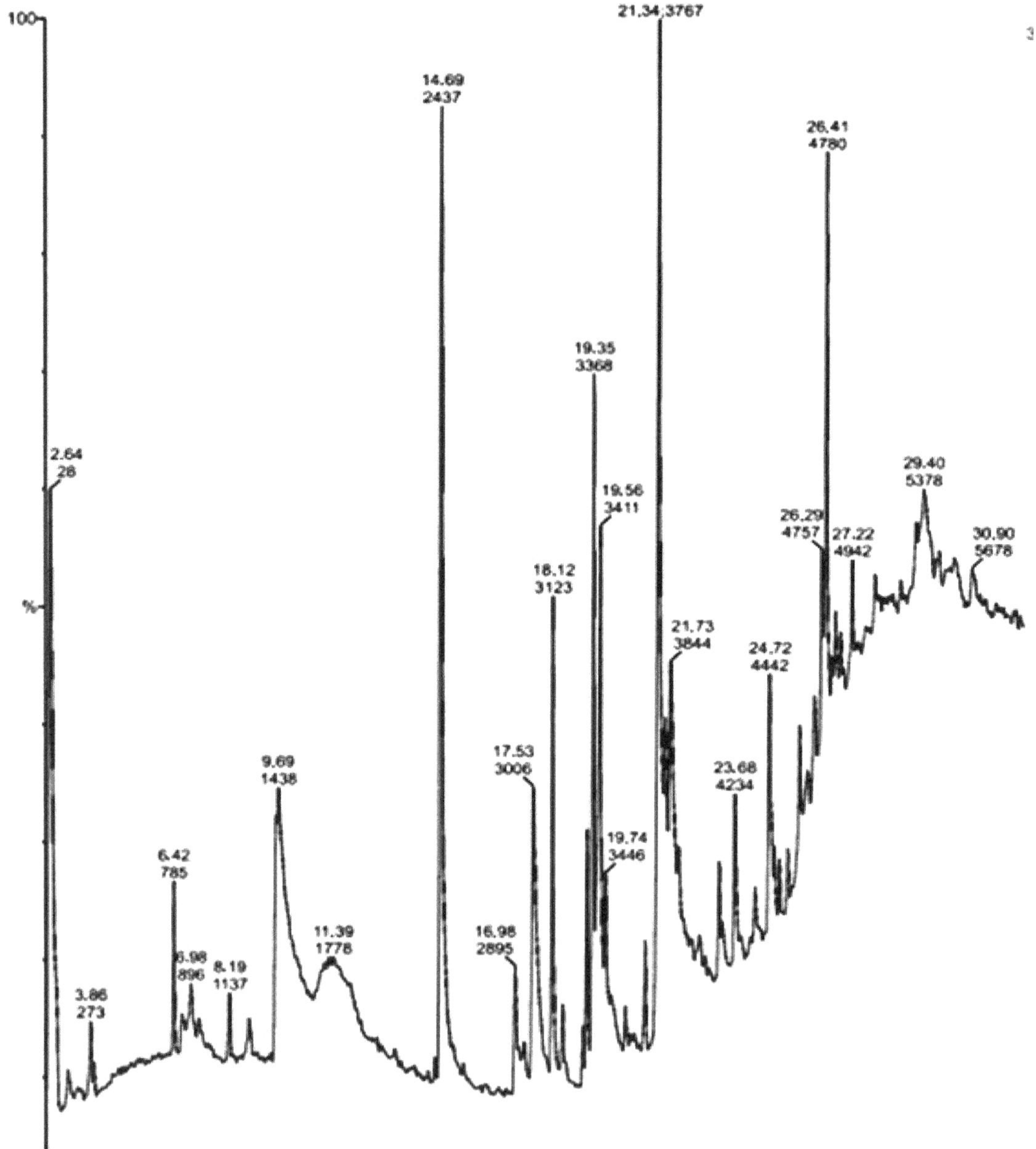

Figure 4. GC-MS chromatogram of ethyl acetate extract of *Beijerinickia fluminensis*.

Table 4. Activity of bioactive compounds found in ethyl acetate extract of *Beijerinickia fluminensis*.

S.No	Compounds	Major Peak Area (%)	Retention Time (RT)	Major Activities * Present in Sample
1	N, N-Dimethylheptanamide	28	2.64	Antibacterial, Antitumor
2	Butanamide,3,N-Dihydroxy	273	3.86	Antitumor
3	3-O-Acetyl-Exo-1,2-O-Ethylidene-Alpha-D-Erythrof	785	6.42	Anticancer
4	5-Aminovaleric acid	1438	9.69	Anticarcinogenic
5	Glycine,N-Octyl-,Ethyl Ester	2437	14.69	Antibacterial, Antitumor
6	As-Triazine-3,5(2H,4H)-Dione, 6-(Dimethylamino)	2895	16.98	
7	Glycyl-L-Proline	3006	17.53	Antibacterial
8	Propyl Aldoxime, 2-Methyl-, Syn	3123	18.12	DNA synthesis inhibitor
9	Actinomycin C2	3368	19.35	Antibacterial, Anticancer
10	(S)-3,4-Dimethylpentanol	3411	19.56	Antibacterial, Anticancer, Antidiabetic, DNA synthesis inhibitor
11	7-Tetradecene, (E)	3767	21.34	Antibacterial, Anticancer
12	2-Decenioc acid	3844	21.73	Anticarcinogenic
13	1-Decene, 8-Methyl	4234	23.68	Methyl-guanidine inhibitor
14	Heneicosane, 11-Phenyl	4442	24.72	-
15	Dodecane, 1-ChloroHeneicosane, 11-Phenyl	4757	26.29	-
16	Dichloroacetic Acid, 2-Ethylhexyl Ester	4780	26.41	-
17	Pterin-6-Carboxylic acid	4942	27.22	Anticarcinogenic
18	2,3-Anhydro-D-Galactosan	5378	29.40	Anticancer

* Dr. Duke's Phytochemical and Ethnobotanical Database [25].

3. Discussion

Actinomycetes have proven to be one of the best sources of secondary metabolites, and they have potent bioactive compounds [26]. Currently, the incidence of infectious diseases in the aquaculture industry is increasing, leading to significant loss in the aquaculture industry. Hence, there is a dire need to develop novel bioactive compounds that can be effective against various pathogens. While screening for *Actinomycetes*, various fast growing microbial colonies, which have the ability to inhibit the colonization of *Actinomycetes* on their particular isolation medium, were detected. In order to isolate *Actinomycetes* and prevent the growth of unwanted as well as ubiquitous bacteria, pre-treatment of the sediment samples was preferred. Gebreyohannes et al. [27] reported that potent antibiotic-producing actinomycetes isolated from the marine sediment were subjected to physical pre-treatment methods to stimulate the growth of *Actinomycetes*. In the present study, similar aseptic methods were followed to prevent the growth of unwanted bacteria.

In recent times, there has been a growing interest in the isolation of various novel and potent bioactive *Actinomycetes* from the marine environment, since they contain bioactive compounds. Many culture-independent studies [28] have reported that the marine environment contains a vast and rich diversity of rare *Actinomycetes.* They have been useful in designing successful isolation schemes to isolate a wide variety of marine *Actinomycetes.* Oggerin et al. [29] isolated *Beijerinickia fluminensis* strains UQM 1685^T and CIP 106281^T from soil. Similarly, a recent study was conducted by Shwaaiman et al. [30] who isolated a novel bacterium *Beijerinickia fluminensis* BFC-33 from soil. In the present study, we isolated *Beijerinickia fluminensis* from marine sediments. **Based on the literature survey, this might be the first report on the isolation of *Beijerinickia fluminensis* from marine sediments.**

Trakunjae et al. [31] reported that they isolated a rare *Actinomycete, Rhodococcus* so. BSRT1-1, which was Gram-positive, non-motile, with short rods that later converted to cocci. Oggerin et al. [29] isolated *Beijerinickia fluminensis* from a soil sample. In the present work, the bacterium *Beijerinickia fluminensis* was isolated from marine sediments. The present work revealed that the isolated bacterium is light orange colored. It is Gram positive, with cocci in clusters, and non-motile. This corroborated the study of Oggerin et al. [29] who reported that a white and cream colored isolate was non-motile. Several studies reported on the biochemical and physiological characteristics of *Actinomycetes.* A study by Ref [32] reported the biochemical and physiological characteristics of *Actinomycetes.* They reported the presence of starch and gelatin. Casein hydrolysis was also reported. They also reported negative results for indole, Voges–Proskauer, urease and lipase, and positive results for methyl-red and citrate. In the present study, *B. fluminensis* gave positive results for starch, casein and gelatin hydrolysis, citrate and urease, while negative results were obtained for methyl-red, Voges–Proskauer and indole tests. Rajkumar et al. [33] isolated actinomycetes from different environments. They reported that the isolated *Actinomycetes* showed negative Indole, MR and VP. They also reported that the isolates from *Actinomycetes* CAHSH-2 and *Actinomycetes* CAHSH-3 were positive for citrate, urease, starch, casein and gelatin. This is in agreement with our report. In the case of physiological characterization, Sivanandhini et al. [32] revealed that the *Actinomycetes* were able to survive in high salt concentrations. Similarly, our present work showed that *B. fluminensis* was able to tolerate up to 7% salt concentration. Another study reported by Ref [27] on the *Actinomycetes* isolated from water and sediments showed starch and urea hydrolysis. It was also able to tolerate NaCl concentration up to 5%. Our present study showed that both starch and urea were able to hydrolyze. It was also able to survive in 7% salt concentration. Similarly, Undabarrena et al. [34] reported the physiological characteristics of several species of Actinobacteria. They reported that they were able to survive at 1%, 3.5%, 5%, 7% and 10% NaCl concentration. In the present work, we observed that they were able to tolerate up to 7% salt concentration. Rajkumar et al. [33] also reported that the *Actinomycetes* were able to survive in 9% NaCl. In the present study, equally similar biochemical and physiological characteristics, such as NaCl tolerance (salinity), temperature tolerance, as well as biochemical characteristics of *Actinomycetes,* were reported.

Primary and secondary screening was performed to evaluate the antimicrobial potential of the microorganism against the pathogens. Singh et al. [35] screened thirty-six *Actinomycete* isolates against both Gram-positive and Gram-negative bacterial pathogens, such as *Staphylococcus aureus* MTCC 96, *Bacillus subtilis* MTCC 441 and *Escherichia coli* MTCC 64, and a fungal pathogen *Candida albicans* MTCC 183. Among them, fifteen showed strong and moderate antimicrobial activity. Later, they were subjected to secondary screening, where out of fifteen active isolates, thirteen exhibited strong antimicrobial activity. Another study by Ref. [36] reported that different strains of *Streptomyces* sp. showed antibacterial activity against *Vibrio* spp. In the present study, good antibacterial activity was seen in the primary screening of the isolate, which exhibited a zone of inhibition of 11 mm and 13 against pathogens *A. hydrophila* and *V. parahemolyticus*, respectively, followed by secondary screening.

Norouzi et al., 2018 [37] reported the antibacterial activity potential of several *Actinomycete* isolates against many bacterial pathogens in comparison to antibiotics such as Gentamicin using the Kirby–Bauer disc diffusion method. Antibiotic sensitivity test and disc diffusion were carried out to compare the activity of the crude extract of the *Actinomycete* isolate with the antibiotic Gentamycin. The disc diffusion revealed that, among the three isolates, two potent marine *Actinomycetes* isolates MN2 and MN39 were able to produce biomolecules with antibacterial activity against multidrug-resistant (MDR) bacteria. In our present study, we observed that our isolate gave a better result compared to Gentamicin. Gentamicin exhibited zones of inhibition of 11 mm and 12 mm in the case of both pathogens, *V. parahemolyticus* and *A. hydrophila*. The zones of inhibition observed in the isolate against pathogens *A. hydrophila* and *V. parahemolyticus* were 11 mm and 13 mm, respectively.

Currently, the number of many novel secondary metabolites reported from marine *Actinomycetes* has surpassed the number of their terrestrial counterparts [38]. Dholakiya et al. [39] reported the results of a study where the bacteria *Streptomyces variabilis* exhibited a wide spectrum of antibacterial activities against several Gram-negative and Gram-positive bacteria. Merochlorin E and F isolated from a marine bacterium *Streptomyces* sp. displayed antibacterial activities against several bacteria, such as *Staphylococcus. aureus, Bacillus. subtilis* and *Kocuria. rhizophila* [40,41]. Voon et al. [42] reported the FTIR spectrum of biocellulose from *Beijerinickia fluminensis* WAUPM53 in different media. Another study by Ref [43] reported the FTIR spectra of melanin produced by *Beijerinickia fluminensis.* In the current study, several potent antibacterial compounds were found in the FTIR and GC-MS analyses of *Actinomycetes.*

4. Materials and Methods

4.1. Sample Collection

The samples for screening were collected in the month of May 2021, off the coast of Digha, West Bengal, India. The marine sediment samples were collected from Digha (21°37′35.82″ N latitude and 87°30′26.75″ E longitude) and transferred to autoclavable bags and then stored in cell frost at 4 °C [27,34].

4.2. Media Used

Media

The Actinomycetes Isolation Agar (AIA) medium (Hi Media) was used for the isolation of *Actinomycetes.* Nalidixic acid (50 µg/mL) and Mycostatin (100 µg/mL) were added to the medium. The International Streptomyces Project (ISP-4) medium (TM media, Delhi, India) was used for purification of the isolates [44]. The Mueller–Hinton Agar (MHA) medium (HiMedia, Mumbai, India) was used for the determination of antimicrobial activity of the *Actinomycete* isolates.

4.3. Pre-Treatment of Sediment Samples

To enable the isolation of *Actinomycetes*, samples were subjected to pre-treatment. Sediment samples were kept overnight at 70 °C in a hot air oven for drying [45]. They were crushed and ground aseptically using mortar and pestle under aseptic condition.

4.4. Isolation of Samples

From the pre-treated sediment sample, 1 g was weighed and dispersed into a 100 mL conical flask containing distilled water with 0.9% saline and shaken continuously at room temperature using an orbital shaker for 10 min. This was considered as the stock culture for the sediment samples. A volume of 1 mL was taken from the stock solution and transferred to 9 mL sterile test tube containing water and mixed vigorously to reach a dilution factor of 10^{-1}. Afterward, serial dilutions from 10^{-2} to 10^{-6} were carried out.

After serial dilution, a 0.1 mL aliquot sample of each dilution from 10^{-3} to 10^{-6} was plated in the AIA medium (HiMedia) containing 0.9% NaCl using the spread plate technique. The medium was amended with 50 µg/mL of Nalidixic acid (HiMedia) and Mycostatin

(100 μg/mL) (HiMedia) to prevent the growth of other microorganisms. The Petri plates were incubated at 30 °C for 7–14 days until visible colonies were observed [39,46].

4.5. Screening of Samples

The pure colonies were counted using a colony counter. Their morphological characteristics and pigmentation were recorded.

After isolation, the isolates were screened and sub-cultured by individual streaking into AIA medium prepared with 0.9% NaCl. They were then transferred to new Petri plates containing AIA medium to obtain pure colonies. In addition, they were also sub-cultured and maintained in AIA slants at 4 °C [47]. The isolated bacteria were later stored at −20 °C in 20% glycerol [34,48].

4.6. Characterization of Isolates

Strains were identified based on morphological, cultural, biochemical characteristics and molecular analysis using 16S rRNA sequencing.

4.6.1. Morphological Characterization

The selected isolates were streaked into an ISP4 medium to study their morphological characters by the macroscopic method. The colonies, morphology of the substrate and spores, aerial hyphae, branching and pigment production on the plates were observed in the preliminary stage. Gram staining, motility test and capsule staining were carried out to check the structure and motility of the isolated colonies.

Gram Staining

Gram staining was performed to identify the isolates, i.e., whether they are Gram positive or Gram negative. Later, the organism was first observed under 10× and 40× magnification and then under 100× with oil immersion [49].

Motility Test

A motility test was carried out by inoculating the isolate using the Sulphide Indole Motility (SIM) medium with an inoculation loop up to half or one-third inch above the bottom in a straight line. It was then incubated at 30 °C for 7 days. In addition, it was also observed under 10× and 40× magnification through the microscope to check its motility.

Capsule Staining

For capsule staining, Nigrosin was used as the primary stain and crystal violet as a counterstain. It was observed under 100× oil immersion microscope.

Field Emission Scanning Electron Microscope (FESEM) Analysis

The isolates selected were examined microscopically for spore chain morphology under 3000×, 5000×, 8000×, 10,000× and 13,000×. FESEM analysis was performed by two different methods to check the arrangement of the spores, the sporulating structures and the external surface morphology of the isolates. In the first method, the isolate was inoculated in a broth medium and incubated in a rotary shaker at 30 °C for 7 days until there was growth of bacteria. The broth culture was centrifuged at 7000 rpm for 10 min, and the supernatant was discarded. An amount of 1 mL autoclaved 1× Phosphate Buffer Saline (PBS) was added to the cell-free supernatant (CFS) and mixed well. Later, 10–20 μL of this culture was transferred to a clean grease-free slide. An amount of 20 μL of 0.25% Glutaraldehyde solution was dispensed into the slide containing the culture and oven dried overnight at 40 °C. In the second method, a clean, grease-free slide was taken, and a thin smear of the isolate was made in the center of it, air dried completely and kept overnight in a hot air oven at +40 °C.

4.6.2. Biochemical Characterization

Biochemical Tests

Biochemical tests, such as ONPG, Ornithine, Lysine, Phenylalanine, Urease, H_2S Production, Nitrate Reduction, Methyl Red–Voges–Proskauer's Tests (MR-VP), Indole Production, Esculin Hydrolysis, Malonate, Catalase, Oxidase, Citrate Utilization, Crystal violet, Casein Utilization and Growth in MacConkey Agar, were carried out using a readily prepared KB003 Hi 25 Kit (HiMedia). Both Strip-I and Strip-II reaction wells of the kit were inoculated with *Actinomycete* culture using an inoculation loop incubated at 30 °C for 7–14 days.

In addition to the above-conducted tests, readily prepared KB009 Hi Carbo^{TM} Kit B009C HiMedia kit was used to check the capability of the isolated *Actinomycetes* to use several carbon compounds as an energy source. The kit consisted of 11 sugars, including Melezitose, Xylitol, Cellobiose, D-Arabinose, Sorbose, Rhamnose, alpha-Methyl-D-Mannoside, Malonate, Esculin, Citrate, O-Nitrophenyl-β-D-galactopyranoside (ONPG) and one control. Parts C of the wells in the kit were inoculated similarly to the *Actinomycetes* and incubated at 30 °C for 7–14 days [33].

Starch Utilization Test

A starch hydrolysis test was performed in which the isolates were inoculated on sterile starch agar plates and incubated at 30 °C for seven days. Following the incubation period, iodine solution was flooded into the plates. Starch hydrolysis was confirmed by the presence of a clear zone of hydrolysis around the bacterial growth [27].

Gelatin Hydrolysis Test

In the gelatin hydrolysis test, the isolates were inoculated on gelatin agar plates and incubated at 30 °C for seven days to check the gelatin hydrolysis. At the end of the incubation period, 1 mL mercuric chloride solution was dispersed into the agar medium to observe the zone of hydrolysis [27].

Casein Hydrolysis Test

The isolates were similarly streaked on skimmed milk agar medium plates. They were then incubated at 30 °C for seven days to see whether there was a zone of hydrolysis [27].

Citrate Utilization Test

The isolates were also inoculated on Simon's citrate slant agar test tubes incubated at 30 °C for seven days until a color change was found [27].

4.6.3. Physiological Characterization

To test the NaCl resistance, the AIA medium was prepared in seven batches and amended with 0%, 1%, 3.5%, 5%, 7%, 10% and 20% NaCl. The isolates were streaked on the media and incubated at 30 °C for seven days. The growth of the isolates at the highest salt concentration was observed and recorded.

The sterile AIA plates were inoculated with the isolates and incubated at 4 °C, 20 °C, 25 °C, 30 °C, 37 °C and 40 °C for seven days. The maximum growth at the optimal temperature was recorded by the visual observation of growth [50].

4.6.4. Molecular Characterization

Based on the biochemical characteristics and morphology of the isolate, the isolate that showed good antibacterial activity was chosen for further study. The 16S ribosomal RNA (rRNA) sequence analysis was carried out to confirm the isolated bacterium.

4.7. Bacterial Cultures Used

The two bacterial fish pathogens selected for this study were *Vibrio parahemolyticus* (MTCC 451) and *Aeromonas hydrophila* (MTCC 1739). The stock and slant cultures of these test organisms were obtained from the Microbial Type Culture Collection (MTCC). Both cultures were inoculated into the nutrient broth and sub-cultured on nutrient agar plates before performing the test.

4.8. In Vitro Antibacterial Activity

4.8.1. Primary Screening

Preliminary screening was carried out by the cross-streak method in which the isolates were screened against the selected test pathogens. Mueller–Hinton Agar (MHA) plates were prepared and supplemented with 2% sodium chloride. The *Actinomycete* isolates were inoculated with a single streak in the center of MHA plates and incubated at 30 °C for 7 days. Later, the sub-cultured test cultures were streaked perpendicular to the *Actinomycetes* isolates at a 90° angle and incubated at 37 °C for 24 h. The zone of inhibition was recorded [35]. Based on the zone of inhibition observed against the test organism, the potential isolates of *Actinomycetes* were selected for secondary screening.

4.8.2. Secondary Screening

The *Actinomycete* isolates that showed prospective antibacterial activities in primary screening were subjected to solvent extraction to obtain the crude extracts. The isolates were inoculated into a Starch Casein broth/ISP2 broth and kept in a rotary shaker incubator for seven days at 30 °C and 120 rpm until growth was visible. Centrifugation was performed at 4 °C at 10,000 rpm for 20 min, and the filtrate was aseptically transferred into conical flasks. Equal volumes of organic solvents with ethanol and an aqueous extract with distilled water were added to the cell-free supernatant (CFS) and shaken well for two hours to extract the antibacterial compounds [51].

Agar Well Diffusion Method

The antibacterial activities of the crude extracts were tested at different concentrations, ranging from 50 µg/mL to 250 µg/mL, by the agar well diffusion method. Lawn cultures of the pathogenic organisms were spread on the solidified MHA agar plates using sterile cotton swabs. The wells were prepared on MHA plates using a sterile cork borer. The different concentrations of both ethanolic and aqueous extracts of *Actinomycete* isolates were poured into each well and allowed to diffuse completely. Culture plates were incubated at 37 °C for 24 h. Experiments were carried out three times, and the mean of the zone of inhibition (in mm) was recorded [52].

Disc Diffusion Method

Gentamycin (10 mcg) was taken for the antibiotic sensitivity test, which was performed for both pathogens. The *Actinomycete* isolate showing the maximum zone of inhibition was chosen for further study by the Kirby–Bauer agar disc diffusion method [36] to check the antibacterial activity against pathogens. In this method, the pathogens were first swabbed entirely from the surface of sterile MHA plates. Then, the concentration of the isolate showing positive results was added inside the disc and placed on the surface of the medium and incubated at 37 °C for 24 h. The zone of inhibition was recorded in triplicate to calculate the mean value of the inhibition zones.

4.9. Extraction of the Active Compound

The inoculum was prepared by inoculating the selected *Actinomycete* isolate into a Starch Casein Broth/ISP-2 Broth kept in a rotary shaker incubator at 30 °C and harvested after 7 days. The broth culture was then filtered through Whatman No. 1 filter paper. The filtrate obtained was centrifuged at 4 °C, 10,000 rpm, for 20 min. The CFS collected was further used for the extraction of metabolites. The CFS was extracted three times with an equal volume of ethyl acetate by shaking manually at a 1:1 ratio by using a separating funnel. Then, it was kept in a shaker for 48 h, after which it was removed and the solvent layer collected. The concentration of the solvent layer was carried out using a rotary evaporator to obtain the crude metabolites [33]. As per some earlier reports, studies were conducted using various solvents, which gave better results in the extraction of bioactive compounds from the CFS of *Actinomycetes*. Among them, ethyl acetate was found to be a good polar solvent [53,54].

4.10. Identification of the Crude Bioactive Compound

Compounds obtained from the *Actinomycete* isolate were dissolved in ethyl acetate and identified using gas chromatography–mass spectroscopy (GC-MS) and Fourier transform infrared spectroscopy (FTIR) analyses.

A 13 mm potassium bromide (KBr) powder was mixed and ground thoroughly using a mortar and pestle with the crude bioactive compounds isolated from potential *Actinomycete* isolates. The sample was placed inside a pellet dye in a quantity just sufficient to cover its bottom. It was pressed at 5000–10,000 psi using the pellet press to form a pellet disc and placed in the sample holder. It was then analyzed by FTIR to obtain the measurement in the spectrum wave number range between 4 and 4000/cm at a resolution of 4 cm^{-1} using a FTIR spectrometer (Jasco, Victoria, BC, Canada).

GC-MS was performed for the identification of the active components present in the ethanolic extract of *Actinomycetes*. The unknown components were identified using a GC-MS (Agilent 6890/Hewlett-Packard 5975 (Agilent, Palo Alto, CA, USA/Hewlett Packard Labs, Milpitas, CA, USA) under the electron impact (EI) mode. Several acquisition parameters were followed with respect to the analysis. The interpretation of the mass spectrum was carried out using the National Institute Standard and Technology (NIST) database. The NIST library was used for searching for the spectrum of unknown compounds from the *Actinomycetes* [55].

5. Conclusions

Sixteen *Actinomycetes* were successfully isolated from marine sediments. Among them, two strains displayed antibacterial activity, but one strain, *Beijerinickia fluminensis*, was reported to have potent antibacterial activity and was hence selected for further analysis. FTIR and GC-MS revealed several antibacterial compounds that can be used as an alternative to antibiotics for the treatment of several pathogens in aquaculture. Based on the literature survey, this might be the first report on the isolation of *Beijerinickia fluminensis* from marine sediments.

Author Contributions: All authors contributed to the study conceptualization and design. Formal analysis, methodology, validation and writing of the original draft were performed by H.M. Conceptualization, investigation, supervision, validation, review and editing of the manuscript were performed by J.T. All authors commented on previous versions of the manuscript. All authors have read and agreed to the published version of the manuscript.

Funding: This research received no external funding.

Institutional Review Board Statement: Not applicable.

Informed Consent Statement: Not applicable.

Data Availability Statement: Not applicable.

Acknowledgments: The authors are thankful to VIT Vellore for providing the required facilities to carry out this work.

Conflicts of Interest: The authors declare no conflict of interest.

References

1. Mo, F.; Wa, F.; Gm, D. Marine Actinomycetes the Past, the Present and the Future. *Pharm. Res.* **2021**, *5*, 000241. [CrossRef]
2. Arbat, A.B.; Zodpe, S.N. Biodiversity of Actinomycetes Species Isolated from Saline Belt of Akola District. *Indian J. Appl. Res.* **2011**, *4*, 450–452. [CrossRef]
3. Rashad, F.M.; Fathy, H.M.; El-Zayat, A.S.; Elghonaimy, A.M. Isolation and Characterization of Multifunctional Streptomyces Species with Antimicrobial, Nematicidal and Phytohormone Activities from Marine Environments in Egypt. *Microbiol. Res.* **2015**, *175*, 34–47. [CrossRef]
4. Jagannathan, S.V.; Manemann, E.M.; Rowe, S.E.; Callender, M.C.; Soto, W. Marine Actinomycetes, New Sources of Biotechnological Products. *Mar. Drugs* **2021**, *19*, 365. [CrossRef] [PubMed]
5. Rasool, U.; Hemalatha, S. Diversified Role and Applications of Marine Actinomycetes in the Field of Biology. *World Sci. News* **2017**, *74*, 1–14.
6. Imada, C.; Koseki, N.; Kamata, M.; Kobayashi, T.; Hamada-Sato, N. Isolation and Characterization of Antibacterial Substances Produced by Marine Actinomycetes in the Presence of Seawater. *Actinomycetologica* **2007**, *21*, 27–31. [CrossRef]
7. Valli, S.; Sugasini, S.S.; Aysha, O.S.; Nirmala, P.; Vinoth Kumar, P.; Reena, A. Antimicrobial Potential of Actinomycetes Species Isolated from Marine Environment. *Asian Pac. J. Trop. Biomed.* **2012**, *2*, 469–473. [CrossRef]
8. Helmke, E.; Weyland, H. Rhodococcus Marinonascens. *Int. J. Syst. Bacteriol.* **1984**, *34*, 127–138. [CrossRef]
9. Colquhoun, J.A.; Mexson, J.; Goodfellow, M.; Ward, A.C.; Horikoshi, K.; Bull, A.T. Novel Rhodococci and Other Mycolate Actinomycetes from the Deep Sea. *Antonie Van Leeuwenhoek Int. J. Gen. Mol. Microbiol.* **1998**, *74*, 27–40. [CrossRef]
10. Mincer, T.J.; Jensen, P.R.; Kauffman, C.A.; Fenical, W. Widespread and Persistent Populations of a Major New Marine Actinomycete Taxon in Ocean Sediments. *Appl. Environ. Microbiol.* **2002**, *68*, 5005–5011. [CrossRef]
11. Yi, H.; Schumann, P.; Sohn, K.; Chun, J. *Serinicoccus Marinus* Gen. Nov., Sp. Nov., a Novel Actinomycete with L-Ornithine and L-Serine in the Peptidoglycan. *Int. J. Syst. Evol. Microbiol.* **2004**, *54*, 1585–1589. [CrossRef] [PubMed]
12. Das, S.; Lyla, P.S.; Khan, S.A. Distribution and Generic Composition of Culturable Marine Actinomycetes from the Sediments of Indian Continental Slope of Bay of Bengal. *Chin. J. Oceanol. Limnol.* **2008**, *26*, 166–177. [CrossRef]
13. Freel, K.C.; Edlund, A.; Jensen, P.R. Microdiversity and Evidence for High Dispersal Rates in the Marine Actinomycete "Salinispora Pacifica". *Environ. Microbiol.* **2012**, *14*, 480–493. [CrossRef] [PubMed]
14. Ghanem, N.B.; Sabry, S.A.; El-Sherif, Z.M.; Abu El-Ela, G.A. Isolation and Enumeration of Marine Actinomycetes from Seawater and Sediments in Alexandria. *J. Gen. Appl. Microbiol.* **2000**, *46*, 105–111. [CrossRef]
15. Sharma, M.; Dangi, P.; Choudhary, M. Actinomycetes: Source, identification, and their applications. *Int. J. Curr. Microbiol. Appl. Sci.* **2014**, *3*, 801–832.
16. Solanki, R.; Khanna, M.; Lal, R. Bioactive Compounds from Marine Actinomycetes. *Indian J. Microbiol.* **2008**, *48*, 410–431. [CrossRef]
17. Okami, Y.; Okazaki, T. *Nocardia and Streptomyces*; Gaustav Fischer: Stuttgart, Germany, 1978; pp. 145–151.
18. Magarvey, N.A.; Keller, J.M.; Bernan, V.; Dworkin, M.; Sherman, D.H. Isolation and Characterization of Novel Marine-Derived Actinomycete Taxa Rich in Bioactive Metabolites. *Appl. Environ. Microbiol.* **2004**, *70*, 7520–7529. [CrossRef]
19. Dilip, C.; Mulaje, S.S.; Mohalkar, R.Y. A Review on Actinomycetes and Their Biotechnological Application. *Int. J. Pharm. Sci. Res.* **2013**, *4*, 1730–1742. [CrossRef]
20. Riedlinger, J.; Reicke, A.; Zähner, H.; Krismer, B.; Bull, A.T.; Maldonado, L.A.; Ward, A.C.; Goodfellow, M.; Bister, B.; Bischoff, D.; et al. Abyssomicins, Inhibitors of the Para-Aminobenzoic Acid Pathway Produced by the Marine Verrucosispora Strain AB-18-032. *J. Antibiot.* **2004**, *57*, 271–279. [CrossRef]
21. Soria-Mercado, I.E.; Prieto-Davo, A.; Jensen, P.R.; Fenical, W. Antibiotic Terpenoid Chloro-Dihydroquinones from a New Marine Actinomycete. *J. Nat. Prod.* **2005**, *68*, 904–910. [CrossRef]
22. El-Gendy, M.M.A.; Shaaban, M.; Shaaban, K.A.; El-Bondkly, A.M.; Laatsch, H. Essramycin: A First Triazolopyrimidine Antibiotic Isolated from Nature. *J. Antibiot.* **2008**, *61*, 149–157. [CrossRef] [PubMed]
23. Manivasagan, P.; Venkatesan, J.; Sivakumar, K.; Kim, S.K. Pharmaceutically Active Secondary Metabolites of Marine Actinobacteria. *Microbiol. Res.* **2014**, *169*, 262–278. [CrossRef] [PubMed]
24. Newman, S.G. Antibiotics in Aquaculture. Is Responsible Use Possible? 2007. Available online: https://www.globalseafood.org/advocate/antibiotics-in-aquaculture-is-responsible-use-possible/ (accessed on 28 September 2022).
25. Duke, J.U.S. *Dr. Duke's Phytochem. Ethanobotanical Databases*; Department of Agriculture, Agricultural Research Service: Washington, DC, USA, 1992. [CrossRef]
26. Suthindhiran, K.; Kannabiran, K. Cytotoxic and Antimicrobial Potential of Actinomycete Species Saccharopolyspora Salina VITSDK4 Isolated from the Bay of Bengal Coast of India. *Am. J. Infect. Dis.* **2009**, *5*, 90–98. [CrossRef]
27. Gebreyohannes, G.; Moges, F.; Sahile, S.; Raja, N. Isolation and Characterization of Potential Antibiotic Producing Actinomycetes from Water and Sediments of Lake Tana, Ethiopia. *Asian Pac. J. Trop. Biomed.* **2013**, *3*, 426–435. [CrossRef]
28. Chen, Y.; Zeng, D.; Ding, L.; Li, X.L.; Liu, X.T.; Li, W.J.; Wei, T.; Yan, S.; Xie, J.H.; Wei, L.; et al. Three-Dimensional Poly-(ε-Caprolactone) Nanofibrous Scaffolds Directly Promote the Cardiomyocyte Differentiation of Murine-Induced Pluripotent Stem Cells through Wnt/β-Catenin Signaling. *BMC Cell Biol.* **2015**, *16*, 1–13. [CrossRef]

29. Oggerin, M.; Arahal, D.R.; Rubio, V.; Marín, I. Identification of Beijerinckia Fluminensis Strains CIP 106281T and UQM 1685T as Rhizobium Radiobacter Strains, and Proposal of Beijerinckia Doebereinerae Sp. Nov. to Accommodate Beijerinckia Fluminensis LMG 2819. *Int. J. Syst. Evol. Microbiol.* **2009**, *59*, 2323–2328. [CrossRef]
30. Shwaaiman, H.A.; Shahid, M.; Elgorban, A.M.; Siddique, K.H.M.; Syed, A. Beijerinckia Fluminensis BFC-33, a Novel Multi-Stress-Tolerant Soil Bacterium: Deciphering the Stress Amelioration, Phytopathogenic Inhibition and Growth Promotion in Triticum Aestivum L. (Wheat). *Chemosphere* **2022**, *295*, 133843. [CrossRef]
31. Trakunjae, C.; Boondaeng, A.; Apiwatanapiwat, W.; Kosugi, A.; Arai, T.; Sudesh, K.; Vaithanomsat, P. Enhanced Polyhydroxybutyrate (PHB) Production by Newly Isolated Rare Actinomycetes Rhodococcus Sp. Strain BSRT1-1 Using Response Surface Methodology. *Sci. Rep.* **2021**, *11*, 1896. [CrossRef]
32. Sivanandhini, T.; Subbaiya, R.; Gopinath, M.; Mahavinod Angrasan, J.K.V.; Kabilan, T.; Masilamani Selvam, M. An Investigation on Morphological Characterization of Actinomycetes Isolated from Marine Sediments. *Res. J. Pharm. Biol. Chem. Sci.* **2015**, *6*, 1234–1243.
33. Rajkumar, T.; Manimaran, M.; Taju, G.; Vimal, S.; Majeed, S.A.; Kannabiran, K.; Sivakumar, S.; Kumar, K.M.; Madhan, S.; Hameed, A.S. Antiviral Viral Compound from Streptomyces Ghanaensis like Strain against White Spot 2 Syndrome Virus (WSSV) of Shrimp. *bioRxiv* **2018**. [CrossRef]
34. Undabarrena, A.; Beltrametti, F.; Claverías, F.P.; González, M.; Moore, E.R.B.; Seeger, M.; Cámara, B. Exploring the Diversity and Antimicrobial Potential of Marine Actinobacteria from the Comau Fjord in Northern Patagonia, Chile. *Front. Microbiol.* **2016**, *7*, 1135. [CrossRef] [PubMed]
35. Singh, V.; Haque, S.; Singh, H.; Verma, J.; Vibha, K.; Singh, R.; Jawed, A.; Tripathi, C.K.M. Isolation, Screening, and Identification of Novel Isolates of Actinomycetes from India for Antimicrobial Applications. *Front. Microbiol.* **2016**, *7*, 1921. [CrossRef] [PubMed]
36. Azhar, N.; Yudiati, E.; Subagiyo; Alghazeer, R. Producing Active Secondary Metabolite Against Pathogenic Vibrio Spp. by Actinobacteria-Sodium Alginate Co-Culture. *Ilmu Kelaut. Indones. J. Mar. Sci.* **2021**, *26*, 254–264. [CrossRef]
37. Norouzi, H.; Danesh, A.; Mohseni, M.; Khorasgani, M.R. Marine Actinomycetes with Probiotic Potential and Bioactivity against Multidrugdrug-Resistant Bacteria. *Int. J. Mol. Cell. Med.* **2018**, *7*, 44.
38. Subramani, R.; Aalbersberg, W. Marine Actinomycetes: An Ongoing Source of Novel Bioactive Metabolites. *Microbiol. Res.* **2012**, *167*, 571–580. [CrossRef]
39. Dholakiya, R.N.; Kumar, R.; Mishra, A.; Mody, K.H.; Jha, B. Antibacterial and Antioxidant Activities of Novel Actinobacteria Strain Isolated from Gulf of Khambhat, Gujarat. *Front. Microbiol.* **2017**, *8*, 2420. [CrossRef]
40. Ryu, M.J.; Hwang, S.; Kim, S.; Yang, I.; Oh, D.C.; Nam, S.J.; Fenical, W. Meroindenon and Merochlorins e and F, Antibacterial Meroterpenoids from a Marine-Derived Sediment Bacterium of the Genus Streptomyces. *Org. Lett.* **2019**, *21*, 5779–5783. [CrossRef]
41. Wang, C.; Lu, Y.; Cao, S. Antimicrobial Compounds from Marine Actinomycetes. *Arch. Pharm. Res.* **2020**, *43*, 677–704. [CrossRef]
42. Voon, W.W.Y.; Muhialdin, B.J.; Yusof, N.L.; Rukayadi, Y.; Meor Hussin, A.S. Bio-Cellulose Production by Beijerinckia Fluminensis WAUPM53 and Gluconacetobacter Xylinus 0416 in Sago By-Product Medium. *Appl. Biochem. Biotechnol.* **2019**, *187*, 211–220. [CrossRef]
43. Joshi, M.H.; Patil, A.A.; Adivarekar, R.V. Characterization of Brown-Black Pigment Isolated from Soil Bacteria, Beijerinckia Fluminensis. *bioRxiv Microbiol.* **2021**. [CrossRef]
44. Imada, C.; Masuda, S.; Kobayashi, T.; Hamada-Sato, N.; Nakashima, T. Isolation and Characterization of Marine and Terrestrial Actinomycetes Using a Medium Supplemented with NaCl. *Actinomycetologica* **2010**, *24*, 12–17. [CrossRef]
45. Manimaran, M.; Rajkumar, T.; Vimal, S.; Taju, G.; Majeed, S.A.; Hameed, A.S.S.; Kannabiran, K. Antiviral Activity of 9(10H)-Acridanone Extracted from Marine Streptomyces Fradiae Strain VITMK2 in Litopenaeus Vannamei Infected with White Spot Syndrome Virus. *Aquaculture* **2018**, *488*, 66–73. [CrossRef]
46. Selvan, G.P.; Ravikumar, S.; Ramu, A.; Neelakandan, P. Antagonistic Activity of Marine Sponge Associated Streptomyces Sp. Against Isolated Fish Pathogens. *Asian Pac. J. Trop. Dis.* **2012**, *2*, S724–S728. [CrossRef]
47. Malviya, M.K.; Pandey, A.; Sharma, A.; Tiwari, S.C. Characterization and Identification of Actinomycetes Isolated from "fired Plots" under Shifting Cultivation in Northeast Himalaya, India. *Ann. Microbiol.* **2013**, *63*, 561–569. [CrossRef]
48. El Karkouri, A.; Assou, S.A.; El Hassouni, M. Isolation and Screening of Actinomycetes Producing Antimicrobial Substances from an Extreme Moroccan Biotope. *Pan Afr. Med. J.* **2019**, *33*, 1–9. [CrossRef] [PubMed]
49. Cheesbrough, M. *Adaptive or Specific Immune Response, District Laboratory in Tropical Countries*; Cambridge University Press: Cape Town, South Africa, 2010; p. 11.
50. Muiru, W.M.; Mutitu, E.W.; Mukunya, D.M. Identification of Selected Actinomycetes Isolates and Characterization of Their Antibiotic Metabolites. *J. Biol. Sci.* **2008**, *8*, 1021–1026. [CrossRef]
51. Thirumurugan, D.; Vijayakumar, R. A Potent Fish Pathogenic Bacterial Killer Streptomyces Sp. Isolated from the Soils of East Coast Region, South India. *J. Coast. Life Med.* **2013**, *1*, 175–180. [CrossRef]
52. Sapkota, A.; Thapa, A.; Budhathoki, A.; Sainju, M.; Shrestha, P.; Aryal, S. Isolation, Characterization, and Screening of Antimicrobial-Producing Actinomycetes from Soil Samples. *Int. J. Microbiol.* **2020**, *2020*, 2716584. [CrossRef]
53. Barakat, K.M.; Beltagy, E.A. Bioactive Phthalate from Marine Streptomyces Ruber EKH2 against Virulent Fish Pathogens. *Egypt. J. Aquat. Res.* **2015**, *41*, 49–56. [CrossRef]

54. Ramalingam, V.; Rajaram, R. Antioxidant Activity of 1-Hydroxy-1-Norresistomycin Derived from Streptomyces Variabilis KP149559 and Evaluation of Its Toxicity against Zebra Fish Danio Rerio. *RSC Adv.* **2016**, *6*, 16615–16623. [CrossRef]
55. Thanigaivel, S.; Vijayakumar, S.; Mukherjee, A.; Chandrasekaran, N.; Thomas, J. Antioxidant and Antibacterial Activity of Chaetomorpha Antennina against Shrimp Pathogen Vibrio Parahaemolyticus. *Aquaculture* **2014**, *433*, 467–475. [CrossRef]

antibiotics MDPI

Article

In Vitro Antibacterial Activity of Green Synthesized Silver Nanoparticles Using *Mangifera indica* Aqueous Leaf Extract against Multidrug-Resistant Pathogens

Yahya S. Alqahtani [1], Amal Bahafi [2], Kiran K. Mirajkar [3], Rakshith Rudrapura Basavaraju [3], Susweta Mitra [4], Shailaja S [4], Sunil S. More [4,*], Uday M. Muddapur [5], Aejaz Abdullatif Khan [6], P. Renuka Sudarshan [3] and Ibrahim Ahmed Shaikh [7,*]

1 Department of Pharmaceutical Chemistry, College of Pharmacy, Najran University, Najran 66462, Saudi Arabia
2 Department of Pharmaceutical Chemistry, Ibn Sina National College for Medical Studies, Jeddah 21418, Saudi Arabia
3 Department of Biochemistry, University of Agricultural Sciences, Dharwad 580005, India
4 School of Basic and Applied Sciences, Dayananda Sagar University, Bangalore 560111, India
5 Department of Biotechnology, KLE Technological University, BVB Campus, Hubli 580031, India
6 Department of General Science, Ibn Sina National College for Medical Studies, Jeddah 21418, Saudi Arabia
7 Department of Pharmacology, College of Pharmacy, Najran University, Najran 66462, Saudi Arabia
* Correspondence: sunilsmorey@gmail.com (S.S.M.); i.ibrahimshaikh09@gmail.com (I.A.S.)

Citation: Alqahtani, Y.S.; Bahafi, A.; Mirajkar, K.K.; Basavaraju, R.R.; Mitra, S.; S, S.; More, S.S.; Muddapur, U.M.; Khan, A.A.; Sudarshan, P.R.; et al. In Vitro Antibacterial Activity of Green Synthesized Silver Nanoparticles Using *Mangifera indica* Aqueous Leaf Extract against Multidrug-Resistant Pathogens. *Antibiotics* **2022**, *11*, 1503. https://doi.org/10.3390/antibiotics11111503

Academic Editors: Valério Monteiro-Neto and Elizabeth S. Fernandes

Received: 29 September 2022
Accepted: 26 October 2022
Published: 28 October 2022

Abstract: An estimated 35% of the world's population depends on wheat as their primary crop. One fifth of the world's wheat is utilized as animal feed, while more than two thirds are used for human consumption. Each year, 17–18% of the world's wheat is consumed by China and India. In wheat, spot blotch caused by *Bipolaris sorokiniana* is one of the major diseases which affects the wheat crop growth and yield in warmer and humid regions of the world. The present work was conducted to evaluate the effect of green synthesized silver nanoparticles on the biochemical constituents of wheat crops infected with spot blotch disease. Silver nanoparticles (AgNPs) were synthesized using *Mangifera indica* leaf extract and their characterization was performed using UV-visible spectroscopy, SEM, XRD, and PSA. Characterization techniques confirm the presence of crystalline, spherical silver nanoparticles with an average size of 52 nm. The effect of green synthesized nanoparticles on antioxidative enzymes, e.g., Superoxide dismutase (SOD), Catalase (CAT), Glutathione Reductase (GR), Peroxidase (POX), and phytochemical precursor enzyme Phenylalanine Ammonia-Lyase (PAL), and on primary and secondary metabolites, e.g., reducing sugar and total phenol, in *Bipolaris sorokiniana* infected wheat crop were studied. Inoculation of fungal spores was conducted after 40 days of sowing. Subsequently, diseased plants were treated with silver nanoparticles at different concentrations. Elevation in all biochemical constituents was recorded under silver nanoparticle application. The treatment with a concentration of nanoparticles at 50 pp min diseased plants showed the highest resistance towards the pathogen. The efficacy of the green synthesized AgNPs as antibacterial agents was evaluated against multi drug resistant (MDR) bacteria comprising Gram-negative bacteria *Escherichia coli* (n = 6) and *Klebsiella pneumoniae* (n = 7) and Gram-positive bacteria Methicillin resistant *Staphylococcus aureus* (n = 2). The results show promising antibacterial activity with significant inhibition zones observed with the disc diffusion method, thus indicating green synthesized *M. indica* AgNPs as an active antibacterial agent against MDR pathogens.

Keywords: MDR pathogens; silver nanoparticles; antibacterial activity; *Mangifera indica* leaf extract; *Triticum* spp.

1. Introduction

Wheat (*Triticum* spp.), the so-called "King of cereals" is considered one of the oldest staple food crops grown worldwide, with an annual production of 734 million tonnes [1].

Cultivation of wheat started as a part of the 'Neolithic Revolution' 10,000 years ago. The wheat kernel contains an average of 12% moisture, 70% carbohydrates, 12% proteins, 2% fat, 2.2% crude fibre, and 1.8% minerals. Niacin, riboflavin, and thiamine are the major vitamins present in wheat [2]. *Triticum aestivum*, *T. durum*, *T. dicoccum*, and *T. sphaerococcum* are the four important wheat species grown globally. Agriculture production is greatly affected by abiotic and biotic stresses, leading to huge economic losses. Spot blotch disease of wheat caused by *Bipolaris sorokiniana* is one of the major diseases which causes severe yield loss in wheat crop grown in the warm and humid regions of the world [3]. Although this disease was reported in the early 1990s, its importance was recognized only after the green revolution, when more numbers of semi-dwarf wheat cultivars became susceptible to spot blotch. When a crop reaches the late post-anthesis stage, which coincides with warm and high humid conditions, disease severity is higher.

Nanomaterials are used efficiently for the safe administration of pesticides, herbicides, and fertilizers at low concentrations. Pesticides affect human health and pollinating insects. To reduce the harmful effects of these chemicals, the use of green nanoparticles can be beneficial. Green nanoparticles play an important role in decreasing toxicity and, in turn, increase efficiency [4,5].

Consequently, new techniques are implemented by the modern agriculture system to minimize yield losses by several crop diseases. Among these technologies, nanotechnology has assumed a prominent place by virtue of its vast range of applications in various fields, e.g., agriculture, pharmaceuticals, electronics, etc. The development and application of biosynthesized nanotechnology in agricultural research initiated the development of an eco-friendly and effective biosynthesized nanoparticles to control diseases. The biological manufacture of silver nanoparticles (AgNPs) has been demonstrated as a reliable, nontoxic, and environmentally acceptable strategy for plant disease treatment [6]. Previous research has demonstrated the toxicity of silver nanoparticles on the growth of fungal hyphae and conidia, while AgNPs made from cow milk have also shown potent antifungal action against a variety of phytopathogens.

Plants contain a multitude of structurally divergent phytochemicals which are together termed as secondary metabolites. These metabolites act as natural pesticides, antibiotics and protective agents which are normally toxic to microbes. Plants can activate different metabolic pathways to control pathogen attacks. Phytoalexins, lignin, and phytoanticipins are the different compounds produced by the plant to overcome pathogen attacks. Biochemical resistance against diseases depends on pre-existing and induced substances present in the plant. Elicitors are the prime molecules capable to induce defense responses. Both exogenous (pathogen origin) and endogenous (plant origin) elicitors act as defense inducers [7]. The programmed cell death mechanism is a complex defense system shown by the plant to overcome the pathogen attack. Superoxide anion (O_2^-), hydroxyl radical (OH•) and Hydrogen peroxide (H_2O_2) are the major reactive oxygen species (ROS) produced by the plant under these conditions and are strong oxidizing agents which attack all the biomolecules causing cellular damage. Plant cells contain oxygen radical detoxifying enzymes, e.g., catalase, peroxidase, superoxide dismutase, ascorbate peroxidase, and glutathione reductase, and non-enzymatic antioxidants, such as ascorbic acid and phenolic derivatives, to overcome the ROS.

The SOD, which catalyzes the dismutation of superoxide anion to hydrogen peroxide (H_2O_2), appears to play an important role in the emergence and progression of the necrotic reaction. The induction of SOD activity in plants has been widely observed in response to pathogen invasion, and it is a reaction that is frequently related with plant resistance. Catalase is the most important enzyme in the elimination of hydrogen peroxide. In addition to serving as a substrate for the POD and CAT, the hydrogen peroxide produced by the SOD plays an important function in the synthesis of lignin [8]. Furthermore, due to its diffusible nature, hydrogen peroxide increases the expression of defense genes, such as phenylalanine ammonia-lyase, chalcone synthase, endochitinases, and phytoalexin biosynthetic enzymes, and is a major indicator of the related necrotic reaction. The hydrogen peroxide also induces the expression of genes that code for antioxidant proteins, such as glutathione-S-transferase, glutathione peroxidase, and polyubiquitins, which prevent the death of healthy cells with hypersensitivity [9,10]. The phenylpropanoid pathway leads to the biosynthesis of a wide range of phenolics. In many circumstances, the activation of the phenylpropanoid metabolism, with the main enzyme phenylalanine ammonia-lyase (PAL), is involved in the initial disease resistance reactions of plants, leading to the creation of several defense-related chemicals such as antimicrobial phytoalexins and lignin [11]. PAL has been linked to barley and wheat defense responses to the necrotrophic fungal disease caused by *Bipolaris sorokiniana* (Sacc.) [12]. After inoculating barley and wheat leaves with highly or weakly aggressive isolates of *Bipolaris sorokiniana*, PAL activity was strongly induced [13].

The main objective of this study was to evaluate the effect of green synthesized *M. indica* AgNPs on antioxidative enzymes and secondary metabolites, such as reducing sugar and total phenol, in *Bipolaris sorokiniana* infected wheat crop.

2. Materials and Methods

2.1. Source of Plant Material (Mangifera indica) and Chemicals

M. indica fresh leaves were collected, thoroughly washed several times with tap water followed by distilled water to remove the dust particles, and air-dried at room temperature for six to seven days. Dried leaves powder was prepared and stored at room temperature for further use. In this study, standard biochemical reagents and chemicals, including potassium phosphate, H_2O_2, $AgNO_3$, etc., were of analytical grade, procured from Sigma Aldrich, Himedia, and Sisco Research Laboratories Pvt. Ltd.

2.2. Preparation of the M. indica Aqueous Leaf Extract

The aqueous *M. indica* leaf extract was prepared following the protocol described by Narayana et al. with minor modifications. Thus, 10 g of *M. indica* leaves powder was boiled in 100 mL of distilled water for 30 min (10% aqueous plant extract) and filtered through Whatman No. 1 filter paper. The filtrate was collected and stored in the refrigerator at 4 °C until further use.

2.3. Green Synthesis of Silver Nanoparticles (AgNPs) Using M. indica Aqueous Leaf Extract

The $AgNO_3$ solution was used as a precursor for the synthesis of silver nanoparticles (AgNPs). Hence, 0.02 M solutions were prepared by the addition of 1.7 g of $AgNO_3$ to 500 mL of distilled water. For the green synthesis of AgNPs we used 0.02 M silver nitrate ($AgNO_3$) as a precursor and 10% *M. indica* aqueous leaf extract as both reducing and capping agent [14]. Leaf extract and $AgNO_3$ (0.02 M) solutions were mixed in the ratio of 9:1 and kept for reaction in the dark for approximately 24 h. The plant extract reduced the silver nitrate and led to the formation of AgNPs. The change in solution color from brown to blackish indicated the formation of AgNPs. The reduction process was monitored by using UV-Visible spectroscopy.

2.4. Characterization of Green Synthesized Silver Nanoparticles (AgNPs)

The synthesized nanoparticles were characterized in terms of shape, size, and morphology using UV-Visible spectrophotometry (UV-Vis), particles size analyzer (PSA), scanning electron microscope (SEM), and X-ray diffraction (XRD). Spectral analysis of AgNPs was performed using an ELICO UV1900, Hyderabad, India, Double Beam Spectrophotometer with spectrum scanning in the wavelength range of 190–1100 nm with a band width of 2 nm. Green nanoparticles were analyzed at a wavelength ranging from 200 to 700 nm. Characterization of AgNPs in terms of mean diameter and distribution was performed using PSA (Nicomp, NANOZ Z3000 PSS, Billerica, MA, USA). To assess the topology of green synthesized nanoparticles, SEM (Carl Zeiss-EVO-18- Cambridge, UK) was used. An X-ray Diffractometer (Powder) (Model: SmartLab SE, Tokyo, Japan) with fully automated alignment under computer control SAXS capabilities, with Optional D/teX Ultra-high-speed, position-sensitive detector system with 2D: 2–150°, was used to understand the structural information of AgNPs.

2.5. Biochemical Analysis

The seeds were procured from the Durum Wheat Main Agricultural Research Station of the AICRP (All India Coordinated Research Projects), at the University of Agricultural Sciences in Dharwad. The tests were performed in a factorial complete-replicate design with three replicates. The variety of durum wheat employed in this study was Bijaga Yellow which is a tall durum wheat variety cultivated in rain fed regions.

Wheat plant was evaluated by pot culture experiments in a polyhouse. Blast disease was inoculated artificially after 40 days of sowing. The AgNPs treatment was given after the development of disease as described in the study design (Table 1). Leaves samples were collected after 2–3 days of treatment for determining the reducing sugar and total phenol and antioxidant enzyme activities, such as superoxide dismutase (SOD), catalase (CAT), glutathione reductase (GR), peroxidase (POX), and phenylalanine ammonia lyase (PAL).

Table 1. Details of the treatment groups used in the study.

Variety of Durum Wheat	Treatment	Replication
Bijaga Yellow	C: Control leaves T_1: Diseased + 0 ppm AgNPs (Disease control) T_2: Diseased + 5 ppm AgNPs T_3: Diseased + 10 ppm AgNPs T_4: Diseased + 15 ppm AgNPs T_5: Diseased + 20 ppm AgNPs T_6: Diseased + 30 ppm AgNPs T_7: Diseased + 50 ppm AgNPs	Three

2.6. Extraction of Enzyme and Determination of Protein Content

In the present work, enzymes were extracted from freshly collected leaf tissues at 0–4 °C. The antioxidant enzyme activities of SOD, CAT, POX, GR, and PAL were determined by extracting 0.5 g of leaf tissues with 2 mL of respective buffers (0.05 M sodium phosphate buffer of pH 7.8 for SOD, pH 7.0 for CAT and POX, 0.1 M Tris-HCl buffer of pH 7.8 containing 2 mM dithiothreitol for GR and 0.1 M Tris buffer of pH 8.5 for PAL). The homogenates were centrifuged at 14,000 rpm for 20-min at 4 °C. The resulting supernatant was used as an enzyme source for determining antioxidant activities and the protein content was measured using Lowry's method.

2.7. Antioxidant Enzyme Assays

2.7.1. Catalase

CAT (EC 1.11.1.6) activity of Wheat leaves was analyzed using the Beers and Sizers spectrophotometric technique [15]. To commence the reaction, 20 μL of enzyme extract was added to a reaction mixture containing 2.98 mL of 16.65 mM hydrogen peroxide in 50 mM phosphate buffer, pH 7.0. The decrease in absorbance was monitored at 240 nm against substrate blank for a period of 3 min. Specific activity was expressed as μM min^{-1} mg^{-1} protein and one unit of CAT was defined as 1 μM of H_2O_2 degraded per minute at pH 7.0 and 25 °C.

2.7.2. Superoxide Dismutase

Using the Beauchamp and Fridovich method [16], the SOD activity in wheat leaves was measured photochemically at 560 nm. The composition of 3 mL of reaction mixture included 50 mM phosphate buffer (pH 7.8), 20 μL enzyme extract, 10 mM L-methionine, 33 μM p-nitroblue tetrazolium chloride (NBT), 0.66 μM ethylene diamine tetra acetic acid (EDTA), and 3.3 μM riboflavin. The addition of riboflavin was immediately followed by illuminating the glass tube with the light of a 15 W fluorescent lamp at 25 °C for 20 min. Absorbance at 560 nm was read to determine the concentration of blue formazan formed by NBT photoreduction against a blank which comprised of same reaction mixture excluding the exposure to light. The specific activity of SOD is measured in international units (IU) per milligram of protein, and one unit of SOD is defined as the quantity of enzyme needed to inhibit 50% of the NBT photo-reduction per minute.

2.7.3. Peroxidase Activity

Peroxidase activity (POX, EC 1.11.1.7) in wheat leaves was measured kinetically by the method of Chance and Maehly [17]. The addition of 20 µL of the enzyme extract to reaction mixture, containing 2.88 mL of 0.1 M potassium phosphate buffer (pH 7.0), 50 µL of 0.02 M guaiacol, and 50 µL of 0.042% hydrogen peroxide, initiated the reaction and the increase in optical density at 436 nm was recorded for 5 min. The specific activity of POX was quantified as µM of oxidized guaiacol produced per minute per milligram of protein.

2.7.4. Glutathione Reductase

The spectrophotometric quantification of GR (EC 1.8.1.7) activity in wheat leaves was performed following the method of Mavis and Stellwagen [18]. Composition of reaction mixture included 0.1 mL of 30 mM oxidized glutathione, 1.5 mL of 100 mM potassium phosphate buffer with 3.4 mM EDTA, pH 7.6, 0.35 mL of 0.8 mM reduced ß-nicotinamide adenine dinucleotide phosphate (NADPH), and 0.95 mL of distilled water. The addition of 100 µL of enzyme extract commenced the reaction and the reduction in absorbance at 340 nm was monitored for 5 min. GR specific activity is represented as µM min^{-1} mg^{-1} protein, and one GR unit is equal to the quantity of enzyme required to oxidize 1 µM of NADPH per minute at pH 7.6 and 25 °C.

2.7.5. Phenylalanine Ammonia Lyase

Using the spectrophotometric approach developed by Paltonen and Karjalainen [12], the PAL (EC.4.3.1.5.) activity in wheat leaves was measured. The assay mixture containing 0.5 mL of enzyme extract and 2.5 mL of 0.2% L-phenylalanine in 0.1 mL of Tris buffer, pH 8.5 was incubated for 1-h at 40 °C and the reaction was terminated with 0.2 M HCl. The absorbance was measured at 290 nm against a substrate blank. The specific activity of PAL was reported in units of µM of trans-cinnamate released per minute per milligram of protein. One unit of PAL is defined as the quantity of enzyme necessary to liberate 1 µM of trans-cinnamate from L-phenylalanine per minute.

2.8. Estimation of Reducing the Sugar

For the quantitative analysis of reducing sugars in wheat leaves, 1 g of dried leaf material was introduced in hot ethanol (80%) for 15 min and then homogenized with pestle and mortar. The homogenate thus obtained was filtered using muslin cloth, the residue was re-extracted two or three times, and the filtrate were made up to final volume of 10 mL with 80% ethanol. Reducing sugars were quantified by employing Nelson Somogi's method (Norton Nelson, 1944) in which the reducing sugars in alkaline medium reduce cupric ions to cuprous ions which form a chromophore with arsenomolybdate reagent and have a λmax at 510 nm. The reducing sugar was expressed as mg per gram dry weight.

2.9. Estimation of Total Phenols

The Folin-Ciocalteau reagent (FCR) method was used to estimate the total phenols present in wheat leaf samples. The reaction between phenols and the oxidizing agent phosphomolybdo-phosphotungstate in FC reagent results in the formation of blue-colored chromophore with a maximum absorption at 660 nm. The amount of phenol has a direct correlation with the development of color with the reagent. Total phenols were extracted using 10.0 mL of hot 80% alcohol with 1 g of dried leaf as the starting material (Sadasivam and Manikam, 1992). The colorimetric method of (Bray and Thorpe, 1954.) was used for the determination of total phenols using the Folin–Ciocalteau reagent. The phenol content was expressed as mg per gram dry weight

2.10. *Bacterial Strains Used in This Study*

We used 15 multidrug-resistant (MDR) bacteria comprising gram negative bacteria- *Escherichia coli* (n = 6), *Klebsiella pneumoniae* (n = 7) and gram-positive bacteria- Methicillin Resistant *Staphylococcus aureus* (n = 2). These clinical strains were phenotypically and genotypically characterized in our previous study and found to possess wide spectrum of beta lactamase genes responsible for antibiotic resistance [17]. ATCC 25922 *E. coli* was used as a control strain.

2.11. *In Vitro Antibiotic Susceptibility Test of* M. indica *Nano Particles (MNPs) against MDR Bacteria*

Disc Diffusion Method

The antibacterial activity of MNPs against the selected MDR bacterial (n = 15) strains was carried out using the Kirby–Bauer disc diffusion method [18]. The bacterial strains were grown until attaining McFarland standard O.D of 0.5. The inoculum was spread in Muller Hinton Agar (MHA) (Himedia, India) using sterile cotton swabs and sterile antimicrobial susceptibility disks were loaded on the plates with 10 µL of MNP. The plates were incubated at 37 °C for 48 h and zone of inhibition was recorded.

2.12. *Statistical Analysis*

The data of the experiment were analyzed statistically by the procedure described by Gomez and Gomez [19].

3. Results and Discussion

3.1. *Characterization of Green Synthesized AgNPs*

3.1.1. The UV-Visible Absorption Spectrum of Green Synthesized AgNPs

The addition of *M. indica* leaf extract to the silver nitrate solution led to the change in color from brown to dark blackish-brown, indicating the formation of AgNPs due to the reduction of silver ions by the reducing agents present in the *M. indica* leaf extract. Further confirmation of AgNPs formation was performed using the UV- visible spectrophotometer. Surface plasmonic resonance absorbance range between 410 and 480 nm was, in particular, used as an indicator to confirm the reduction of Ag^+ to metallic Ag^o [20]. The strong and broad surface plasmonic resonance (SPR) peak centered at 475 nm and 410 nm confirmed the formation of AgNPs (Figure 1).

3.1.2. AgNPs Particle Size Distribution Study by Particle Size Analyzer (PSA)

Nanoparticles size distribution and mean diameter of green synthesized AgNPs using *M. indica* aqueous leaf extract were characterized using PSA (Nicomp NANOZ Z3000 PSS). PSA showed a mean diameter at 52.8 nm and nanoparticle size distribution from 5 nm to 220 nm (Figure 2).

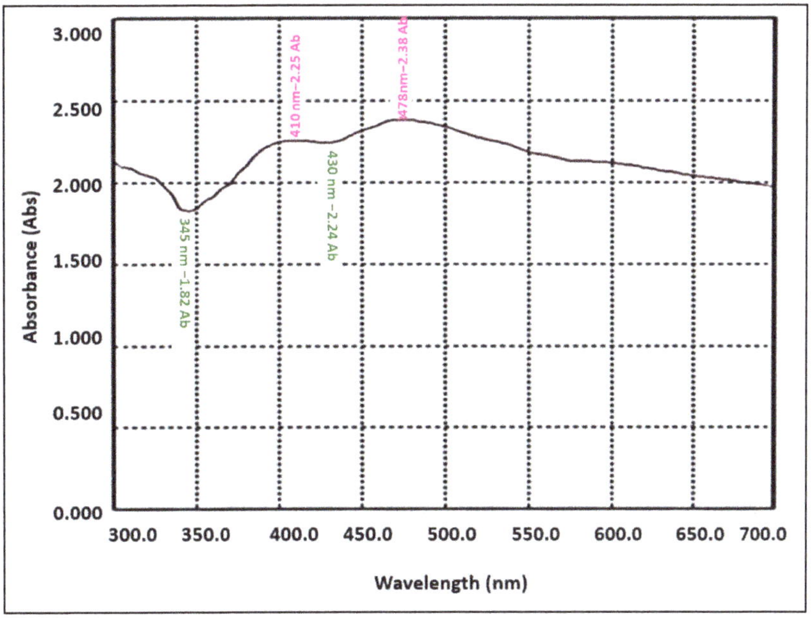

Figure 1. UV–Visible absorption spectrum of green synthesized AgNPs using *M. indica* aqueous leaf extract.

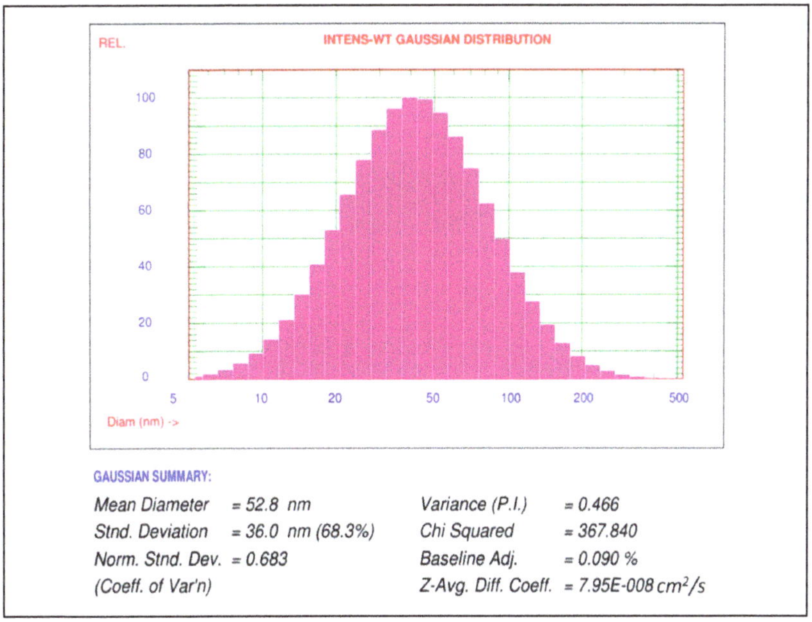

Figure 2. Particle size distribution of AgNPs synthesized using *M. indica* aqueous leaf extract.

3.1.3. SEM Micrographs of Green Synthesized AgNPs

Scanning electron microscopy imaging was carried out to view the morphology and size of the silver nanoparticles. SEM micrograph obtained for the AgNPs synthesized using *M. indica* leaf extract showed high-density nanoscale particles (52 nm) with spherical shape (Figure 3). SEM is a surface imaging technique that can resolve micro- and nano-scale differences in particle size, particle size distribution, nanomaterial type, and particle surface morphology. Through scanning electron microscopy (SEM), we can analyze particle morphology and create a histogram from images, either manually by measuring and counting the particles or automatically using specialized software. Our findings are in line with previous studies involving nanoparticles synthesized using *Raphanus sativus* and *Vitex leucoxylon* extract [21,22].

Figure 3. SEM micrograph of green synthesized AgNPs using *M. indica* aqueous leaf extract.

3.1.4. XRD Spectrum of AgNPs Synthesized by Using *M. indica* Leaf Extract

Characterization of AgNPs synthesized by using *M. indica* leaf extract was carried out by performing XRD spectrum analysis. The result indicated peaks angle of 2θ at 27.98°, 32.42°, 38.31°, 44.46°, 46.34°, 54.93°, 57.61°, 64.63°, and 77.53°. The highest intensity counts were at the angle of 38.31°, followed by 44.46°, 77.58°, 64.63°, 32.42°, 46.34°, and 27.98° for leaf extracts which can be assigned the plane of silver crystals (Figure 4). When compared with the standard (JCPDS No. 87-0579), the obtained XRD spectrum confirmed that the synthesized silver nanoparticles were in nanocrystal form and crystalline in nature. The peaks can be assigned to the planes (122), (111), (200), (220), and (311) facet of silver crystal, respectively. These results corroborate the findings of earlier studies [23,24].

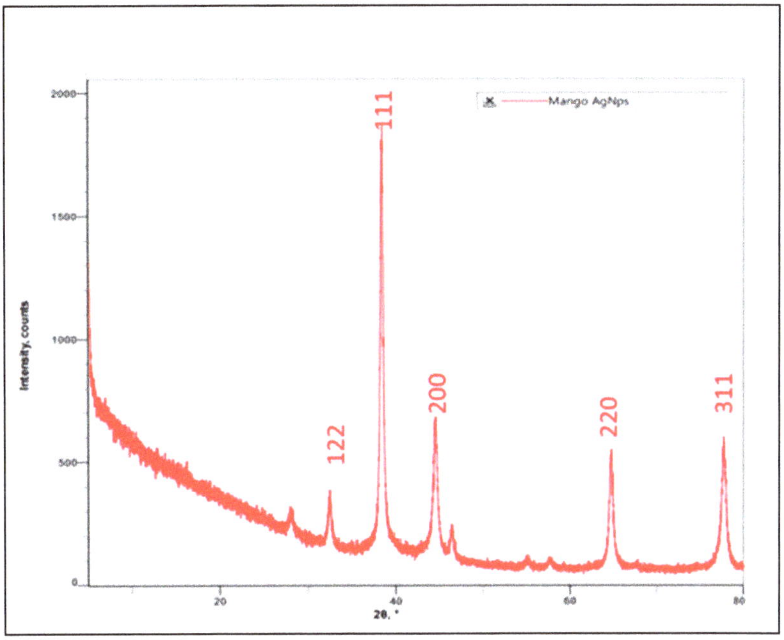

Figure 4. XRD graph of green synthesized AgNPs using *M. indica aqueous* leaf extract.

3.2. Estimation of Primary and Secondary Metabolites in AgNPs Treated Wheat Plants Infected with Spot Blotch Infection (Bipolaris sorokiniana)

3.2.1. Reducing Sugar

Levels of reducing sugar in the AgNPs -treated plants showed a slight increase when compared to the diseased wheat plant (2.73 g % dry weight) (Table 2). Among the treatments, plants treated with a higher concentration of silver nanomaterials (30 ppm: 6.94 g % and 50 ppm: 7.82 g % dry weight) showed a phenomenal increase in reducing sugar content compared to plants treated with lower concentration (5 ppm: 4.44 g %, 10 ppm: 5.00 g %, 15 ppm: 5.45 g %, and 20 ppm: 5.87 g % dry weight) and diseased plant (2.34 g % dry weight). In comparison to control, the disease inoculated wheat plant leaves recorded a reduction in reducing sugar content (2.73 g % dry weight) as compared to control leaves (10.61 g % dry weight).

3.2.2. Total Phenol

Total phenol content in all nano-treated plants showed a slight increase when compared to the diseased wheat plants (1.65 g % dry weight). Among the treatments, plants treated with a higher concentration of AgNPs (30 ppm: 1.98 g %, 50 ppm: 2.09 g % dry weight) showed a phenomenal increase in total phenol content compared to low concentration (5 ppm: 1.72 g %, 10 ppm: 1.76 g %, 15 ppm: 1.81 g %, and 20 ppm: 1.85 g % dry weight) and diseased plant (1.65 g % dry weight). In comparison to control, the disease inoculated wheat plant leaves recorded a higher total phenol content (1.65 g % dry weight) as compared to control leaves (1.2 g % dry weight) (Table 3).

Table 2. Estimation of reducing sugar in spot blotch infected wheat plant leaves in response to nanoparticle treatment.

Treatments	Reducing Sugar	
	g % Dry Weight	% Variation
Control	10.61	-
T_1: 0 ppm AgNPs	2.73	−288.645
T_2: 5 ppm AgNPs	4.44	−138.964
T_3: 10 ppm AgNPs	5.00	−112.20
T_4: 15 ppm AgNPs	5.45	−94.67
T_5: 20 ppm AgNPs	5.87	−80.74
T_6: 30 ppm AgNPs	6.94	−52.88
T_7: 50 ppm AgNPs	7.82	−35.67
SE (m)	0.22	
C.D.	0.67	

Table 3. Estimation of total phenol in spot blotch infected wheat plant leaves in response to nanoparticle treatment.

Treatments	Total Phenol	
	g % Dry Weight	% Variation
Control	1.2	-
T_1: 0 ppm AgNPs	1.65	27.27
T_2: 5 ppm AgNPs	1.72	30.23
T_3: 10 ppm AgNPs	1.76	31.81
T_4: 15 ppm AgNPs	1.81	33.70
T_5: 20 ppm AgNPs	1.85	35.13
T_6: 30 ppm AgNPs	1.98	39.39
T_7: 50 ppm AgNPs	2.09	42.58
SE (m)	0.05	
C.D.	0.16	

3.3. Estimation of Stress Response Enzyme Activities in AgNPs Treated Wheat Plants Infected with Bipolaris sorokiniana

3.3.1. Estimation of SOD Activity

SOD activity increased by nearly 35.53% in the diseased plants compared to the control plants. All AgNPs treated plants showed elevation in SOD activity in all different treatments (30 ppm: 3.91 U/mg protein and 50 ppm: 4.42 U/mg protein). The AgNPs treated plants showed significant elevation in enzyme activity and a reduction in disease severity. In T2, T3, T4, and T5 treatments, crops showed elevation in enzyme activity but no significant reduction in disease severity. At 5 ppm: 38.98%, 10 ppm: 38.80%, 15 ppm: 42.41%, 20 ppm: 42.41% elevation in enzyme activity was recorded (Table 4).

Table 4. Estimation of Superoxide dismutase activity in spot blotch infected wheat plant leaves in response to nanoparticle treatment.

Treatments	**Superoxide Dismutase**	
	U/mg Protein	**% Variation**
Control	2.05	-
T_1: 0 ppm AgNPs	3.18	35.53
T_2: 5 ppm AgNPs	3.36	38.98
T_3: 10 ppm AgNPs	3.35	38.80
T_4: 15 ppm AgNPs	3.54	42.09
T_5: 20 ppm AgNPs	3.56	42.41
T_6: 30 ppm AgNPs	3.91	47.57
T_7: 50 ppm AgNPs	4.42	53.61
SE (m)	0.11	
C.D.	0.34	

3.3.2. Estimation of CAT Activity

CAT activity increased by nearly 22.22% in the diseased plants compared to the control plants. All AgNPs treated plants showed elevation in CAT activity in all different treatments. At 30 ppm: 492 U/mg protein and 50 ppm: 583 U/mg protein, the AgNPs treated plants showed significant elevation in enzyme activity and a reduction in disease severity. In T2, T3, T4, and T5, crops showed elevation in enzyme activity but no significant reduction in disease severity. At 5 ppm: 24.49%, 10 ppm: 22.76%, 15 ppm: 30.29%, and 20 ppm: 28.05% elevation in enzyme activity was recorded (Table 5).

Table 5. Estimation of catalase activity in spot blotch infected wheat plant leaves in response to nanoparticle treatment.

Treatments	**Catalase**	
	U/mg Protein	**% Variation**
Control	336	-
T_1: 0 ppm AgNPs	432	22.22
T_2: 5 ppm AgNPs	445	24.49
T_3: 10 ppm AgNPs	435	22.76
T_4: 15 ppm AgNPs	482	30.29
T_5: 20 ppm AgNPs	461	28.05
T_6: 30 ppm AgNPs	492	31.71
T_7: 50 ppm AgNPs	583	42.37
SE (m)	14.77	
C.D.	44.68	

3.3.3. Estimation of POX Activity

POX activity increased by nearly 40.61% in the diseased plants compared to the control plants. All AgNPs treated plants showed elevation in POX activity in all different treatments. At 30 ppm: 3.00 U/mg protein (61.00%) and 50 ppm: 3.24 U/mg protein (63.89%), the nanoparticle treated plants showed significant elevation in enzyme activity and a reduction in disease severity. In T2, T3, T4, and T5, plants showed elevation in enzyme activity but no significant reduction in disease severity. At 5 ppm: 52.24%, 10 ppm:

56.5%, 15 ppm: 57.30%, and 20 ppm: 59.52% elevation in enzyme activity was recorded (Table 6).

Table 6. Estimation of Peroxidase activity in spot blotch infected wheat plant leaves in response to nanoparticle treatment.

Treatments	Peroxidase	
	U/mg Protein	% Variation
Control	1.17	-
T_1: 0 ppm AgNPs	1.97	40.61
T_2: 5 ppm AgNPs	2.45	52.24
T_3: 10 ppm AgNPs	2.69	56.51
T_4: 15 ppm AgNPs	2.74	57.30
T_5: 20 ppm AgNPs	2.89	59.52
T_6: 30 ppm AgNPs	3.00	61.00
T_7: 50 ppm AgNPs	3.24	63.89
SE (m)	0.07	
C.D.	0.22	

3.3.4. Estimation of GR Activity

GR activity increased by nearly 39.37% in the diseased plant compared to the control plant. All AgNPs treated plants showed elevation in GR activity in all different treatments. At 30 ppm: 3.79 U/mg protein and 50 ppm: 4.00 U/mg protein, the nanoparticle treated plant showed significant elevation in enzyme activity and a reduction in disease severity. In T2, T3, T4, and T5, wheat plants showed elevation in enzyme activity but no significant reduction in disease severity. At 5 ppm: 49.00%, 10%: 49.50%, 15 ppm: 53.0%, and 20 ppm: 54.70% elevation in enzyme activity was recorded (Table 7).

Table 7. Estimation of Glutathione reductase activity in spot blotch infected wheat plant leaves in response to nanoparticle treatment.

Treatments	Glutathione Reductase	
	U/mg Protein	% Variation
Control	1.54	-
T_1: 0 ppm AgNPs	2.54	39.37
T_2: 5 ppm AgNPs	3.02	49.00
T_3: 10 ppm AgNPs	3.05	49.50
T_4: 15 ppm AgNPs	3.28	53.04
T_5: 20 ppm AgNPs	3.40	54.70
T_6: 30 ppm AgNPs	3.79	59.36
T_7: 50 ppm AgNPs	4.00	61.50
SE (m)	0.12	
C.D.	0.38	

3.3.5. Estimation of PAL Activity

PAL activity increased by nearly 32.98% in the diseased plant than in the control plant. All AgNPs treated plants showed elevation in PAL activity in all different treatments. At 30 ppm: 6.36 U/mg protein and 50 ppm: 6.78 U/mg protein, the nanoparticle treated plants showed significant elevation in enzyme activity and a reduction in disease severity. In T2,

T3, T4, and T5, crops showed elevation in enzyme activity but no significant reduction in disease severity. At 5 ppm: 34.92%, 10 ppm: 34.92%, 15 ppm: 33.45%, and 20 ppm: 39.37% elevation in enzyme activity was recorded (Table 8).

Table 8. Estimation of Phenylalanine ammonia lyase activity in spot blotch infected wheat plant leaves in response to nanoparticle treatment.

Treatments	Phenylalanine Ammonia Lyase	
	U/mg Protein	% Variation
Control	3.82	-
T_1: 0 ppm AgNPs	5.70	32.98
T_2: 5 ppm AgNPs	5.87	34.92
T_3: 10 ppm AgNPs	5.74	33.45
T_4: 15 ppm AgNPs	6.13	37.68
T_5: 20 ppm AgNPs	6.20	39.37
T_6: 30 ppm AgNPs	6.36	39.94
T_7: 50 ppm AgNPs	6.78	43.66
SE (m)	0.17	
C.D.	0.51	

3.4. Antibacterial Activity against MDRs

The AgNPs showed antibacterial activity against all the MDR pathogens included in this study. The results of disc diffusion assay are summarized in Table 9. The clearly visible zone of inhibition exhibited by MNP against the Gram-negative and -positive MDR pathogens is depicted in Figure 5. The scientific name of the bacterial strain used in the study is presented in Table 10.

Table 9. Diameter of zone of inhibition (mm) observed for MNP against MDR pathogens.

	Bacteria	Diameter of Inhibition Zone (mm)
		MNP
1.	*ATCC 25922 E. coli*	12
2.	*E. coli Ec1*	11
3.	*E. coli Ec2*	13
4.	*E. coli Ec3*	11
5.	*E. coli Ec4*	12
6.	*E. coli Ec5*	12
7.	*E. coli Ec6*	11
8.	*K. pneumoniae KP-1*	10
9.	*K. pneumoniae KP-2*	11
10.	*K. pneumoniae KP-3*	12
11.	*K. pneumoniae KP-4*	11
12.	*K. pneumoniae KP-5*	11
13.	*K. pneumoniae KP-6*	8
14.	*K. pneumoniae KP-7*	11
15.	*MRSA-1*	15
16.	*MRSA-2*	16

Table 10. Scientific name/ strain for the organisms used in the study.

S. No.	Bacterial Strain	Gram Stain	ATCC Number
1	*Klebsiella pneumoniae*-KP1	Gram negative	ATCC 13883
2	*Escherichia coli*	Gram negative	ATCC 25922
3	*Methicillin-resistant Staphylococcus aureus* MRSA-SA1	Gram positive	ATCC BAA-41

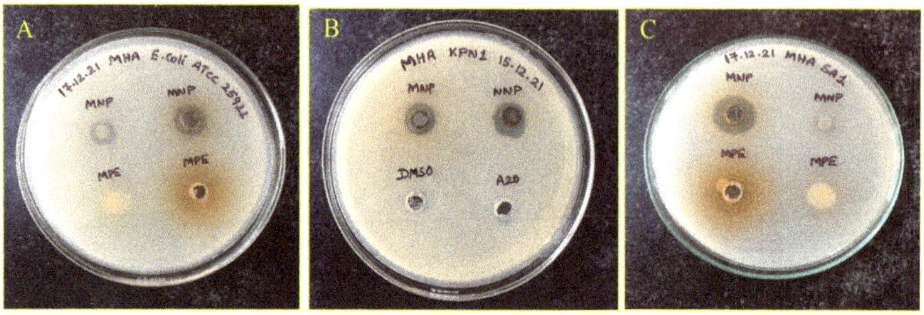

Figure 5. Zone of Inhibition (visible clear zones) produced by MNP against clinical MDR pathogens: (**A**): *E. coli* ATCC 25922 (**B**) *K. pneumoniae* KP1 and (**C**): Methicillin Resistant *S. aureus* (MRSA-SA1).

3.5. Green Synthesis of AgNPs by Using M. indica Leaf Extract and Their Characterization

The present study reports the successful synthesis and characterization of green synthesized AgNPs using *M. indica* aqueous leaf extract. Green synthesis of AgNPs using different plant sources has been well documented in the literature [14,25]. In the present study, we chose *M. indica* leaf extract for the green synthesis of AgNPs because it has a significant amount of reducing activity and capping properties due to the presence of phytochemicals, e.g., alkaloids, phenols, flavonoids, alkaloids, fats, amino acids, saponins, and oils. These bioactive compounds efficiently reduce silver salts (Ag^+) to Ag^0 [26]. Another reason for *M. indica* leaf selection is that some of the phytochemicals present in the plant extracts exhibit antimicrobial properties. In addition, *M. indica* leaf extracts not only efficiently synthesize silver nanoparticles, but also do so through capping.

Green synthesized AgNPs were characterized for their mean size, distribution, shape, morphology, and topography of nanoparticles. In this study, phytosynthesized AgNPs were initially identified by the change in color from light brown to blackish, indicating the formation of AgNPs. The present results were reported from colorless to brown grey in previous studies [27,28].

Further confirmation of AgNPs formation was ascertained by a surface plasmon resonance peak centered at 475 nm with a median nanoparticle size of 52 nm. A similar AgNPs size was observed in other studies. For instance, *M. indica* leaf extract based AgNPs size was reported to be 20–55 nm with plasmon resonance peak at 452 nm, 5–40 nm with plasmon resonance peak at 470 nm [29], 10–26 nm with surface plasmon resonance peak at 450 nm [30], and 32 nm with a surface plasmon peak at 393 nm [31]. *Catharanthus roseus* based silver nanoparticles were evaluated as 20–50 nm with a peak absorbance at 500 nm [27], 49 nm with surface plasmon resonance absorption peak at 425 nm [32], and 40–60 nm [33]. AgNPs other than the spherical shape can also contribute to the shift in surface plasmon resonance peak towards 500 nm and above.

SEM and XRD results confirmed the presence of AgNPs. SEM micrographs showed spherical shape AgNPs and the presence of crystalline AgNPs through the XRD spectrum analysis [34,35]. The SEM micrograph revealed phytosynthesized AgNPs using *Catharanthus roseus* leaf extract present as a bunch form in shape [33].

3.6. Studies on Biochemical Changes in Wheat Plants Infected with Spot Blotch (Bipolaris sorokiniana)

3.6.1. Reducing Sugars

Sugars constitute the primary biomolecule providing energy and structural material for defense responses in plants, and also act as signal molecules interacting with the plant immune system. Sugars cause oxidative burst at the early stages of infection, increasing the lignification of cell walls, stimulating the synthesis of flavonoids, and inducing certain PR proteins. Some sugars act as priming agents, inducing higher plant resistance to pathogens reported [36].

In the present study, the response of AgNPs applied to wheat crops infected with spot blotch was studied to correlate the role of reducing sugar content upon spot blotch infection and nanoparticle application. As per the results obtained (Table 1), the control plant recorded the highest reducing sugar content, and the lowest reducing sugar content was observed in the spot blotch infected plant(T_1). Spot blotch induced a decrease in sugar content. All other treatments showed elevation in reducing sugar content after the application of silver nanoparticles.

Carbohydrates and proteins are the primary metabolites that are exploited by biotic stress inducers for their growth and development. These primary metabolites also function as a precursor for secondary substances, which play a major role in plant defense [37]. Alberto (2014) [38] observed a decrease in reducing sugar content in the disease-infected plant compared to the healthy plants in their study on biochemical changes in *Theium-cepa L.* infected with anthracnose. Kumar et al. (2011) [39] reported that reducing sugar was decreased after infection of spot blotch infection in barley. Yanik F and Vardar F (2019) [40] observed the elevation in reducing sugar content in wheat plants applied with silver nanoparticles at different concentrations. Krishnaraj et al. (2012) in their study on biologically synthesized AgNPs found that these particles exerted a slight stress condition on the growth and metabolism of *B. monnieri* [41] accompanied by an elevation in total sugar content based on concentration applied.

3.6.2. Total Phenol

Phenols play an important role in the plant defense system against plant pathogens, insects, and the cyclic reduction of reactive oxygen species (ROS), such as superoxide anion and hydroxide radicals, H_2O_2, and singlet oxygen [42].

In the present study, the response of AgNPs applied to wheat plants infected with spot blotch was studied to correlate the role of total phenol content upon spot blotch infection and nanoparticle application. As per the results obtained (Table 2) among control and disease infected wheat crops, the healthy plant showed a low amount of total phenol. AgNPs treated spot blotch infected wheat plants showed elevation in total phenol content correlating the concentration of nanoparticles applied, and reduction in severity was also observed at higher concentration nanoparticle application.

Similar results were reported earlier whereby rust-resistant sorghum genotypes recorded higher phenols as the age of the crop increased [43]. A similar study on spot blotch (*Bipolaris sorokiniana*) in healthy and infected leaves of barley (*Hordeum vulgare*) indicated that resistant genotypes had more amount of total phenols content than susceptible genotypes [44]. Phenols inhibit disease development through different mechanisms like inhibition of extracellular fungal enzymes (cellulase, pectinase, laccase and xylanase), fungal oxidative phosphorylation, nutrition deprivation and antioxidant activity in plant tissue [45]. A similar study on the effect of nanoparticle application on spot blotch infected barley biochemical systems also showed an elevation in total phenol content associated to nanoparticle application concentration [39].

3.6.3. Superoxide Dismutase

SOD catalyzes the dismutation of superoxide radicals to hydrogen peroxide and oxygen, and constitutes the most important enzyme in cellular defense, because its activation directly modulates the amount of superoxide anion⁻ and H_2O_2 [46]. SOD activity may reduce the risk of hydroxyl radical formation. SODs are classified into three major groups based on their metal cofactor, and they are iron SOD (Fe-SOD) and copper-zinc SOD (Cu/Zn-SOD) are localized in the chloroplast, cytosol; manganese SOD (Mn-SOD) is found mainly in mitochondria and peroxisomes [47].

In the present study, biochemical changes in AgNPs treated wheat plants infected with spot blotch disease were analyzed. The wheat plants showed elevation in the SOD activity in all silver nanoparticles treated conditions when compared with the control. As per the results obtained (Table 4), all the AgNPs treated wheat plants showed an elevation in SOD activity, whereas only the 30 ppm and 50 ppm silver AgNPs treatment plants showed a reduction in spot blotch severity and elevation in SOD activity.

Similar results were observed [41] in studies on the effect of AgNPs application on *B. monnieri* biochemical activity and the treated plants showed an elevation in SOD activity at 100 ppm AgNPs treatment. Similar results were observed by Du et al. (2015), who studied the effect of AgNPs application on wheat plants and recorded an elevation in SOD activity at 100 mg/kg nanoparticle application [48].

3.6.4. Catalase

Plants possess several antioxidant enzymes that eliminate reactive oxygen species (ROS). CAT enzymes have a defensive role against ROS. They are responsible for the removal of toxic H_2O_2 in the cells, thereby protecting the cells from getting damaged. Up to a certain level, ROS production under stress may work as a signal for triggering defense responses via transduction pathways [49,50]. A high amount of ROS production in the cell causes cell death.

In the present study, the wheat plants showed elevation in the catalase activity in all AgNPs treated conditions when compared with a control. As per the results obtained (Table 4), all the silver nanoparticle treated wheat plants showed elevation in CAT activity but there was no reduction in disease severity in other treatments, except at 30 ppm and 50 ppm AgNPs treatment, where the infected plants showed a reduction in spot blotch severity and elevation in catalase activity.

Similar results were obtained in the study on AgNPs induced morphological, physiological, and biochemical changes in the wheat plant. Elevation in all of the antioxidative enzymes and phenol content was recorded [51]. A similar study conducted by Mohamed et al., in 2017, which also showed corresponding biochemical changes in a crop treated with nanoparticles [51,52].

3.6.5. Peroxidase (POX)

A heme-containing protein called peroxidase (POX) preferentially oxidizes aromatic electron donors at the expense of H_2O_2. By using oxidative polymerization, POX facilitates the conversion of cinnamyl alcohol to lignin. Vascular plants have many POX genes, and each class of these genes may have a unique physiological function in preventing oxidative damage to the cell membrane [53].

In the present study, the wheat plants showed elevation in the peroxidase activity in all of the AgNPs treated conditions when compared with a control. As per the results obtained (Table 5), all the AgNPs treated wheat plants showed elevation in POX activity, but there was no reduction in disease severity in other treatments except at 30 ppm and 50 ppm AgNPs treatment. With 30 ppm and 50 ppm silver nanoparticle treatment, crops showed a reduction in spot blotch severity and elevation in peroxidase activity was recorded. The present study indicates the crop showed more resistance towards disease at 30 and 50 ppm silver nanoparticles, but all treatments showed elevation in all plant biochemical constituents.

Numerous fungal species, including *Trichosporium vesiculosum*, *Macrophomin phaseolina*, *Coprinus comatus*, *Mycophaerella arachidicola*, *Fusarium exosporium*, and *Botrytis cenere*, have been shown to be resistant to POX in some studies [53,54]. Following inoculation with stem and leaf rust pathogens, low infection type isogenic lines of wheat showed a quicker increase in POX activity than sensitive isogenic lines of wheat [55].

3.6.6. Glutathione Reductase

Another important pathway involved in the control of ROS levels in the plant tissue is the ascorbate/glutathione cycle in which Glutathione reductase plays a major role [56]. GR is a flavoprotein oxidoreductase that catalyzes the reduction of glutathione disulphide to the sulfhydryl form GSH. This enzyme employs NADPH as a reductant.

As per the results obtained (Table 6), all treatments showed elevation in GR activity, but a reduction in disease severity was observed in only 30 ppm and 50 ppm AgNPs treatment conditions. In the present study, the wheat plant showed elevation in the GR activity in all AgNPs treated conditions when compared with the control.

Similarly, a rapid elevation in GR content was observed by Yanik F and Vardar F, 2019, in their study on silver nanoparticle-induced morphological, physiological, and biochemical changes in the wheat plant [40]. An elevation in all antioxidant enzyme and phenol content was recorded.

3.6.7. Phenylalanine Ammonia-Lyase

The primary enzyme in the plant phenylpropanoid pathway, phenylalanine ammonia-lyase catalyzes the synthesis of secondary metabolites from L-phenylalanine, such as lignin, flavonoids, and phytoalexins. L-phenylalanine is non-oxidatively deaminated by the PAL enzyme, resulting in the formation of trans-cinnamic acid and a free ammonium ion. Phenylalanine, an amino acid, is converted to trans-cinnamic acid in plants, which is the first step in the transfer of carbon from primary metabolism to phenylpropanoid secondary metabolism.

As per the results obtained (Table 7) among all the treatments, T6 and T7 recorded the highest PAL activity, whereas the lowest PAL activity was observed in the control plant. Spot blotch infected plants also showed a small elevation in PAL activity. In the present study, biochemical changes in AgNPs treated wheat crops infected with spot blotch disease were analyzed. The wheat plant showed elevation in the peroxidase activity in all AgNPs treated conditions when compared with the control condition. As per the results obtained, all the AgNPs treated wheat plants showed elevation in PAL, but no reduction in disease severity was observed. With 30 ppm and 50 ppm silver nanoparticle treatment, crops showed a reduction in spot blotch severity and elevation in PAL activity was recorded. The present study indicates the crop showed more resistance towards disease at 30 ppm and 50 ppm, but all treatments showed elevation in all plant biochemical constituents.

It is possible that the rapid induction of PAL genes in resistant interactions between groundnut and its pathogen/elicitor is caused by the involvement of a signal transduction mechanism. This mechanism is triggered specifically as a result of the interaction between elicitor and receptor molecules, thereby showing differential transcriptional rates of PAL compatible and incompatible interactions. A mechanism quite similar to this one was postulated earlier in several plants [56,57]

4. Conclusions

Silver nanoparticles were successfully synthesized by using *Mangifera indica* leaf extract with a size below 100 nm. The formation of AgNPs was confirmed using various analytical techniques, such as UV-visible spectrophotometer, particle size analyzer (PSA), scanning electron microscopy (SEM), and X-ray diffraction. Biochemical analysis of silver nanoparticle treated wheat crop infected with spot blotch showed elevation in all biochemical constituents, e.g., reducing sugar, total phenol, superoxide dismutase (SOD), catalase (CAT), peroxidase (POX), glutathione reductase (GR), and phenylalanine ammonia-lyase (PAL). Elevation in all biochemical constituents was recorded under all treatment conditions but a reduction in disease severity was recorded only in 30 ppm and 50 ppm concentration AgNPs treatments. The overall study revealed that the higher amount of reducing sugar, total phenols, antioxidant enzyme activity, and phytochemical precursor enzyme PAL plays an important role in the defense mechanism of plants against wheat spot blotch infection. Moreover, this study revealed that silver nanoparticles application on wheat crops causes the elevation in antioxidative enzyme activity and the amount of total phenol, reducing sugar content, and promising antibacterial activity against MDR strains.

Author Contributions: Conceptualization, R.R.B., K.K.M., P.R.S. and S.M.; methodology, Y.S.A., S.S.M. and S.S.; soft-ware, U.M.M.; validation, A.A.K., I.A.S. and P.R.S.; formal analysis, S.M.; investigation, K.K.M., A.B. and S.M.; resources, I.A.S. and Y.S.A.; data curation, A.A.K.; writing—original draft preparation, K.K.M., S.S.M. and A.A.K.; writing—review and editing, I.A.S. and A.A.K.; visualization, R.R.B., and A.B.; supervision, P.R.S., and S.S.M.; project administration, S.S., A.A.K., I.A.S. and P.R.S.; funding acquisition, Y.S.A.. All authors have read and agreed to the published version of the manuscript.

Funding: The authors are thankful to the Deanship of Scientific Research, Najran University, Najran, Saudi Arabia, for funding this research through grant research code NU/NRP/MRC/11/5.

Institutional Review Board Statement: Not applicable.

Informed Consent Statement: Not applicable.

Data Availability Statement: All the data has been presented in this article.

Acknowledgments: The authors are thankful to Najran University, Najran, KSA; UAS Dharwad, Dayanand Sagar University, Bangalore, India; for supporting steps of this work.

Conflicts of Interest: The authors declare no conflict of interest.

References

1. Food & Agriculture Organization of the United Nations (FAO). *Faostat: FAO Statistical Databases*; Food & Agriculture Organization of the United Nations (FAO): Rome, Italy, 2000; ISBN 9780119861105.
2. Mielgo-Ayuso, J.; Aparicio-Ugarriza, R.; Olza, J.; Aranceta-Bartrina, J.; Gil, A.; Ortega, R.M.; Serra-Majem, L.; Varela-Moreiras, G.; González-Gross, M. Dietary Intake and Food Sources of Niacin, Riboflavin, Thiamin and Vitamin B6 in a Representative Sample of the Spanish Population. The ANthropometry, Intake, and Energy Balance in Spain (ANIBES) Study dagger. *Nutrients* **2018**, *10*, 846. [CrossRef] [PubMed]
3. Joshi, A.K.; Chand, R. Variation and Inheritance of Leaf Angle, and Its Association with Spot Blotch (*B. Sorokiniana*) Severity in Wheat (*Triticum Aestivum*). *Euphytica* **2002**, *124*, 283–290. [CrossRef]
4. Rajan, M.S. Nano: The next Revolution. *Natl. Book Trust* **2004**, 15–72.
5. Mousavi, S.R.; Rezaei, M. Nanotechnology in Agriculture and Food Production. *J. App Environ. Biol. Sci.* **2011**, *1*, 414–419.
6. Mishra, S.; Singh, H. Silver nanoparticles mediated altered gene expression of melanin biosynthesis genes in *Bipolaris sorokiniana*. *Microbiol. Res.* **2015**, *172*, 16–18. [CrossRef]
7. Boller, T. Chemoperception of Microbial Signals in Plant Cells. *Annu. Rev. Plant Biol.* **1995**, *46*, 189–214. [CrossRef]
8. Kobayashi, I.; Murdoch, L.J.; Hardham, A.R.; Kunoh, H. Cell biology of early events in the plant resistance response to infection by pathogenic fungi. *Can. J. Bot.* **1995**, *73*, 418–425. [CrossRef]
9. Brisson, L.F.; Tenhaken, R.; Lamb, C. Function of Oxidative Cross-Linking of Cell Wall Structural Proteins in Plant Disease Resistance. *Plant Cell* **1994**, *6*, 1703. [CrossRef]
10. Bolwell, G.P. Role of active oxygen species and NO in plant defence responses. *Curr. Opin. Plant Biol.* **1999**, *2*, 287–294. [CrossRef]
11. Hahlbrock, K.; Scheel, D. Physiology and Molecular Biology of Phenylpropanoid Metabolism. *Annu. Rev. Plant Biol.* **1989**, *40*, 347–369. [CrossRef]
12. Peltonen, S.; Karjalainen, R. Phenylalanine Ammonia Lyase Activity in Barlay after Infection with *Bipolaris Sorokiniana* or Treatment with Its Purified Baralay. *J. Phytopathol.* **1995**, *143*, 239–245. [CrossRef]
13. Peltonen, S. Response of Barley and Wheat to Pathogen, Non- Pathogens and Wounding as Indicates by Induced Phenylalanine Ammonia Lyase Activity. *Acta Agric. Scand Sect. B Soil Plant Sci.* **1998**, *48*, 184–191.
14. Narayan, A.; Jacob, R.; Paulsonb, J.M.; Dinesh, K.B.; Narayan, C.R. Mango Leaf Extract Synthesized Silver Nanorods Exert Anticancer Activity on Breast Cancer and Colorectal Carcinoma Cells. *J. Drug. Deliv. Sci. Technol.* **2018**, *44*, 8–12.
15. Beauchamp, C.; Fridovich, I. Superoxide dismutase: Improved assays and an assay applicable to acrylamide gels. *Anal. Biochem.* **1971**, *44*, 276–287. [CrossRef]
16. Mavis, R.D.; Stellwagen, E. Purification and Subunit Structure of Glutathione Reductase from Bakers' Yeast. *J. Biol. Chem.* **1968**, *243*, 809–814. [CrossRef]
17. Tewari, R.; Ganaie, F.; Venugopal, N.; Mitra, S.; Shome, R.; Shome, B.R. Occurrence and characterization of genetic determinants of β-lactam-resistance in Escherichia coli clinical isolates. *Infect. Genet. Evol.* **2022**, *100*, 105257. [CrossRef]

18. Bauer, M.A.W.; Kirby, M.W.M.M.; Sherris, M.J.C.; Turck, M.M. Antibiotic Susceptibility Testing by a Standardized Single Disk Method. *Am. J. Clin. Pathol.* **1966**, *45*, 493–496. [CrossRef]
19. Gomez, G.I.; Gonez, H.K. Analysis of CRD and DMRT's Methods. *Electron. J. Appl. Stat. Anal.* **1984**.
20. Kora, A.J.; Mounika, J.; Jagadeeshwar, R. Rice leaf extract synthesized silver nanoparticles: An in vitro fungicidal evaluation against Rhizoctonia solani, the causative agent of sheath blight disease in rice. *Fungal Biol.* **2020**, *124*, 671–681. [CrossRef]
21. Al Awadh, A.A.; Shet, A.R.; Patil, L.R.; Shaikh, I.A.; Alshahrani, M.M.; Nadaf, R.; Mahnashi, M.H.; Desai, S.V.; Muddapur, U.M.; Achappa, S.; et al. Sustainable Synthesis and Characterization of Zinc Oxide Nanoparticles Using *Raphanus sativus* Extract and Its Biomedical Applications. *Crystals* **2022**, *12*, 1142. [CrossRef]
22. Nahari, M.H.; Al Ali, A.; Asiri, A.; Mahnashi, M.H.; Shaikh, I.A.; Shettar, A.K.; Hoskeri, J. Green Synthesis and Characterization of Iron Nanoparticles Synthesized from Aqueous Leaf Extract of *Vitex leucoxylon* and Its Biomedical Applications. *Nanomaterials* **2022**, *12*, 2404. [CrossRef] [PubMed]
23. Zainab; Saeed, K.; Ammara; Ahmad, S.; Ahmad, H.; Ullah, F.; Sadiq, A.; Uddin, A.; Khan, I.; Ahmad, M. Green Synthesis, Characterization and Cholinesterase Inhibitory Potential of Gold Nanoparticles. *J. Mex. Chem. Soc.* **2021**, *65*, 416–423. [CrossRef]
24. Khan, I.; Saeed, K.; Khan, I. Nanoparticles: Properties, applications and toxicities. *Arab. J. Chem.* **2019**, *12*, 908–931. [CrossRef]
25. Bharathi, V.; Firdous, J.; Muhamad, N.; Mona, R. Green Synthesis of Mangifera Indica Silver Nanoparticles and Its Analysis Using Fourier Transform Infrared and Scanning Electron Microscopy. *Natl. J. Physiol. Pharm. Pharm.* **2017**, *7*, 83–94.
26. Mishra, J.N.; Verma, N.K. A Brief Study on Catharanthus Roseus: A Review. *Int. J. Res. Pharma Pharm. Sci.* **2017**, *2*, 20–23.
27. Ghozali, S.Z.; Vuanghao, L.; Ahmad, N.H. Biosynthesis and Characterization of Silver Nanoparticles Using Catharanthus Roseus Leaf Extract and Its Proliferative Effects on Cancer Cell Lines. *J. Nanomed. Nanotechnol.* **2015**, *6*, 4–11.
28. Bhusnure, O.G.; Kuthar, V.S.; Gholve, S.B.; Giram, P.S.; Shembekar, V.S.; Zingade, S.G.; Jadhav, P.P. Green Synthesis of Silver Nanoparticle Using Catharanthus Roseus Extract for Pharmacological Activity. *Int. J. Pharma Pharm. Res* **2017**, *10*, 77–88.
29. Banala, R.R.; Nagati, V.B.; Karnati, P.R. Green synthesis and characterization of Carica papaya leaf extract coated silver nanoparticles through X-ray diffraction, electron microscopy and evaluation of bactericidal properties. *Saudi J. Biol. Sci.* **2015**, *22*, 637–644. [CrossRef]
30. Ajani, A.; Dada, E.; Olu-Arotiowa, O.; Okeowo, I.; Alade, A.; Afolabi, T. Adsorption of Methylene Blue from Aqueous Solution using Microwave-Assisted $BaCl_2$ Modified Activated Carbon Produced from Mango Seed Shell. *LAUTECH J. Civ. Environ. Stud.* **2019**, *3*, 72–82. [CrossRef]
31. Sundeep, D.; Vijaya Kumar, T.; Rao, P.S.S.; Ravikumar, R.V.S.S.N.; Gopala Krishna, A. Green synthesis and characterization of Ag nanoparticles from Mangifera indica leaves for dental restoration and antibacterial applications. *Prog. Biomater.* **2017**, *6*, 57–66. [CrossRef]
32. Hanady, S.A.; Wasnaa, H.; Ghassan, M.; Saadoon, A.H. Biosynthesis of Silver Nanoparticles from CatharanthusRoseus Leaf Extract and Assessing Their Antioxidant, Antimicrobial, and Wound-Healing Activities. *Artif. Cells Nanomed. Biotechnol.* **2016**, *45*, 1234–1240.
33. Ahmed, S.; Saifullah; Ahmad, M.; Swami, B.L.; Ikram, S. Green synthesis of silver nanoparticles using Azadirachta indica aqueous leaf extract. *J. Radiat. Res. Appl. Sci.* **2016**, *9*, 1–7. [CrossRef]
34. Ponarulselvam, S.; Panneerselvam, C.; Murugan, K.; Aarthi, N.; Kalimuthu, K.; Thangamani, S. Synthesis of Silver Nanoparticles Using Leaves of *Catharanthus Roseus* L. (G) Don. and Their Antiplasmodial Activities. *Asian. Pac. J. Trop. Biomed.* **2012**, *2*, 574–580. [CrossRef]
35. Rajagopal, T.; Jemimah, I.A.A.; Ponmanickam, P.; Ayyanar, M. Synthesis of Silver Nanoparticles Using Catharanthus Roseus Root Extract and Its Larvicidal Activities. *J. Environ. Biol.* **2015**, *36*, 1283–1289.
36. Morkunas, I.; Ratajczak, L. The Role of Suger Signalling in Plant Defence Response against Fungal Pathogen. *Acta Physiol. Plant* **2014**, *36*, 1607–1619. [CrossRef]
37. Kazemi-Shahandashti, S.-S.; Maali-Amiri, R. Global insights of protein responses to cold stress in plants: Signaling, defence, and degradation. *J. Plant Physiol.* **2018**, *226*, 123–135. [CrossRef]
38. Alberto, R. Pathological response and biochemical changes in *Allium cepa* L. (bulb onions) infected with anthracnose-twister disease. *Plant Pathol. Quar.* **2014**, *4*, 23–31. [CrossRef]
39. Kumar, P.; Kalappanavar, I.K.; Arukumar, G.S.; Pradeep, P.E. Studies on Selected Physiological and Biochemical Parameters Responsible for Resistance to Spot Blotch of Barlay. *J. Wheat Res.* **2011**, *3*, 25–29.
40. Yanik, F.; Vardar, F. Assessment of Silver Nanoparticle-Induced Morphological, Biochemical and Physiological Alterations in Wheat Roots. *Ann. Bot.* **2019**, *9*, 83–94.
41. Krishnaraj, C.; Jagan, E.; Ramachandran, R.; Abirami, S.; Mohan, N.; Kalaichelvan, P. Effect of biologically synthesized silver nanoparticles on Bacopa monnieri (Linn) Wettst. plant growth metabolism. *Process Biochem.* **2012**, *47*, 651–658. [CrossRef]
42. Maffei, M.E.; Mithöfer, A.; Boland, W. Before gene expression: Early events in plant–insect interaction. *Trends Plant Sci.* **2007**, *12*, 310–316. [CrossRef] [PubMed]
43. Naik, S.T.; Anahousur, K.H.; Hegde, R.K. Role of Sugar Phenol and Amino Acid in Rust Resistance in Sorghum. *J. Agril. Sci.* **1981**, *15*, 282–288.
44. Reddy, M.N. Changes in Phenolic Acids in Groundnut Leaves Infected with Rust. *J. Phytopathol.* **1984**, *110*, 78–81. [CrossRef]
45. Scalbert, A. Antimicrobial properties of tannins. *Phytochemistry* **1991**, *30*, 3875–3883. [CrossRef]

46. Foyer, C.; Noctor, G. Oxidant and Antioxidant Signaling in Plants, a Reevaluation of the Concept of Oxidative Stress in a Physiological Context. *Plant Cell Environ.* **2005**, *28*, 1056–1071. [CrossRef]
47. Alscher, R.G.; Erturk, N.; Heath, L.S. Role of superoxide dismutases (SODs) in controlling oxidative stress in plants. *J. Exp. Bot.* **2002**, *53*, 1331–1341. [CrossRef]
48. Du, W.; Gardea-Torresdey, J.L.; Ji, R.; Yin, Y.; Zhu, J.; Peralta-Videa, J.R.; Guo, H. Physiological and Biochemical Changes Imposed by CeO_2 Nanoparticles on Wheat: A Life Cycle Field Study. *Environ. Sci. Technol.* **2015**, *49*, 11884–11893. [CrossRef]
49. Cruz, C.M.H. Drought Stress and Reactive Oxygen Species: Production, Scavenging and Signaling. *Plant Signal. Behav.* **2008**, *3*, 156–165. [CrossRef]
50. Miller, G.; Suzuki, N.; Ciftci-Yilmaz, S.; Mittler, R. Reactive oxygen species homeostasis and signalling during drought and salinity stresses. *Plant Cell Environ.* **2010**, *33*, 453–467. [CrossRef]
51. Mohamed, A.K.S.H.; Qayyum, M.F.; Abdel-Hadi, A.M.; Rehman, R.A.; Ali, S.; Rizwan, M. Interactive effect of salinity and silver nanoparticles on photosynthetic and biochemical parameters of wheat. *Arch. Agron. Soil Sci.* **2017**, *63*, 1736–1747. [CrossRef]
52. Luthaje, S.; MeisrimLer, C.; Moller, B. Phylogenetic, Topology, Structural and Functions of Membrane Bound Peroxidase in Vascular Plant. *Phytochemistry* **2011**, *72*, 1124–1135. [CrossRef]
53. Ye, X.; Ng, T. Isolation of a novel peroxidase from French bean legumes and first demonstration of antifungal activity of a non-milk peroxidase. *Life Sci.* **2002**, *71*, 1667–1680. [CrossRef]
54. Ghosh, M. Antifungal Properties of Haem Peroxidase from Acorus calamus. *Ann. Bot.* **2006**, *98*, 1145–1153. [CrossRef]
55. Johnson, L.; Lee, R. Peroxidase changes in wheat isolines with compatible and incompatible leaf rust infections. *Physiol. Plant Pathol.* **1978**, *13*, 173–181. [CrossRef]
56. Dixon, R.A.; Dey, P.M.; Lamb, C.J. Phytoalexins: Enzymology and Molecular Biology. *Adv. Enzymol. Relat. Areas Mol. Biol.* **1983**, *55*, 1–136. [CrossRef]
57. Lawton, M.A.; Lamb, C.L. Transcriptional Activation of Plant Defense Gene by Fungus Elicitor, Wonding, and Infection. *Mol. Cell Biol.* **1987**, *7*, 335–341.

antibiotics

Article

Potentiation of Antibiotic Action and Efflux Pump Inhibitory Effect on *Staphylococcus aureus* Strains by Solasodine

Ana Raquel Pereira da Silva [1], Maria do Socorro Costa [1], Nara Juliana Santos Araújo [2], Thiago Sampaio de Freitas [3], Ray Silva de Almeida [3], José Maria Barbosa Filho [4], Josean Fechine Tavares [5], Erlânio Oliveira de Souza [6], Pablo Antonio Maia de Farias [7], Jacqueline Cosmo Andrade Pinheiro [2] and Henrique Douglas Melo Coutinho [3,*]

1 Graduate Program in Biotechnology, Universidade Estadual do Ceará—UECE, Fortaleza 60714-903, Brazil
2 Graduate Program in Health Sciences, Universidade Federal do Cariri—UFCA, Barbalha 63180-000, Brazil
3 Department of Biological Chemistry, Universidade Regional do Cariri—URCA, Crato 63105-000, Brazil
4 Laboratory of Pharmaceutical Technology, Universidade Federal da Paraíba—UFPB, João Pessoa 58033-455, Brazil
5 Department of Pharmaceutical Sciences, Universidade Federal da Paraíba—UFPB, João Pessoa 58051-900, Brazil
6 Faculty of Technology of Cariri, Juazeiro do Norte 63040-000, Brazil
7 College of Dentistry—CECAPE, Juazeiro do Norte 63024-015, Brazil
* Correspondence: hdmcoutinho@gmail.com

Abstract: A worrisome fact is the increase in microbial resistance, which has as its main cause the indiscriminate use of antibiotics. Scientific studies have investigated bioactive compounds such as steroidal sapogenins, in the perspective of new beneficial alternatives for the control of bacterial resistance. Therefore, the objective of this work was to verify the antibacterial activity as well as the modifying action of antibiotics associated with solasodine and its ability to inhibit the efflux pump mechanism in strains of *Staphylococcus aureus*. Tests were performed to verify the minimum inhibitory concentration (MIC). In addition, the action-modifying potential of antibiotics and the inhibitory capacity of the efflux pump NorA and MepA through synergistic effects on the antibiotic and ethidium bromide were evaluated. Solasodine showed significant results for the standard bacteria with an MIC of 512 μg/mL, and when associated with the antibiotics gentamicin and nofloxacin for the multidrug-resistant bacteria *S. aureus* 10, *Escherichia coli* 06, and *Pseudomonas aeruginosa* 24, it showed a 50% reduction in MIC. The association of solasodine with the antibiotic ciprofloxacin against *S. aureus* K2068 (MepA) showed synergism, with a reduction in the MIC of the antibiotic from 64 μg/mL to 40 μg/mL, and also a reduction in the MIC when the antibiotic was used in conjunction with the efflux pump inhibitors. Solasodine may be acting on the mechanism of action of the antibiotic, as it has shown a potentiating effect when associated with antibiotics, inducing a reduction in the MIC against Gram-positive and Gram-negative bacteria. Therefore, this study demonstrated significant results for the potentiating action of solasodine when associated with antibiotics of clinical importance.

Keywords: antibacterial activity; bacterial resistance; sapogenin

Citation: da Silva, A.R.P.; Costa, M.d.S.; Araújo, N.J.S.; de Freitas, T.S.; de Almeida, R.S.; Barbosa Filho, J.M.; Tavares, J.F.; de Souza, E.O.; de Farias, P.A.M.; Pinheiro, J.C.A.; et al. Potentiation of Antibiotic Action and Efflux Pump Inhibitory Effect on *Staphylococcus aureus* Strains by Solasodine. *Antibiotics* **2022**, *11*, 1309. https://doi.org/10.3390/antibiotics11101309

Academic Editor: Carlos M. Franco

Received: 18 August 2022
Accepted: 23 September 2022
Published: 27 September 2022

1. Introduction

Antibiotics are drugs that have revolutionized the treatment of diseases caused by bacteria. However, infectious diseases are currently one of the leading causes of death in the world, due to the increasing number of microorganisms that are less susceptible to drugs, thus resulting in a worrying increase in acquired bacterial resistance, either by mechanisms of mutation, transformation, or conjunction; these processes of bacterial adaptation that were supposed to be natural have resulted in a consequent phenomenon of resistance, characterized as a serious health problem that has as its main cause the indiscriminate use of antibiotics that consequently causes a succession of adverse events to the human microbiota [1–3].

The mechanism of action, such as the efflux pump present in some bacteria, is one of the causes of antibiotic resistance, since it promotes active pumping of antimicrobials from the intracellular to the extracellular environment, in which its active flux produces resistance to certain antimicrobials [4]. NorA is a multidrug efflux pump and one of the most studied systems; it is a pump present in the bacterium *Staphylococcus aureus* [5]. MepA is another multidrug efflux pump belonging to the multidrug and toxic compound extrusion (MATE) family also present in *S. aureus* [6].

Therefore, antimicrobial resistance is an issue of worldwide concern and scientific studies have investigated natural products as a source of new drugs [7], and affordable and beneficial alternatives that can contribute to the control of microbial resistance. They have also investigated the use of antibiotics of conventional use associated with bioactive compounds of biotechnological interest, in the perspective of a new potentiating agent of the effect of drugs, contributing in the alteration of the permeability of the cellular membrane of the bacteria or a possible inhibiting effect of the efflux pump mechanism [8,9].

Steroidal compounds have been valuable precursors to complementary sources for the synthesis of various drugs. Sapogenins are already known for their multiple structural variations and pharmacological properties that can be used in industry, both in human and veterinary medicinal applications [10–12].

Solasodine is a steroidal alkaloid, found in nature mainly in plants of the genus *Solanum*, being widely studied for its pharmacological applications in view of the presence of steroidal sapogenins that have pharmaceutical and industrial importance [11,13,14]. Thus, the interest in promoting this work is due to the activities already described in previous studies, in which solasodine demonstrates anticonvulsant [15], anti-inflammatory [16], antioxidant [17], antiviral [18], antifungal [19,20], anticancer [21–23], and antibacterial activity [24–26].

This study aimed to verify the antibacterial activity and the modifying action of antibiotics by the broth microdilution method, as well as the ability for solasodine to inhibit the NorA and MepA efflux pump mechanism.

2. Results

2.1. Antibacterial Activity

With the minimum inhibitory concentration (MIC) analysis, solasodine was found to have no direct antibacterial activity for the bacteria *S. aureus* 10, *S. aureus* K2068, *P. aeruginosa* 24, and *E. coli* 06, showing an MIC greater than or equal to 1024 μg/mL.

However, for the bacteria *S. aureus* 1199B, *S. aureus* ATCC 25923, *P. aeruginosa* ATCC 9027, and *E. coli* ATCC 25922, solasodine showed an MIC of 512 μg/mL, as presented in Table 1.

Table 1. Determination of the minimum inhibitory concentration (μg/mL).

Bacteria	S.A 10	P.A 24	E.C 06	S.A K2068	S.A 1199B	S.A ATCC 25923	P.A ATCC 9027	E.C ATCC 25922
Solasodine	≥1024	≥1024	≥1024	≥1024	512	512	512	512

Abbreviations: S.A: *Staphylococcus aureus*; P.A: *Pseudomonas aeruginosa*, E.C: *Escherichia coli*.

2.2. Antibiotic Activity Modifier Action

Solasodine in association with gentamicin demonstrated an antibiotic potentiation effect against *S. aureus* 10 with MIC reduction from 50 μg/mL to 6 μg/mL, *Escherichia coli* 06 with MIC reduction from 40 μg/mL to 5 μg/mL, and *Pseudomonas aeruginosa* 24 with MIC reduction from 40 μg/mL to 25 μg/mL. The association of solasodine with norfloxacin also showed an antibiotic potentiation effect against *S. aureus* 10 with MIC reduction from 128 μg/mL to 64, *E. coli* 06 with MIC reduction from 322 μg/mL to 80 μg/mL, and *P. aeruginosa* 24 with MIC reduction from 50 μg/mL to 6 μg/mL (Figure 1).

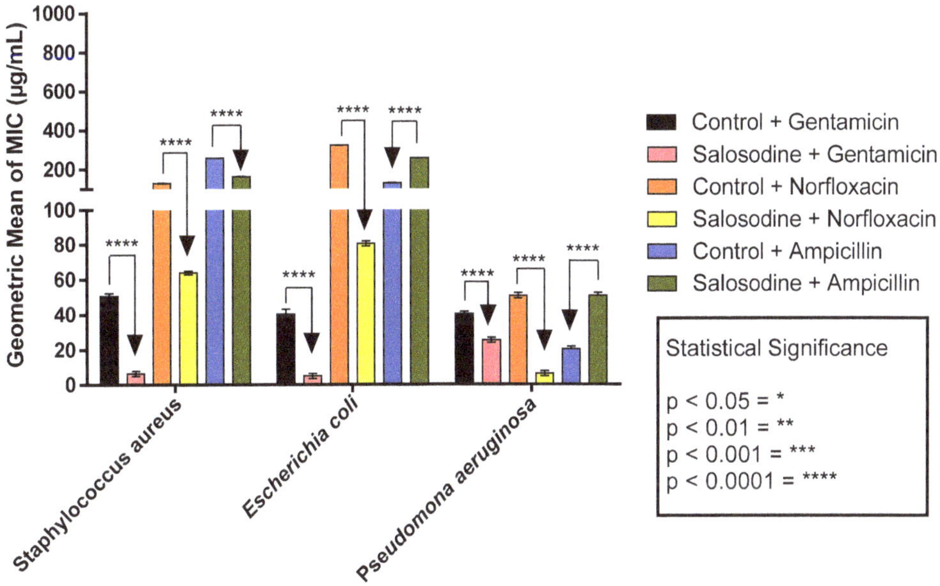

Figure 1. Modifying effect of antibiotics by solasodine against *S. aureus* 10, *E. coli* 06, and *P. aeruginosa* 24. The control refers to antibiotics. **** Statistically significant value when $p < 0.0001$; not statistically significant when $p > 0.05$.

For solasodine in association with ampicillin, antibiotic potentiation was only observed against *S. aureus* 10 with a reduction in the MIC from 256 µg/mL to 161 µg/mL. Against *E. coli* 06 and *P. aeruginosa* 24, the association of sapogenin with the antibiotic expressed itself in an antagonistic effect with an increase in the MIC (Figure 1).

2.3. Efflux Pump Inhibitory Effect Nora E Mepa

The association of solasodine with the antibiotic norfloxacin against *S. aureus* 1199B resulted in an antagonistic effect with an increase in the MIC. However, when norfloxacin was tested together with the efflux pump inhibitors chlorpromazine and CCCP, it showed a reduction of the MIC (Figure 2).

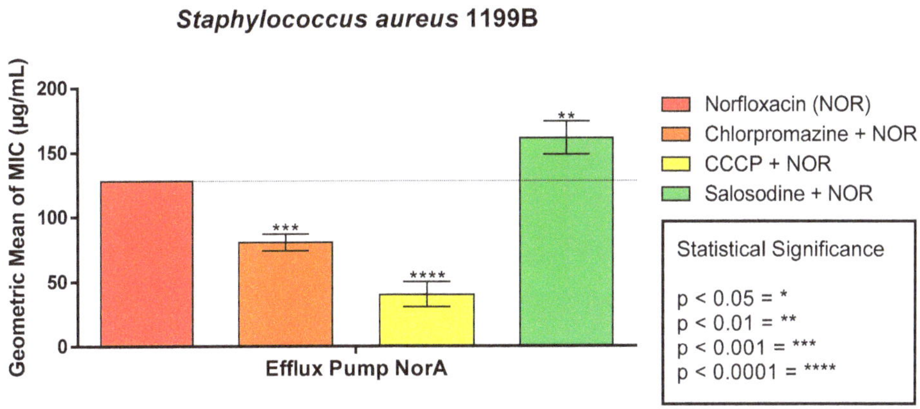

Figure 2. Modifying effect of norfloxacin antibiotic action by solasodine on *S. aureus* strain 1199B.

When solasodine was associated with ethidium bromide, it did not show a statistically significant result. However, there was a reduction in the MIC of the bromide in conjunction with the efflux pump inhibitors (Figure 3).

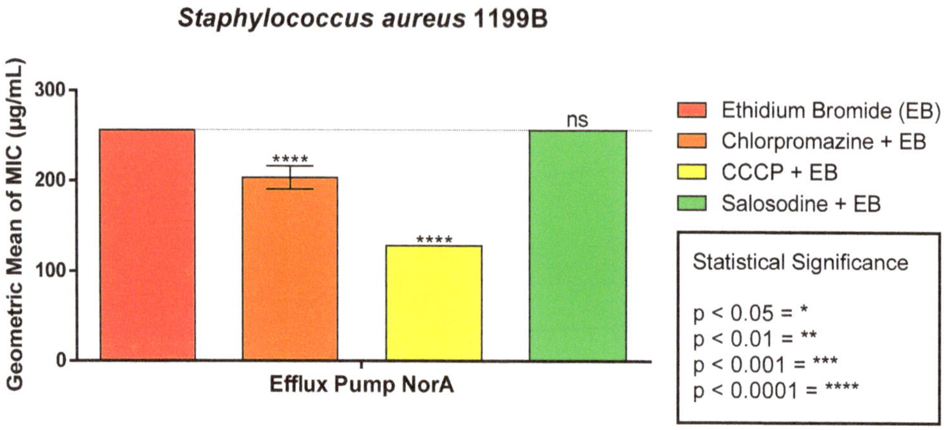

Figure 3. Modifying effect of ethidium bromide by solasodine on *S. aureus* strain 1199B.

The association of solasodine with the antibiotic ciprofloxacin against *S. aureus* K2068 demonstrated antibiotic potentiation with a reduction in the MIC from 64 µg/mL to 40 µg/mL. It also resulted in a reduction in the MIC of ciprofloxacin when tested in conjunction with the efflux pump inhibitors (Figure 4). This suggests that solasodine may be acting on the mechanism of action of the antibiotic, increasing its effect against the bacteria tested.

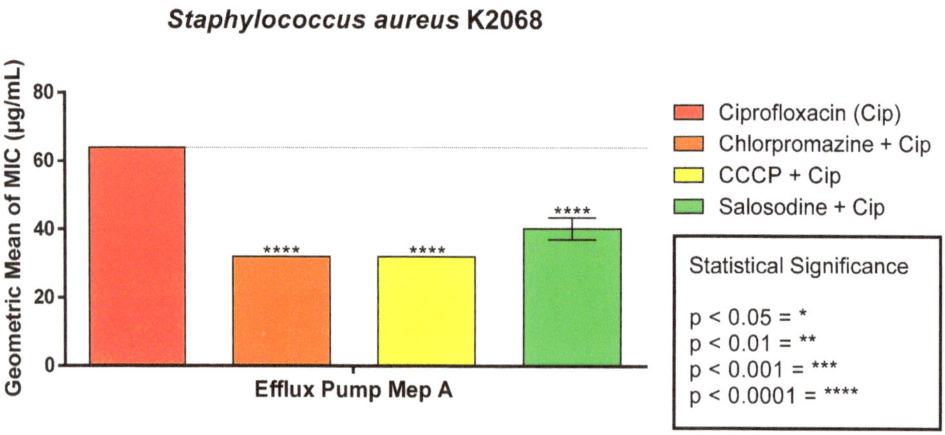

Figure 4. Modifying effect of norfloxacin antibiotic action by solasodine on *S. aureus* strain K2068.

The solasodine associated with ethidium bromide expressed an antagonistic effect, having a high increase in the MIC. The bromide in conjunction with the standard inhibitors showed a reduction in the MIC (Figure 5).

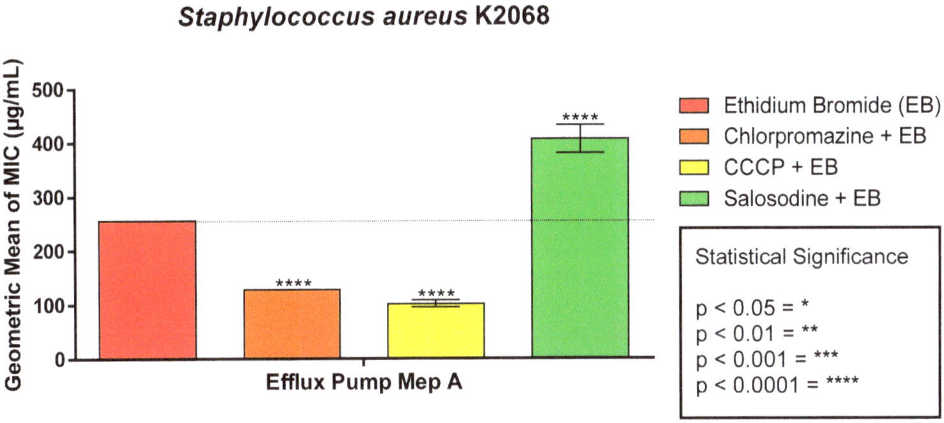

Figure 5. Modifying effect of ethidium bromide by solasodine on *S. aureus* strain K206B.

3. Discussion

Bacteria have the ability to adapt easily to the environment, in which they can develop several complex mechanisms, such as altering membrane permeability, producing enzymes capable of inactivating the action of antibiotics, and modifying the molecular structure; these factors cause resistance [27]. However, rather than replacing antibiotics that are ineffective for MDR bacteria, combining them with bioactive compounds is a new alternative for antimicrobial therapy. The combination of steroidal alkaloids with antibiotics can contribute to bacterial inhibition and the restoration of antibiotic efficacy [28,29], since the main antibacterial mechanisms of alkaloids include alteration in cell membrane permeability, inhibition of bacterial cell wall synthesis, inhibition of bacterial metabolism, efflux pump inhibition, and inhibition of nucleic acid and protein synthesis [30,31].

The analysis of the antibacterial activity of solasodine demonstrated a reduction in the MIC for the standard bacteria *S. aureus*, *P. aeruginosa*, and *E. coli*. This result is in agreement with the antibacterial effect reported by [24,25], where solasodine is shown to have an effect against Gram-positive and Gram-negative bacteria. Therefore, the intrinsic activity of solasodine may be related to the interaction of this alkaloid with the bacterial cell wall structure.

Niño et al. [25] evaluated steroidal alkaloids isolated from *Solanum* and reported that the antibacterial activity of solasodine against *S. aureus* is probably through the inhibition of DNA topoisomerase II; this may be attributed to the fact that alkaloids possess mechanisms for inhibiting bacterial nucleic acid and protein synthesis [31]. In addition, alkaloids and steroids present in other plant species are also associated with antimicrobial activity against Gram-positive bacteria [32], which has a thicker peptidoglycan layer in its cell wall, and the synthesis of this layer involves the participation of a beta-lactam antibiotic [33].

Bibon [26] evaluated solasodine isolated from a species of the genus *Solanum* against *E. coli*; he classified the antibacterial efficacy of this sapogenin as lethal, due to solasodine damaging the DNA of the bacteria, proving that solasodine indeed presents a damaging activity to the bacterial DNA. This is in agreement with the results of this study, in which the Gram-negative bacteria *E. coli* 06 and *P. aeruginosa* 24 also showed significant results.

In the antibiotic potentiation activity where solasodine was associated with antibiotics from the aminoglycoside group, flouroquinolones and a beta-lactam, a reduction in the MIC for the three MDR bacteria tested was shown. The solasodine may be acting to increase the effect of these antibiotics against the strains tested. This effect may be alternatively related to synergism, which is a positive interaction of a compound when associated with an antibiotic, resulting in a reduction in the MIC of the antibiotic used, as well as a decrease in the appearance of resistant bacterial cells [34]. It can also be attributed to the fact that the solasodine alkaloid contributes to the alteration of cell membrane permeability with the uptake of molecules through porin channels [27,31].

The antagonistic effect was also presented by the association of solasodine with the antibiotics ampicillin and norfloxacin, having resulted in the decrease of the effect of both combinations. It can be explained by the fact that synergistic and antagonistic interactions are dependent on the mechanism of bacterial inhibition by the class of antibiotic [34], as well as the possibility of an action arising from the chelation mechanism of the antibiotic when associated with solasodine against the bacteria tested [8]. Additionally, the antagonism often coincides with the increase of ethidium bromide of the bacteria [35].

Therefore, for this work, the MIC of ethidium bromide showed a reduction in conjunction with the inhibitors, thus demonstrating that the bacteria tested have an active efflux pump mechanism from the intracellular to the extracellular medium [36].

4. Materials and Methods

4.1. Substances, Antibiotics and Reagents

Salosodine (Figure 6) was obtained from the species *Solanum paludosum* Moric., was first dissolved in dimethyl sulfoxide (DMSO) and then diluted in sterile water to obtain the concentrations of (1024 μg/mL), (512 μg/ML) and (128 μg/mL).

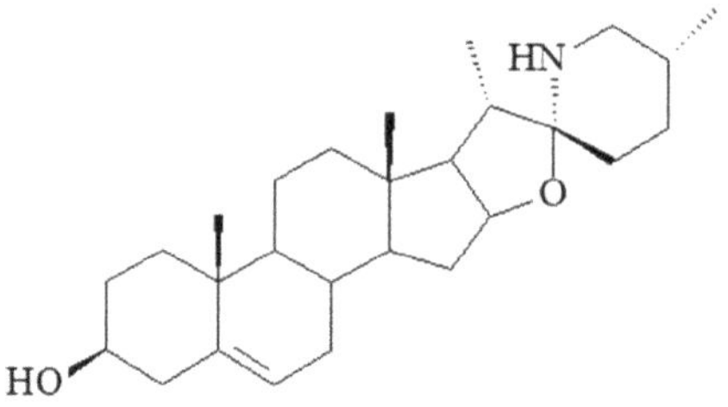

$C_{27}H_{43}NO_2$

PM 413.64

Figure 6. Solasodine.

In the test to evaluate the action-modifying activity of antibiotics, gentamicin, norfloxacin, and ampicillin (Sigma Aldrich Co. Ltd., São Paulo, Brazil) were used at a concentration of (1024 μg/mL).

In the efflux pump inhibitory effect test, the following were used: chlorpromazine (CPZ), carbonyl-m-chlorophenyl hydrazone cyanide (CCCP), and ethidium bromide (Sigma Aldrich Co. Ltd., São Paulo, Brazil) at a concentration of (1024 μg/mL). Norfloxacin and ciprofloxacin antibiotics specific for NorA and MepA were also used at a concentration of (1024 μg/mL). Ethidium bromide and the antibiotics were diluted in sterile water, except for norfloxacin and CPZ which followed the same dilution process as solasodine. CCCP was dissolved in a 1:1 methanol/water solution to obtain the concentration (1024 μg/mL).

4.2. Bacterial Strains

In the test to verify antibacterial activity, the standard strains used were *Escherichia coli* ATCC 25922, *Staphylococcus aureus* ATCC 25923, *Pseudomonas aeruginosa* ATCC 902, and the multidrug-resistant strains *Escherichia coli* 06, *Staphylococcus aureus* 10, *Pseudomonas aeruginosa* 24, obtained through the Microbiology and Molecular Biology Laboratory (LMBM) University of Region of Cariri (URCA, Crato, Brazil). Before performing each test, they were cultured for 24 h at 37 °C in Brain Heart infusion agar (BHI) (Laboratorios cond S.A., Madrid, Spain).

Staphylococcus aureus strains SA-1199B (NorA overexpressed) and SA-K2068 (MepA expressed), provided by Prof. S. Gibbons (University of London, London, England), were used to test the inhibitory effect of the efflux pump. Initially, they were maintained in blood agar (Laboratory Difco Ltda., Franklin Lakes, NJ, USA), but before testing they were cultured for 24 h at 37 °C in Brain Heart infusion agar (BHI) (Laboratorios cond S.A., Madrid, Spain). All culture media were prepared according to the manufacturer's instructions.

4.3. Determination of the Minimum Inhibitory Concentration (MIC)

To determine the minimum inhibitory concentration, the methodology used was CLSI [37] with modifications. Suspensions of the microorganisms were prepared in tubes containing 4 mL of sterile solution (NaCl a 0.9 %). These suspensions were then stirred with the aid of a vortex apparatus and the turbidity was compared and adjusted to that presented by the barium sulfate suspension in the 0.5 tube of the McFarland scale, which corresponds to an inoculum of approximately 10^5 colony forming units/mL (CFU/mL). A solution containing 900 µL of 10% BHI broth and 100 µL of the inoculum was prepared in each eppendorf, and 100 µL of the solution was dispensed into each well of the 96-well microdilution plate. We also added 100 µL of solasodine at the initial concentration of (1024 µg/mL) and it was passed to the other wells through successive dilutions in a 1:1 ratio, ranging from (512 µg/mL) in the first well to (4 µg/mL) in the penultimate well, with the last well reserved for the test control.

To analyze the MIC, after incubating the plates for 24 h at 37 °C, an indicator solution of sodium resazurin in sterile distilled water at a concentration of 0.01% was prepared and 20 µL of this indicator solution was added to each well and the plates were incubated for 1 hour at room temperature. Color change from blue to pink was identified as proof of bacterial growth. The MIC is defined as the lowest concentration capable of inhibiting bacterial growth, which was evidenced by the unchanged blue color.

4.4. Evaluation of Modifying Antibiotic Activity Action

For the antibiotic action modifier activity, solasodine was tested at sub-inhibitory concentrations (MIC/8); the methodology proposed by Coutinho et al. [38] was used.

For the test, eppendorfs containing the solasodine at a concentration of (128 µg/mL) were prepared and suspensions of 10^5 CFU/mL of the microorganisms were deposited together with BHI medium. A control for the antibiotic was prepared with the same amount of bacterial inoculum corresponding to the 10% volume in the eppendorf and 1350 µL of BHI. We dispensed 100 µL of these solutions into the cavities of the 96-well plate. Then 100 µL of each antibiotic was added at a concentration of (1024 µg/mL) following with successive microdilutions at a 1:1 ratio, ranging from (512 µg/mL) in the first well to (0.5 µg/mL) in the penultimate well, with the last well reserved for the test control. The plates were incubated for 24 h at 37 °C and then read using the colorimetric indicator resazurin.

4.5. Evaluation of Efflux Pump Inhibitory Effect Nora E Mepa

In the efflux pump inhibition test, solasodine was tested at sub-inhibitory concentrations (MIC/8). The inhibitors CPZ, CCCP, and ethidium bromide were used to verify the effect. The methodology used was Tintino et al. [39] with modifications.

For analysis of efflux pump inhibition, solasodine was tested at concentrations of 128 µg/mL and 64 µg/mL. Eppendorfs were prepared containing solasodine and the bacterial inoculum for each strain tested, and the final volume was filled with BHI medium. In preparing the control group, 150 µL of the bacterial inoculum was transferred to Eppendorfs and 1350 mL of BHI was added. We distributed 100 µL of these solutions in 96-well microplates following with successive dilutions at a 1:1 ratio, with 100 µL of the antibiotic corresponding to each strain, or ethidium bromide, both at an initial concentration of 1024 µg/mL, with final concentrations varying from 512 µg/mL to 0.5 µg/mL. After 24 h of incubation of the plates, the reading was performed using the colorimetric indicator for microbiological tests.

4.6. Statistical Analysis

The data were analyzed using the two-way ANOVA test, in which the results were expressed as geometric mean and standard deviation using the GraphPad Prisma software version 5.0. It was subjected to the Bonferrini post hoc test, and the results were considered significant when $p < 0.05$, $p < 0.0001$, and not significant when $p > 0.05$.

5. Conclusions

In the evaluation of antibacterial activity by the broth microdilution method, solasodine showed no clinically relevant intrinsic activity for four of the multidrug-resistant bacteria tested, showing a reduction in MIC only for *S. aureus* 1199B. However, as a modifier for the action of antibiotics, significant results against Gram-positive and Gram-negative strains were identified for association with the antibiotics gentamicin and norfloxacin, having a reduction of MIC by 50%. For association with ampicillin, only *S. aureus* 10 showed synergism.

Despite not showing direct antibacterial activity, solasodine demonstrated a synergistic effect when associated with the ciprofloxacin, inducing a reduction in the minimum inhibitory concentration, indicating an MePA efflux pump inhibitory effect.

Therefore, the data obtained in this study are significant and may stimulate future research on the use of sapogenins as associated compounds to enhance drug activity, thus reducing possible mechanisms of bacterial resistance to antibiotics.

Author Contributions: Conceptualization, H.D.M.C. and J.C.A.P.; methodology, A.R.P.d.S., R.S.d.A., M.d.S.C. and N.J.S.A.; software, T.S.d.F., E.O.d.S. and P.A.M.d.F.; validation, H.D.M.C., J.C.A.P., J.F.T. and J.M.B.F.; formal analysis, A.R.P.d.S.; investigation, A.R.P.d.S. and M.d.S.C.; data curation, H.D.M.C. and J.C.A.P.; writing—original draft preparation, A.R.P.d.S.; writing—review and editing, A.R.P.d.S. and J.C.A.P.; supervision, H.D.M.C. All authors have read and agreed to the published version of the manuscript.

Funding: CNPq: CAPES, and FUNCAP for the financial.

Conflicts of Interest: The authors declare no conflict of interest.

References

1. Yu, H.; Han, X.; Quiñones Pérez, D. La humanidad enfrenta un desastre: La resistencia antimicrobiana. *Rev. Habanera Cienc. Médicas* **2021**, *20*, e3850.
2. Brito, G.B.; Trevisan, M. O uso indevido de antibióticos e o eminente risco de resistência bacteriana. *Rev. Artigos. Com.* **2021**, *30*, e7902.
3. Alves, J.W.S. Antibióticos e mecanismos de resistência bacteriana: Uma questão de saúde pública. *Rev. Multidiscip. Saúde* **2021**, *2*, 38.
4. Bhardwaj, K.A.; Mohanty, P. Bacterial efflux pumps involved in multidrug resistance and their inhibitors: Rejuvinating the antimicrobial chemotherapy. *Recent Pat. Antiinfect. Drug Discov.* **2012**, *7*, 73–89. [CrossRef] [PubMed]
5. Schindler, B.D.; Jacinto, P.; Kaatz, G.W. Inhibition of drug efflux pumps in *Staphylococcus aureus*: Current status of potentiating existing antibiotics. *Future Microbiol.* **2013**, *8*, 491–507. [CrossRef] [PubMed]
6. Kaatz, G.W.; Mcaleese, F.; Seo, S.M. Multidrug resistance in *Staphylococcus aureus* due to overexpression of a novel multidrug and toxin extrusion (MATE) transport protein. *Antimicrob. Agents Chemother.* **2005**, *49*, 1857–1864. [CrossRef]
7. Bhattarai, K.; Baral, B. Next-generation microbial drugs developed from microbiome's natural products. *Adv. Genet.* **2021**, *108*, 341–382.
8. Coutinho, H.D.M.; Brito, S.M.O.; Leite, N.F.; Vandesmet, V.C.S.; Oliveira, M.T.A.; Martins, G.M.A.B.; Silva, A.R.P.; Costa, M.S. Avaliação comparativa da modulação de antibióticos, frente às cepas bacterianas de *Escherichia coli*, *Staphylococcus aureus*. *Rev. Cienc. Salud* **2015**, *13*, 345–354. [CrossRef]
9. Tintino, S.R.; Oliveira-Tintino, C.D.; Campina, F.F.; Limaverde, P.W.; Pereira, P.S.; Siqueira-Junior, J.P.; Coutinho, H.D.M.; Quintans-Júnior, L.J.; Silva, T.G.; Leal-Balbino, T.C.; et al. Vitamin K enhances the effect of antibiotics inhibiting the efflux pumps of *Staphylococcus aureus* strains. *Med. Chem. Res.* **2018**, *27*, 261–267. [CrossRef]
10. Jastrzebska, I.; Morzycki, J.M. Some observations on solasodine reactivity. *Steroids* **2017**, *127*, 13–17. [CrossRef]

11. Kumar, R.; Khan, M.I.; Prasad, M. Solasodine: A Perspective on their roles in Health and Disease. *Res. J. Pharm. Technol.* **2019**, *12*, 2571–2576. [CrossRef]
12. Qu, M.; Xeu, P.; Zhang, Q.; Lu, T.; Liu, K.; Hu, B.; Pang, J.; Xiao, Q.; Xu, T.; Wang, Q.; et al. Pharmacokinetics, oral bioavailability and metabolic analysis of solasodine in mice by dried blood spotLC-MS/MS and UHPLC-Q-Exactive MS. *J. Pharm. Biom. Anal.* **2022**, *210*, 114–542. [CrossRef] [PubMed]
13. Hernández, A.E.F.; Oliveira, N.C.C.; Ferrer, M.E.A.; Ferrer, M.T. Estudio de las hojas y frutos de *Solanum monachophyllum* Dunal.(jurubebina) como una fuente de compuestos esteroidales. *Rev. Cuba. Plantas Med.* **2022**, *26*, 968–982.
14. Pendharkar, G.B.; Benerjee, T.; Patil, S.; Dalal, K.S.; Chaudhari, B.L. Biotransformation of Industrially Important Steroid Drug Precursors. In *Industrial Microbiology and Biotechnology*; Springer: Singapore, 2022; pp. 307–333.
15. Chauhan, K.; Sheth, N.; Ranpariya, V.; Parmar, S. Anticonvulsant activity of solasodine isolated from *Solanum sisymbriifolium* fruits in rodents. *Pharm. Biol.* **2011**, *49*, 194–199. [CrossRef]
16. Pandurangan, A.; Khosa, R.L.; Hemalatha, S. Anti-inflammatory activity of an alkaloid from *Solanum trilobatum* on acute and chronic inflammation models. *Nat. Prod. Res.* **2011**, *25*, 1132–1141. [CrossRef]
17. Sharma, T.; Airao, V.; Panara, N.; Vaishnav, D.; Ranpariya, V.; Sheth, N.; Parmar, S. Solasodine protects rat brain against ischemia/reperfusion injury through its antioxidant activity. *Eur. J. Pharmacol.* **2014**, *15*, 40–46. [CrossRef]
18. Ashal, H.A. Biosynthesis and biological activities of in vitro derived solasodine glycoalkaloids from *Solanum laciniatum*. *Nat. Prod. J.* **2017**, *7*, 199–207.
19. Li, Y.; Chang, W.; Zhang, M.; Lou, H. Natural Product Solasodine3-O-ß-glucopyranoside inhibits the virulence factors of *Candida albicans*. *FEMS Yeast Res.* **2015**, *15*, fov060. [CrossRef]
20. Chang, W.; Li, Y.; Zhang, M.; Zheng, S.; Li, Y.; Lou, H. Solasodine-3-O-β-d-glucopyranoside kills *Candida albicans* by disrupting the intracellular vacuole. *Food Chem. Toxicol.* **2017**, *106*, 139–146. [CrossRef]
21. Hameed, A.; Ijaz, S.; Mohammad, I.S.; Muhammad, K.S.; Akhtar, N.; Khan, H.M.S. Aglycone solanidine and solasodine derivatives: A natural approach towards cancer. *Biomed. Pharmacother.* **2017**, *94*, 446–457. [CrossRef]
22. Zhuang, Y.W.; Wu, C.; Zhou, J.Y.; Chen, X.; Wu, J.; Jiang, S.; Peng, H.Y.; Zou, X.; Liu, J.Y.; Wu, D.P.; et al. Solasodine inhibits human colorectal cancer cells through suppression of the AKT/glycogen synthase kinase-3β/β-catenin pathway. *Câncer Sci.* **2017**, *108*, 2248–2264. [CrossRef] [PubMed]
23. Fan, Y.; Li, Z.; Wu, L.; Lin, F.; Shao, J.; Ma, X.; Yao, Y.; Zhuang, W.; Wang, Y. Solasodine, Isolated from *Solanum sisymbriifolium* Fruits, has a Potent Anti-Tumor Activity Against Pancreatic Cancer. *Drug Des. Devel. Ther.* **2021**, *13*, 1509–1519. [CrossRef] [PubMed]
24. Almazinia, M.A.I.; Abbas, H.G.; Abdul-Amer, A. Antibacterial activity of the solasodine of *Solanum nigrum* against bacterial isolates from the wounds. *Basrah J. Vet. Res.* **2009**, *8*, 137–147.
25. Niño, J.; Correa, Y.M.; Mosqueram, O.M. Biological activities of steroidal alkaloids isolated from *Solanum leucocarpum*. *Pharm. Biol.* **2009**, *47*, 255–259. [CrossRef]
26. Bibon, M.B. Antibacterial in-vitro evaluation of phenotypically screened solasodine from *Solanum nigrum* Linn. against enterohemorrhagic *Escherichia coli* (0157:H7). *Int. J. Sci. Res. Multidiscip. Stud.* **2021**, *7*, 43–48.
27. Tortora, G.J.; Funke, B.R.; Case, C.L. *Microbiologia*, 10th ed.; Artmed: Porto Alegre, Brazil, 2012; pp. 1–939.
28. Soltani, R.; Fazeli, H.; Najafi, R.B.; Jelokhanian, A. Evaluation of the synergistic effect of tomatidine with several antibiotics against standard and clinical isolates of *Staphylococcus aureus*, *Enterococcus faecalis*, *Pseudomonas aeruginosa* and *Escherichia coli*. *Iran. J. Pharm. Res.* **2017**, *16*, 290–296.
29. Siriyong, T.; Voravuthikunchai, S.P.; Coote, P.J. Steroidal alkaloids and conessine from the medicinal plant *Holarrhena antidysenterica* restore antibiotic efficacy in a *Galleria mellonella* model of multidrug-resistant *Pseudomonas aeruginosa* infection. *BMC Comp. Altern. Med.* **2018**, *18*, 285. [CrossRef]
30. Casciaro, B.; Mangiardi, L.; Cappiello, F.; Romeo, I.; Loffredo, M.R.; Iazzetti, A.; Calcaterra, A.; Goggiamani, A.; Ghirga, F.; Mangoni, M.L.; et al. Naturally-Occurring Alkaloids of Plant Origin as Potential Antimicrobials against Antibiotic-Resistant Infections. *Molecules* **2020**, *25*, 3619. [CrossRef]
31. Yan, Y.; Li, X.; Zhang, C.; Lv, L.; Gao, B.; Li, M. Research Progress on Antibacterial Activities and Mechanisms of Natural Alkaloids: A Review. *Antibiotics* **2021**, *10*, 318. [CrossRef]
32. Braquehais, I.D.; Vasconcelos, F.R.; Ribeiro, A.R.C.; Da Silva, A.R.A.; Franca, M.G.A.; De Lima, D.R.; De Paiva, C.F.; Guedes, M.I.F.; Magalhães, F.E.A. Toxicological, antibacterial, and phytochemical preliminary study of the ethanolic extract of *Jatropha mollissima* (Pohl) Baill (pinhão-bravo, *Euphorbiaceae*) leaves, collected in Tauá, Ceará, Northeastern Brazil. *Rev. Bras. De Plantas Med.* **2016**, *18*, 582–587. [CrossRef]
33. Araújo, L.L.C.; Azevedo, F.H.C. Study of the beta-lactamase enzyme and its relationship with antibiotic resistance. *Res. Soc. Dev.* **2020**, *9*, e663974594. [CrossRef]
34. Guliu, C.; Green, S.I.; Min, L.; Clark, J.R.; Salazar, C.K.; Terwilliger, A.L.; Kaplan, H.B.; Trautner, R.F.; Ramig, R.F.; Maresso, A.w. Phage-antibiotic synergy is driven by a unique combination of antibacterial mechanism of action and stoichiometry. *MBio* **2020**, *11*, e01462-20.
35. Shrestha, P.; Ni, J.; Wong, T.Y. Synergistic and antagonistic interactions of triclosan with various antibiotics in bacteria. *J. Environ. Sci. Health Toxicol. Carcinog.* **2020**, *38*, 187–203. [CrossRef]

36. Coêlho, M.L.; Ferreira, J.H.L.; Siqueira-Júnior, J.P.; Kaatz, W.G.; Barreto, H.M.; Cavalcante, A.A.C.M. Inhibition of the NorA multi-drug transporter by oxygenated monoterpenes. *Microb. Pathog.* **2016**, *99*, 173–177. [CrossRef] [PubMed]
37. CLSI. Performance Standards for Antimicrobial Susceptibility Testing; Twenty Fifth Informational Supplement. In *CLSI Document M100-S25*; Clinical and Laboratory Standards Institute: Wayne, PA, USA, 2015; Volume 35.
38. Coutinho, H.D.M.; Falcão-Silva, V.S.; Gonçalves, F.G. Pulmonary bacterial pathogens in cystic fibrosis patients and antibiotic therapy: A tool for the health workers. *Int. Arch. Med.* **2008**, *1*, 24. [CrossRef]
39. Tintino, S.R.; Oliveira-Tintino, C.D.; Campina, F.F.; Silva, R.L.P.; Costa, M.S.; Menezes, I.R.A.; Calixto-Júnior, J.T.; Siqueira-Junior, J.P.; Coutinho, H.D.M.; Leal-Balbino, T.C.; et al. Evaluation of the tannic acid inhibitory effect against the NorA efflux pump of *Staphylococcus aureus*. *Microb. Pathog.* **2016**, *97*, 9–13. [CrossRef] [PubMed]

antibiotics

Article

In Vitro Biological Activity of Natural Products from the Endophytic Fungus *Paraboeremia selaginellae* against *Toxoplasma gondii*

Flaminia Mazzone [1,†], Viktor E. Simons [2,†], Lasse van Geelen [2], Marian Frank [2], Attila Mándi [3], Tibor Kurtán [3], Klaus Pfeffer [1,*] and Rainer Kalscheuer [2,*]

1 Institute of Medical Microbiology and Hospital Hygiene, Heinrich Heine University, 40225 Duesseldorf, Germany
2 Institute of Pharmaceutical Biology and Biotechnology, Heinrich Heine University, 40225 Duesseldorf, Germany
3 Department of Organic Chemistry, University of Debrecen, 4002 Debrecen, Hungary
* Correspondence: klaus.pfeffer@hhu.de (K.P.); rainer.kalscheuer@hhu.de (R.K.); Tel.: +49-211-8112459 (K.P.); +49-211-8114180 (R.K.)
† These authors contributed equally to this work.

Abstract: *Toxoplasma gondii* is an apicomplexan pathogen able to infect a wide range of warm-blooded animals, including humans, leading to toxoplasmosis. Current treatments for toxoplasmosis are associated with severe side-effects and a lack efficacy to eradicate chronic infection. Thus, there is an urgent need for developing novel, highly efficient agents against toxoplasmosis with low toxicity. For decades, natural products have been a useful source of novel bioactive compounds for the treatment of infectious pathogens. In the present study, we isolated eight natural products from the crude extract of the endophytic fungus *Paraboeremia selaginellae* obtained from the leaves of the plant *Philodendron monstera*. The natural products were tested for inhibiting *Toxoplasma gondii* proliferation, and their cytotoxicity was evaluated in different human cell lines. Six natural products showed antitoxoplasma activity with low or no cytotoxicity in human cell lines. Together, these findings indicate that biphenyl ethers, bioxanthracenes, and 5*S*,6*S*-phomalactone from *P. selaginellae* are potential candidates for novel anti-toxoplasma drugs.

Keywords: *Toxoplasma gondii*; *Paraboeremia selaginellae*; endophytic fungi; natural products; bioactivity; biphenyl ether; bioxanthracene; phomalactone

Citation: Mazzone, F.; Simons, V.E.; van Geelen, L.; Frank, M.; Mándi, A.; Kurtán, T.; Pfeffer, K.; Kalscheuer, R. In Vitro Biological Activity of Natural Products from the Endophytic Fungus *Paraboeremia selaginellae* against *Toxoplasma gondii*. *Antibiotics* **2022**, *11*, 1176. https://doi.org/10.3390/antibiotics11091176

Academic Editors: Valério Monteiro-Neto and Elizabeth S. Fernandes

Received: 12 August 2022
Accepted: 29 August 2022
Published: 31 August 2022

1. Introduction

Toxoplasma gondii is an obligate intracellular protozoan parasite member of the phylum Apicomplexa, which includes known human pathogens such as *Plasmodium* sp., *Eimeria* sp., *Neospora*, *Babesia*, *Theileria*, and *Cryptosporidium* spp., with which it shares significant biological similarities [1]. Beyond these organisms, the study of *T. gondii* has experimental advantages since its basic biology and the methodology for the genetic manipulation and quantification of its different stages are well established. Thus, *T. gondii* is considered a major model for the study of apicomplexan biology and for anti-apicomplexan drug target validation [2]. *T. gondii* infections are among the most common human zoonoses, leading to toxoplasmosis disease [3]. *T. gondii* is considered one of the world's most successful parasites due its ability to infect a wide range of warm-blooded vertebrate intermediate hosts [4]. *T. gondii* is estimated to chronically infect one-third of the world's human population and is acquired mainly through two ways: by ingesting oocysts shed from feline hosts (the definitive hosts) in contaminated food or water and by the consumption of raw or undercooked meat containing viable tissue cysts [5]. Waterborne and food-borne outbreaks of toxoplasmosis have been reported from countries with diverse cultural, social, and ethnic backgrounds [6]. In immunocompetent individuals, infection with *T. gondii* is usually asymptomatic or has a subclinical course with mild symptoms. In contrast, immunocompromised (i.e., acquired immune deficiency syndrome (AIDS), organ transplant or cancer) patients can develop the disease, leading to life-threatening cerebral and ocular toxoplasmosis due to a reactivation of the latent infection. Additionally, primary infection in pregnant women may result in fetal death, spontaneous abortion, and birth defects [7–9]. Although many gaps have been filled in the epidemiological, diagnostic, and biological fields to understand of the interaction of the parasite with the host, little progress has been made in drug discovery for the treatment of toxoplasmosis.

Current treatments of acute toxoplasmosis are largely limited to anti-folate therapy. Pyrimethamine and sulfadiazine, the current gold-standards for the treatment of toxoplasmosis, can suppress the parasite growth in the active stage of the infection by targeting the tachyzoite stage, but they have no effect in the bradyzoites stage. Additionally, they have been found to have high rates of toxic side effects, leading to discontinuation of therapy. Thus, there is an urgent need to identify novel potent candidates that would be well-tolerated to eradicate latency as well as to treat the acute infection [10,11]. Natural products profoundly impact the history of drug discovery, especially in the research of novel anti-cancer, anti-bacterial, and anti-parasitic treatments. Nature continues to provide diverse and unique chemical sources of bioactive lead compounds that inspire novel drug discoveries [12]. The antiparasitic bioactivity of natural products from various sources, especially plant-derived secondary metabolites, has been deeply investigated in in vitro and in vivo studies [13]. Many fungal metabolites have also been reported to exhibit antimicrobial properties against parasitic pathogens. However, most of these studies focused on bioactivity against *Plasmodium falciparum*, whereas there is a scarcity of investigations to explore the potential of fungi as a source of novel anti-toxoplasma agents [14].

In this study, we extracted and purified eight natural products from the crude extract of *Paraboeremia selaginellae*, an endophytic fungus isolated from the ornamental plant *Philodendron monstera*. Isolated compounds were structurally characterized and evaluated for their anti-toxoplasma activities. Biphenyl ethers, bioxanthracenes, and phomalactone showed substantial activity against *T. gondii* proliferation. Therefore, we suggest these compounds as promising candidates for novel anti-parasitic therapies.

2. Results

2.1. Isolation of Compounds from Paraboeremia selaginella

We isolated an endophytic fungus from fresh surface-sterilized leaves of the ornamental plant *Philodendron monstera*. The isolated strain was identified as *Paraboeremia selaginella* by the internal transcribed spacer (ITS) sequence with 99.56% identity in comparison with the ITS database of the National Center for Biotechnology Information. From the crude extract of a culture of *Paraboeremia selaginella* grown on solid rice medium, eight compounds were isolated by chromatographic methods and structurally elucidated by complementary spectroscopic analyses (Figure 1). All eight compounds have previously been reported from other sources but are reported here for the first time as natural products occurring in the genus *Paraboeremia*.

Different stereoisomers of phomalactone (**F**) were isolated and reported previously from various sources [15–18], and some papers did not specify the absolute configuration [19,20], while others assigned the (5*R*,6*R*) absolute configuration to the large positive specific rotation [21], which was opposite to previous studies [15–18,22]. In order to determine the absolute configuration of phomalactone independently and unambiguously, we performed TDDFT-ECD, TDDFT-OR, and DFT-VCD studies, which consistently confirmed the (+)-*cis*-(5*S*,6*S*) absolute configuration (see Supplementary Materials) [23,24]. Comparisons of the experimental and computed VCD spectra of *cis*-(5*S*,6*S*) are shown in Figure 2, which produced good agreement. Other computational results are shown in the Supplementary Materials.

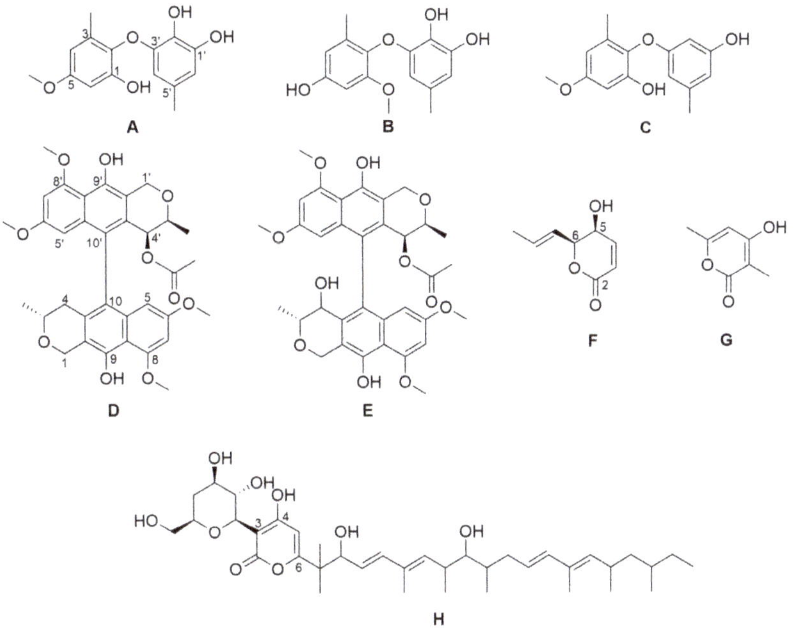

Figure 1. Chemical structures of the isolated compounds. NK-A 17e233 (**A**); 1,2-benzenediol, 3-(4-hydroxy-2-methoxy-6-methylphenoxy)-5-methyl-(ACI) (**B**); cyperin (**C**); ES-242-1 (**D**); ES-242-3 (**E**); 5*S*,6*S*-phomalactone (**F**); methyltriaceticlactone (**G**); S 39163/F-1 (**H**).

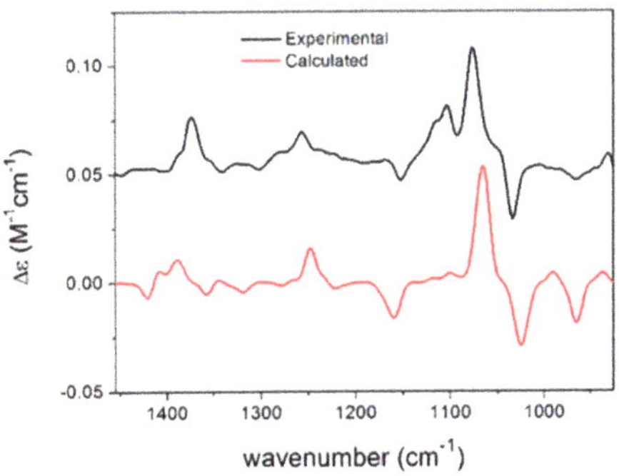

Figure 2. Comparison of the experimental VCD spectrum of **F** measured in $CDCl_3$ and the calculated VCD spectrum of *cis*-(5*S*,6*S*)-**F** computed at the B3LYP/TZVP PCM/$CHCl_3$ level for the eight lowest-energy conformers gained from the DFT optimization performed at the same level.

2.2. Anti-*T. gondii* Activity

The eight natural products isolated from *P. selaginellae* were tested for anti-*T. gondii* activity. Interestingly, **A**–**F** showed activity against *T. gondii* growth, with IC_{50} values of 5.75, 22.16, 27.22, 7.38, 17.99, and 5.13 μM, respectively (Table 1 and Figure 3). Therefore, we further explored the in vitro cytotoxicity of the natural compounds in different human cell lines.

Table 1. In vitro activity (IC_{50} values) of the natural compounds (**A**–**H**) from *P. selaginellae* against the *T. gondii* strain ME49. All experiments were conducted in triplicate.

Compound	IC_{50} (μM)
NK-A 17e233 (**A**)	5.75
1,2-benzenediol, 3-(4-hydroxy-2-methoxy-6-methylphenoxy)-5-methyl-(ACI) (**B**)	22.16
cyperin (**C**)	27.22
ES-242-1 (**D**)	7.38
ES-242-3 (**E**)	17.99
5*S*,6*S*-phomalactone (**F**)	5.13
methyltriaceticlactone (**G**)	Not active
S 39163/F-1 (**H**)	Not active
pyrimethamine	0.06

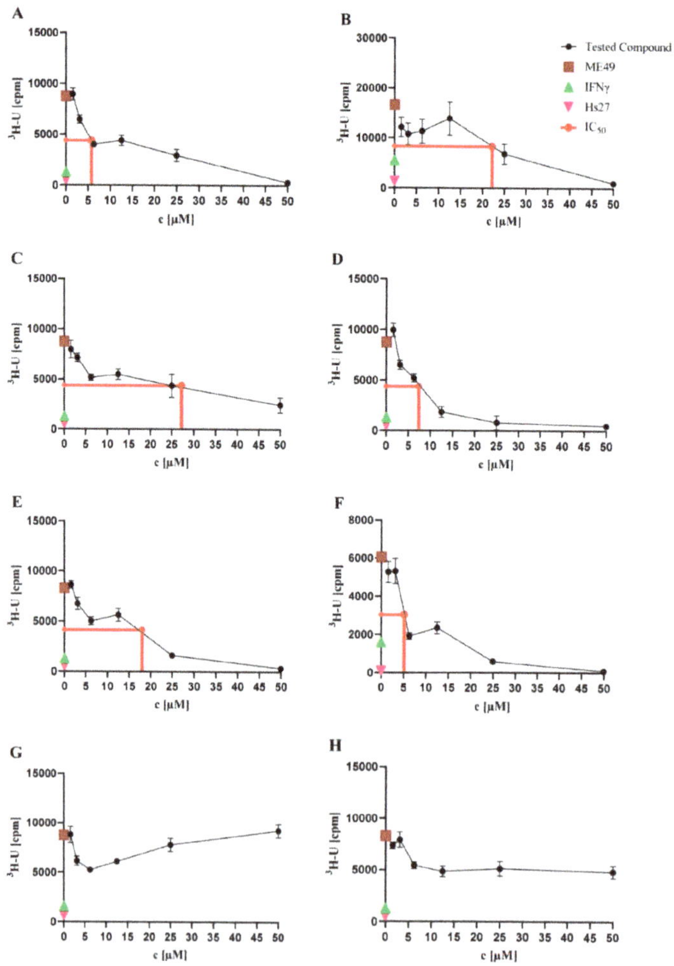

Figure 3. Toxoplasma proliferation assays. Toxoplasma proliferation assays were performed to investigate the activity of the natural products against *T. gondii* strain ME49. Hs27 cells were cultured in a monolayer in 96-well plates and infected with *T. gondii* (3×10^4). Cultures were treated with the

natural products at the concentration range of 0.56–50.00 μM for 48 h at 37 °C. Afterwards, the cultures were labelled with ^{3}H-U (5 mCi, diluted 1:30) for 28–30 h at 37 °C. Based on the incorporation of ^{3}H-U into the parasite nucleic acid, the parasite growth was quantified. As controls, uninfected Hs27 cells without treatment (pink triangles) and IFNγ pre-stimulated infected Hs27 cells (green triangles) were used, while untreated *T. gondii*-infected Hs27 cells (red squares) served as a negative control. Values shown in (**A–H**) represent the means of three independent experiments each done in duplicate (n = 6) ± SEM. The mean of the IC_{50} values (red line) of each compound is shown. Activity of NK-A 17e233 (**A**); 1,2-benzenediol, 3-(4-hydroxy-2-methoxy-6-methylphenoxy)-5-methyl-(ACI) (**B**); cyperin (**C**); ES-242-1 (**D**); ES-242-3 (**E**); 5*S*,6*S*-phomalactone (**F**); methyltriaceticlactone (**G**); S 39163/F-1 (**H**).

2.3. Cytotoxicity Assays

First, we evaluated the cytotoxicity of compounds **A** to **F** in an MTT assay against Hs27 human fibroblasts (same cell type used for the *T. gondii* proliferation assay). The results of the MTT assay are shown in Figure 4 and Table 2. **A–E** showed no cytotoxicity at 100 μM against Hs27 cells. Only **F** showed moderate cytotoxicity with a cytotoxic concentration CC_{50} = 81 ± 2.16 μM.

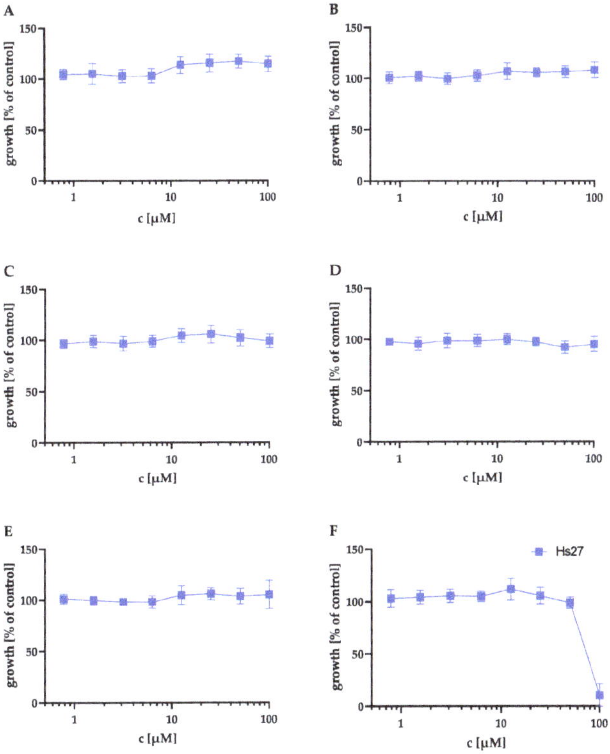

Figure 4. Effect of the natural products on the metabolic activity of Hs27 cells via MTT assay. Hs27 cells were plated in 96-well plates and grown to confluence prior to incubation at 37 °C for 24 h with the natural products in the concentration range of 0.56–100.00 μM. The cultures were incubated with 10 μL of the 12 mM MTT stock solution for approximately 4 h. Afterwards, 100 μL of SDS dissolved in HCl was added to each well and incubated again for 4 h at 37 °C. Finally, the absorbance was measured at 570 nm by spectrophotometry. Values shown in (**A–F**) represent the means of three independent experiments each done in duplicate (n = 6) ± SEM. Cytotoxicity in Hs27 cells of NK-A 17e233 (**A**); 1,2-benzenediol, 3-(4-hydroxy-2-methoxy-6-methylphenoxy)-5-methyl-(ACI) (**B**); cyperin (**C**); ES-242-1 (**D**); ES-242-3 (**E**); 5*S*,6*S*-phomalactone (**F**).

Table 2. In vitro cytotoxicity (CC_{50} values) of the natural compounds (**A–F**) from *P. selaginellae* against human fibroblasts Hs27. Concentration >100 µM indicates no activity in the experimental setup. All experiments were conducted in triplicate.

Compound	CC_{50} (µM)
A	>100
B	>100
C	>100
D	>100
E	>100
F	81
Pyrimethamine	44

These compounds also were tested against the THP-1, Huh-7, and Hek 293 cell lines in a Resazurin assay. The mean IC_{50} values of the Resazurin assay are shown in Table 3. While compounds **A–C** showed no cytotoxic effect in concentrations < 100 µM against any of the tested cell lines, **D** had only a weak cytotoxic activity against the Hek 293 cell line with an IC_{50} of 93.8 µM. **E** showed moderate cytotoxic activity against all of the three tested cell lines and thus was the most cytotoxic of the tested compounds. **F** showed no or weak cytotoxic effects against the Huh-7 and Hek-293 cell lines. The cytotoxic effect against the THP-1 cell line was higher, with an IC_{50} of 24.3 µM. The graphs for the Resazurin assay are shown in Figure 5.

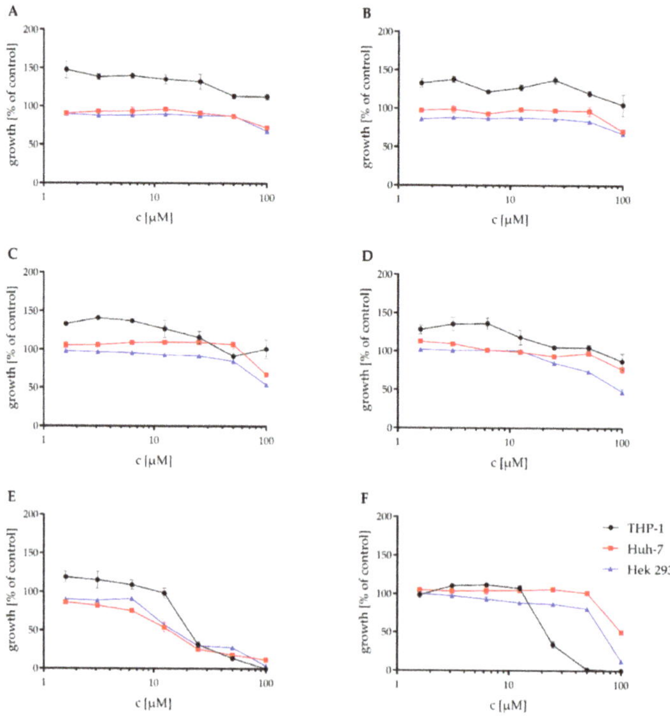

Figure 5. Effect of the natural products on the viability of THP-1, Huh-7, and HEK-293 cells. Cytotoxic effect of NK-A17e233 (**A**); 1,2-benzenediol,3-(4-hydroxy-2-methoxy-6-methylphenoxy)-5-methyl-(ACI) (**B**); cyperin (**C**); ES-242-1 (**D**); ES-242-3 (**E**); 5*S*,6*S*-phomalactone (**F**) against the human cell lines THP-1, Huh-7, and HEK-293 as determined by resazurin assay. 100% growth control DMSO, 0% growth control cycloheximide. Values represent the means of triplicates ± SEM.

Table 3. Mean IC_{50} values of compounds **A–F** against human cell lines THP-1, Huh-7, and Hek293. All concentrations are shown in μM. Concentration >100 μM indicates no activity in the experimental setup. All experiments were conducted in triplicate. The IC_{50} values were calculated using GraphPad Prism 7.

Compound	Mean IC_{50} [μM]		
	THP1	**Huh-7**	**HEK-293**
A	>100	>100	>100
B	>100	>100	>100
C	>100	>100	>100
D	>100	>100	93.8
E	21.9	13	16.95
F	24.3	100	66.9

2.4. Determination of Anti-Bacterial Activity

In our ongoing research for antibacterial and particularly antitubercular compounds, **A** to **H** were also tested in a minimal inhibitory concentration assay against *S. aureus* ATCC 700699, *P. aeruginosa* ATCC 87110, and *M. tuberculosis* H37Rv. Compounds **A** to **H** had no inhibitory effect on *S. aureus* ATCC 700699 and *P. aeruginosa* ATCC 87110, except for compound **F**, which showed a weak inhibitory effect on *P. aeruginosa* ATCC 87110 with an MIC_{90} of 100 μM. Compounds **A**, **B**, **C**, **F**, **G**, and **H** had no inhibitory effect on the growth of *M. tuberculosis*, and compounds **D** and **E** showed a weak inhibitory effect with an MIC_{90} of 50 and 100 μM, respectively. The results are shown in Table 4, highlighting that compounds **A** to **F** had a specific anti-toxoplasma effect and were devoid of broad, unspecific antimicrobial activity.

Table 4. MIC_{90} against *S. aureus* ATCC 700699, *P. aeruginosa* ATCC 87110, and *M. tuberculosis* H37Rv. All concentrations are shown in μM. Concentration >100 μM indicates no activity in the experimental setup. All experiments were conducted in triplicate.

Compound	MIC_{90} [μM]		
	***S. aureus* ATCC 700699**	***P. aeruginosa* ATCC 87110**	***M. tuberculosis* H37Rv**
A	>100	>100	>100
B	>100	>100	>100
C	>100	>100	>100
D	>100	>100	50
E	>100	>100	100
F	>100	100	>100
G	>100	>100	>100
H	>100	>100	>100

3. Discussion

Natural products have played an important role in the history of drug discovery for infectious disease. In the quest for new anti-*T. gondii* drugs, natural products have been proven to exhibit high potential for the discovery and development of new lead compounds with strong anti-*T. gondii* activity [25,26]. In this study, we isolated eight natural products from the crude extract of the endophytic fungus *P. selaginellae*. A previous report on the inhibitory activity of one of these compounds (phomalactone, **F**) against the apicomplexan parasite *Plasmodium falciparum* with an IC_{50} of 84.32 μM [27] prompted us to test the natural products for anti-*T. gondii* activity. Interestingly, six compounds showed activity against *T. gondii* proliferation with no or low cytotoxicity in different human cell lines and no or low antibacterial activity against a gram-positive, a gram-negative, and a mycobacterial representative, revealing reasonable anti-*T. gondii* specificity and promising therapeutic windows. These results establish diphenyl ethers, bioxanthracenes, and lactones from *P. selaginellae* as potential candidates for further preclinical development of novel anti-toxoplasma therapeutics.

Some of the isolated compounds share similar structural elements, which give insights into a structure activity relationship of the natural products against the tested *T. gondii* strain ME49. Compounds **A**, **B**, and **C** are biphenyl ether derivatives that differ either in the position of the methoxy group or in the number of substituted hydroxyl groups. The most potent of these compounds is **A** (IC_{50} = 5.75 μM), followed by **B** (IC_{50} = 19.35 μM) and **C** (IC_{50} = 27.22 μM). While **A** only differs from **B** by a switch in the position of the methoxylated hydroxyl group from position 2 to 4, it differs from **C** only by an additional hydroxyl group in position 2′, which it shares with **B**. Because of the higher potency of **A** in the toxoplasma proliferation assay compared to **B** and **C**, the position of the methoxy group in 4 and the amount and position of hydroxyl groups in 2′ and 3′ both are likely to have an influence on the antitoxoplasma activity of these derivatives. This suggestion, nevertheless, needs further experimental evidence. Furthermore, diphenyl ethers **A**, **B**, and **C** are structurally related to triclosan, a well-known broad spectrum antifungal and antibacterial agent targeting lipid synthesis [28]. It has been shown that triclosan also inhibits the growth of apicomplexans by inhibition of the enoyl reductase ENR (FabI) enzyme, the second reductive step in the type II fatty acid biosynthesis pathway. Nevertheless, due the poor solubility of triclosan, there is considerable interest in finding novel potent triclosan analogs with improved properties such as solubility, activity, and toxicity [29,30]. The mechanism of action of **A**, **B**, and **C** may be similar to that of triclosan, but further studies are necessary to explore and confirm their mode of action and cellular target. Furthermore, in vitro and in vivo pharmacokinetic characterization is needed to reveal whether any of the compounds reported here has superior properties compared to triclosan.

Compounds **D** and **E** represent bioxanthracenes belonging to the ES-242 class and share the same structure, differing only in position 4′ by the hydroxyl group that is present only in **E**. The IC_{50} values in the toxoplasma proliferation assay were 7.38 μM and 17.99 μM for **D** and **E**, respectively, suggesting a reduction in the antitoxoplasma activity if position 4′ is substituted by a hydroxyl group. The bioxanthracenes **D** and **E** were previously isolated from *Verticillium* spp. and are well-known to act as *N*-methyl-*D*-aspartate receptor antagonists [31]. Both compounds were also found to be active against the apicomplexan parasite *P. falciparum* with IC_{50} values of 8.44 and 13.22 μM, respectively [32]. Interestingly, the activities of **D** and **E** against *T. gondii* in this study were comparable to their reported activity in *P. falciparum* with IC_{50} values of 7.38 μM and 17.99 μM (see Table 1). Nevertheless, the mechanism of action of **D** and **E** on apicomplexans is still unknown and is probably independent from their activity as NMDA receptor antagonists [32].

Compounds **F** and **G** are small δ-lactonic molecules; 5*S*,6*S*-phomalactone (**F**) differs from methyltriaceticlactone (**G**) in the length of the sidechain in position 6, the hydroxyl group in position 5, and in the absence of the methyl group that is present in methyltriaceticlactone in position 2. Interestingly, antitoxoplasmal activity was observed for **F**, but not for **G**, suggesting that one or more of these structural differences and not only the presence of the δ-lactonic base structure plays a crucial role in the bioactivity against *T. gondii*. Phomalactone (**F**) is a frequent fungal metabolite and was first isolated from the plant-pathogenic fungus *Nigrospora* sp. [16]. It has a wide range of activities such as antifungal, immunomodulating, insecticide, nematocidal, and phytotoxic activity [15,19,33–35]. In addition, it has been found to be active against the apicomplexan parasite *P. falciparum*, with an IC_{50} of 84.32 μM [27]. In the present study, we tested **F** for inhibition of *T. gondii* proliferation and, interestingly, it showed a more potent activity with an IC_{50} of 5.13 μM (see Table 1). No target or mode of action has been suggested for phomalactone in *P. falciparum*, and the target of this compound in *T. gondii* also remains elusive and has to be determined in the future. Importantly, the newly identified natural products with inhibitory activity against *T. gondii* showed very little in vitro toxicity and should be evaluated in in vivo infection model systems in the future. In general, this study highlights the potential of endophytic fungi as a promising source for novel antitoxoplasma compounds.

4. Materials and Methods

4.1. General Experimental Procedures

Optical rotations were measured on a Jasco P 2000 polarimeter (Jasco, Pfungstadt, Germany). UV-spectra were obtained by the use of a Dionex P580 system in combination with a diode array detector (UVD340S) and an Eurosphere 10 C18 column (125 mm × 4 mm). ECD spectra were measured on a JASCO J-810 spectropolarimeter. VCD spectra were recorded on a BioTools Chiral-IR-2X at a resolution of 4 cm^{-1} under ambient temperature for 18 × 3000 scans. Samples were dissolved in $CDCl_3$, and the solution was placed in a 100 μm BaF_2 cell. 1D and 2D NMR spectra were recorded on a Bruker Avance III (^1H, 600 MHz; ^{13}C 150 MHz) spectrometer. Mass spectra were measured on a Finnigan LCQ Deca (Thermo Quest, Egelsbach, Germany) mass spectrometer and for HRESIMS, on a UHR-QTOF maXis 4G (Bruker Daltonics, Bremen, Germany) mass spectrometer. Semipreperative HPLC was performed on a Lachrom-Merck Hitachi system (pump L7100, UV-detector L7400, Eurospher 100 C18 column 300 mm × 8 mm, Knauer, Everswinkel, Germany). VLC and non-vacuum-column chromatography were accomplished using Macherey Nagel silica gel 60M (0.04–0.063 mm). Precoated TLC silica gel 60 F254 plates (Merck, Darmstadt, Germany) were used for tracking separation using detection under UV light at 254 and 365 nm wavelengths or spraying anisaldehyde–sulfuric acid reagent. Sephadex LH20 (GE Healthcare Bio.Sciences AB, Uppsala, Sweden) was used as a stationary phase for column chromatography. The measurement of optical rotations was accomplished by using spectral grade solvents.

4.2. Fungal Material

The fungus was obtained from the leaves of the plant *Philodendron monstera* as an endophyte. A single leaf was surface sterilized by soaking it with 70% ethanol for 30 s and letting it dry under sterile conditions. With a heat-sterilized scalpel, the leaf was cut into pieces and put onto a YPD agar plate, which was enriched with 100 mg/L chloramphenicol to suppress bacterial growth. After seven days of incubation at room temperature, distinct fungal growth was observed on the plate. A 1 cm^2 piece of the fungus was cut out of the agar medium using a heat-sterilized scalpel under sterile conditions and was transferred onto a new sterile YPD agar plate to isolate a pure organism. The isolated strain was identified as *Paraboeremia selaginella* by the internal transcribed spacer (ITS) sequence with 99.56% identity in comparison with the ITS database of the National Center for Biotechnology Information (GenBank Accession ON231784).

4.3. Fermentation and Extraction

The fungus was fermented on solid rice medium. Ten Erlenmeyer flasks were used; 100 g of rice and 100 mL of demineralized water were added to each flask and autoclaved at 121 °C for 15 min. Under sterile conditions, 1 cm^2 of fungal material was cut out of an agar plate using a sterile scalpel and transferred onto the autoclaved rice medium. The fungus was grown for 4 weeks under static conditions at room temperature. Each flask was soaked with 250 mL of ethylacetate for at least 12 h. The rice medium was then cut into small pieces and shaken for 8 h at 150 rpm. The liquid crude extract was filtrated into round flasks and evaporated using a rotary evaporator to yield 14.66 g of dry crude extract.

4.4. Isolation

The crude extract (14.66 g) obtained from the fermentation was separated using vacuum liquid chromatography with silica gel as a stationary phase. A step gradient from 100% hexane to 100% ethylacetate followed by a step gradient from 100% dichloromethane to 100% methanol gave 18 fractions (V1–V18). Two fractions (V4 and V6) were chosen based on initial bioactivity observed against *Candida albicans*. However, this bioactivity was lost during the purification process. Fraction V4 (200.7 mg) was further separated using a Sephadex LH20 column with MeOH as eluent to give five subfractions (V4-S1–S5). Fraction V4S3 (47.3 mg) was subjected to semipreparative HPLC using a MeOH-H_2O step gradient from 50% to 80% MeOH followed by a washing step with 100% MeOH to yield **A** (20.7 mg), **B** (2.8 mg), and **C** (7.7 mg). Fraction V6 (1010 mg) was purified using a Sephadex LH20 column with CH_2Cl_2 and MeOH (50/50) as eluent to yield six subfractions (V6-S1–S6). Subfraction S2 (72.0 mg) was purified using a silica column with 40% hexane and 60% ethylacetate to elute **D** (25.0 mg) and **E** (8.6 mg) as pure compounds. Subfraction V6-S4 (516 mg) was further purified by using a Sephadex LH20 column with MeOH as eluent to yield five subfractions (V6S4-S1–S5). Subfraction V6S4S2 (496 mg) was subjected to a silica column with a mixture of CH_2Cl_2 and MeOH (95/5) as eluent to give four subfractions (V6S4S2-K1–K4). Silica subfraction V6S4S2K2 (47 mg) was then purified by semipreparative HPLC using a MeOH-H_2O step gradient from 10% to 30% MeOH followed by a washing step with 100% MeOH to yield **F** (22.2 mg) and **G** (4.5 mg). Fraction V12 (744.1 mg) was separated using a Sephadex LH20 column with 50% MeOH and 50% CH_2Cl_2 to yield six subfractions (V12-S1–S6). Subfraction S2 (180.3 mg) was then further separated using a silica column with 10% MeOH and 90% CH_2Cl_2 as eluent to give seven subfractions (V12S2-K1–K7). Silica subfraction K7 (56.1 mg) was subjected to semipreparative HPLC using a step gradient from 70% to 100% MeOH to yield **H** (5.0 mg).

NK-A 17e233 (A): Brown oil; UV (MeOH) λ_{max} 220.0, 234.3, 279.7 nm; ^{1}H NMR (DMSO-d6) see Supplementary Materials Figure S1; HRESIMS *m/z* 277.1075 $[M + H]^+$ (calcd. for $C_{15}H_{17}O_5$ 277.1071 *m/z*).

3-(4-Hydroxy-2-methoxy-6-methylphenoxy)-5-methylbenzene-1,2-diol (B): Brown oil; UV (MeOH) λ_{max} 211.7, 286.3 nm; ^{1}H NMR ($CDCl_3$), see Supplementary Materials Figure S6; HRESIMS *m/z* 277.1065 $[M + H]^+$ (calcd. for $C_{15}H_{17}O_5$ 277.1071 *m/z*).

Cyperin (C): Brown oil; UV (MeOH) λ_{max} 212.1, 279.8 nm; ^{1}H NMR ($CDCl_3$), see Supplementary Materials Figure S10; HRESIMS *m/z* 261.1126 $[M + H]^+$ (calcd. for $C_{15}H_{17}O_4$ 261.1121 *m/z*).

ES-242-1 (D): Brown amorphous powder; $[\alpha]^{24}_D$ *+18* (c 1.0, MeOH); UV (MeOH) λ_{max} 239.0, 309.8, 345.8 nm; ^{1}H NMR ($CDCl_3$) and ^{13}C NMR ($CDCl_3$), see Supplementary Materials Figures S14 and S15; HRESIMS *m/z* 622.2644 $[M + NH_4]^+$ (calcd. for $C_{34}H_{40}NO_{10}$ 622.2647 *m/z*).

ES-242-3 (E): Brown amorphous powder; $[\alpha]^{24}_D$ +66 (c 1.0, $CHCl_3$); UV (MeOH) λ_{max} 239.2, 298.6, 309.4 nm; ^{1}H NMR ($CDCl_3$) and ^{13}C NMR ($CDCl_3$), see Supplementary Materials Figures S19 and S20; HRESIMS *m/z* 638.2588 $[M + NH_4]^+$ (calcd. for $C_{34}H_{40}NO_{11}$ 638.2596 *m/z*).

Phomalactone (F): light yellowish oil; $[\alpha]^{24}_D$ *+172* (c 1.0, EtOH); UV (MeOH) λ_{max} 216.0 nm; For the details of VCD, ECD, and OR calculations, see Supplementary Materials; ^{1}H NMR ($CDCl_3$) and ^{13}C NMR ($CDCl_3$), see Supplementary Materials Figures S24 and S25; HRESIMS *m/z* 155.0702 $[M + H]^+$ (calcd. for $C_8H_{11}O_3$ 155.0703 *m/z*) and *m/z* 137.0597 $[M - OH^-]^+$ (calcd. for $C_8H_9O_2$ 137.0597 *m/z*).

Methyltriaceticlactone (G): White amorphous powder; UV (MeOH) λ_{max} 290.5 nm; ^{1}H NMR (DMSO-d6) and ^{13}C NMR (DMSO-d6), see Supplementary Materials Figures S29 and S30; HRESIMS *m/z* 141.0549 $[M + H]^+$ (calcd. for $C_7H_9O_3$ 141.0546 *m/z*).

S 39163/F-1 (H): Brown amorphous gum; $[\alpha]^{24}_D$ *−11* (c 1.0, MeOH); UV (MeOH) λ_{max} 218.2, 238.8, 291.9 nm; ^{1}H NMR ($CDCl_3$), see Supplementary Materials Figure S34; HRESIMS *m/z* 661.4312 $[M + H]^+$ (calcd. for $C_{38}H_{61}O_9$ 661.4310 *m/z*).

4.5. Preparation of Compounds for T. gondii Proliferation Assay

The purified natural products **A**–**F** and pyrimethamine [36] were dissolved in DMSO as 10 mM stocks and stored at −20 °C. The compounds were diluted in Iscove's Modified Dulbecco's medium (Gibco–Thermo Fisher Scientific, Braunschweig, Germany) immediately prior to use.

4.6. Parasites and Cell Culture for T. gondii Proliferation Assay

T. gondii ME49 tachyzoites (ATCC/LGC Standards GmbH, Wesel, Germany) were cultured in human foreskin fibroblast Hs27 cells (ATCC/LGC Standards GmbH, Wesel, Germany) as host cells as described previously [37]. The cells were maintained in Iscove's modified Dulbecco's medium (Gibco–Thermo Fisher Scientific, Braunschweig, Germany) supplemented with 10% heat-inactivated fetal bovine serum (Invitrogen, Karlsruhe, Germany) and 50 mM 2-mercaptoethanol (Gibco–Thermo Fisher Scientific, Braunschweig, Germany) and were grown in a humidified incubator at 37 °C with 5% CO_2 in air atmosphere. For toxoplasma propagation, 25 cm^2 cell culture flasks, containing a confluent monolayer of Hs27 cells, were infected with 5×10^6 *T. gondii* tachyzoites after medium change. After three days, the supernatant of the cell culture containing parasites was harvested and transferred to a 15 mL centrifuge tube and centrifuged at 700 rpm for five minutes and resuspended in cell culture medium. The number of parasites was counted using a hemocytometer.

4.7. *T. gondii Proliferation Assay*

Microtiter plates (96-well) with a final volume of 200 µL per well were used for the assay. Hs27 fibroblast monolayers were infected with 3×10^4 freshly harvested tachyzoites per well (MOI = 1) and incubated for 48 h at 37 °C, after which various concentrations of the tested compounds (0.04, 0.09, 0.19, 0.39, 0.78, 1.5, 3.12, 6.25, 12.5, 25, 50 µM) in culture medium were added to the cells. Pyrimethamine (0.007, 0.01, 0.03, 0.06, 0.125, 0.25, 0.5, 1 µM) was added under identical conditions as a positive drug control [37]. Hs27 cells were pre-stimulated for 24 h with IFNγ (300 U/mL) and infected with *T. gondii* cells without further treatment as the growth inhibition control. After 48 h, proliferating toxoplasma parasites were radioactively labelled with tritiated uracil (5 mCi, Hartmann Analytic, Braunschweig, Germany) and diluted 1:30 (10 µL per 200 µL total culture volume per well) in order to determine parasite proliferation [38]. After 28–30 h, the microtiter plates were frozen at −20 °C. To evaluate the assay, the microtiter plates were thawed at room temperature. Cells were transferred to glass-fiber filters (Printed Filtermat A 102 mm × 258 mm, PerkinElmer, Waltham, MA, USA) using a cell harvester (Basic96 Harvester, Zinsser Analytic, Skatron Instruments, Northridge, CA, USA). The filters were dried for 20 min at 130 °C in a drying cabinet and were then soaked in 10 mL of scintillation fluid (Betaplate Scint, PerkinElmer, Waltham, MA, USA) and shrink-wrapped in plastic covers (Sample Bag for Betaplate, PerkinElmer, Waltham, MA, USA). The filters were then clamped in cassettes and evaluated using a beta-counter device (Betaplate Liquid Scintillation Counter 1205, LKB-WALLAK, Melbourne, Australia) to measure the Cherenkov radiation, which refers to the amount of incorporation of tritiated uracil into the RNA of *T. gondii*. IC_{50} values, the concentration of inhibitors necessary to inhibit the growth of tachyzoites by 50%, were determined for each experiment with the use of Prism GraphPad version 9.2.0 software.

4.8. *Cell Viability Assay against Hs27 Cells*

The 3-[4,5-dimethylthiazole-2-yl]-2,5-diphenyltetrazolium bromide (MTT) test was used to assess cell viability of the isolated active compounds against Hs27 cells. The MTT assay is a colorimetric reaction based on the enzymatic reduction of MTT to MTT-formazan, which is catalyzed by mitochondrial succinate dehydrogenase [39].

In brief, Hs27 cells were seeded 96-well plates in a monolayer in Iscove's modified Dulbecco's medium (Gibco–Thermo Fisher Scientific, Braunschweig, Germany) and incubated at 37 °C with different concentrations of the tested natural products (1.56, 3,12, 6.25, 12,5, 25, 50, 100 µM) in the culture media. Staurosporine (0.007, 0.01, 0.03, 0.06, 1.25, 0.25, 0.5, 1 µM), a well-known cytotoxicity-inducing kinase inhibitor [40], untreated Hs27 cells, and DMSO were used as controls. After 24 h, the medium of the culture was removed and replaced with 100 µL of DMEM without red phenol (Gibco–Thermo Fisher Scientific, Braunschweig, Germany) plus 10% heat-inactivated fetal bovine serum (Invitrogen, Karlsruhe, Germany), and 50 mM 2-mercaptoethanol (Gibco–Thermo Fisher Scientific, Braunschweig, Germany). Afterwards, the 12 mM MTT solution was added to each well according to the manufacturer's instruction (Vybrant MTT Cell Proliferation Assay Kit, Thermo Fisher Scientific, Braunschweig, Germany). The OD value of each well was assayed at the wavelength of 570 nm on a microplate reader (TECAN Sunrise, Männedorf, Switzerland). The 50%

cytotoxic concentration (CC_{50} values) of the tested natural products on Hs27 cells was calculated and all data were analyzed using Prism GraphPad version 9.2.0 software.

4.9. Determination of the Minimal Inhibitory Concentration against Different Pathogenic Bacteria

Testing for antibacterial activity was done as described previously [37]. Briefly, a single colony of Methicillin-resistant *Staphylococcus aureus* (MRSA strain Mu50, ATCC 700699) or *Pseudomonas aeruginosa* (strain PAO1, ATCC 87110) were grown in Mueller-Hinton broth (MHB) at 37 °C shaking at 120 rpm to reach an optical density of approx. 0.4. The cell suspension was adjusted to 10^6 CFU/mL, of which 50 µL was seeded into a prepared 96-well polystyrene round-bottom plate containing test compounds diluted in MHB in a 1:1 serial dilution ranging from 100 µM to 0.78 µM. The plates were incubated at 37 °C statically for 24 h, and readout was performed using the BacTiter Glo assay (Promega) following the manufacturer's instructions. Briefly, BacTiter Glo reagent was added to a white flat-bottom 96-well plate, and an equal volume of bacteria suspension was added to each well and mixed carefully. After 5 min, the luminescence was measured using a TECAN plate reader. The growth was calculated in regard to the vehicle (DMSO) and sterile control. Moxifloxacin and cefuroxime were used as a positive and negative control, respectively. All compounds were tested in triplicate.

For the testing against *M. tuberculosis* H37Rv, the Minimal Inhibitory Concentration (MIC) was determined in 96-well microtiter plates containing a total volume of 100 µL employing a resazurin reduction assay [41]. Briefly, a 96-well plate was prepared containing 7H9 medium supplemented with 10% ADS (0.81% NaCl, 5% BSA, 2% dextrose), 0.5% glycerol, and 0.05% tyloxapol. Compounds were two-fold serially diluted with the highest tested concentration of 100 µM. A *M. tuberculosis* culture was pre-grown to an $OD_{600\ nm}$ of approx. 0.4–0.6 by shaking at 37 °C in PETG square bottles (ThermoFisher Scientific, Braunschweig, Germany) containing 10 mL supplemented 7H9 medium. The cell density was adjusted to an $OD_{600\ nm}$ of 0.08 (10^6 CFU/mL, and 5×10^4 CFU were added to each well). Rifampicin and DMSO were used as a positive and solvent control, respectively. The 96-well plates were incubated for 5 days at 37 °C and 5% CO_2 in humidified atmosphere. Afterwards, 10 µL of a 100 mg/mL resazurin solution was added to each well and resuspended carefully. After another 24 h at room temperature, the cells were fixed by adding 100 µL of a 10% formalin solution to each well. The readout was performed using a TECAN plate reader at 535 nm excitation and 590 nm emission. The growth was calculated in relation to the solvent control being 100% growth. The experiment was performed in triplicate.

4.10. Cytotoxicity Assay against Different Human Cell Lines

The cytotoxicity study was carried out using the THP-1 (human monocytic leukemia cell line), Huh-7 (Human liver carcinoma cell line), and HEK293 (human embryonic kidney cell line) cell lines as described before [37]. The THP-1 cells were cultured using RPMI 1640 medium containing 2 mM L-glutamine and supplemented with 10% fetal calf serum (FCS) and 1% sodium pyruvate. Huh-7 cells were cultured using a 1:1 mixture of RPMI 1640 medium containing 2 mM L-glutamine and 10% FCS medium and DMEM containing 10% FCS and 1% sodium pyruvate. The HEK-293 cells were cultured with DMEM including 2 mM L-glutamine and supplemented with 1% NE amino acids, 1% 1.0 mM sodium pyruvate and 10% FCS. All three cell lines were then incubated at 37 °C in an atmosphere of 5% CO_2 under humid conditions for 2 weeks while renewing the medium twice weekly. Subsequently, the cells were suspended and adjusted to a density of 2×10^5 cells/mL. In a 96-well flat-bottom microtiter plate, the cells were adjusted to a total volume of 100 µL containing 2-fold serial dilutions of the tested compounds **A–F** ranging from 100 to 1.56 µM. Cycloheximide (4, 2, 1, 0.5, 0.25, 0.13, 0.06, 0.03 µg/mL) was used as a positive control. After an incubation time of 48 h at 37 °C in an atmosphere of 5% CO_2 under humid conditions, 10 µL resazurin solution (100 µg/mL) was added to each well and incubated for another 4 h. The fluorescence was then quantified using a Tecan Infinite 200pro microplate reader (excitation 540 nm, emission 590 nm). The residual growth was calculated relative to non-inoculated conditions (0% growth) and controls treated with DMSO (100% growth).

Supplementary Materials: The following supporting information can be downloaded at: https://www.mdpi.com/article/10.3390/antibiotics11091176/s1, Figures S1–S58: Spectroscopic data used for the structure elucidation of compounds A–H; Figure S59: Comparison of the experimental ECD spectrum of F measured in MeCN and the calculated ECD spectra of (5S,6S)-F computed at various levels of theory for the 10 lowest-energy ωB97X/TZVP PCM/MeCN conformers; Figure S60: Geometries of the low-energy ωB97X/TZVP PCM/MeCN conformers of (5S,6S)-F; Figure S61: Geometries of the low-energy B3LYP/TZVP PCM/CHCl3 conformers of (5S,6S)-F; Table S1: Boltzmann populations and specific optical rotations of the low-energy conformers of (5S,6S)-F computed at various levels for the low-energy ωB97X conformers. References [42–45] are cited in Supplementary Materials.

Author Contributions: Conceptualization, R.K. and K.P.; methodology, F.M., V.E.S., T.K., R.K. and K.P.; investigation, F.M., V.E.S., L.v.G. and A.M.; data curation, F.M., V.E.S., M.F., T.K., R.K. and K.P.; writing—original draft preparation, F.M. and V.E.S.; writing—review and editing, funding acquisition, T.K., R.K. and K.P. All authors have read and agreed to the published version of the manuscript.

Funding: This work was supported by the Deutsche Forschungsgemeinschaft (DFG, German Research Foundation)–project number 270650915/GRK 2158 (to RK and KP). T.K. and A.M. were supported by the National Research, Development and Innovation Office (K138672 and FK134653).

Institutional Review Board Statement: Not applicable.

Informed Consent Statement: Not applicable.

Data Availability Statement: All data presented in this study are contained within the article and the supplementary materials. The internal transcribed spacer (ITS) sequence for *Paraboeremia selaginellae* has been deposited in the National Center for Biotechnology Information (NCBI) GenBank under accession number ON231784.

Acknowledgments: We thank the CeMSA@HHU (Center for Molecular and Structural Analytics@Heinrich Heine University) for recording the mass-spectrometric and the NMR-spectroscopic data. We thank Heike Goldbach-Gecke for testing and measuring the cytotoxic effect of the isolated compounds against the tested human cell lines. The Governmental Information-Technology Development Agency (KIFÜ) is acknowledged for CPU time. We thank Karin Buchholz for expert technical assistance and Daniel Degrandi as well as Ursula Sorg for scientific advice and discussions.

Conflicts of Interest: The authors declare no conflict of interest. The funders had no role in the design of the study; in the collection, analyses, or interpretation of data; in the writing of the manuscript, or in the decision to publish the results.

References

1. Frolich, S.; Entzeroth, R.; Wallach, M. Comparison of protective immune responses to apicomplexan parasites. *J. Parasitol. Res.* **2012**, *2012*, 852591. [CrossRef] [PubMed]
2. Kim, K.; Weiss, L.M. Toxoplasma gondii: The model apicomplexan. *Int. J. Parasitol.* **2004**, *34*, 423–432. [CrossRef] [PubMed]
3. Toxoplasmosis of Animals and Man. By J.P. Dubey and C. P. Beattie. 220 pages. ISBN 0 8493 4618 5. CRC Press, Boca Raton, 1988. £108.00. *Parasitology* **2009**, *100*, 500–501. [CrossRef]
4. Dubey, J.P. *Toxoplasmosis of Animals and Humans*; CRC Press: Boca Raton, FL, USA, 2016.
5. Saadatnia, G.; Golkar, M. A review on human toxoplasmosis. *Scand. J. Infect. Dis.* **2012**, *44*, 805–814. [CrossRef] [PubMed]
6. Dubey, J.P. Outbreaks of clinical toxoplasmosis in humans: Five decades of personal experience, perspectives and lessons learned. *Parasites Vectors* **2021**, *14*, 263. [CrossRef] [PubMed]
7. Furtado, J.M.; Smith, J.R.; Belfort, R., Jr.; Gattey, D.; Winthrop, K.L. Toxoplasmosis: A global threat. *J. Glob. Infect. Dis.* **2011**, *3*, 281–284. [CrossRef]
8. de Jong, P.T. Ocular toxoplasmosis; common and rare symptoms and signs. *Int. Ophthalmol.* **1989**, *13*, 391–397. [CrossRef]
9. Elbez-Rubinstein, A.; Ajzenberg, D.; Dardé, M.L.; Cohen, R.; Dumètre, A.; Yera, H.; Gondon, E.; Janaud, J.C.; Thulliez, P. Congenital toxoplasmosis and reinfection during pregnancy: Case report, strain characterization, experimental model of reinfection, and review. *J. Infect. Dis.* **2009**, *199*, 280–285. [CrossRef]
10. Dunay, I.R.; Gajurel, K.; Dhakal, R.; Liesenfeld, O.; Montoya, J.G. Treatment of Toxoplasmosis: Historical Perspective, Animal Models, and Current Clinical Practice. *Clin. Microbiol. Rev.* **2018**, *31*, e00057-17. [CrossRef]

11. Gopalakrishnan, A.M.; López-Estraño, C. Comparative analysis of stage specific gene regulation of apicomplexan parasites: Plasmodium falciparum and Toxoplasma gondii. *Infect. Disord. Drug Targets* **2010**, *10*, 303–311. [CrossRef]
12. Newman, D.J.; Cragg, G.M. Natural Products as Sources of New Drugs over the Nearly Four Decades from 01/1981 to 09/2019. *J. Nat. Prod.* **2020**, *83*, 770–803. [CrossRef] [PubMed]
13. Cheraghipour, K.; Masoori, L.; Ezzatpour, B.; Roozbehani, M.; Sheikhian, A.; Malekara, V.; Niazi, M.; Mardanshah, O.; Moradpour, K.; Mahmoudvand, H. The Experimental Role of Medicinal Plants in Treatment of Toxoplasma gondii Infection: A Systematic Review. *Acta Parasitol.* **2021**, *66*, 303–328. [CrossRef] [PubMed]
14. Lenzi, J.; Costa, T.M.; Alberton, M.D.; Goulart, J.A.G.; Tavares, L.B.B. Medicinal fungi: A source of antiparasitic secondary metabolites. *Appl. Microbiol. Biotechnol.* **2018**, *102*, 5791–5810. [CrossRef] [PubMed]
15. Fukushima, T.; Tanaka, M.; Gohbara, M.; Fujimori, T. Phytotoxicity of three lactones from Nigrospora sacchari. *Phytochemistry* **1998**, *48*, 625–630. [CrossRef]
16. Evans, R.H.; Ellestad, G.A.; Kunstmann, M.P. Two new metabolites from an unidentified nigrospora species. *Tetrahedron Lett.* **1969**, *10*, 1791–1794. [CrossRef]
17. Gusmao, A.S.; Abreu, L.S.; Tavares, J.F.; de Freitas, H.F.; Silva da Rocha Pita, S.; Dos Santos, E.G.; Caldas, I.S.; Vieira, A.A.; Silva, E.O. Computer-Guided Trypanocidal Activity of Natural Lactones Produced by Endophytic Fungus of Euphorbia umbellata. *Chem. Biodivers* **2021**, *18*, e2100493. [CrossRef]
18. Hussain, H.; Kock, I.; Al-Harrasi, A.; Al-Rawahi, A.; Abbas, G.; Green, I.R.; Shah, A.; Badshah, A.; Saleem, M.; Draeger, S.; et al. Antimicrobial chemical constituents from endophytic fungus Phoma sp. *Asian Pac. J. Trop. Med.* **2014**, *7*, 699–702. [CrossRef]
19. Khambay, B.P.S.; Bourne, J.M.; Cameron, S.; Kerry, B.R.; Zaki, M.J. A nematicidal metabolite from Verticillium chlamydosporium. *Pest Manag. Sci. Former. Pestic. Sci.* **2000**, *56*, 1098–1099. [CrossRef]
20. Meepagala, K.M.; Johnson, R.D.; Techen, N.; Wedge, D.E.; Duke, S.O. Phomalactone from a Phytopathogenic Fungus Infecting ZINNIA elegans (ASTERACEAE) Leaves. *J. Chem. Ecol.* **2015**, *41*, 602–612. [CrossRef]
21. Trisuwan, K.; Rukachaisirikul, V.; Sukpondma, Y.; Preedanon, S.; Phongpaichit, S.; Sakayaroj, J. Pyrone derivatives from the marine-derived fungus *Nigrospora* sp. PSU-F18. *Phytochemistry* **2009**, *70*, 554–557. [CrossRef]
22. Wu, S.H.; Chen, Y.W.; Shao, S.C.; Wang, L.D.; Yu, Y.; Li, Z.Y.; Yang, L.Y.; Li, S.L.; Huang, R. Two new solanapyrone analogues from the endophytic fungus *Nigrospora* sp. YB-141 of Azadirachta indica. *Chem. Biodivers* **2009**, *6*, 79–85. [CrossRef] [PubMed]
23. Mandi, A.; Kurtan, T. Applications of OR/ECD/VCD to the structure elucidation of natural products. *Nat. Prod. Rep.* **2019**, *36*, 889–918. [CrossRef] [PubMed]
24. Szabo, Z.; Paczal, A.; Kovacs, T.; Mandi, A.; Kotschy, A.; Kurtan, T. Synthesis and Vibrational Circular Dichroism Analysis of N-Heterocyclic Carbene Precursors Containing Remote Chirality Centers. *Int. J. Mol. Sci* **2022**, *23*, 3471. [CrossRef]
25. Atanasov, A.G.; Zotchev, S.B.; Dirsch, V.M.; Orhan, I.E.; Banach, M.; Rollinger, J.M.; Barreca, D.; Weckwerth, W.; Bauer, R.; Bayer, E.A.; et al. Natural products in drug discovery: Advances and opportunities. *Nat. Rev. Drug Discov.* **2021**, *20*, 200–216. [CrossRef] [PubMed]
26. Guo, H.-Y.; Jin, C.; Zhang, H.-M.; Jin, C.-M.; Shen, Q.-K.; Quan, Z.-S. Synthesis and Biological Evaluation of (+)-Usnic Acid Derivatives as Potential Anti-Toxoplasma gondii Agents. *J. Agric. Food Chem.* **2019**, *67*, 9630–9642. [CrossRef] [PubMed]
27. Jiménez-Romero, C.; Ortega-Barría, E.; Arnold, A.E.; Cubilla-Rios, L. Activity against *Plasmodium falciparum* of Lactones Isolated from the Endophytic Fungus *Xylaria* sp. *Pharm. Biol.* **2008**, *46*, 700–703. [CrossRef]
28. McMurry, L.M.; Oethinger, M.; Levy, S.B. Triclosan targets lipid synthesis. *Nature* **1998**, *394*, 531–532. [CrossRef]
29. McLeod, R.; Muench, S.P.; Rafferty, J.B.; Kyle, D.E.; Mui, E.J.; Kirisits, M.J.; Mack, D.G.; Roberts, C.W.; Samuel, B.U.; Lyons, R.E.; et al. Triclosan inhibits the growth of Plasmodium falciparum and Toxoplasma gondii by inhibition of apicomplexan Fab I. *Int. J. Parasitol.* **2001**, *31*, 109–113. [CrossRef]
30. Tipparaju, S.K.; Muench, S.P.; Mui, E.J.; Ruzheinikov, S.N.; Lu, J.Z.; Hutson, S.L.; Kirisits, M.J.; Prigge, S.T.; Roberts, C.W.; Henriquez, F.L.; et al. Identification and development of novel inhibitors of Toxoplasma gondii enoyl reductase. *J. Med. Chem.* **2010**, *53*, 6287–6300. [CrossRef]
31. Toki, S.; Ando, K.; Kawamoto, I.; Sano, H.; Yoshida, M.; Matsuda, Y. ES-242-2, -3, -4, -5, -6, -7, and -8, novel bioxanthracenes produced by *Verticillium* sp., which act on the N-methyl-D-aspartate receptor. *J. Antibiot.* **1992**, *45*, 1047–1054. [CrossRef]
32. Jaturapat, A.; Isaka, M.; Hywel-Jones, N.L.; Lertwerawat, Y.; Kamchonwongpaisan, S.; Kirtikara, K.; Tanticharoen, M.; Thebtaranonth, Y. Bioxanthracenes from the insect pathogenic fungus. Cordyceps pseudomilitaris BCC 1620. I. Taxonomy, fermentation, isolation and antimalarial activity. *J. Antibiot.* **2001**, *54*, 29–35. [CrossRef] [PubMed]
33. Komai, S.-I.; Hosoe, T.; Nozawa, K.; Okada, K.; de Campos Takaki, G.M.; Fukushima, K.; Miyaji, M.; Horie, Y.; Kawai, K.-I. Antifungal activity of pyranone and furanone derivatives, isolated from *Aspergillus* sp. IFM51759, against *Aspergillus fumigatus*. *MYCOTOXINS-TOKYO-* **2003**, *53*, 11–18. [CrossRef]
34. Krasnoff, S.B.; Gupta, S. Identification of the antibiotic phomalactone from the entomopathogenic fungusHirsutella thompsonii var. synnematosa. *J. Chem. Ecol.* **1994**, *20*, 293–302. [CrossRef] [PubMed]
35. Krivobok, S.; Thomasson, F.; Seigle-Murandi, F.; Steiman, R.; Bottex-Gauthier, C. 6-Allyl-5, 6-dihydro-5-hydroxypyran-2-one, a lactone produced by a new Drechslera species: Specified 1H and 13C NMR assignments, mutagenic and immunomodulating testings. *Die Pharm.* **1994**, *49*, 605–607.

36. Kumarihamy, M.; Ferreira, D.; Croom, E.M., Jr.; Sahu, R.; Tekwani, B.L.; Duke, S.O.; Khan, S.; Techen, N.; Nanayakkara, N.P.D. Antiplasmodial and Cytotoxic Cytochalasins from an Endophytic Fungus, *Nemania* sp. UM10M, Isolated from a Diseased *Torreya taxifolia* Leaf. *Molecules* **2019**, *24*, 777. [CrossRef]
37. Meier, D.; Hernandez, M.V.; van Geelen, L.; Muharini, R.; Proksch, P.; Bandow, J.E.; Kalscheuer, R. The plant-derived chalcone Xanthoangelol targets the membrane of Gram-positive bacteria. *Bioorg Med. Chem.* **2019**, *27*, 115151. [CrossRef]
38. Pfefferkorn, E.R.; Pfefferkorn, L.C. Specific Labeling of Intracellular Toxoplasma gondii with Uracil. *J. Protozool.* **1977**, *24*, 449–453. [CrossRef]
39. Mosmann, T. Rapid colorimetric assay for cellular growth and survival: Application to proliferation and cytotoxicity assays. *J. Immunol. Methods* **1983**, *65*, 55–63. [CrossRef]
40. Tamaoki, T.; Nomoto, H.; Takahashi, I.; Kato, Y.; Morimoto, M.; Tomita, F. Staurosporine, a potent inhibitor of phospholipid/Ca++dependent protein kinase. *Biochem. Biophys. Res. Commun.* **1986**, *135*, 397–402. [CrossRef]
41. Rehberg, N.; Akone, H.S.; Ioerger, T.R.; Erlenkamp, G.; Daletos, G.; Gohlke, H.; Proksch, P.; Kalscheuer, R. Chlorflavonin Targets Acetohydroxyacid Synthase Catalytic Subunit IlvB1 for Synergistic Killing of *Mycobacterium tuberculosis*. *ACS Infect. Dis.* **2018**, *4*, 123–134. [CrossRef]
42. MacroModel; Schrödinger LLC. 2015. Available online: http://www.schrodinger.com/MacroModel (accessed on 31 July 2022).
43. Frisch, M.J.; Trucks, G.W.; Schlegel, H.B.; Scuseria, G.E.; Robb, M.A.; Cheeseman, J.R.; Scalmani, V.; Barone, G.; Mennucci, B.; Petersson, G.A.; et al. *Gaussian 09 (Revision E.01)*; Gaussian, Inc.: Wallingford, CT, USA, 2013.
44. Stephens, P.J.; Harada, N. ECD cotton effect approximated by the Gaussian curve and other methods. *Chirality* **2009**, *22*, 229–233. [CrossRef] [PubMed]
45. Varetto, U. *Molekel 5.4*; Swiss National Supercomputing Centre: Manno, Switzerland, 2009.

antibiotics

Review

Surface-Active Compounds Produced by Microorganisms: Promising Molecules for the Development of Antimicrobial, Anti-Inflammatory, and Healing Agents

Jéssica Araujo [1,2,†], Joveliane Monteiro [1,2,†], Douglas Silva [2], Amanda Alencar [2], Kariny Silva [2], Lara Coelho [2], Wallace Pacheco [2], Darlan Silva [3], Maria Silva [3], Luís Silva [1,4] and Andrea Monteiro [1,2,*]

1 Rede de Biodiversidade e Biotecnologia da Amazônia Legal, BIONORTE, São Luís 65055-310, MA, Brazil
2 Laboratório de Microbiologia Aplicada, Universidade CEUMA, São Luís 65075-120, MA, Brazil
3 Laboratório de Ciências do Ambiente, Universidade CEUMA, São Luís 65075-120, MA, Brazil
4 Laboratório de Patogenicidade Microbiana, Universidade CEUMA, São Luís 65075-120, MA, Brazil
* Correspondence: andreasmont@gmail.com
† These authors contributed equally to this work.

Abstract: Surface-active compounds (SACs), biomolecules produced by bacteria, yeasts, and filamentous fungi, have interesting properties, such as the ability to interact with surfaces as well as hydrophobic or hydrophilic interfaces. Because of their advantages over other compounds, such as biodegradability, low toxicity, antimicrobial, and healing properties, SACs are attractive targets for research in various applications in medicine. As a result, a growing number of properties related to SAC production have been the subject of scientific research during the past decade, searching for potential future applications in biomedical, pharmaceutical, and therapeutic fields. This review aims to provide a comprehensive understanding of the potential of biosurfactants and emulsifiers as antimicrobials, modulators of virulence factors, anticancer agents, and wound healing agents in the field of biotechnology and biomedicine, to meet the increasing demand for safer medical and pharmacological therapies.

Keywords: surface-active compounds; biosurfactants; bioemulsifier; biological properties; biotechnology

Citation: Araujo, J.; Monteiro, J.; Silva, D.; Alencar, A.; Silva, K.; Coelho, L.; Pacheco, W.; Silva, D.; Silva, M.; Silva, L.; et al. Surface-Active Compounds Produced by Microorganisms: Promising Molecules for the Development of Antimicrobial, Anti-Inflammatory, and Healing Agents. *Antibiotics* **2022**, *11*, 1106. https://doi.org/10.3390/antibiotics11081106

Academic Editor: Carlos M. Franco

Received: 3 July 2022
Accepted: 2 August 2022
Published: 16 August 2022

1. Introduction

Microorganisms can produce several surface-active compounds (SACs) with hydrophilic and hydrophobic moieties. These structural features allow them to interact with the surface and interfacial tensions, form micelles, and emulsify immiscible substances [1,2].

Biosurfactants (BSs) and bioemulsifiers (BEs) are considered SACs because of their ability to interfere and with modifying surfaces. Because these biomolecules are amphiphilic and are produced by different microorganisms, they have different physicochemical properties and physiological roles, which contribute to their specific functions in nature and biotechnological applications [3].

Recently, the production of SACs has received extensive attention because of their diverse applications, such as dissolving water-insoluble compounds, heavy metal binding, contaminant desorption, inhibiting bacterial pathogenesis, adhesion, and cell aggregation [4–8]. In addition, SACs also have several advantages over synthetic surfactants, such as low toxicity, lower critical micelle concentration (CMC), higher biodegradability, and ecological acceptability [9].

Moreover, these compounds exhibit antibacterial [5,10], antifungal [11], antiviral [12], and antitumor activities [13]. Their antiadhesive properties and antibiofilm activities are also important in inhibiting the adhesion and colonization of pathogenic microorganisms and removing preformed biofilms on silicone discs and other biomedical instruments [14].

The present use of these biomolecules has aroused interest from several sectors because of their numerous functions and sustainable properties, allowing various applications in petroleum, food, medicine, pharmaceuticals, chemicals, pulp and paper, textiles, and cosmetics. Furthermore, because of their application in soil bioremediation, they are considered the "green molecules" of the 21st century [15].

2. Surface-Active Compounds

2.1. Biosurfactants

Biosurfactants (BSs), which are low molecular weight microbial compounds, are synthesized extracellularly or linked to the cell membrane of bacteria [16], yeasts [17], and filamentous fungi [18]. Produced under various extreme environmental conditions, their chemical compositions depend on the microorganism that produces them, raw materials, and process conditions [6].

Surfactants are amphiphilic molecules with a hydrophobic moiety comprising a hydrocarbon chain with saturated or unsaturated and hydroxylated fatty alcohols or fatty acids, and a hydrophilic moiety comprising hydroxyl, phosphate, or carboxyl groups, or carbohydrates (such as mono-, oligo-, or polysaccharides) or peptide fractions [3,19]. Depending on their biochemical nature, these compounds are classified as glycolipids, lipopeptides, lipoproteins or fatty acids, and phospholipid polymers, with glycolipids and lipopeptides being the most abundant [20,21].

Glycolipids consist of mono- or oligosaccharides and lipids, where different sugars (glucose, mannose, galactose, glucuronic acid, or rhamnose) link to form saturated or unsaturated fatty acids, hydroxylated fatty acids, or fatty alcohols. The most studied groups include sophorolipids (SLs), mannosylerythritol lipids, trehalolipids, and rhamnolipids (RLs) [22,23], which are usually produced by the yeast *Starmerella bombicola* [24], *Pseudozyma* sp. [25,26] *Rhodococcus erythropolis* [27] and *Pseudomonas aeruginosa* [28], respectively.

On the other hand, lipopeptides (LP) consist of cyclopeptides with amino acids linked to fatty acids of different chain lengths [29]. The most common among these are surfactin, iturin, and fengycin [29–31] which are produced by different microorganisms, such as the genera *Bacillus* [32], in turn, other lipopeptides have been detected in *Bacillus amyloliquefaciens* [33], *Streptomyces* sp. [34], *Pseudomonas guguanensis* [35], and *Serratia marcescens* [36].

Microorganisms that produce BS inhabit water (fresh, underground, and sea) and land (soil, sediments, and mangroves) and can grow in extreme environments (oil reservoirs) and under different temperatures, pH values, and salinity levels [37–39].

These microorganisms are generally heterotrophs that need carbon, nitrogen, minerals, vitamins, growth factors, and water to grow and produce metabolites. In general, carbon sources (carbohydrates, oils, and fats) and hydrocarbon groups are often consumed during BS production. For example, glucose, a carbon source easily metabolized by microorganisms through glycolysis to generate energy, is commonly reported as a factor in producing higher yields [37,40].

Because of their amphipathic nature, BSs can mix immiscible fluids, reduce surface and interfacial tensions, and promote solubility of polar compounds in nonpolar solvents [41] that help exhibit numerous properties, such as foaming, dispersion, wetting, emulsification, demulsification, and coating, making them suitable for physicochemical and biological remediation technologies of organic and metallic contaminants [42].

Biosurfactants due to their physicochemical properties have industrial applications in pharmaceuticals, textile processing, agriculture, cosmetics, personal care, and the food industry, as well as environmental applications in soil remediation, hydrocarbon degradation, and oil recovery [43–45].

Several BSs have antibacterial, antifungal, antiviral, or antitumor properties, making them potential alternatives to conventional therapeutics in many biomedical applications [45,46].

Despite their versatility and efficiency in terms of applicability in different fields, their production has always been a challenge because of inefficient bioprocessing and high costs due to the expensive substrates used [33]. Therefore, optimizing strategies on cost efficiency and high-yield bioprocessing is critical for low-cost production and mass commercialization.

2.2. Bioemulsifier (BE)

Unlike BSs, bioemulsifiers (BEs) have high molecular weight and can emulsify, even at low concentrations, two immiscible liquids, while not reducing surface or interfacial tension [47]. These comprise complex mixtures of heteropolysaccharides, lipopolysaccharides, proteins, glycoproteins, or lipoproteins, which guarantee better emulsification potential and emulsion stabilization [3,48,49].

Bioemulsifiers, which are synthesized by bacteria, yeasts, and filamentous fungi, can be isolated from contaminated soil, mangroves, seawater, freshwater, and human skin [50–53]. The most studied polymeric BEs include emulsan, alasan, liposan, mannoprotein, and other polysaccharide-protein complexes. Members of the genus *Acinetobacter* sp. are commonly reported to produce BEs [15].

Despite numerous reports on the production of BEs and BSs by different bacteria, the genus *Acinetobacter* spp. received special attention because it is the first known producer of BEs, with emulsan, biodispersan, and alasan as the best examples of BEs commercially produced by the genus. These BEs are mainly used in microbial oil recovery and the biodegradation of toxic compounds [15].

Compared with synthetic surfactants, BEs have many advantages as they are eco-friendly, biocompatible, less toxic with higher biodegradability, and active at extreme temperatures, pH values, and salinity levels. Furthermore, BEs can be produced from low-cost renewable substrates, such as industrial waste, vegetable oils, and hydrocarbons [53].

Various carbon sources are used in BE production, such as ethanol, n-hexadecane, crude oil, glucose, lactic acid, methylnaphthalene, peptone, n-heptadecane, edible oil, olive oil, glycerol, and C-heavy oil [54]. Conventionally, microbial production of BE is still expensive, with the use of synthetic sources as one of the factors contributing most to the high costs. One promising strategy to make the cost economically viable is to include renewable sources from agro-industrial residues and by-products. In this sense, previous research had explored several alternative low-cost substrates, such as residual soybean oil from frying and corn steep liquor, as alternatives to synthetic sources of carbon and nitrogen [53].

Despite their potential advantages, several obstacles hinder practical BE applications, including low yields and high purification costs. To address these issues, researchers have been striving to develop more cost-efficient BEs, which can be used at low concentrations [55].

Bioemulsifiers can form very stable emulsions and dispersions that do not mix, remain attached to the droplet interfaces, and can re-emulsify by adding or replacing a new solvent without diluting. Because of these advantages, BEs are preferred over BSs in the cosmetics and food industries [48].

Because of diverse functions, such as emulsification, wetting, foaming, cleaning, phase separation, surface activity, and hydrocarbon viscosity reduction, BEs are best suited for bioremediation, enhanced oil recovery, cleaning of pipe and vessels contaminated with oil, and more. In addition, emulsifiers are widely used in the food and drug industry [56].

2.3. Microorganisms Producing SACS

For many years, researchers have tirelessly searched for microorganisms that have the potential to produce secondary metabolites with surfactant or emulsifying properties. The amount of BS or BE produced depends on the type of microorganisms and their sources (Table 1).

Table 1. Lists some microorganisms that produce surface-active compounds.

Microorganism	Biosurfactant/Bioemulsifier	Reference
Acinetobacter calcoaceticus RAG-1	Emulsan	[57]
Acinetobacter radioresistant KA53	Alasan	[58]
Acinetobacter junii B6	Surfactin/fengycin	[59]
Acinetobacter junii BD	Rhamnolipids	[60]
Acinetobacter calcoaceticus A2	Biodispersan	[61]
Bacillus nealsonii strain S2MT	Surfactin	[2]
Bacillus subtilis 3NA	Surfactin	[62]
Bacillus thailandensis E264	Rhamnolipids	[63]
Bacillus velezensis	Iturin, surfactin, and fengycin	[64]
Candida keroseneae GBME-IAUF-2	Sophorolipids	[65]
Candida lipolytica UCP 0988	Rufisan	[66]
Lactobacillus sp.	Surfactin, iturin, and lichenysin	[67]
Pseudomonas aeruginosa SG	Rhamnolipids	[68]
Pseudomonas fluorescens SBW25	Viscosine	[69]
Pseudomonas sp. S2WE	Rhamnolipids	[70]
Serratia sp. ZS6 strain	Serrawettina	[71]
Yarrowia lipolytica IMUFRJ50682	Yansan	[72]
Trichosporon mycotoxinivorans CLA2	Lipid-polysaccharide complex	[73]

3. Biological Properties

3.1. Antimicrobial Activities

The discovery of antibiotics in the last century can be considered a major advancement in medicine because the use of these antimicrobial agents significantly reduced morbidity and mortality associated with microbial infections. Antibacterial and antifungal factors reduce and eliminate the viability and growth of microbial populations through several mechanisms: (i) disruption of extracellular membranes and/or their cell wall, (ii) inhibition of gene expression, (iii) DNA damage, or (iv) manipulation of important metabolic pathways [74].

Bacteria become resistant to antimicrobial agents in several ways: through horizontal gene transfer between genetic elements of different strains and the environment that confer resistance and through mutations that interfere with basic cell functions in addition to conferring resistance to microorganisms [75,76].

The most resistant bacteria associated with serious hospital infections include *Enterococcus faecalis*, *Staphylococcus aureus*, *Klebsiella pneumoniae*, *Acinetobacter baumanii*, *P. aeruginosa*, and *Enterobacter* sp., which often result in high mortality rates [77]. Furthermore, other microorganisms such as *Candida* spp. can also be considered a global health threat because of their resistance to antimicrobial agents [78–80].

The increasing rates of antimicrobial resistance and the emergence of new microbial pathogens reinforce the need to find new antimicrobial compounds to fight microbial infections. Among these new strategies, SACs have promising antibiotic and disinfectant potential, as well as antibiotic delivery properties due to their physicochemical properties. Most of these biomolecules can break the outer and inner membranes of pathogens, thereby exploiting their charge and hydrophobicity. The advantages of using SACs as antimicrobials include their broad-spectrum bactericidal action and the absence of pathogen resistance mechanisms [81].

Cationic surfactants comprise the largest class of synthetic surfactants with antimicrobial properties because of their broad spectrum of biostatic and biocidal activities against planktonic pathogens. The hydrophobic chain of cationic surfactants penetrates the microbial cell membrane and interferes with membrane continuity and metabolic processes, leading to cell death [82]. Despite exhibiting antimicrobial efficiency mainly against Gram-positive bacteria (29–32 mm), such as *S. aureus* and *Bacillus subtilis*, these compounds are less biodegradable than natural surfactants [83].

Previous studies reported the antimicrobial efficacy of glycolipid SACs produced by microorganisms. For example, RLs produced by *P. aeruginosa* significantly inhibited the growth of *S. mutans* UA159 and *S. sanguinis* ATCC10556. Furthermore, they completely inhibited the growth of *Aggregatibacter actinomycetemcomitans* Y4 at high concentrations [7].

Similarly, the synergistic action of two RL BSs produced by *P. aeruginosa* C2 and *Bacillus stratosphericus* A15 demonstrated bactericidal activity by rupturing the membrane of gram-positive and gram-negative bacteria, such as *S. aureus* ATCC 25923 and *Escherichia coli* K8813 [84]. Because of these actions, the membrane disintegrates, leading to penetration into the cell wall and plasma membrane through the formation of pores, followed by leakage of internal cytoplasmic materials, leading to cell death [85].

A previous study demonstrated that the synergism between essential oils of oregano, cinnamon tree, and lavender with RLs produced by *P. aeruginosa* increased the antimicrobial effect against *Candida albicans* and *S. aureus* which are resistant to methicillin [86], revealing that SAC activity can be enhanced when they establish a synergistic relationship with other compounds. In addition to RLs, SLs are also easily extracted and are usually produced by *Candida* spp. yeast [87] either in the lactone form or the acid form or as a mixture of both forms [88,89].

A previous study showed that SL produced by *C. albicans* SC5314 and *Candida glabrata* CBS138 showed antibacterial properties against pathogenic bacteria *Bacillus subtilis* and *E. coli* [10]. Besides its antibacterial activity against both Gram-positive and Gram-negative bacteria, this class of BS also exhibited promising antifungal activity against pathogenic fungi including *Colletotrichum gloeosporioides*, *Fusarium verticillioides*, *Fusarium oxysporum*, *Corynespora cassiicola*, and *Trichophyton rubrum* [90].

The antimicrobial activity of SACs glycolipids was found to be dependent on the type of glycolipid and the interaction with the cell membrane. Diaz de Renzo et al. [63] demonstrated that RLs inhibit bacterial growth in the exponential phase while SLs inhibit growth between the exponential and stationary phases.

The antimicrobial potential of lipopeptide SACs has also been recognized; these biomolecules correspond to the most important components of metabolites that are synthesized by many species of the genus *Bacillus* spp., which characterize the strains of this genus as important parts of plant disease control and food safety [91–93].

Antimicrobial lipopeptides, such as iturin, fengycin, and surfactin, have been identified in *Bacillus velezensis* HC6. Surfactin exhibited strong antibacterial effects against *Listeria monocytogenes* and *Bacillus cereus*, while fengycin and iturin inhibited the growth of pathogenic fungi *Aspergillus flavus*, *Aspergillus parasiticus*, *Aspergillus sulphureus*, *Fusarium graminearum*, and *Fusarium oxysporum* [94]. Researchers also found that when *B. velezensis* HC6 is applied to corn, it reduced the levels of aflatoxin and ochratoxin produced by fungi.

Ohadi et al. [95] demonstrated that lipopeptides produced by *Acinetobacter junii* displayed nonselective activity against Gram-positive and Gram-negative bacterial strains. The data showed that this bioproduct had effective antibacterial activity at concentrations almost below the CMC and that the minimal inhibitory concentration (MIC) values were lower than the standard antifungal activity, exhibiting almost 100% inhibition against *Candida utilis*.

Other broad classes of bacterial metabolites with surface-active potential and antimicrobial effects include glycoproteins, peptides, and fatty acids. *Lactobacillus* spp. produced a bioactive glycolipoprotein surfactant with antimicrobial activity against *C. albicans* using sugarcane molasses as substrate, and some pathogenic gram-positive and gram-negative bacteria [96]. A cyclic heptapeptide containing a fatty acid moiety produced by *Bacillus subtilis*, called bacaucin 1, exhibited specific antibacterial activity against methicillin-resistant *S. aureus* (MRSA) by disrupting the membrane without detectable toxicity to mammalian cells or induction of bacterial resistance. In addition, this peptide was found to be efficient in preventing infections in both in vitro and in vivo models [97].

Finally, some microorganisms excrete mixtures of bioactive compounds with surface-reducing ability and emulsifying potential. For example, the actinomycete strains of *Strep-*

tomyces griseoplanus NRRL-ISP5009 produced a BS component that is a complex mixture of proteins, carbohydrates, and lipids that have antimicrobial activity against gram-positive bacteria (*Bacillus subtilis, S. aureus*) and pathogenic fungi (*C. albicans* and *Aspergillus fumigatus*). However, it is only moderately active against Gram-negative bacteria *E. coli* and *Salmonella typhimurium* [37].

3.2. Antiviral Activity

Viruses represent a serious threat to human health at a global level. Previous studies have described secondary metabolites with surface-active properties for their antiviral properties against a variety of viruses. Antiviral activity by SACs was shown to be effective against various viruses, enveloped and nonenveloped (Table 2).

Table 2. Antiviral properties of SACS.

Biosurfactant/ Bioemulsifier	Microorganism	Antiviral Activity	Virus	Reference
Surfactin	*Bacillus subtilis*	Rupturing the viral lipid membrane and part of the capsid	Semliki Forest virus Simplex virus (HSV-1, HSV-2) Suid herpesvirus (SHV-1)	[98]
		Inhibited the proliferation	Simian immunodeficiency (SIV) Feline calicivirus (FCV) Coronaviruses: Epidemic porcine diarrhea (PEDV) Transmissible gastroenteritis virus (TGEV)	[99]
Lipopeptides	-	Inhibited the membrane fusion between the virus and host cells.	Influenza A (H1N1) Human Coronavirus SARS-CoV-2	[100] [101–103]
Sophorolipids	*Candida bombicola*	Virucidal property	Human Immunodeficiency Virus (HIV)	[104,105]
Rhamnolipids	*Pseudomonas* spp.	Inhibits the cytopathic effect	Simplex virus: HSV-1, and HSV-2;	[106]
	Pseudomonas gessardii M15	Inhibited the proliferation	Simplex vírus: HSV-1 and HSV-2, Human coronavírus: HCoV-229E and SARS-CoV-2	[12]

Viral infections represent one of the main causes of human and animal morbidity and mortality that lead to significant healthcare costs [107]. Therefore, secondary metabolites with surface-active properties should be considered promising substances for the development of antiviral compounds.

3.3. Anti-Inflammatory Activity

Inflammatory responses represent a crucial aspect of a tissue's response to certain injuries, chemical irritation, or microbial infections. This complex response involves leukocyte cells, macrophages, neutrophils, and lymphocytes. In response to inflammation, these cells release specialized substances, including amines and vasoactive peptides, eicosanoids, pro-inflammatory cytokines, and acute-phase proteins, which mediate the inflammatory process and prevent additional tissue damage [108].

Currently, studies on SACs are looking into their potential as anti-inflammatory drugs. For example, a recent in vivo study showed that surfactin inhibited the pro-inflammatory response in *Zebrafish larvae* (*Danio rerio*), significantly reducing the expression of interleukin (IL)-1β, IL-8, tumor necrosis factor-α (TNF-α), nitric oxide (NO), nuclear factor kappa-β p65 (NF-κBp65), cyclooxygenase-2 (COX-2), and inducible nitric oxide synthase (iNOS) and increasing the expression of IL-10. Furthermore, the study showed that surfactin reduced neutrophil migration and alleviated liver damage [109].

Other studies showed that surfactin systematically induced CD4 + CD25 + FoxP3 + Tregs in the spleen of mice, which inhibit T cells from producing pro-inflammatory cytokines such as TNF-α and interferon (IFN)-γ. Moreover, surfactin attenuation of chronic inflammation increased IL-10 expression in atherosclerotic lesions of the aorta of mice, demonstrating that BSs can restore the balance in the Th1/Th2 response in mice [110], as well as induce the maturation of dendritic cells (DCs) and increase the expression of MHC-II molecules and other costimulatory factors [111].

Few anti-inflammatory properties related to glycolipid BSs have been reported. Sophorolipids produced by *C. bombicola* reduced lipopolysaccharide-induced expression of TNF-α, COX-2, and IL-6 in RAW 264.7 cells [112], and reduced the level of immunoglobulin E (IgE), TLR-2, IL-6, and STAT3 mRNA expression [113].

In previous in vivo models, SLs reduced sepsis-related mortality and exhibited anti-inflammatory effects in mice by inhibiting nitric oxide and inflammatory cytokine production [114,115]. On the other hand, the glycolipid complex had no significant effect on the proliferative effect of peripheral blood leukocytes because it activated the production of pro-inflammatory cytokines (IL-1β and TNF-α) without affecting the IL-6 production in vitro in the monocyte fraction [116].

3.4. Anticancer Activity

Cancer is considered a multistage disease, the etiology of which is associated with high incidence and mortality rates globally. Chemotherapy drugs, surgery, and radiation remain the most common treatments to fight the disease in humans. However, they are all associated with serious adverse effects, indicating the lack of specificity and the need to discover new antitumor agents to improve the effectiveness of conventional chemotherapy drugs while reducing the adverse effects [74].

For these purposes, several studies have demonstrated the antitumor potential of several SACs (Table 3). Biosurfactants have been proposed as suitable drug candidates for many diseases including cancer [117]. Given their wide applications, the interest in exploring their role in promoting human health continues to grow.

Table 3. Anticancer activity of SACS against cancer cells.

Biosurfactant/ Bioemulsifier	Microorganism	Anticancer Activity	Cancer	Reference
Rhamnolipids: monorhamnolipid and dirhamnolipid	*P. aeruginosa* MR01	Inhibiting cell division at lower concentrations	Human breast cancer MCF-7	[118]
Sophorolipids	*Wickerhamiella domercqiae* Y2A	Increased the apoptosis	HepG2 liver cancer cells	[109]
		Cytotoxicity	Breast cancer MDA-MB-231	[119]
		Inhibited cell proliferation	Liver Lung Leukemia	[120]
Surfactin		Reduced tumor growth and weight; Apoptosis; Elevated levels of immune-boosting mediators	Melanoma skin cancer	[13]
	Bacillus saphensis	Cytotoxic activity against cancer cell lines	Breast cancer Melanoma	[46]
Iturin	*Bacillus megaterium*	Inhibited the growth of cancer cells	Breast cancer	[121]

3.5. Antibiofilm Activity

Biofilms comprise microbial communities attached to the surface and embedded in an extracellular matrix composed of extracellular polymeric substances (EPS) secreted by cells that reside within them. In general, EPS is a mixture of polysaccharides, proteins, extracellular DNA (eDNA), and other smaller components. The biofilm matrix constituents' physical and chemical properties enable the matrix to protect resident cells from desiccation, chemical disturbance, invasion by other bacteria, and death from predators. However, biofilms often cause major medical problems and are the cause of chronic infections because biofilm communities can house bacteria that are tolerant and persistent against antibiotic treatment and are more resistant to antibiotics compared with planktonic bacteria [9,122].

Because of their composition, biofilms cause a wide range of chronic diseases due to the emergence of antibiotic-resistant bacteria that have become difficult to treat effectively. To date, available antibiotics are ineffective in treating these biofilm-related infections because of their higher MIC and minimal bactericidal concentration values, which may lead to in vivo toxicity. Therefore, designing or tracking antibiofilm molecules that can effectively minimize and eradicate biofilm-related infections is important [123].

Because of their antimicrobial, antiadhesive, and antibiofilm properties, SACs can be used to neutralize biofilm formation and the emergence of drug-resistant microorganisms [14]. These biomolecules tend to interact with antimicrobials [124,125], which are usually less effective against biofilms in general and multispecies biofilms associated with extremely complicated polymicrobial infections.

A mixture of lipopeptides (surfactin, iturin, and fengycin), which are synthesized by *B. subtilis*, prevented biofilm formation by inhibiting cell adhesion of *Trichosporon* spp. by up to 96.89% and dispersed mature biofilms (up to 99.2%), reducing their thickness and cell viability. This mixture reduced cell ergosterol content and altered the membrane permeability and surface hydrophobicity of planktonic cells [126].

Another mixture of lipopeptides (surfactin, iturin, and lichenysin) was identified for the first time in *Lactobacillus* spp. vaginal exhibited strong antiadhesive activity (up to 74.4%) against the biofilm producer *C. albicans* [67]. Mixed lipopeptides (iturin, fengycin, and surfactin) with higher surfactin content produced by *B. subtilis* TIM10 and *B. vallismortis* TIM68 inhibited the biofilm formation of *Malassezia* spp., especially TIM10, by about 90% [127].

Meanwhile, surfactin-type BS produced by *B. subtilis* reduced adhesion and stopped the formation of *S. aureus* biofilm on glass, polystyrene, and stainless-steel surfaces. Surfactin significantly decreased biofilm percentage and reduced *icaA* and *icaD* expressions, which are important for staphylococcal biofilm structure. Furthermore, lipopeptides have been shown to affect the quorum sensing (QS) system in *S. aureus* by regulating the autoinducer 2 activity [94].

In terms of the antibiofilm activity of glycolipids, Allegrone et al. [128] reported the effects of different types of RLs. They demonstrated that RL produced by *P. aeruginosa* 89 (R89BS) was 91.4% pure and comprised 70.6% of monorhamnolipids and 20.8% of dirhamnolipids. The pure extract R89BS inhibited S. aureus adhesion (97%) and biofilm formation (85%). Furthermore, purified monorhamnolipids (mR89BS) have been observed to induce dispersion of preformed biofilms at all concentrations (0.06–2 mg/mL) by 80–99%, unlike the pure extract R89BS and purified dirhamnolipids (dR89BS), which depended on the tested concentration.

Ceresa et al. [5] demonstrated that R89BS-coated silicone elastomeric disks significantly neutralized *Staphylococcus* spp. biofilm formation in terms of accumulated biomass (up to 60% inhibition in 72 h) and cellular metabolic activity (up to 68% inhibition in 72 h). The results suggested that RL coatings may be an effective antibiofilm treatment procedure and represent a promising strategy for preventing infections associated with implantable medical devices.

Shen et al. [129] demonstrated that besides inhibiting the formation of new biofilms, RLs were superior in eradicating mature biofilms formed by *Helicobacter pylori*, *E. coli*, and *Streptococcus mutans* in a dose-dependent manner, compared with other antibacterial agents even at concentrations below minimum inhibitory concentrations (MICs). They can enhance the effect of antimicrobial agents. Sidrim et al. [130] observed that these molecules significantly increased the activity of meropenem and amoxicillin-clavulanate against mature *Burkholderia pseudomallei* biofilms.

Rhamnolipids produced by *P. aeruginosa* SS14 also inhibited planktonic cells of filamentous fungi of *Trichophyton rubrum* and *Trichophyton mentagrophytes*. The formation and rupture of mature biofilms were dose-dependent, with the highest activity observed at concentrations of 2 × MIC against both pathogens [131].

Like RLs, SLs exhibited an effective inhibitory activity against biofilm formation. Ceresa et al. [132] obtained three different SL products: SLA (acid congeners), SL18 (lactonic congeners), and SLV (mixture of acid and lactone congeners), which all showed an inhibitory effect of 70%, 75%, and 80% for *S. aureus*, *P. aeruginosa*, and *C. albicans*, respectively. Using 0.8% *w/v* SLA on pre-coated medical silicone disks reduced *S. aureus* biofilm formation by 75%. In co-incubation experiments, 0.05% *w/v* SLA significantly inhibited *S. aureus* and *C. albicans* from forming biofilms and adhering to surfaces by 90–95% at concentrations between 0.025 and 0.1% *w/v*.

Antibiofilm activities were also demonstrated for BSs produced by probiotics of the genus Bacillus sp. that were isolated from cervicovaginal samples. This bioproduct, called BioSa3, was highly effective in the formation of biofilms of different pathogenic and multidrug-resistant strains, such as *S. aureus* and *Staphylococcus haemolyticus*. The anti-biofilm effect may be related to the ability of BioSa3 to alter the membrane physiology of the tested pathogens to cause a significant decrease in surface hydrophobicity [133].

Thus, SACs are good candidates for the emergence of new therapies to better control multidrug-resistant microorganisms and inhibit infections associated with biofilms, protecting surfaces from microbial contamination.

3.6. Wound Healing

Wound healing is an important but complicated process of tissue repair in humans or animals, comprising a multifaceted process organized by sequential and overlapping phases, including hemostasis, inflammation phase, proliferation phase, and remodeling phase [134,135]. Failure of one of these phases caused by a deregulated immune response or insufficient oxygenation impairs the healing process, leading to ulcerative skin defect (chronic wound) or excessive scar tissue formation (hypertrophic or keloid scarring) [136,137].

Treating wounds of different etiologies constitutes an important part of the total health budget, mostly affected by three important cost drivers: curing time, frequency of dressing change, and complications. Moreover, chronic wound infection, one of the leading causes of nonhealing, contributes significantly to rising healthcare costs. Although the treatment of an uncomplicated surgical incision is relatively inexpensive, the costs can increase significantly when infections occur [138].

Biofilms, commonly found in chronic wounds, contribute to infections, causing slower healing. Infections in chronic wounds are usually caused by multiple species [139], with *P. aeruginosa* and *S. aureus* being the most common. Although most microbial communities usually form on the wound's outer layer, some biofilms are also embedded in deeper layers, such as *P. aeruginosa* biofilms, which are difficult to diagnose via traditional wound smear culture [140,141]. Moreover, antibiotic resistance of bacteria in biofilms is a crucial problem in the management and treatment of chronic wounds [139].

For these reasons, physicians and the scientific community consider the management and treatment of wounds, as well as biofilm prevention, a top priority. In this context, SACs recently emerged as promising agents that promote wound healing with low irritation and high compatibility with human skin [14]. Furthermore, these bioproducts promote fibroblast and epithelial cell proliferation, faster re-epithelialization, and collagen deposition, leading to a faster healing process [142,143].

Surfactin A from *B. subtilis* promotes wound healing and scar inhibition. During the healing process, it up-regulates the expression of hypoxia-inducible factor-1α (HIF-1α) and vascular endothelial growth factor, accelerates keratinocyte migration via mitogen-activated protein kinase (MAPK), and factor nuclear-κB (NF-κB) signaling pathways and also regulates pro-inflammatory cytokine secretion and macrophage phenotypic exchange. Furthermore, surfactin A inhibits scar tissue formation by influencing α-smooth muscle actin (α-SMA) and transforming growth factor (TGF-β) expression [144]. Therefore, the healing potency of the lipopeptides *B. subtilis* SPB1 is due to their antioxidant activity potential revealed in vitro [143].

A previously unknown lipopeptide 78 (LP78) from *S. epidermidis* inhibited TLR3-mediated skin inflammation and promoted wound healing. The skin lesion activated TLR3/NF-κB, promoting p65 and PPARγ interaction in the nuclei and initiating the inflammatory response in keratinocytes. Next, LP78 activated the TLR2-SRC, inducing β-catenin phosphorylation in Tyr. Phospho-β-catenin is translocated into the nuclei to bind to PPARγ, thereby interrupting the p65 and PPARγ interaction. Dissociation between p65 and PPARγ reduced TLR3-induced inflammatory cytokine expression in skin wounds of normal and diabetic mice, which correlated with faster wound healing [145].

As an alternative to improve this healing process, the formulation of nanolipopeptide biosurfactant (NLPB) from the lipopeptide biosurfactant (LPB) produced by *Acinetobacter junii* was reported as promising for performing healing activity. The percentage of wound closure of mice treated with NLPB hydrogels at 2 mg/mL was approximately 80% on day 7 and 100% on day 15. The NLPB hydrogel formulation showed better efficacy in wound closure and healing when compared to the control [146].

A BS of glycolipid nature, which was synthesized by *Bacillus licheniformis* SV1, showed good cytocompatibility and increased 3T3/NIH fibroblasts proliferation in vitro. A previous study showed that the application of BS ointment in a skin excision wound in rats promoted re-epithelialization, fibroblast cell proliferation, and faster collagen deposition, demonstrating its potential transdermal properties to improve skin wound healing [147].

A previous study administered an RL-containing ointment (5 g/L) on the back of Wistar mice after creating an excision wound. Histopathological results revealed a significant healing effect of RL, demonstrating increased wound closure, improved collagenases, and reduced inflammation (decreasing the level of TNF-α) without inducing skin irritation [84]. Dirhamnolipid treatment has been suggested for cutaneous scar therapy, demonstrating an antifibrotic function in rabbit ear hypertrophic scar models with a significant reduction in the scar elevation index, type I collagen fibers, and α-SMA expression [148].

A cell culture model has demonstrated the wound healing capacity of SLs by using an in vitro human dermal fibroblast model as a substitute for human skin, revealing that SLs affected the ability of human skin fibroblasts to express collagen I mRNA (Col-I) and elastase inhibition (IC_{50} = 38.5 μg/mL) [112]. In addition, Kwak et al. (2021), using an in vitro wound healing assay in human colorectal adenocarcinoma (HT-29) cell line, showed a significantly increased collagenase-1 expression ($p < 0.05$) 48 h after SL treatment. Moreover, all SL dosages significantly increased occludin and matrilysin-1 (MMP-7) expression [149].

3.7. Other Considerations

We also consider that there are SACs molecules obtained by chemical synthesis processes, such as ultrashort synthetic surface active (USSA) [150,151]. Some of these can be synthesized as C-terminal amides on Rink amide (4-Methylbenzhydrylamine (MBHA) resin using 9-fluorenylmethoxycarbonyl/t-butylcarbamate [151]. The fundamental difference of the USSA, as lipopeptoids (modified SAC) in relation to the natural ones, is their immunomodulatory capacity. As seen in mouse infection models, they reduce the exacerbation of the disease, even if not presenting direct antibacterial activity [151]. This characteristic would be a limiting activity, since many natural ones lead to a disturbance of biological membranes, with antifungal and antibacterial actions [151].

New possibilities can be obtained for the SACs, as transformation systems applying recombinant plasmids have been employed to substantially increase the productivity of microbial biosurfactants, e.g., the engineered strain *Pseudozyma* sp. SY16, which increases the production of mannosylerythritol lipids (MELs) by up to 31.5%, suggesting that genetic engineering can improve the industrial application of yeast [152].

4. Conclusions

The BS and BE surface-active compounds have drawn the attention of the scientific community as a new generation of products with high potential in the biomedical and pharmaceutical fields. Their use, whether alone or in combination with other antimicrobial or chemotherapeutic agents, opens paths for new strategies to prevent and combat infections caused by bacteria, fungi, and viruses, as well as the formation and proliferation of biofilms. Furthermore, new anticancer treatments and wound healing applications can be explored in future studies.

These molecules affect various biological activities, making them suitable candidates for use in new generations of agents in the biotechnological, biomedical, and pharmaceutical fields. However, it is necessary to investigate their applications, cost-effectiveness, and availability further.

Author Contributions: Conceptualization, J.A. and A.M.; writing—original draft preparation, J.A., J.M., D.S. (Douglas Silva), A.A., K.S., L.C., D.S. (Darlan Silva) and W.P.; writing—review and editing, J.M., M.S., L.S. and A.M.; visualization, L.S., M.S. and A.M.; supervision, A.M.; project administration, J.M. and A.M.; funding acquisition, A.M. All authors have read and agreed to the published version of the manuscript.

Funding: This study was financed in part by Fundação de Amparo à Pesquisa e o Desenvolvimento Científico e Tecnológico do Maranhão (FAPEMA, BD-01413/19); and Conselho Nacional de Desenvolvimento Científico e Tecnológico (CNPQ, 434149/2018-7).

Institutional Review Board Statement: Not applicable.

Informed Consent Statement: Not applicable.

Data Availability Statement: Not applicable.

Acknowledgments: The authors would like to thank Hélio Euclides S. dos Santos and Marinaldo for technical assistance.

Conflicts of Interest: The authors declare no conflict of interest.

Abbreviations

ACE2	angiotensin-converting enzyme 2
BE	bioemulsifier
BS	biosurfactant
CD4	differentiation cluster 4
CD25	differentiation cluster 25
CD31	differentiation cluster 31
COX-2	cyclooxygenase-2
FOXP3	Forkhead box protein P3
HIF-1α	hypoxia-inducible factor-1α
HT-29	human colorectal adenocarcinoma
IL-2	interleukin 2
IL-10	interleukin 10
IL1-1β	interleukin 10-1β
IL-8	interleukin 8
IgE	immunoglobulin E
iNOS	inducible nitric oxide synthase
LP78	lipopeptide 78
MBHA	(4-Methylbenzhydrylamine)

MAPK	mitogen-activated protein kinase
MELs	mannosylerythritol lipids
MIC	minimum inhibitory concentration
MICs	minimum inhibitory concentrations
MRSA	methicillin-resistant *S. aureus*
NF	κB factor nuclear-κB
NF	κBp65- nuclear factor kappa- β p65
NLPB	nanolipopeptide biosurfactant
NO	nitric oxide
P-Akt	P-mitogen-activated
P-GSK3β	protein kinase syntoxic glycogen-3 beta
PPARγ	peroxisome proliferator-activated receptor γ
QS	quorum sensing
RL	rhamnolipid
SACs	surface-active compounds
TGF-β	transforming growth factor
TLR3	Toll-like receptor 3
TNF-α	tumor necrosis factor α
USSA	ultrashort synthetic surface active
α-SMA	influencing α-smooth muscle actin

References

1. Franzetti, A.; Tamburini, E.; Banat, I.M. Applications of Biological Surface-Active Compounds in Remediation Technologies. *Adv. Exp. Med. Biol.* **2010**, *672*, 121–134. [CrossRef] [PubMed]
2. Phulpoto, I.A.; Yu, Z.; Hu, B.; Wang, Y.; Ndayisenga, F.; Li, J.; Liang, H.; Qazi, M.A. Production and Characterization of Surfactin-like Biosurfactant Produced by Novel Strain *Bacillus nealsonii* S2MT and It's Potential for Oil Contaminated Soil Remediation. *Microb. Cell Fact.* **2020**, *19*, 145. [CrossRef] [PubMed]
3. Uzoigwe, C.; Burgess, J.G.; Ennis, C.J.; Rahman, P.K.S.M. Bioemulsifiers Are Not Biosurfactants and Require Different Screening Approaches. *Front. Microbiol.* **2015**, *6*, 245. [CrossRef] [PubMed]
4. Durval, I.J.B.; Mendonça, A.H.R.; Rocha, I.V.; Luna, J.M.; Rufino, R.D.; Converti, A.; Sarubbo, L.A. Production, Characterization, Evaluation and Toxicity Assessment of a *Bacillus cereus* UCP 1615 Biosurfactant for Marine Oil Spills Bioremediation. *Mar. Pollut. Bull.* **2020**, *157*, 111357. [CrossRef]
5. Ceresa, C.; Tessarolo, F.; Maniglio, D.; Tambone, E.; Carmagnola, I.; Fedeli, E.; Caola, I.; Nollo, G.; Chiono, V.; Allegrone, G.; et al. Medical-Grade Silicone Coated with Rhamnolipid R89 Is Effective against *Staphylococcus* spp. Biofilms. *Molecules* **2019**, *24*, 3843. [CrossRef]
6. Md Badrul Hisham, N.H.; Ibrahim, M.F.; Ramli, N.; Abd-Aziz, S. Production of Biosurfactant Produced from Used Cooking Oil by *Bacillus* Sp. HIP3 for Heavy Metals Removal. *Molecules* **2019**, *24*, 2617. [CrossRef]
7. Yamasaki, R.; Kawano, A.; Yoshioka, Y.; Ariyoshi, W. Rhamnolipids and Surfactin Inhibit the Growth or Formation of Oral Bacterial Biofilm. *BMC Microbiol.* **2020**, *20*, 358. [CrossRef]
8. Sun, W.; Zhu, B.; Yang, F.; Dai, M.; Sehar, S.; Peng, C.; Ali, I.; Naz, I. Optimization of Biosurfactant Production from *Pseudomonas* sp. CQ2 and Its Application for Remediation of Heavy Metal Contaminated Soil. *Chemosphere* **2021**, *265*, 129090. [CrossRef]
9. Sharma, D.; Misba, L.; Khan, A.U. Antibiotics versus Biofilm: An Emerging Battleground in Microbial Communities. *Antimicrob. Resist. Infect. Control* **2019**, *8*, 76. [CrossRef]
10. Gaur, V.K.; Regar, R.K.; Dhiman, N.; Gautam, K.; Srivastava, J.K.; Patnaik, S.; Kamthan, M.; Manickam, N. Biosynthesis and Characterization of Sophorolipid Biosurfactant by *Candida* spp.: Application as Food Emulsifier and Antibacterial Agent. *Bioresour. Technol.* **2019**, *285*, 121314. [CrossRef]
11. Kumari, A.; Kumari, S.; Prasad, G.S.; Pinnaka, A.K. Production of Sophorolipid Biosurfactant by Insect Derived Novel Yeast *Metschnikowia churdharensis* f.A., Sp. Nov., and Its Antifungal Activity against Plant and Human Pathogens. *Front. Microbiol.* **2021**, *12*, 678668. [CrossRef] [PubMed]
12. Giugliano, R.; Buonocore, C.; Zannella, C.; Chianese, A.; Palma Esposito, F.; Tedesco, P.; De Filippis, A.; Galdiero, M.; Franci, G.; de Pascale, D. Antiviral Activity of the Rhamnolipids Mixture from the Antarctic Bacterium *Pseudomonas gessardii* M15 against *Herpes simplex* Viruses and Coronaviruses. *Pharmaceutics* **2021**, *13*, 2121. [CrossRef] [PubMed]
13. Kim, H.-Y.; Jung, H.; Kim, H.-M.; Jeong, H.-J. Surfactin Exerts an Anti-Cancer Effect through Inducing Allergic Reactions in Melanoma Skin Cancer. *Int. Immunopharmacol.* **2021**, *99*, 107934. [CrossRef] [PubMed]
14. Ceresa, C.; Fracchia, L.; Fedeli, E.; Porta, C.; Banat, I.M. Recent Advances in Biomedical, Therapeutic and Pharmaceutical Applications of Microbial Surfactants. *Pharmaceutics* **2021**, *13*, 466. [CrossRef]
15. Mujumdar, S.; Joshi, P.; Karve, N. Production, Characterization, and Applications of Bioemulsifiers (BE) and Biosurfactants (BS) Produced by *Acinetobacter* Spp.: A Review. *J. Basic Microbiol.* **2019**, *59*, 277–287. [CrossRef]

16. Cazals, F.; Huguenot, D.; Crampon, M.; Colombano, S.; Betelu, S.; Galopin, N.; Perrault, A.; Simonnot, M.-O.; Ignatiadis, I.; Rossano, S. Production of Biosurfactant Using the Endemic Bacterial Community of a PAHs Contaminated Soil, and Its Potential Use for PAHs Remobilization. *Sci. Total Environ.* **2020**, *709*, 136143. [CrossRef]
17. Loeto, D.; Jongman, M.; Lekote, L.; Muzila, M.; Mokomane, M.; Motlhanka, K.; Ndlovu, T.; Zhou, N. Biosurfactant Production by Halophilic Yeasts Isolated from Extreme Environments in Botswana. *FEMS Microbiol. Lett.* **2021**, *368*, 146. [CrossRef]
18. Silva, M.E.T.; Duvoisin, S., Jr.; Oliveira, R.L.; Banhos, E.F.; Souza, A.Q.L.; Albuquerque, P.M. Biosurfactant Production of *Piper hispidum* Endophytic Fungi. *J. Appl. Microbiol.* **2021**, *130*, 561–569. [CrossRef]
19. Nayarisseri, A.; Singh, P.; Singh, S.K. Screening, Isolation and Characterization of Biosurfactant-Producing *Bacillus tequilensis* Strain ANSKLAB04 from Brackish River Water. *Int. J. Environ. Sci. Technol.* **2019**, *16*, 7103–7112. [CrossRef]
20. Desai, J.D.; Banat, I.M. Microbial Production of Surfactants and Their Commercial Potential. *Fuel Energy Abstr.* **1997**, *38*, 221. [CrossRef]
21. Chen, W.-C.; Juang, R.-S.; Wei, Y.-H. Applications of a Lipopeptide Biosurfactant, Surfactin, Produced by Microorganisms. *Biochem. Eng. J.* **2015**, *103*, 158–169. [CrossRef]
22. Otzen, D.E. Biosurfactants and Surfactants Interacting with Membranes and Proteins: Same but Different? *Biochim. Biophys. Acta* **2017**, *1859*, 639–649. [CrossRef] [PubMed]
23. Hogan, D.E.; Tian, F.; Malm, S.W.; Olivares, C.; Palos Pacheco, R.; Simonich, M.T.; Hunjan, A.S.; Tanguay, R.L.; Klimecki, W.T.; Polt, R.; et al. Biodegradability and Toxicity of Monorhamnolipid Biosurfactant Diastereomers. *J. Hazard. Mater.* **2019**, *364*, 600–607. [CrossRef]
24. Qazi, M.A.; Wang, Q.; Dai, Z. Sophorolipids Bioproduction in the Yeast *Starmerella bombicola*: Current Trends and Perspectives. *Bioresour. Technol.* **2022**, *346*, 126593. [CrossRef]
25. Ceresa, C.; Hutton, S.; Lajarin-Cuesta, M.; Heaton, R.; Hargreaves, I.; Fracchia, L.; De Rienzo, M.A.D. Production of Mannosylerythritol Lipids (MELs) to Be Used as Antimicrobial Agents against *S. aureus* ATCC 6538. *Curr. Microbiol.* **2020**, *77*, 1373–1380. [CrossRef] [PubMed]
26. Niu, Y.; Wu, J.; Wang, W.; Chen, Q. Production and Characterization of a New Glycolipid, Mannosylerythritol Lipid, from Waste Cooking Oil Biotransformation by *Pseudozyma aphidis* ZJUDM34. *Food Sci. Nutr.* **2019**, *7*, 937–948. [CrossRef] [PubMed]
27. Luong, T.M.; Ponamoreva, O.N.; Nechaeva, I.A.; Petrikov, K.V.; Delegan, Y.A.; Surin, A.K.; Linklater, D.; Filonov, A.E. Characterization of Biosurfactants Produced by the Oil-Degrading Bacterium *Rhodococcus erythropolis* S67 at Low Temperature. *World J. Microbiol. Biotechnol.* **2018**, *34*, 20. [CrossRef]
28. Kubicki, S.; Bollinger, A.; Katzke, N.; Jaeger, K.-E.; Loeschcke, A.; Thies, S. Marine Biosurfactants: Biosynthesis, Structural Diversity and Biotechnological Applications. *Mar. Drugs* **2019**, *17*, 408. [CrossRef]
29. Hu, F.; Liu, Y.; Li, S. Rational Strain Improvement for Surfactin Production: Enhancing the Yield and Generating Novel Structures. *Microb. Cell Fact.* **2019**, *18*, 42. [CrossRef]
30. Habe, H.; Taira, T.; Sato, Y.; Imura, T.; Ano, T. Evaluation of Yield and Surface Tension-Lowering Activity of Iturin a Produced by *Bacillus subtilis* RB14. *J. Oleo Sci.* **2019**, *68*, 1157–1162. [CrossRef]
31. Sarwar, A.; Brader, G.; Corretto, E.; Aleti, G.; Abaidullah, M.; Sessitsch, A.; Hafeez, F.Y. Qualitative Analysis of Biosurfactants from *Bacillus* Species Exhibiting Antifungal Activity. *PLoS ONE* **2018**, *13*, e0198107. [CrossRef]
32. Denoirjean, T.; Doury, G.; Poli, P.; Coutte, F.; Ameline, A. Effects of Bacillus Lipopeptides on the Survival and Behavior of the Rosy Apple Aphid Dysaphis Plantaginea. *Ecotoxicol. Environ. Saf.* **2021**, *226*, 112840. [CrossRef] [PubMed]
33. Singh, P.; Patil, Y.; Rale, V. Biosurfactant Production: Emerging Trends and Promising Strategies. *J. Appl. Microbiol.* **2019**, *126*, 2–13. [CrossRef]
34. Zambry, N.S.; Rusly, N.S.; Awang, M.S.; Md Noh, N.A.; Yahya, A.R.M. Production of Lipopeptide Biosurfactant in Batch and Fed-Batch *Streptomyces* sp. PBD-410L Cultures Growing on Palm Oil. *Bioprocess Biosyst. Eng.* **2021**, *44*, 1577–1592. [CrossRef] [PubMed]
35. Pardhi, D.S.; Panchal, R.R.; Raval, V.H.; Rajput, K.N. Statistical Optimization of Medium Components for Biosurfactant Production by *Pseudomonas guguanensis* D30. *Prep. Biochem. Biotechnol.* **2022**, *52*, 171–180. [CrossRef] [PubMed]
36. Dos Santos, R.A.; Rodríguez, D.M.; Ferreira, I.N.d.S.; de Almeida, S.M.; Takaki, G.M.d.C.; de Lima, M.A.B. Novel Production of Biodispersant by *Serratia marcescens* UCP 1549 in Solid-State Fermentation and Application for Oil Spill Bioremediation. *Environ. Technol.* **2021**, *43*, 1–12. [CrossRef]
37. Elkhawaga, M.A. Optimization and Characterization of Biosurfactant from *Streptomyces griseoplanus* NRRL-ISP5009 (MS1). *J. Appl. Microbiol.* **2018**, *124*, 691–707. [CrossRef]
38. Bezza, F.A.; Chirwa, E.M.N. Biosurfactant-Enhanced Bioremediation of Aged Polycyclic Aromatic Hydrocarbons (PAHs) in Creosote Contaminated Soil. *Chemosphere* **2016**, *144*, 635–644. [CrossRef]
39. Elakkiya, V.T.; SureshKumar, P.; Alharbi, N.S.; Kadaikunnan, S.; Khaled, J.M.; Govindarajan, M. Swift Production of Rhamnolipid Biosurfactant, Biopolymer and Synthesis of Biosurfactant-Wrapped Silver Nanoparticles and Its Enhanced Oil Recovery. *Saudi J. Biol. Sci.* **2020**, *27*, 1892–1899. [CrossRef]
40. Nurfarahin, A.H.; Mohamed, M.S.; Phang, L.Y. Culture Medium Development for Microbial-Derived Surfactants Production-an Overview. *Molecules* **2018**, *23*, 1049. [CrossRef]

41. Liu, J.-F.; Mbadinga, S.M.; Yang, S.-Z.; Gu, J.-D.; Mu, B.-Z. Chemical Structure, Property and Potential Applications of Biosurfactants Produced by *Bacillus subtilis* in Petroleum Recovery and Spill Mitigation. *Int. J. Mol. Sci.* **2015**, *16*, 4814–4837. [CrossRef] [PubMed]
42. De Almeida, D.G.; Soares Da Silva, R.d.C.F.; Luna, J.M.; Rufino, R.D.; Santos, V.A.; Banat, I.M.; Sarubbo, L.A. Biosurfactants: Promising Molecules for Petroleum Biotechnology Advances. *Front. Microbiol.* **2016**, *7*, 1718. [CrossRef] [PubMed]
43. Ostendorf, T.A.; Silva, I.A.; Converti, A.; Sarubbo, L.A. Production and Formulation of a New Low-Cost Biosurfactant to Remediate Oil-Contaminated Seawater. *J. Biotechnol.* **2019**, *295*, 71–79. [CrossRef] [PubMed]
44. Silva, I.A.; Veras, B.O.; Ribeiro, B.G.; Aguiar, J.S.; Campos Guerra, J.M.; Luna, J.M.; Sarubbo, L.A. Production of Cupcake-like Dessert Containing Microbial Biosurfactant as an Emulsifier. *PeerJ* **2020**, *8*, e9064. [CrossRef]
45. Ohadi, M.; Shahravan, A.; Dehghannoudeh, N.; Eslaminejad, T.; Banat, I.M.; Dehghannoudeh, G. Potential Use of Microbial Surfactant in Microemulsion Drug Delivery System: A Systematic Review. *Drug Des. Dev. Ther.* **2020**, *14*, 541–550. [CrossRef]
46. Abdelli, F.; Jardak, M.; Elloumi, J.; Stien, D.; Cherif, S.; Mnif, S.; Aifa, S. Antibacterial, Anti-Adherent and Cytotoxic Activities of Surfactin(s) from a Lipolytic Strain *Bacillus safensis* F4. *Biodegradation* **2019**, *30*, 287–300. [CrossRef] [PubMed]
47. Bhaumik, M.; Dhanarajan, G.; Chopra, J.; Kumar, R.; Hazra, C.; Sen, R. Production, Partial Purification and Characterization of a Proteoglycan Bioemulsifier from an Oleaginous Yeast. *Bioprocess Biosyst. Eng.* **2020**, *43*, 1747–1759. [CrossRef]
48. Rosenberg, E.; Ron, E.Z. High- and Low-Molecular-Mass Microbial Surfactants. *Appl. Microbiol. Biotechnol.* **1999**, *52*, 154–162. [CrossRef]
49. Ortega-de la Rosa, N.D.; Vázquez-Vázquez, J.L.; Huerta-Ochoa, S.; Gimeno, M.; Gutiérrez-Rojas, M. Stable Bioemulsifiers Are Produced by *Acinetobacter bouvetii* UAM25 Growing in Different Carbon Sources. *Bioprocess Biosyst. Eng.* **2018**, *41*, 859–869. [CrossRef]
50. Patil, J.R.; Chopade, B.A. Studies on Bioemulsifier Production by *Acinetobacter* Strains Isolated from Healthy Human Skin. *J. Appl. Microbiol.* **2001**, *91*, 290–298. [CrossRef]
51. Fan, Y.; Tao, W.; Huang, H.; Li, S. Characterization of a Novel Bioemulsifier from *Pseudomonas stutzeri*. *World J. Microbiol. Biotechnol.* **2017**, *33*, 161. [CrossRef] [PubMed]
52. Adetunji, A.I.; Olaniran, A.O. Production and Characterization of Bioemulsifiers from *Acinetobacter* strains Isolated from Lipid-Rich Wastewater. *3 Biotech* **2019**, *9*, 151. [CrossRef] [PubMed]
53. Marques, N.S.A.A.; Silva, I.G.S.d.; Cavalcanti, D.L.; Maia, P.C.S.V.; Santos, V.P.; Andrade, R.F.S.; Campos-Takaki, G.M. Eco-Friendly Bioemulsifier Production by Mucor Circinelloides UCP0001 Isolated from Mangrove Sediments Using Renewable Substrates for Environmental Applications. *Biomolecules* **2020**, *10*, 365. [CrossRef] [PubMed]
54. Gudiña, E.J.; Pereira, J.F.B.; Costa, R.; Evtuguin, D.V.; Coutinho, J.A.P.; Teixeira, J.A.; Rodrigues, L.R. Novel Bioemulsifier Produced by a Paenibacillus Strain Isolated from Crude Oil. *Microb. Cell Fact.* **2015**, *14*, 14. [CrossRef]
55. Tao, W.; Lin, J.; Wang, W.; Huang, H.; Li, S. Designer Bioemulsifiers Based on Combinations of Different Polysaccharides with the Novel Emulsifying Esterase AXE from Bacillus Subtilis CICC 20034. *Microb. Cell Fact.* **2019**, *18*, 173. [CrossRef] [PubMed]
56. Dastgheib, S.M.M.; Amoozegar, M.A.; Elahi, E.; Asad, S.; Banat, I.M. Bioemulsifier Production by a Halothermophilic *Bacillus* Strain with Potential Applications in Microbially Enhanced Oil Recovery. *Biotechnol. Lett.* **2008**, *30*, 263–270. [CrossRef]
57. Goldman, S.; Shabtai, Y.; Rubinovitz, C.; Rosenberg, E.; Gutnick, D.L. Emulsan in *Acinetobacter calcoaceticus* RAG-1: Distribution of Cell-Free and Cell-Associated Cross-Reacting Material. *Appl. Environ. Microbiol.* **1982**, *44*, 165–170. [CrossRef]
58. Toren, A.; Navon-Venezia, S.; Ron, E.Z.; Rosenberg, E. Emulsifying Activities of Purified Alasan Proteins from *Acinetobacter radioresistens* KA53. *Appl. Environ. Microbiol.* **2001**, *67*, 1102–1106. [CrossRef]
59. Ohadi, M.; Dehghannoudeh, G.; Forootanfar, H.; Shakibaie, M.; Rajaee, M. Investigation of the Structural, Physicochemical Properties, and Aggregation Behavior of Lipopeptide Biosurfactant Produced by *Acinetobacter junii* B6. *Int. J. Biol. Macromol.* **2018**, *112*, 712–719. [CrossRef]
60. Dong, H.; Xia, W.; Dong, H.; She, Y.; Zhu, P.; Liang, K.; Zhang, Z.; Liang, C.; Song, Z.; Sun, S.; et al. Rhamnolipids Produced by Indigenous *Acinetobacter junii* from Petroleum Reservoir and Its Potential in Enhanced Oil Recovery. *Front. Microbiol.* **2016**, *7*, 1710. [CrossRef]
61. Rosenberg, E.; Rubinovitz, C.; Gottlieb, A.; Rosenhak, S.; Ron, E.Z. Production of Biodispersan by *Acinetobacter calcoaceticus* A2. *Appl. Environ. Microbiol.* **1988**, *54*, 317–322. [CrossRef] [PubMed]
62. Klausmann, P.; Hennemann, K.; Hoffmann, M.; Treinen, C.; Aschern, M.; Lilge, L.; Morabbi Heravi, K.; Henkel, M.; Hausmann, R. *Bacillus subtilis* High Cell Density Fermentation Using a Sporulation-Deficient Strain for the Production of Surfactin. *Appl. Microbiol. Biotechnol.* **2021**, *105*, 4141–4151. [CrossRef] [PubMed]
63. Díaz De Rienzo, M.A.; Kamalanathan, I.D.; Martin, P.J. Comparative Study of the Production of Rhamnolipid Biosurfactants by *B. thailandensis* E264 and *P. aeruginosa* ATCC 9027 Using Foam Fractionation. *Process Biochem.* **2016**, *51*, 820–827. [CrossRef]
64. Kim, Y.T.; Kim, S.E.; Lee, W.J.; Fumei, Z.; Cho, M.S.; Moon, J.S.; Oh, H.-W.; Park, H.-Y.; Kim, S.U. Isolation and Characterization of a High Iturin Yielding Bacillus Velezensis UV Mutant with Improved Antifungal Activity. *PLoS ONE* **2020**, *15*, e0234177. [CrossRef]
65. Ganji, Z.; Beheshti-Maal, K.; Massah, A.; Emami-Karvani, Z. A Novel Sophorolipid-Producing *Candida keroseneae* GBME-IAUF-2 as a Potential Agent in Microbial Enhanced Oil Recovery (MEOR). *FEMS Microbiol. Lett.* **2020**, *367*, fnaa144. [CrossRef]

66. Rufino, R.D.; Luna, J.M.; Sarubbo, L.A.; Rodrigues, L.R.M.; Teixeira, J.A.C.; Campos-Takaki, G.M. Antimicrobial and Anti-Adhesive Potential of a Biosurfactant Rufisan Produced by *Candida lipolytica* UCP 0988. *Colloids Surf. B Biointerfaces* **2011**, *84*, 1–5. [CrossRef]
67. Nelson, J.; El-Gendy, A.O.; Mansy, M.S.; Ramadan, M.A.; Aziz, R.K. The Biosurfactants Iturin, Lichenysin and Surfactin, from Vaginally Isolated Lactobacilli, Prevent Biofilm Formation by Pathogenic *Candida*. *FEMS Microbiol. Lett.* **2020**, *367*, fnaa126. [CrossRef]
68. Zhao, F.; Han, S.; Zhang, Y. Comparative Studies on the Structural Composition, Surface/Interface Activity and Application Potential of Rhamnolipids Produced by *Pseudomonas aeruginosa* Using Hydrophobic or Hydrophilic Substrates. *Bioresour. Technol.* **2020**, *295*, 122269. [CrossRef] [PubMed]
69. Bonnichsen, L.; Bygvraa Svenningsen, N.; Rybtke, M.; de Bruijn, I.; Raaijmakers, J.M.; Tolker-Nielsen, T.; Nybroe, O. Lipopeptide Biosurfactant Viscosin Enhances Dispersal of *Pseudomonas fluorescens* SBW25 Biofilms. *Microbiology* **2015**, *161*, 2289–2297. [CrossRef]
70. Phulpoto, I.A.; Wang, Y.; Qazi, M.A.; Hu, B.; Ndayisenga, F.; Yu, Z. Bioprospecting of Rhamnolipids Production and Optimization by an Oil-Degrading *Pseudomonas* sp. S2WE Isolated from Freshwater Lake. *Bioresour. Technol.* **2021**, *323*, 124601. [CrossRef]
71. Hu, X.; Cheng, T.; Liu, J. A Novel Serratia Sp. ZS6 Isolate Derived from Petroleum Sludge Secretes Biosurfactant and Lipase in Medium with Olive Oil as Sole Carbon Source. *AMB Express* **2018**, *8*, 165. [CrossRef] [PubMed]
72. Amaral, P.F.F.; da Silva, J.M.; Lehocky, M.; Barros-Timmons, A.M.V.; Coelho, M.A.Z.; Marrucho, I.M.; Coutinho, J.A.P. Production and Characterization of a Bioemulsifier from *Yarrowia lipolytica*. *Process Biochem.* **2006**, *41*, 1894–1898. [CrossRef]
73. de Souza Monteiro, A.; Domingues, V.S.; Souza, M.V.; Lula, I.; Gonçalves, D.B.; de Siqueira, E.P.; Dos Santos, V.L. Bioconversion of Biodiesel Refinery Waste in the Bioemulsifier by *Trichosporon Mycotoxinivorans* CLA2. *Biotechnol. Biofuels* **2012**, *5*, 29. [CrossRef] [PubMed]
74. Anestopoulos, I.; Kiousi, D.E.; Klavaris, A.; Galanis, A.; Salek, K.; Euston, S.R.; Pappa, A.; Panayiotidis, M.I. Surface Active Agents and Their Health-Promoting Properties: Molecules of Multifunctional Significance. *Pharmaceutics* **2020**, *12*, 688. [CrossRef]
75. Lerminiaux, N.A.; Cameron, A.D.S. Horizontal Transfer of Antibiotic Resistance Genes in Clinical Environments. *Can. J. Microbiol.* **2019**, *65*, 34–44. [CrossRef]
76. Melnyk, A.H.; Wong, A.; Kassen, R. The Fitness Costs of Antibiotic Resistance Mutations. *Evol. Appl.* **2015**, *8*, 273–283. [CrossRef]
77. Tacconelli, E.; Carrara, E.; Savoldi, A.; Harbarth, S.; Mendelson, M.; Monnet, D.L.; Pulcini, C.; Kahlmeter, G.; Kluytmans, J.; Carmeli, Y.; et al. WHO Pathogens Priority List Working Group. Discovery, research, and development of new antibiotics: The WHO priority list of antibiotic-resistant bacteria and tuberculosis. *Lancet Infect Dis.* **2018**, *18*, 318–327. [CrossRef]
78. Colombo, A.L.; Júnior, J.N.d.A.; Guinea, J. Emerging Multidrug-Resistant *Candida* Species. *Curr. Opin. Infect. Dis.* **2017**, *30*, 528–538. [CrossRef]
79. Kean, R.; Ramage, G. Combined Antifungal Resistance and Biofilm Tolerance: The Global Threat of *Candida Auris*. *mSphere* **2019**, *4*, e00458-19. [CrossRef]
80. Kumar, A.; Nair, R.; Kumar, M.; Banerjee, A.; Chakrabarti, A.; Rudramurthy, S.M.; Bagga, R.; Gaur, N.A.; Mondal, A.K.; Prasad, R. Assessment of Antifungal Resistance and Associated Molecular Mechanism in *Candida albicans* Isolates from Different Cohorts of Patients in North Indian State of Haryana. *Folia Microbiol.* **2020**, *65*, 747–754. [CrossRef]
81. Anestopoulos, I.; Kiousi, D.-E.; Klavaris, A.; Maijo, M.; Serpico, A.; Suarez, A.; Sanchez, G.; Salek, K.; Chasapi, S.A.; Zompra, A.A.; et al. Marine-Derived Surface Active Agents: Health-Promoting Properties and Blue Biotechnology-Based Applications. *Biomolecules* **2020**, *10*, 885. [CrossRef] [PubMed]
82. Zhou, C.; Wang, F.; Chen, H.; Li, M.; Qiao, F.; Liu, Z.; Hou, Y.; Wu, C.; Fan, Y.; Liu, L.; et al. Selective Antimicrobial Activities and Action Mechanism of Micelles Self-Assembled by Cationic Oligomeric Surfactants. *ACS Appl. Mater. Interfaces* **2016**, *8*, 4242–4249. [CrossRef] [PubMed]
83. Labena, A.; Hegazy, M.A.; Sami, R.M.; Hozzein, W.N. Multiple Applications of a Novel Cationic Gemini Surfactant: Anti-Microbial, Anti-Biofilm, Biocide, Salinity Corrosion Inhibitor, and Biofilm Dispersion (Part II). *Molecules* **2020**, *25*, 1348. [CrossRef] [PubMed]
84. Sana, S.; Datta, S.; Biswas, D.; Sengupta, D. Assessment of Synergistic Antibacterial Activity of Combined Biosurfactants Revealed by Bacterial Cell Envelop Damage. *Biochim. Biophys. Acta Biomembr.* **2018**, *1860*, 579–585. [CrossRef]
85. Naughton, P.J.; Marchant, R.; Naughton, V.; Banat, I.M. Microbial Biosurfactants: Current Trends and Applications in Agricultural and Biomedical Industries. *J. Appl. Microbiol.* **2019**, *127*, 12–28. [CrossRef] [PubMed]
86. Haba, E.; Bouhdid, S.; Torrego-Solana, N.; Marqués, A.M.; Espuny, M.J.; García-Celma, M.J.; Manresa, A. Rhamnolipids as Emulsifying Agents for Essential Oil Formulations: Antimicrobial Effect against *Candida albicans* and Methicillin-Resistant *Staphylococcus aureus*. *Int. J. Pharm.* **2014**, *476*, 134–141. [CrossRef]
87. Kurtzman, C.P.; Price, N.P.J.; Ray, K.J.; Kuo, T.-M. Production of Sophorolipid Biosurfactants by Multiple Species of the Starmerella (*Candida*) *bombicola* Yeast Clade: Sophorolipids from Yeasts. *FEMS Microbiol. Lett.* **2010**, *311*, 140–146. [CrossRef]
88. Callaghan, B.; Lydon, H.; Roelants, S.L.K.W.; Van Bogaert, I.N.A.; Marchant, R.; Banat, I.M.; Mitchell, C.A. Lactonic Sophorolipids Increase Tumor Burden in Apcmin+/− Mice. *PLoS ONE* **2016**, *11*, e0156845. [CrossRef]
89. Van Bogaert, I.N.A.; Zhang, J.; Soetaert, W. Microbial Synthesis of Sophorolipids. *Process Biochem.* **2011**, *46*, 821–833. [CrossRef]
90. Sen, S.; Borah, S.N.; Bora, A.; Deka, S. Production, Characterization, and Antifungal Activity of a Biosurfactant Produced by Rhodotorula Babjevae YS3. *Microb. Cell Fact.* **2017**, *16*, 95. [CrossRef]

91. Chen, M.; Wang, J.; Liu, B.; Zhu, Y.; Xiao, R.; Yang, W.; Ge, C.; Chen, Z. Biocontrol of Tomato Bacterial Wilt by the New Strain *Bacillus velezensis* FJAT-46737 and Its Lipopeptides. *BMC Microbiol.* **2020**, *20*, 160. [CrossRef] [PubMed]
92. Wu, S.; Liu, G.; Zhou, S.; Sha, Z.; Sun, C. Characterization of Antifungal Lipopeptide Biosurfactants Produced by Marine Bacterium Bacillus Sp. CS30. *Mar. Drugs* **2019**, *17*, 199. [CrossRef] [PubMed]
93. Perez, K.J.; Viana, J.D.S.; Lopes, F.C.; Pereira, J.Q.; Dos Santos, D.M.; Oliveira, J.S.; Velho, R.V.; Crispim, S.M.; Nicoli, J.R.; Brandelli, A.; et al. *Bacillus* Spp. Isolated from Puba as a Source of Biosurfactants and Antimicrobial Lipopeptides. *Front. Microbiol.* **2017**, *8*, 61. [CrossRef] [PubMed]
94. Liu, J.; Li, W.; Zhu, X.; Zhao, H.; Lu, Y.; Zhang, C.; Lu, Z. Surfactin Effectively Inhibits *Staphylococcus aureus* Adhesion and Biofilm Formation on Surfaces. *Appl. Microbiol. Biotechnol.* **2019**, *103*, 4565–4574. [CrossRef]
95. Ohadi, M.; Forootanfar, H.; Dehghannoudeh, G.; Eslaminejad, T.; Ameri, A.; Shakibaie, M.; Adeli-Sardou, M. Antimicrobial, Anti-Biofilm, and Anti-Proliferative Activities of Lipopeptide Biosurfactant Produced by *Acinetobacter junii* B6. *Microb. Pathog.* **2020**, *138*, 103806. [CrossRef] [PubMed]
96. Mouafo, T.H.; Mbawala, A.; Ndjouenkeu, R. Effect of Different Carbon Sources on Biosurfactants' Production by Three Strains of *Lactobacillus* spp. *Biomed Res. Int.* **2018**, *2018*, 5034783. [CrossRef]
97. Liu, Y.; Ding, S.; Dietrich, R.; Märtlbauer, E.; Zhu, K. A Biosurfactant-Inspired Heptapeptide with Improved Specificity to Kill MRSA. *Angew. Chem. Int. Ed. Engl.* **2017**, *56*, 1486–1490. [CrossRef]
98. Vollenbroich, D.; Ozel, M.; Vater, J.; Kamp, R.M.; Pauli, G. Mechanism of Inactivation of Enveloped Viruses by the Biosurfactant Surfactin from *Bacillus subtilis*. *Biologicals* **1997**, *25*, 289–297. [CrossRef]
99. Yuan, L.; Zhang, S.; Wang, Y.; Li, Y.; Wang, X.; Yang, Q. Surfactin Inhibits Membrane Fusion during Invasion of Epithelial Cells by Enveloped Viruses. *J. Virol.* **2018**, *92*, e00809-18. [CrossRef]
100. Wu, W.; Wang, J.; Lin, D.; Chen, L.; Xie, X.; Shen, X.; Yang, Q.; Wu, Q.; Yang, J.; He, J.; et al. Super Short Membrane-Active Lipopeptides Inhibiting the Entry of Influenza A Virus. *Biochim. Biophys. Acta* **2015**, *1848*, 2344–2350. [CrossRef]
101. Chowdhury, T.; Baindara, P.; Mandal, S.M. LPD-12: A Promising Lipopeptide to Control COVID-19. *Int. J. Antimicrob. Agents* **2021**, *57*, 106218. [CrossRef] [PubMed]
102. Outlaw, V.K.; Bovier, F.T.; Mears, M.C.; Cajimat, M.N.; Zhu, Y.; Lin, M.J.; Addetia, A.; Lieberman, N.A.P.; Peddu, V.; Xie, X.; et al. Inhibition of Coronavirus Entry in vitro and ex vivo by a Lipid-Conjugated Peptide Derived from the SARS-CoV-2 Spike Glycoprotein HRC Domain. *MBio* **2020**, *11*, e01935-20. [CrossRef] [PubMed]
103. Subramaniam, M.D.; Venkatesan, D.; Iyer, M.; Subbarayan, S.; Govindasami, V.; Roy, A.; Narayanasamy, A.; Kamalakannan, S.; Gopalakrishnan, A.V.; Thangarasu, R.; et al. Biosurfactants and Anti-Inflammatory Activity: A Potential New Approach towards COVID-19. *Curr. Opin. Environ. Sci. Health* **2020**, *17*, 72–81. [CrossRef] [PubMed]
104. Shah, V.; Doncel, G.F.; Seyoum, T.; Eaton, K.M.; Zalenskaya, I.; Hagver, R.; Azim, A.; Gross, R. Sophorolipids, Microbial Glycolipids with Anti-Human Immunodeficiency Virus and Sperm-Immobilizing Activities. *Antimicrob. Agents Chemother.* **2005**, *49*, 4093–4100. [CrossRef] [PubMed]
105. Borsanyiova, M.; Patil, A.; Mukherji, R.; Prabhune, A.; Bopegamage, S. Biological Activity of Sophorolipids and Their Possible Use as Antiviral Agents. *Folia Microbiol.* **2016**, *61*, 85–89. [CrossRef]
106. Remichkova, M.; Galabova, D.; Roeva, I.; Karpenko, E.; Shulga, A.; Galabov, A.S. Anti-Herpesvirus Activities of *Pseudomonas* sp. S-17 Rhamnolipid and Its Complex with Alginate. *Z. Naturforsch. C* **2008**, *63*, 75–81. [CrossRef]
107. Lozach, P.-Y. Cell Biology of Viral Infections. *Cells* **2020**, *9*, 2431. [CrossRef]
108. Abdulkhaleq, L.A.; Assi, M.A.; Abdullah, R.; Zamri-Saad, M.; Taufiq-Yap, Y.H.; Hezmee, M.N.M. The Crucial Roles of Inflammatory Mediators in Inflammation: A Review. *Vet. World* **2018**, *11*, 627–635. [CrossRef]
109. Wang, Y.; Tian, J.; Shi, F.; Li, X.; Hu, Z.; Chu, J. Protective Effect of Surfactin on Copper Sulfate-Induced Inflammation, Oxidative Stress, and Hepatic Injury in Zebrafish. *Microbiol. Immunol.* **2021**, *65*, 410–421. [CrossRef]
110. Gan, P.; Jin, D.; Zhao, X.; Gao, Z.; Wang, S.; Du, P.; Qi, G. Bacillus-Produced Surfactin Attenuates Chronic Inflammation in Atherosclerotic Lesions of ApoE(−/−) Mice. *Int. Immunopharmacol.* **2016**, *35*, 226–234. [CrossRef]
111. Xu, W.; Liu, H.; Wang, X.; Yang, Q. Surfactin Induces Maturation of Dendritic Cells in Vitro. *Biosci. Rep.* **2016**, *36*, e00387. [CrossRef] [PubMed]
112. Maeng, Y.; Kim, K.T.; Zhou, X.; Jin, L.; Kim, K.S.; Kim, Y.H.; Lee, S.; Park, J.H.; Chen, X.; Kong, M.; et al. A Novel Microbial Technique for Producing High-Quality Sophorolipids from Horse Oil Suitable for Cosmetic Applications. *Microb. Biotechnol.* **2018**, *11*, 917–929. [CrossRef] [PubMed]
113. Hagler, M.; Smith-Norowitz, T.A.; Chice, S.; Wallner, S.R.; Viterbo, D.; Mueller, C.M.; Gross, R.; Nowakowski, M.; Schulze, R.; Zenilman, M.E.; et al. Sophorolipids Decrease IgE Production in U266 Cells by Downregulation of BSAP (Pax5), TLR-2, STAT3 and IL-6. *J. Allergy Clin. Immunol.* **2007**, *119*, S263. [CrossRef]
114. Hardin, R.; Pierre, J.; Schulze, R.; Mueller, C.M.; Fu, S.L.; Wallner, S.R.; Stanek, A.; Shah, V.; Gross, R.A.; Weedon, J.; et al. Sophorolipids Improve Sepsis Survival: Effects of Dosing and Derivatives. *J. Surg. Res.* **2007**, *142*, 314–319. [CrossRef]
115. Mueller, C.M.; Lin, Y.-Y.; Viterbo, D.; Pierre, J.; Murray, S.A.; Shah, V.; Gross, R.; Schulze, R.; Zenilman, M.E.; Bluth, M.H. Sophorolipid Treatment Decreases Inflammatory Cytokine Expression in an in Vitro Model of Experimental Sepsis. *FASEB J.* **2006**, *20*, A204. [CrossRef]
116. Kuyukina, M.S.; Ivshina, I.B.; Gein, S.V.; Baeva, T.A.; Chereshnev, V.A. In Vitro Immunomodulating Activity of Biosurfactant Glycolipid Complex from *Rhodococcus ruber*. *Bull. Exp. Biol. Med.* **2007**, *144*, 326–330. [CrossRef]

117. Thakur, P.; Saini, N.K.; Thakur, V.K.; Gupta, V.K.; Saini, R.V.; Saini, A.K. Rhamnolipid the Glycolipid Biosurfactant: Emerging Trends and Promising Strategies in the Field of Biotechnology and Biomedicine. *Microb. Cell Fact.* **2021**, *20*, 1. [CrossRef]
118. Rahimi, K.; Lotfabad, T.B.; Jabeen, F.; Mohammad Ganji, S. Cytotoxic Effects of Mono- and Di-Rhamnolipids from *Pseudomonas aeruginosa* MR01 on MCF-7 Human Breast Cancer Cells. *Colloids Surf. B Biointerfaces* **2019**, *181*, 943–952. [CrossRef]
119. Ribeiro, I.A.C.; Faustino, C.M.C.; Guerreiro, P.S.; Frade, R.F.M.; Bronze, M.R.; Castro, M.F.; Ribeiro, M.H.L. Development of Novel Sophorolipids with Improved Cytotoxic Activity toward MDA-MB-231 Breast Cancer Cells: Development of Novel Sophorolipids With Cytotoxic Activity. *J. Mol. Recognit.* **2015**, *28*, 155–165. [CrossRef]
120. Chen, J.; Song, X.; Zhang, H.; Qu, Y. Production, Structure Elucidation and Anticancer Properties of Sophorolipid from Wickerhamiella Domercqiae. *Enzyme Microb. Technol.* **2006**, *39*, 501–506. [CrossRef]
121. Dey, G.; Bharti, R.; Dhanarajan, G.; Das, S.; Dey, K.K.; Kumar, B.N.P.; Sen, R.; Mandal, M. Marine Lipopeptide Iturin a Inhibits Akt Mediated GSK3β and FoxO3a Signaling and Triggers Apoptosis in Breast Cancer. *Sci. Rep.* **2015**, *5*, 10316. [CrossRef] [PubMed]
122. Yan, J.; Bassler, B.L. Surviving as a Community: Antibiotic Tolerance and Persistence in Bacterial Biofilms. *Cell Host Microbe* **2019**, *26*, 15–21. [CrossRef] [PubMed]
123. Roy, R.; Tiwari, M.; Donelli, G.; Tiwari, V. Strategies for Combating Bacterial Biofilms: A Focus on Anti-Biofilm Agents and Their Mechanisms of Action. *Virulence* **2018**, *9*, 522–554. [CrossRef] [PubMed]
124. Rivardo, F.; Martinotti, M.G.; Turner, R.J.; Ceri, H. Synergistic Effect of Lipopeptide Biosurfactant with Antibiotics against *Escherichia coli* CFT073 Biofilm. *Int. J. Antimicrob. Agents* **2011**, *37*, 324–331. [CrossRef]
125. Ceresa, C.; Rinaldi, M.; Fracchia, L. Synergistic Activity of Antifungal Drugs and Lipopeptide AC7 against *Candida albicans* Biofilm on Silicone. *AIMS Bioeng.* **2017**, *4*, 318–334. [CrossRef]
126. Cordeiro, R.d.A.; Weslley Caracas Cedro, E.; Raquel Colares Andrade, A.; Serpa, R.; José de Jesus Evangelista, A.; Sales de Oliveira, J.; Santos Pereira, V.; Pereira Alencar, L.; Bruna Leite Mendes, P.; Cibelle Soares Farias, B.; et al. Inhibitory Effect of a Lipopeptide Biosurfactant Produced by *Bacillus subtilis* on Planktonic and Sessile Cells of *Trichosporon* spp. *Biofouling* **2018**, *34*, 309–319. [CrossRef]
127. Da Silva, G.O.; Farias, B.C.S.; da Silva, R.B.; Teixeira, E.H.; Cordeiro, R.d.A.; Hissa, D.C.; Melo, V.M.M. Effects of Lipopeptide Biosurfactants on Clinical Strains of *Malassezia furfur* Growth and Biofilm Formation. *Med. Mycol.* **2021**, *59*, 1191–1201. [CrossRef]
128. Allegrone, G.; Ceresa, C.; Rinaldi, M.; Fracchia, L. Diverse Effects of Natural and Synthetic Surfactants on the Inhibition of *Staphylococcus aureus* Biofilm. *Pharmaceutics* **2021**, *13*, 1172. [CrossRef]
129. Shen, Y.; Li, P.; Chen, X.; Zou, Y.; Li, H.; Yuan, G.; Hu, H. Activity of Sodium Lauryl Sulfate, Rhamnolipids, and N-Acetylcysteine against Biofilms of Five Common Pathogens. *Microb. Drug Resist.* **2020**, *26*, 290–299. [CrossRef]
130. Sidrim, J.J.; Ocadaque, C.J.; Amando, B.R.; de Guedes, G.M.; Costa, C.L.; Brilhante, R.S.; de Cordeiro, A.R.; Rocha, M.F.; Scm Castelo-Branco, D. Rhamnolipid Enhances *Burkholderia pseudomallei* Biofilm Susceptibility, Disassembly and Production of Virulence Factors. *Future Microbiol.* **2020**, *15*, 1109–1121. [CrossRef]
131. Sen, S.; Borah, S.N.; Bora, A.; Deka, S. Rhamnolipid Exhibits Anti-Biofilm Activity against the Dermatophytic Fungi Trichophyton Rubrum and *Trichophyton mentagrophytes*. *Biotechnol. Rep.* **2020**, *27*, e00516. [CrossRef] [PubMed]
132. Ceresa, C.; Fracchia, L.; Williams, M.; Banat, I.M.; Díaz De Rienzo, M.A. The Effect of Sophorolipids against Microbial Biofilms on Medical-Grade Silicone. *J. Biotechnol.* **2020**, *309*, 34–43. [CrossRef] [PubMed]
133. Haddaji, N.; Ncib, K.; Bahia, W.; Ghorbel, M.; Leban, N.; Bouali, N.; Bechambi, O.; Mzoughi, R.; Mahdhi, A. Control of Multidrug-Resistant Pathogenic Staphylococci Associated with Vaginal Infection Using Biosurfactants Derived from Potential Probiotic Bacillus Strain. *Fermentation* **2022**, *8*, 19. [CrossRef]
134. Wang, P.-H.; Huang, B.-S.; Horng, H.-C.; Yeh, C.-C.; Chen, Y.-J. Wound Healing. *J. Chin. Med. Assoc.* **2018**, *81*, 94–101. [CrossRef]
135. Martin, P.; Nunan, R. Cellular and Molecular Mechanisms of Repair in Acute and Chronic Wound Healing. *Br. J. Dermatol.* **2015**, *173*, 370–378. [CrossRef] [PubMed]
136. Tatara, A.M.; Kontoyiannis, D.P.; Mikos, A.G. Drug Delivery and Tissue Engineering to Promote Wound Healing in the Immunocompromised Host: Current Challenges and Future Directions. *Adv. Drug Deliv. Rev.* **2018**, *129*, 319–329. [CrossRef] [PubMed]
137. Yip, W.L. Influence of Oxygen on Wound Healing: Oxygen and Wound Healing. *Int. Wound J.* **2015**, *12*, 620–624. [CrossRef]
138. Lindholm, C.; Searle, R. Wound Management for the 21st Century: Combining Effectiveness and Efficiency: Wound Management for the 21st Century. *Int. Wound J.* **2016**, *13* (Suppl. 2), 5–15. [CrossRef]
139. Rahim, K.; Saleha, S.; Zhu, X.; Huo, L.; Basit, A.; Franco, O.L. Bacterial Contribution in Chronicity of Wounds. *Microb. Ecol.* **2017**, *73*, 710–721. [CrossRef]
140. Pastar, I.; Nusbaum, A.G.; Gil, J.; Patel, S.B.; Chen, J.; Valdes, J.; Stojadinovic, O.; Plano, L.R.; Tomic-Canic, M.; Davis, S.C. Interactions of Methicillin Resistant *Staphylococcus aureus* USA300 and *Pseudomonas aeruginosa* in Polymicrobial Wound Infection. *PLoS ONE* **2013**, *8*, e56846. [CrossRef]
141. Banu, A.; Noorul Hassan, M.M.; Rajkumar, J.; Srinivasa, S. Spectrum of Bacteria Associated with Diabetic Foot Ulcer and Biofilm Formation: A Prospective Study. *Aust. Med. J.* **2015**, *8*, 280–285. [CrossRef] [PubMed]
142. Ohadi, M.; Forootanfar, H.; Rahimi, H.R.; Jafari, E.; Shakibaie, M.; Eslaminejad, T.; Dehghannoudeh, G. Antioxidant Potential and Wound Healing Activity of Biosurfactant Produced by *Acinetobacter junii* B6. *Curr. Pharm. Biotechnol.* **2017**, *18*, 900–908. [CrossRef]

143. Zouari, R.; Moalla-Rekik, D.; Sahnoun, Z.; Rebai, T.; Ellouze-Chaabouni, S.; Ghribi-Aydi, D. Evaluation of Dermal Wound Healing and in vitro Antioxidant Efficiency of Bacillus Subtilis SPB1 Biosurfactant. *Biomed. Pharmacother.* **2016**, *84*, 878–891. [CrossRef] [PubMed]
144. Yan, L.; Liu, G.; Zhao, B.; Pang, B.; Wu, W.; Ai, C.; Zhao, X.; Wang, X.; Jiang, C.; Shao, D.; et al. Novel Biomedical Functions of Surfactin A from *Bacillus subtilis* in Wound Healing Promotion and Scar Inhibition. *J. Agric. Food Chem.* **2020**, *68*, 6987–6997. [CrossRef] [PubMed]
145. Li, D.; Wang, W.; Wu, Y.; Ma, X.; Zhou, W.; Lai, Y. Lipopeptide 78 from *Staphylococcus epidermidis* Activates β-Catenin to Inhibit Skin Inflammation. *J. Immunol.* **2019**, *202*, 1219–1228. [CrossRef] [PubMed]
146. Afsharipour, S.; Asadi, A.; Ohadi, M.; Ranjbar, M.; Forootanfar, H.; Jafari, E.; Dehghannoudeh, G. Preparation and Characterization of Nano-Lipopeptide Biosurfactant Hydrogel and Evaluation of Wound-Healing Properties. *Bionanoscience* **2021**, *11*, 1061–1069. [CrossRef]
147. Gupta, S.; Raghuwanshi, N.; Varshney, R.; Banat, I.M.; Srivastava, A.K.; Pruthi, P.A.; Pruthi, V. Accelerated in Vivo Wound Healing Evaluation of Microbial Glycolipid Containing Ointment as a Transdermal Substitute. *Biomed. Pharmacother.* **2017**, *94*, 1186–1196. [CrossRef]
148. Shen, C.; Jiang, L.; Shao, H.; You, C.; Zhang, G.; Ding, S.; Bian, T.; Han, C.; Meng, Q. Targeted Killing of Myofibroblasts by Biosurfactant Di-Rhamnolipid Suggests a Therapy against Scar Formation. *Sci. Rep.* **2016**, *6*, 37553. [CrossRef]
149. Kwak, M.-J.; Park, M.-Y.; Kim, J.; Lee, H.; Whang, K.-Y. Curative Effects of Sophorolipid on Physical Wounds: In Vitro and in Vivo Studies. *Vet. Med. Sci.* **2021**, *7*, 1400–1408. [CrossRef]
150. Domalaon, R.; Findlay, B.; Ogunsina, M.; Arthur, G.; Schweizer, F. Ultrashort cationic lipopeptides and lipopeptoids: Evaluation and mechanistic insights against epithelial cancer cells. *Peptides* **2016**, *84*, 58–67. [CrossRef]
151. Findlay, B.; Mookherjee, N.; Schweizer, F. Ultrashort Cationic Lipopeptides and Lipopeptoids Selectively Induce Cytokine Production in Macrophages. *PLoS ONE* **2013**, *8*, e54280. [CrossRef] [PubMed]
152. Tran, Q.G.; Ryu, A.J.; Choi, Y.J.; Jeong, K.J.; Kim, H.S.; Lee, Y.J. Enhanced production of biosurfactants through genetic engineering of *Pseudozyma* sp. SY16. *Korean J. Chem. Eng.* **2022**, *39*, 997–1003. [CrossRef]

antibiotics

MDPI

Review

Marine Actinobacteria a New Source of Antibacterial Metabolites to Treat Acne Vulgaris Disease—A Systematic Literature Review

Maria Clara De La Hoz-Romo [1,2], Luis Díaz [1,2] and Luisa Villamil [1,*]

1 Doctoral Program of Biosciences, School of Engineering, Universidad de La Sabana, Chía 140013, Colombia; mariaderom@unisabana.edu.co (M.C.D.L.H.-R.); luis.diaz1@unisabana.edu.co (L.D.)
2 Bioprospecting Research Group, School of Engineering, Universidad de La Sabana, Chía 140013, Colombia
* Correspondence: luisa.villamil@unisabana.edu.co; Tel.: +57-313-534-6269

Abstract: Acne vulgaris is a multifactorial disease that remains under-explored; up to date it is known that the bacterium *Cutibacterium acnes* is involved in the disease occurrence, also associated with a microbial dysbiosis. Antibiotics have become a mainstay treatment generating the emergence of antibiotic-resistant bacteria. In addition, there are some reported side effects of alternative treatments, which indicate the need to investigate a different therapeutic approach. Natural products continue to be an excellent option, especially those extracted from actinobacteria, which represent a prominent source of metabolites with a wide range of biological activities, particularly the marine actinobacteria, which have been less studied than their terrestrial counterparts. Therefore, this systematic review aimed to identify and evaluate the potential anti-infective activity of metabolites isolated from marine actinobacteria strains against bacteria related to the development of acne vulgaris disease. It was found that there is a variety of compounds with anti-infective activity against *Staphylococcus aureus* and *Staphylococcus epidermidis*, bacteria closely related to acne vulgaris development; nevertheless, there is no report of a compound with antibacterial activity or quorum-sensing inhibition toward *C. acnes*, which is a surprising result. Since two of the most widely used antibiotics for the treatment of acne targeting *C. acnes* were obtained from actinobacteria of the genus *Streptomyces*, this demonstrates a great opportunity to pursue further studies in this field, considering the potential of marine actinobacteria to produce new anti-infective compounds.

Keywords: biotechnology; marine actinobacteria; antibacterial activity; anti-biofilm activity; quorum-quenching activity; natural compounds; extracts; biosynthetic gene clusters (BGCs); acne vulgaris; *Cutibacterium acnes*; *Staphylococcus aureus*; *Staphylococcus epidermidis*

Citation: De La Hoz-Romo, M.C.; Díaz, L.; Villamil, L. Marine Actinobacteria a New Source of Antibacterial Metabolites to Treat Acne Vulgaris Disease—A Systematic Literature Review. *Antibiotics* **2022**, *11*, 965. https://doi.org/10.3390/antibiotics11070965

Academic Editors: Valério Monteiro-Neto and Elizabeth S. Fernandes

Received: 16 June 2022
Accepted: 13 July 2022
Published: 18 July 2022

1. Introduction

Acne vulgaris is an inflammatory disease of the pilosebaceous unit that includes the hair follicle, hair shaft, and sebaceous gland. It is classified as a chronic condition due to the prolonged course and physical manifestations [1]. Furthermore, acne causes profound negative psychological and social effects on the quality of life of patients [1], affecting 85% of adolescents, and more than 10% of adults, and the Global Burden of Disease Project estimates that the prevalence of acne at 9.4%, placing it as the eighth most prevalent disease worldwide [2,3].

The pathophysiology of acne is related to the bacteria *Cutibacterium acnes*; this is one of the most abundant microorganisms found on human skin, accounting for up to 87% of the microorganisms in pilosebaceous units [4,5] along with *Staphylococcus epidermidis*, which are also major inhabitant Gram-positive bacteria of the skin microbiota. However, these bacteria adapt to changing skin microenvironments and can shift to being opportunistic pathogens, forming biofilms, and thus are involved in common skin dysbiosis, generating the loss of phylotype diversity of *C. acnes*, with the increase in pathogenic *Staphylococcus aureus* and commensal *S. epidermidis* [6].

Currently, all treatments available for the management of acne, topical or systemic, generate prominent side effects in patients such as psychiatric events, inflammatory bowel disease, hepatotoxicity, lupus-like syndrome, drug hypersensitivity syndrome, and so on [7]. Moreover, antibiotics have been a mainstay in the treatment of the disease; however, this has generated the emergence of antibiotic-resistant strains of *C. acnes*, which in turn exert selective pressure on other host bacteria such as *S. aureus* and *S. epidermidis*, allowing the emergence of antibiotic-resistant bacteria and contributing to bacterial drug resistance [8].

Thus, there is the need for new acne treatments to alleviate bacterial drug resistance, which has become a serious global threat to human health, food safety, animal production, and economic and agricultural development. Antibiotic resistance compromises the efficacy of prevention and treatment of infectious diseases, which are the number one cause of death in tropical countries, accounting for half of all fatalities. In addition, infectious disease mortality rates are also increasing in developed countries [9,10], and the lack of new antimicrobials against the spread of drug-resistant bacteria could generate 10 million deaths in the next 35 years. The prediction of losses of the order of 100 trillion USD is expected in 2050 if nothing is done to reverse the trend [11].

Since 2020, the World Health Organization, WHO (World Health Organization) has warned about the shortage of innovative antibiotics and their danger in treating drug-resistant infections [12,13]. As a result, novel approaches have emerged such as the use of bacteriophages, probiotics [14,15], and anti-biofilm agents/quorum-sensing inhibitors [16] a recommended alternative, since to date it is a mechanism against which bacteria have not shown resistance.

Historically, natural products isolated from a variety of sources such as terrestrial plants, animals, marine organisms, microorganisms, terrestrial vertebrates, and invertebrates have been a prolific source for numerous medical agents. In the early 20th century, approximately 80% of all medicines were obtained from plant sources. Nevertheless, since the discovery of penicillin from *Penicillium notatum* by Alexander Fleming in 1928, a significant shift from plants to microorganisms as a source of natural products has arisen [17].

Consequently, microorganism-derived compounds have been used based on a wide variety of biological activities. Among the bacteria, the actinobacteria phylum represents a noteworthy source of commercially important products and 70% of the known antibiotics are produced by actinobacteria, by the genus *Streptomyces* [18].

Most of these compounds were isolated from terrestrial sources [19]. Nevertheless, in the last 20 years, the re-discovery of previously characterized bioactive compounds and strain redundancy has decreased the interest in these soil-dwelling bacteria as a source of novel bioactive compounds. Therefore, Actinobacteria living in other niches, such as the marine environment (sea sediments, coral reefs, invertebrates, etc.), have gained value because of their chemo-diversity [18–21] influenced by their complex environment with extreme variations in pressure, salinity, light, and temperature [20].

It has been shown that marine actinobacteria exhibit more diverse and superior properties when compared to terrestrial actinobacteria in terms of antifouling, antibacterial, antibiofilm, anticoagulant, antiviral and antibacterial effects [19,22,23].

Since the marine actinobacteria have been less explored, here we did a systematic literature review of metabolites and extracts produced by marine actinobacteria with antimicrobial, anti-biofilm, and quorum-sensing inhibition activities (quorum quenching, QQ), as therapeutic alternatives treatment of acne vulgaris, some skin diseases, and infectious diseases.

Here we sort the reported metabolites regarding the type evaluated for their structure–activity relationship (SAR) and associated each family compound with some of their corresponding biosynthetic genes cluster. This systematic review aimed to assess the anti-infective potential of metabolites or compounds isolated from marine actinobacteria strains as an alternative treatment for acne vulgaris disease.

2. Results

2.1. General Information

2962 articles were collected in this study and 1930 articles were identified after duplicate removal. Out of these, 1678 were excluded during the screening phase by title and abstract reading, and by applying the inclusion and exclusion criteria. Starting with this screening, 252 papers were selected for full-text reading, and they were assessed for eligibility. Finally, 177 papers were included for data extraction as shown in Figure 1.

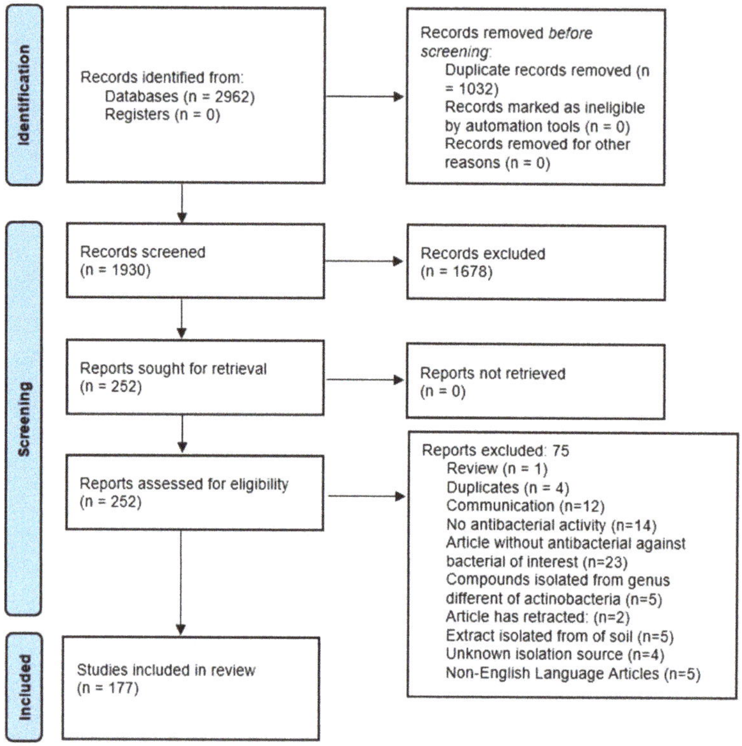

Figure 1. PRISMA flow diagram. Flowchart of systematic literature search according to PRISMA guidelines. Modified from [24]. The systematic review was done following the PRISMA guidelines, the complete checklist can be reviewed in Supplementary Table S1.

2.2. Isolation Sources

The marine actinobacteria with anti-infective activity were collected worldwide, with a higher number of reports from China, (47), followed by India (39), and Egypt (10). In America, the United States was the most predominant country with 13 reported studies, followed by Mexico and Chile with 3 reports, Peru with 2, and Panama with 1. In the Caribbean, only one study was reported in the Bahamas and interestingly, some strains were isolated from oceans such as the Caspian Sea, the Baltic Sea, and the Cantabrian Sea, but not from the Bahamas maritime ecosystems (Figure 2A). Marine sediment was the most prevalent source with 99 of 177 studies reported, which was followed by isolation from sponges, with 30, and other marine invertebrates with 15 studies, such as sea squirts, corals, echinoderm-derived, mollusks, and jellyfish, as well as marine algae with 9 studies, water with 7, mangroves with 7, seagrasses with 2 and fishes with 2 (Figure 2B).

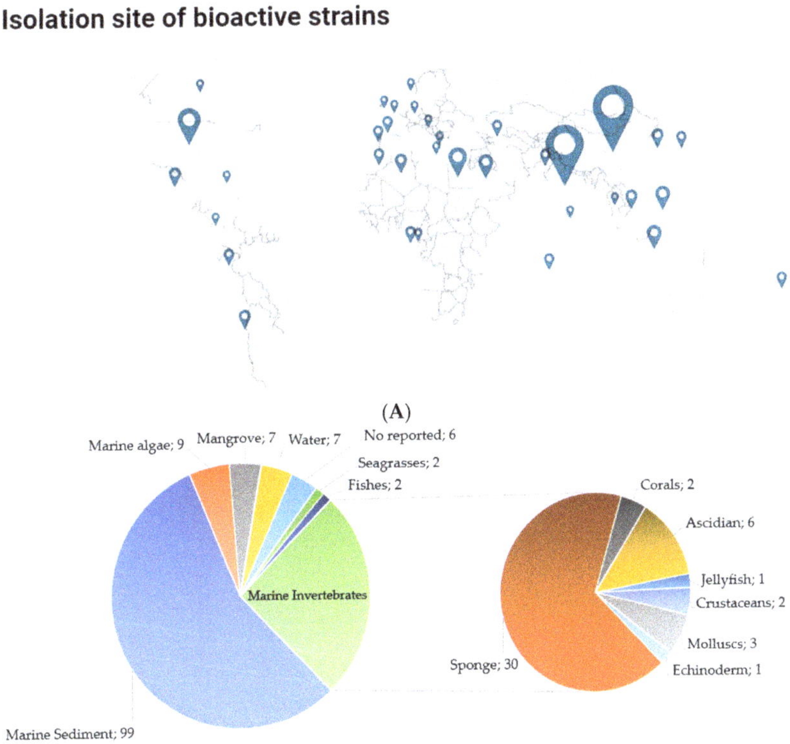

Figure 2. (**A**). World map showing the countries where marine actinobacteria with anti-infective activity were obtained. Max symbol size represents the number of reports. (**B**). Marine actinobacteria with anti-infective activity isolation sources.

Likewise, when analyzing the number of articles published per year, a growth trend was evident; although the search of papers was not restricted by date, the oldest article was from 2002 and the most productive period was 2019 to 2022 (Figure 3A).

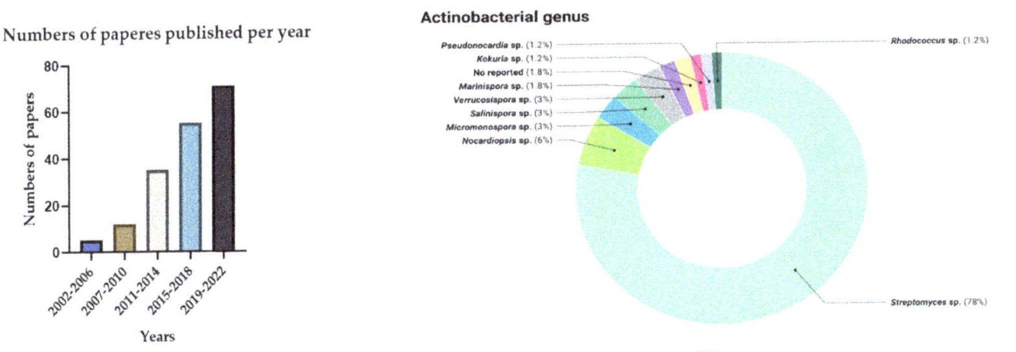

Figure 3. (**A**). Number of works related to marine actinobacteria per year organized by four-year period. (**B**) Actinobacterial genus reported in the studies of marine actinobacteria.

Streptomyces was the genus most reported with anti-infective activity in 78% of the published articles, followed by *Nocardiopsis*, *Micromonospora*, *Salinospora*, and *Verruscosispora* (Figure 3B). Likewise, some genera only were reported in 1% of papers such as *Actinomadura*, *Microbacterium*, *Micrococcus*, *Rothia kristinae*, *Brachybacterium*, *Serinicoccus* and *Solwaraspora* as presented in Figure 3B.

In addition to axenic culture, obtaining compounds from co-cultures has also been described. In total, 4 of 177 papers reported the use of co-cultures, where the most used genus was *Streptomyces*, followed by *Micromonospora* and *Actinokinespora* [25–28].

2.3. Organic Solvents Used to Obtain Anti-Infective Extracts

Extracts and compounds with anti-infective activity have been isolated with different organic solvents; among them, ethyl acetate (EtOAc) is the most frequent. It was used in 66% of studies included in this review, followed by acetone, methanol (MeOH), butanone, butanol, and to a lesser extent dichloromethane, chloroform, ammonium sulfate, etc. Likewise, some solvents combinations have been reported. The most common are butanone–acetone and EtOAc–MeOH, among others. Figure 4 shows this distribution.

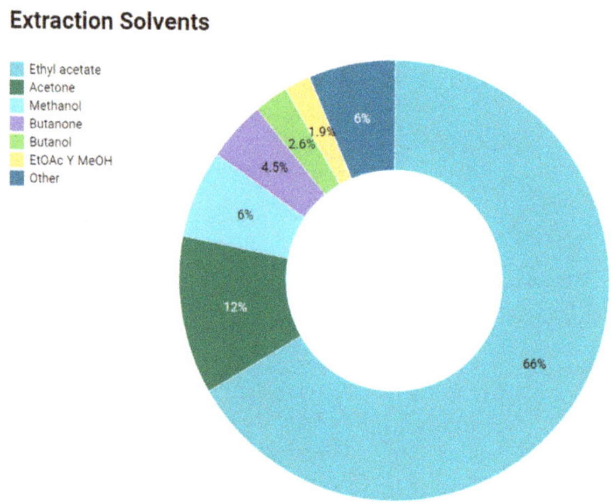

Figure 4. Organic solvents used to obtain actinobacterial extracts and compounds. EtOAc—MeOH: Ethyl acetate—Methanol.

2.4. Anti-Infective Metabolites Derived from Marine Actinobacteria

It is necessary to indicate that here the anti-infective activity refers to a term that includes bacteriostatic, bactericidal, and quorum-quenching activity, which may interfere with virulence factors production, as well as biofilm formation. QQ is not involved in the pathogen elimination or reduction of planktonic cell growth, which may reduce drug resistance and the possibility of bacterial mutation in a high-stress environment [29].

This review focuses on the potential of marine actinobacteria to produce compounds with antibacterial, antifungal (against some fungi such as *Candida albicans* and *Aspergillus fumigatus*) [30–33], antibiofilm activity, and QQ, [34] which inhibits or disrupts an important chemical communication system in bacteria. This involves pathogenic gene expression and metabolism regulation in response to the density of bacterial populations through the production and sensing of some small signal molecules called auto-inductors, both in the same species (intraspecies) as well as among different species (interspecies) [30].

Of the biological activities studied, the antibacterial activity was the most frequently reported in the articles included, with a prevalence of 64% approximately (as shown in Supplementary Table S2), providing the minimum inhibitory concentration (MIC) in some cases (as is presented in Table 1).

Furthermore, this activity was mostly found in compounds obtained from actinobacteria isolated from sediments and marine invertebrates. Likewise, antibiofilm activity and QQ were reported in bacteria of these two sources, and also found in compounds isolated from seawater and mangroves, water actinobacteria, as shown in Figure 5.

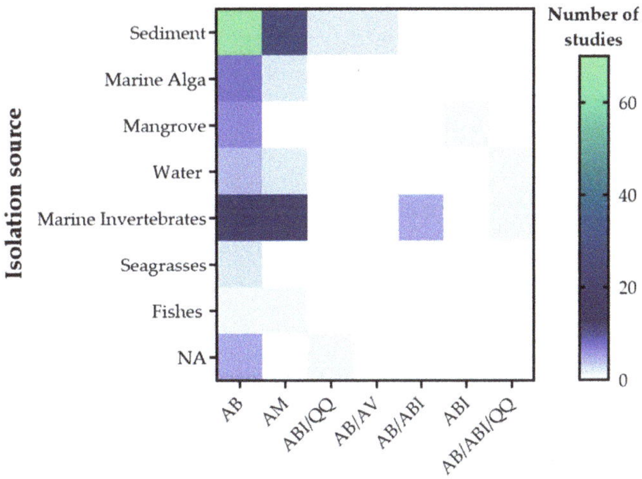

Figure 5. Heatmap of the number of articles included in this study that reported the isolation source of compounds with the biological activity of interest. Isolation sources are arranged from top to bottom, starting with the largest number at the top left. Bioactivities are shown at the bottom from left to right by the largest number of papers reported. The color bar represents the number of studies that reported the source of isolation of bioactive metabolites, from white to blue (lower values), blue to green (medium values), and green (high values). AB: antibacterial activity; AM: antimicrobial activity (activity against bacteria, fungus, parasites); ABI/QQ: antibiofilm and QQ activity; AB/AV: antibacterial and antiviral activity; AB/ABI: antibacterial and antibiofilm activity; ABI: antibiofilm; AB/ABI/QQ: antibacterial, antibiofilm and QQ activity.

Tables 1 and 2 shows the antibacterial and antimicrobial activity, respectively, of crude extract or compounds, obtained from marine actinobacteria expressed in MIC, in which compounds/extracts/fractions with MICs from 0.01–0.02 up to 100, 128, 256, 500 and 1000 μg/mL are reported. Compounds that present activity through the inhibition zone are shown in Tables S3 and S4. Likewise, Tables 1 and 2 present the pathogenic bacteria's target. It is evident that the actinobacteria metabolites exhibit activity towards two of three interesting bacteria that are related to the development of acne vulgaris disease, including MRSA (methicillin-resistant *Staphylococcus aureus*), *S. aureus*, *S. epidermidis* and MRSE (methicillin-resistant *Staphylococcus epidermidis*), but not against *C. acnes*.

Table 1. Antibacterial capacity of actinobacterial crude extracts or compounds.

Genus	Pathogen Target	Compounds/Extracts	MIC (μg/mL)	Ref.
Streptomyces sp.	MRSA [1]	Napyradiomycins 1	0.016	[35]
		Napyradiomycins 8	0.002	[35]
Streptomyces sp.	MRSA [1]	Marinopyrrole A	3.24	[36]
		Marinopyrrole B	3.24	[36]
Streptomyces sp.	*S aureus* ATCC NR-46171	4-methoxyacetanilide	32.4	[18]
Streptomyces sp.	*S. aureus*	Flaviogeranin D	9.2	[37]
		Flaviogeranin C2	8.1	[37]

Table 1. Cont.

Genus	Pathogen Target	Compounds/Extracts	MIC (μg/mL)	Ref.
Streptomyces sp.	*S. aureus*	1-hydroxy-1-norresistomycin	>40	[38]
Streptomyces sp.	MRSA [1]	Fridamycin A	500	[23]
		Fridamycin D	62.5	[23]
Streptomyces sp.	MRSA [1]	Chromomycin A_3	0.698	[39]
Streptomyces sp.	MRSA [1]	Extract	2	[40]
Streptomyces sp.	MRSA ATCC 33591	Actinomycins D_1	0.125	[41]
		Actinomycins D_2	0.25	[41]
		Actinomycins D_3	0.5	[41]
		Actinomycins D_4	0.25	[41]
		Actinomycins D	0.25	[41]
Streptomyces sp.	*S. aureus* CCARM 3090	Grincamycin L	6.25	[42]
Streptomyces sp.	MRSA [1]	Compound [2]	2	[43]
Streptomyces sp.	*S. aureus* (ATCC 29213)	2,4-dichloro-5-sulfamoyl benzoic acid	0.8–4	[44]
Streptomyces sp.	*S. aureus* (ATCC 25923).	Dionemycin	0.5–2	[45]
Streptomyces sp.	*S. aureus* ATCC 43300	Extract	7.9	[46]
Streptomyces sp.	*S. aureus* ATCC 43300	Extract	12.5	[47]
	S. epidermidis (ATCC 12228)	Extract	25	[47]
Streptomyces sp.	*S. aureus*	Aborycin	8.0~64	[48]
	MRSA [1]		16~64	[48]
	MRSE [3]		128	[48]
Streptomyces sp.	MRSA [1]	Supernatant	0.78	[49]
Streptomyces sp.	MRSE [3]	Dehydroxyaquayamycin	16.0	[50]
Streptomyces sp.	MRSA [1]	Medermycin	2	[51]
		G15-F	4	[51]
Streptomyces sp.	MRSA [1] ATCC BAA-44	Bisanhydroaklavinone	6.25	[19]
		1-Hydroxybisanhydroaklavinone	50	[19]
Streptomyces sp.	MRSA [1]	11′,12′-dehydroelaiophylin	1−4	[52]
	MRSA [1], MRSE [3]	Elaiophylin	1−4	[52]
		11-monomethoxylated derivative	2−16	[52]
		Compound 6 [4]	2−16	[52]
Streptomyces sp.	MRSA [1]	Lactoquinomycin A	0.25–0.5	[53]
	MRSA [1]	Stremycin A	16	[54]
		Stremycin B	16	[54]
Streptomyces sp.	MRSA [1] MRSE [3] MSSE [5]	Quinomycin G	16–64	[55]
Streptomyces sp.	*S. aureus* (ATCC 6538)	Actinomycins X2	0.394	[56]
	MRSA [1] (ATCC 43300)	Actinomycins X2	0.190	[56]
	S. aureus (ATCC 6538)	Actinomycins D	0.389	[56]
	MRSA [1] (ATCC 43300)	Actinomycins D	0.188	[56]
Streptomyces sp.	MRSA [1]	Extract [7]	6.25	[25]
	MSSA [6]	Extract [7]	12.5	[25]
Streptomyces sp.	MRSA [1] MSSA [6]	Extract [8]	12.5	[25]
Streptomyces sp.	MRSA [1]	Borrelidins J	0.195	[28]
Streptomyces sp.	*S. aureus*	Extract	256	[57]
Streptomyces sp.	*S. epidermidis*	Extract	128	[57]
Streptomyces sp.	MRSA [1]	Streptopertusacin A	40	[58]
		21,22-en-bafilomycin D	12.5	[58]
		21,22-en-9-hydroxybafilomycin D	12.5	[58]
Streptomyces sp.	*S. aureus* ATCC 29213	Lobophorins E	32	[59]
		Lobophorins F	8	[59]
Streptomyces sp.	MRSA [1]	Pyrrole-derivative	2.8	[60]

Table 1. *Cont.*

Genus	Pathogen Target	Compounds/Extracts	MIC (μg/mL)	Ref.
Streptomyces sp.	MRSA [1]	Julichromes Q_{11}	16–64	[61]
	S. aureus ATCC 29213	Julichromes Q_{10}	16–64	[61]
		Julichromes $Q_{6.6}$	16–64	[61]
		Julichromes Q_6	16–64	[61]
Streptomyces sp.	MRSA [1], *S. aureus*	Lobophorin-like spirotetronate	64	[62]
Streptomyces sp.	MRSA [1], *S. aureus*	Ansamycins	32	[63]
	MRSA [1]	(-)-Streptophenazine B	4.2	[63]
Streptomyces sp.	MRSA [1]	Neo-actinomycin A	16–64	[64]
Streptomyces sp.	*S. aureus* ATCC 29213	MarfuraquinocinsA	8.0	[65]
Streptomyces sp.	MRSE [3] shhs-E1	Marfuraquinocins C	8.0	[65]
	S. aureus ATCC 29213	Marfuraquinocins D	8.0	[65]
Streptomyces sp.	MRSA [1] ATCC 43300	7,8-dideoxygriseorhodin C	0.08–0.12	[66]
		Oxacillin and 7,8-dideoxygriseorhodin C	0.01–0.02	[66]
Streptomyces sp.	MSSA [6] 11497	Desertomycin G	4.0	[67]
	MRSA [1] ATCC 43300	Desertomycin G	4.0	[67]
	MRSA [1] ATCC 25923	Desertomycin G	4.0	[67]
Streptomyces sp.	*S. aureus* ATCC 6518, MTCC 3160 MRSA [1]	Aromatic polyketide	32.40	[68]
Streptomyces sp.	*S. aureus* ATCC 29213 MRSA [1]	Napyradiomycins 1–8 [9]	0.5 to 32	[69]
Streptomyces sp.	*S. aureus* ATCC 29213	Marinopyrroles A–C	<1	[70]
		Marinopyrroles F	3.1	[70]
Streptomyces sp.	MRSA [1]-ATCC33591	A80915A [10]	1–4	[71]
Streptomyces sp.	MRSA [1] ATCC 43300	Polyketide 13 [11]	2	[72]
Streptomyces sp.	MRSA [1]	Fijimycins A–C Etamycin A	4–16	[73]
Streptomyces sp.	*S. aureus* HA- and CA-	Etamycin	1–2	[74]
Streptomyces sp.	MRSA [1]	Lydicamycin congeners	1.56–12.5	[75]
Streptomyces sp.	MRSA [1]	Salinamide F	100	[76]
	S. aureus (ATCC 12600)			
Streptomyces sp.	*S. aureus*	Antimycin B1	32	[77]
Streptomyces sp.	*S. aureus*	Merochlorins G	16	[78]
		Merochlorins J	2	[78]
Streptomyces sp.	*S. aureus*	cyclo(L-Pro-L-Tyr)	160	[79]
		cyclo(L-Pro-L-Phe)	180	[79]
Streptomyces sp.	MRSA [1]	Actinomycin X2	3.125–12.5	[80]
		Actinomycin D	12.5–25	[80]
Streptomyces sp.	*S. aureus*	1,3-Benzodioxole	256	[81]
Streptomyces sp.	*S. aureus ATCC* 29213	Desotamide, Desotamide B	16	[82]
	MRSE		32	[82]
Streptomyces sp.	*S. epidermidis*	Streptophenazines G	3.68	[83]
		Streptophenazines F	6.77	[83]
Streptomyces sp.	MRSA [1]	Citreamicin θ A	0.25	[84]
	ATCC43300	Citreamicin θ B	0.25	[84]
		Citreaglycon A	8.0	[84]
	S. aureus UST950701-005	Dehydrocitreaglycon A	16	[84]
Streptomyces sp.	*S. aureus* DSM346	Alageninthiocin	15	[85]
		Geninthiocin	4	[85]
		Val-geninthiocin	8	[85]
		Indolocarbazole staurosporine	19	[85]
Streptomyces sp.	MRSA [1]	Anthraquinone derivatives	6.25	[86]
Streptomyces sp.	MRSA [1]	Extract	1000	[87]
Streptomyces sp.	*S. aureus*	Extracts	$312–2.5 \times 10^2$	[88]
Streptomyces sp.	*S. aureus*	Extract	400	[89]
Streptomyces sp.	*S. aureus* ATCC 25923	Extract AIA12	2.5×10^2	[90]

Table 1. *Cont.*

Genus	Pathogen Target	Compounds/Extracts	MIC (μg/mL)	Ref.
		Extract AIA17	310	[90]
Streptomyces sp.	MRSA [1]	1-Acetyl-β-Carbonile	128–256	[91]
	MSSA [6]	1-Acetyl-β-Carbonile	64	[91]
Streptomyces sp.	MRSA [1]	Chlororesistoflavins A	0.25	[92]
	MRSA [1]	Chlororesistoflavins B	2.0	[92]
Streptomyces sp.	*S. aureus*	Ligiamycin A	16	[26]
	S. aureus	Ligiamycin B	64	[26]
Verrucosispora sp.	*S. aureus ATCC* 33591	Active fraction	16–32	[93]
Verrucosispora sp	*S. aureus* ATCC29213 MRSA shhs-A1	Proximicins B	16	[94]
Verrucosispora sp.	MRSA [1]	1-hydroxy-2,5-dimethyl benzoate	12.5	[95]
Verrucosispora sp.	MRSA [1]	Proximicin B	3.125	[95]
Micromonospora sp.	*S. aureus* ATCC 29213	Kendomycins B	0.5–2	[96]
	S. aureus 745524	Kendomycins C	0.5–1	[96]
	MRSA [1] shhs-A1	Kendomycins D	1–4	[96]
Micromonospora sp.	MRSA [1]	2-ethylhexyl 1H-imidazole-4- carboxylate	16	[97]
Micromonospora sp.	*S. aureus* ATCC 29213	Micromonohalimanes B	40	[98]
Micromonospora sp.	*S. aureus* ATCC 29213	Rabelomycin	1	[99]
		Phenanthroviridone	0.25	[99]
Micromonospora sp.	*S. aureus* ATCC 29213	homo-dehydrorabelomycin E	1	[100]
Nocardiopsis sp.	MRSA [1]	Bis (2-ethylhexyl) phthalate	7.81	[101]
	MRSA [1] ATCC NR-46071	4-bromophenol	15.62	[101]
Nocardiopsis sp.	MRSA [1]	Nocardiopsistin A	12.5	[102]
		Nocardiopsistin B	3.12	[102]
		Nocardiopsistin C	12.5	[102]
Nocardiopsis sp.	MRSA [1]	α-Pyrone	12.5	[103]
Nocardiopsis sp.	MRSA [1]	Extracts	115–125	[104]
Marinispora sp.	MSSA [6] MRSA [1]	Lipoxazolidinone A	1–2	[105]
Marinispora sp.	MRSA [1] MRSE [3] ATCC 700578c	Lynamicins A–E	2.2–45	[106]
Pseudonocardia carboxydivorans	*S. aureus* ATCC 6538P	Branimycins C	32	[107]
	MRSA [1] MB5393	Branimycins C	20–40	[107]
Kocuria sp.	MRSA [1] ATCC 43300-	Kocurin	0.25–0.5	[108]
Solwaraspora sp.	MRSA [1]	Solwaric acids A	32	[109]
		Solwaric acids B	32	[109]
	MSSA [6]	Solwaric acids A	64	[109]
		Solwaric acids B	64	[109]
Salinispora sp.	MRSA [1]	Rifamycin W	15.62	[110]

[1] MRSA: Methicillin-resistant *Staphylococcus aureus*. [2] Compound: 1 [2-hydroxy-5-((6-hydroxy-4-oxo-4Hpyran-2-yl) methyl)-2-propylchroman-4-one]. [3] MRSE: Methicillin-resistant *Staphylococcus epidermidis*. [4] Compound 6: Compound name no reported. [5] MSSE: Methicillin-susceptible *Staphylococcus epidermidis*. [6] MSSA: Methicillin-susceptible *Staphylococcus aureus*. [7] Extract: Extract Co-culture (MRSA). [8] Extract: Extract Co-culture (*Pseudomonas aeruginosa*). [9] Napyradiomycins 1–8: Except compound 3. [10] A80915A: Napyradiomycin derivatives. [11] Polyketide 13: [=2-hydroxy-5-((6-hydroxy-4-oxo-4H-pyran-2-yl) methyl)-2- propylchroman-4-one].

Table 2. Antimicrobial activity of actinobacterial crude extracts or compounds.

Genus	Pathogen Target	Compounds/Extracts	MIC (μg/mL)	Ref.
Streptomyces sp.	*S. aureus* FDA209P JC-1	Chlorinated α-lapachone	12.5	[31]
Streptomyces sp.	MRSA [1]	Streptoindoles A	25	[32]
		Streptoindoles B	7	[32]
		Streptoindoles D	25	[32]
Streptomyces sp.	MRSA [1]	Streptoglutarimides A–J	9–11	[111]

Table 2. *Cont.*

Genus	Pathogen Target	Compounds/Extracts	MIC (µg/mL)	Ref.
Streptomyces sp.	*S. aureus*	Nitricquinomycin C	17	[112]
Streptomyces sp.	MRSA [1]	Napyradiomycin D1	12–24	[113]
Streptomyces sp.	*S. aureus* ATCC 33591	Polyketide antibiotic SBR-22	64	[114]
Streptomyces sp.	*S. aureus* ATCC 29213	Lobophorins F	6.25	[115]
Streptomyces sp.	*S. aureus*	Polyketide related antibiotic	37.5	[30]
Streptomyces sp.	MRSA [1]	Actinomycin D	0.08	[116]
		Actinomycin V	0.08	[116]
		Actinomycin $X_0\beta$	0.61	[116]
Streptomyces sp.	MRSA [1]	Niphimycins C	4–32	[117]
	MRSE [2]	Niphimycin Iα	4–32	[117]
Streptomyces sp.	*S. aureus* ATCC 25923	Trihydroxylflavanone [3]	32	[118]
		Tetrahydroxylchalcone [4]	1	[118]
Streptomyces sp.	*S. aureus*	Anthracycline analogues	20	[119]
		β-rhodomycin-II	40	[119]
Streptomyces sp.	*S. aureus*	DMBPO [5]	>1000	[120]
Streptomyces sp.	*S. aureus* ATCC 25923	Chromomycin A9	0.03	[121]
		Chromomycin Ap	0.13	[121]
		Chromomycin A2	0.06	[121]
		Chromomycin A3	0.13	[121]
Streptomyces sp.	MRSA [1]	Streptopyrazinones A–D	58–65	[122]
		N-acetyl-L-isoleucine-L-leucinamide	65	[122]
Streptomyces sp.	MRSA [1]	4-dehydro-4a-dechloronapyradiomycin A1	4–8	[123]
		Napyradiomycin A1	0.5–1	[123]
Streptomyces sp.	*S. aureus*	3-propanoic acid [6]	32	[124]
		Propanoic acid methyl ester [7]	64	[124]
		3-(3-chloro-4-hydroxyphenyl) propanoic acid	32	[124]
Streptomyces sp.	*S. aureus* (ATCC 6538)	Natural cyclic peptide	1.25	[125]
	MRSA [1]		12.5	[125]
	S. aureus (ATCC 6538)	Cyclic peptides	0.025–0.156	[125]
	MRSA [1]	Cyclic peptides	0.1–0.78	[125]
Streptomyces sp.	*S. aureus*	Extracts A758	6.25	[126]
		Extracts A759	500	[126]
		Extracts A760	100	[126]
		Extracts A765	3.125	[126]
Streptomyces sp.	MRSA [1]	Novobiocin	0.25	[127]
		Desmethylnovobiocin	16	[127]
		5-Hydroxynovobiocin	8	[127]
Kocuria marina	*S. aureus*	Kocumarin	10	[128]
	MRSA [1]	Kocumarin	10	[128]
Rhodococcus sp.	*S. aureus*	n-butanol fraction	9.3	[34]
		EtOAc fraction	12.6	[34]
Marinispora sp.	MRSA [1]	Marinomycin A	0.130	[129]
		Marinomycin B–C	0.49	[129]
		Marinomycin D	2.43	[129]
Verrucosispora sp.	*S. aureus*	(2-(hydroxymethyl)-3-(2-(hydroxymethyl)-3-methylaziridin-1-yl) (2-hydroxyphenyl) methanone	3.4	[130]

[1] MRSA: Methicillin-resistant *Staphylococcus aureus*. [2] MRSE: Methicillin-resistant *Staphylococcus epidermidis*. [3] Trihydroxylflavanone: lavandulyl-7-methoxy-5,20,40-trihydroxylflavanone. [4] Tetrahydroxylchal-cone 50-lavandulyl-40-methoxy-2,4,20,60-tetrahydroxylchalcone. [5] DMBPO: 5-(2,4-dimethylbenzyl) pyrrolidin-2-one Information no reported. [6] 3-propanoic acid: 3-(3,5-dichloro-4-hydroxyphenyl) propanoic acid. [7] Propanoic acid methyl ester: 3-(3,5-dichloro-4-hydroxyphenyl) propanoic acid methyl ester.

It is well known that actinobacteria are a phylum with the potential to produce molecules with innumerable bioactivities and with multiple applications. Santos et al. [33] studied actinobacteria strains isolated from a marine sponge, in which antimicrobial activity was previously reported (due to this it is not described in Table 2) against MRSA (methicillin-resistant *S. aureus* MB 5393), which could also be involved in skin infections, as well as the fungus *Aspergillus fumigatus* ATCC46645, demonstrating that it also had the capacity to induced lipid reduction on the larvae of zebrafish [33], revealing its potential use in anti-obesity treatments.

Other compounds or strains were reported with activity, but this was presented in growth inhibition percentage for Aa3_DN216_4B10_1, which showed a significant growth inhibition (61%) against MRSA [22].

A wide variety of compounds with antimicrobial activity were reported in plants, such as flavonoids. Interestingly, in this systematic review, some studies reported flavonoids from sponge-derived actinobacteria. Flavonoids are a group of natural substances with variable phenolic structures; they are found in fruits, vegetables, grains, bark, roots, stems, flowers, and wine. These are an important class of natural products; particularly, they belong to a class of plant secondary metabolites having a polyphenolic structure [84].

Historically, flavonoids have been recognized with a broad spectrum of health-promoting effects because of their antioxidative, anti-inflammatory, anti-mutagenic, and anti-carcinogenic properties with their application in various diseases such as cancer, Alzheimer's disease (AD), atherosclerosis, etc. [131]. Cao et al. [118] reported two new lavandulylated flavonoids, 6-lavandulyl-7-methoxy-5,20,40-trihydroxylflavanone and 50-lavandulyl-40-methoxy-2,4,20,60-tetrahydroxylchalcone (Table 2), which had a broad-spectrum of antimicrobial activity against both Gram-positive and Gram-negative bacteria and fungus such as *Candida albicans* [118]. Other compounds with antibacterial activity included in this systematic review are citreamicins, which are polycyclic xanthones (belong to flavonoids class) obtained from marine-derived *Streptomyces caelestis*, isolated in the coastal water of the Red Sea [84]. This *S. caelestis* showed antibacterial activity against a variety of Gram-positive bacteria, including MRSA and vancomycin-resistant *Enterococcus faecalis* (VRE). Four compounds were isolated from *S. caelestis* (Table 2), with antibacterial activity against *S. aureus* UST950701-005 with a MIC from 1 to 16 μg mL^{-1} and three had antibacterial activity against MRSA with a range of MIC between 0.25 and 8 μg mL^{-1} [84].

On the other hand, anthracycline compounds with antibacterial and antimicrobial activity have also been reported among the metabolites derived from marine actinobacteria. Anthracyclines are known as an important class of anticancer compounds used for many years in the treatment of leukemia, breast carcinoma, and other solid tumors. However, their application in cancer treatment has been decreased due to their toxic, dose-related side effects such as stomatitis, gastrointestinal disorders, and cumulative cardiotoxicity. Anthracyclines belong to the group of tetramic acids and have been reported to have antibacterial activity toward Gram-positive bacteria such as vancomycin-resistant *Enterococcus* (VRE). Cong et al. discovered novel anticancer and anti-infective natural products from marine *Streptomyces* sp. SCSIO 41399 which were isolated from coral *Porites* sp. These compounds were isotirandamycin B and two known tirandamycin derivatives. This study is one of the two that in this systematic review that reported a coral with anti-infective activity toward *Streptococcus agalactiae* and *S. aureus*, which may be useful in the control of acne-related bacteria [119].

Other compounds reported with antimicrobial activity in this systematic review were Chromomycins, Napyradiomycins, Marinomycins, and Kokumarin.

Chromomycins are members of the aureolic acid family, and they are polyketides with a tricyclic aglycone core with two aliphatic side chains at C-3 and C-7 and two sugar chains at C-2 and C-6, similar to other aureolic acid family members. Chromomycins interact with the DNA helix minor groove in regions with high GC (guanine–cytosine) content and in a non-intercalative way with Mg^{2+} cations, causing DNA damage in treated cells [121].

Napyradiomycins (NPDs) for their part constitute an interesting family of halogenated natural compounds NPDs that consist of a naphthoquinone core, a prenyl unit attached at C-4a, a monoterpenoid substituent at C-10a, and some congeners have a methyl group at C-7 [123].

Marinomycins possess unique polyene–polyol structures and have unique photoreactivities and chiroptical properties [129]. Finally, Kokumarin was the only compound reported to have antimicrobial activity against MRSA isolated from skin infections [128].

In addition to the antibacterial and antimicrobial activity exhibited by extracts or isolated compounds of marine actinobacteria recovered in this systematic review, some bacteria, especially of the *Streptomyces* and *Nocardiopsis* genera, have been shown to have more than one biological activity such as both antibacterial and antibiofilm activity, as shown in Table 3, with high activity against *S. aureus* and methicillin-resistant *S. epidermidis*, related with the development of acne vulgaris.

Table 3. Antibacterial and anti-biofilm activity of actinobacterial crude extracts or compounds from *Streptomyces* genus.

Genus	Pathogen Target	Compounds/Extracts	MIC (μg/mL)	Ref.
Streptomyces sp.	MRSA [1]	Compound PVI331	1	[132]
Streptomyces sp.	MRSA [1]	8-O-metyltetrangomycin	2	[10]
Streptomyces sp.	MRSE [2] RP62A	Compound (SKC3)	31.25	[133]
Streptomyces sp.	MRSA [1]	PVI401	0.5	[134]
		PVI402	2	

[1] Methicillin-resistant *Staphylococcus aureus*. [2] Methicillin-resistant *Staphylococcus epidermidis*.

Four of five compounds reported to have antibacterial and antibiofilm activity were isolated from the *Streptomyces* genus and these activities were reported against MRSA; only one unidentified compound had an effect against MRSE. This compound, named SKC3, exhibited an antagonistic effect against growth and biofilm formation of the methicillin-resistant *S. epidermidis* at a concentration below the MICs (Table 3). Interestingly, the biofilm inhibitory concentration (BIC_{90}) of SKC3 was 3.95 μg/mL, and this had no considerable influence on bacterial growth. In addition, SKC3 also had an effect in inhibiting the growth and biofilm formation of other strains such as MSSA, MRSA, and VRSA, however, was ineffective against the tested Gram-negative *P. aeruginosa* strains [133].

The compound PVI331 had a prominent antibacterial activity with a MIC of 1 μg/mL, (Table 3) and showed biofilm inhibition at a 92.17 ± 1.67% at 4 μg/mL, concentration against MRSA and it was more effective than the anti-MRSA antibiotic vancomycin, which was used at a concentration of 8 mg/mL, and the biofilm inhibition was 32.58 ± 2.52% [132].

8-O-metyltetrangomycin is an angucycline antibiotic that showed a significant antibiofilm activity toward MRSA, ranging from 52.85 to 86.64% inhibition. Similar to compound PVI331, this angucycline compound exhibited more antibiofilm potential than vancomycin and the highest range of inhibition was observed at 4× MIC, suggesting the stronger potential to reduce biofilm formation that possesses these compounds [10].

Compounds PVI401 and PVI402 exhibited antibacterial activity against MRSA, (Table 3), however, only PVI401 showed antibiofilm activity toward *S. aureus* ATCC25923; this effect was dependent on the concentration obtained in the antibacterial assay of compound PVI401, with poor biofilm formation when compared to controls when the pathogen was treated with a 4× MIC concentration at 2 μg/mL, of PVI401 [134].

Regarding antibiofilm activity, the same compounds were reported by Hifnawy et al. [27] to have antibacterial activity against Gram-negative and Gram-positive bacteria and antibiofilm activity, and the compounds tubermycin and p-anisamide had potent antibiofilm activity against *P. aeruginosa* with inhibition rates of 94 and 73% respectively. On the contrary, compounds **1**, **2** and **9** had antibiofilm activity against *E. coli* with inhibition ranges of 34–54%, and only Compounds **1** and **2** showed a potent to moderate inhibition against *S. aureus* with a percentage of inhibition rates of 50 and 75% respectively [27].

Concerning antibacterial, antibiofilm activity, and QQ, only 2 of 177 papers described compounds with three activities and two more articles reported antibiofilm and QQ (Table 4). These papers evaluated the QQ ability of actinobacteria-derived metabolites. All studies evaluated the AI-1 (Autoinducer 1) system of quorum sensing, using the reporter strain, *Cromobacterium violaceum*. This strain produces a visible purple pigment called violacein, which is under positive regulation by the N-acyl-homoserine lactone CviI/R quorum-sensing system. This system has been reported in Gram-negative bacteria mainly [135–137]. Moreover, one of the studies also evaluated the inhibition of the LuxS/AI-2 quorum-sensing system. In this system, the signal molecule is regulated by the *luxS* gene [138] and it has been reported that is utilized by more than 40 species of Gram-positive and Gram-negative bacteria for communication and transmission [136]. This system has been reported in *C. acnes*, *S. aureus*, and *S. epidermidis*, bacteria under this study, nonetheless, both *S. epidermidis* and *S. aureus* also have been reported to use peptides autoinducers (AIP), regulated by *agr* system for quorum sensing [139]. Nevertheless, any article included in this systematic review that reported inhibition in this system could further be investigated as an effective treatment of acne vulgaris.

From these investigations, two compounds are described as having antibacterial and antibiofilm activity, as well as QQ: one of these is butenolide, which is a compound isolated from marine actinobacteria with antifouling activity studied previously; this compound inhibited quorum sensing and is an unspecific inhibitor due to having the ability to inhibit the AHL system through the inhibition of the violet pigment of two *C. violaceum* strains, *CV026* (short-chain AHLs) and VIR24 (long-chain AHLs), inhibiting short-chain AHLs at a concentration of 100 μg/mL and long-chain AHLs at 50 μg/mL, and with growth inhibition being observed at concentrations of 25–50 μg/mL. This compound also inhibits the AI-2 system through bioluminescence of indicator strains *Vibrio harveyi* BB170, at concentrations of 5, 12.5, and 25 μg/mL with a reduction of luminescence of ~25, ~50, and over 70%, respectively. However, at concentrations above 12.5 μg/mL, it caused growth inhibition against the bacterial cells (Table 4) [138]. Despite this, it is considered to have low antibacterial activity against diverse types of pathogens (both Gram-positive and Gram-negative bacteria) [138].

Likewise, it was found that butenolide not only effectively inhibited the biofilm formation but also eradicated pre-formed biofilms of tested bacteria and it also had a synergistic effect with tetracycline; it was a potential tetracycline enhancer against biofilm-associated infection-producing bacteria such as *E. coli*, *P. aeruginosa*, and MRSA [138].

Another compound with antibacterial, antibiofilm, and QQ activity is a melanin pigment. It was discovered from marine *Nocardiopsis* sp., which exhibited antibacterial activity toward *Bacillus* sp. from extract JN2 with growth inhibition of 68 and >40% against *S. aureus* at a concentration of 150 μg mL^{-1}. Respecting its antibiofilm activity, both the pigments (JN1M and JN2M) inhibited the growth of quorum-sensing bacteria *C. violaceum* MTCC 2656 (Table 4) [137].

Table 4. Anti-biofilm, antibacterial, and quorum-quenching activity of crude extracts or compounds from marine actinobacterial.

Genus	Target Bacteria in Antibiofilm Activity	MBIC [1]	Compounds/ Extracts	Percentage Decreased Biofilm	QS System	QQ Activity (IC50)	Biosensor Strain	Ref.
Streptomyces sp.	MRSA [2]	200	Butenolide	>70	AI-2 up to 70%	NA [3]	*Vibrio harveyi BB170*	[138]
					AHL inhibition up to 97%		*C. violaceum*	[138]
	S. aureus	100	Extract	78.9	AHL	NA [3]	*C. violaceum* 12472	[140]
Nocardiopsis sp.	*S. aureus*	NA [3]	Melanin JN1M	64.2	AHL	NA [3]	*C. violaceum MTCC 26563*	[137]
			Melanin JN2M	65.9	AHL	NA [3]	*C. violaceum MTCC 26563*	[137]

Table 4. *Cont.*

Genus	Target Bacteria in Antibiofilm Activity	MBIC [1]	Compounds/ Extracts	Percentage Decreased Biofilm	QS System	QQ Activity (IC50)	Biosensor Strain	Ref.
Nocardiopsis sp.	*S. aureus*	20 vol % [4]	Culture liquid of JS106	77.94	AHL	NA [3]	*C. violaceum* 12472	[29]
	NA [3]	NA [3]	Questiomycin A	NA [3]	AHL	6.82	*C. violaceum* 12472	[29]
	NA [3]	NA [3]	2-hydroxyacetate-3-hydroxyacetamido-phenoxazine (HHP)	NA [3]	AHL	23.59	*C. violaceum* 12472	[29]

[1] MBIC: The minimum biofilm inhibitory concentration. [2] MRSA: Methicillin-resistant *Staphylococcus aureus*. [2] MRSE: Methicillin-resistant *Staphylococcus epidermidis*. [3] NA: Information not reported. [4] 20 vol %: Concentration expressed in percentage.

In addition, there are compounds reported with antibiofilm and QQ activity. The liquid culture and crude extract of *Nocardiopsis* sp. displayed a decreased antibiofilm activity against *S. aureus* and QQ by inhibiting the violacein production of strain *C. violaceum* 12472, respectively. Likewise, the compounds Questiomycin A and 2-hydroxyacetate-3-hydroxyacetam-ido-phenoxazine (HHP) isolated from this liquid culture also showed QQ activity against *C. violaceum* 12472 at a concentration of 40 μg/mL (Table 4). This compound belongs to the phenoxazinones group and is a structurally unique natural product containing a tricyclic core heterocyclized by nitrogen and oxygen atoms [29].

2.5. Actinobacteria Producing Quorum Quenching Metabolites

Regarding the QQ activity, it was evaluated in only 2.8% of the papers, and the mechanism of inhibition used was the AHL (acyl-homoserine lactone) autoinducer (AI-1), through the indicator strain *C. violaceum*; one only study reported the effect of the extract of marine actinobacteria against mechanism two, the LuxS enzyme autoinducers 2 (AI-2), through the bioluminescence of *V. harveyi* BB170 [138]. Table 5 shows the papers with quorum-quenching activity.

Table 5. Marine actinobacteria with Quorum Quenching (QQ) activity.

Source	Genus	Disrupter QS System	Biosensor Strains	Ref.
Gut of marine fishes	*Streptomyces* sp.	AI-1: AHL	*C. violaceum* and *Serratia marcescens.*	[141]
NA [1]	*Streptomyces* sp.	AI-1: AHL, AI-2: LuxS	*C. violaceum CV026* and *Vibrio harveyi BB170*	[138]
Marine Sponge	*Streptomyces* sp.	AI-1: AHL: LasI	*Pseudomona-* Molecular docking.	[140]
Marine sediment	*Nocardiopsis* sp.	AI-1: AHL	*C. violaceum 12472*	[29]
Seawater	*Nocardiopsis* sp.	AI-1: AHL	*C. violaceum (MTCC 2656)*	[137]

[1] Information no reported.

2.6. Strategies to Maximize Anti-Infective Metabolites Activity and Yield

2.6.1. Culture Conditions to Anti-Infective Production Metabolites

Actinobacteria fermentations often do not generate a high yield of active compounds [51]. It is well known that the culture conditions significantly affect bacterial metabolism. Likewise, the composition of the culture medium is related to the metabolic capacities of the producing organism, influencing the biosynthesis of antibiotics [114]. Some studies included in this systematic review (48 of 177) carried out the identification of the variables that are related to the increase in the production of compounds with anti-infective activity, through some biostatistical methods such as the Placket–Burman design and the response surface method [60]. These analyses revealed that carbon and nitrogen sources played a key role, with the nitrogen source in some cases being more prominent [142] in addition to pH, temperature, and agitation speed.

Starch was described as a carbon source used to achieve maximum production of the anti-infective compound as reported by Djini et al. [43] as presented in Figure 6A. Likewise, Norouzi et al. revealed a significant effect of starch in combination with Peptone (as nitrogen source) and pH, and calcium carbonate, reaching up to a 218% increase in production yield of anti-MRSA compounds [60], and Mohamedin et al. reported antagonistic activity produced from the optimized culture conditions against multidrug-resistant *Staphylococcus epidermidis*, which showed about a 1.37-fold increase using starch as the carbon source and potassium nitrate and yeast extract as the nitrogen source [143].

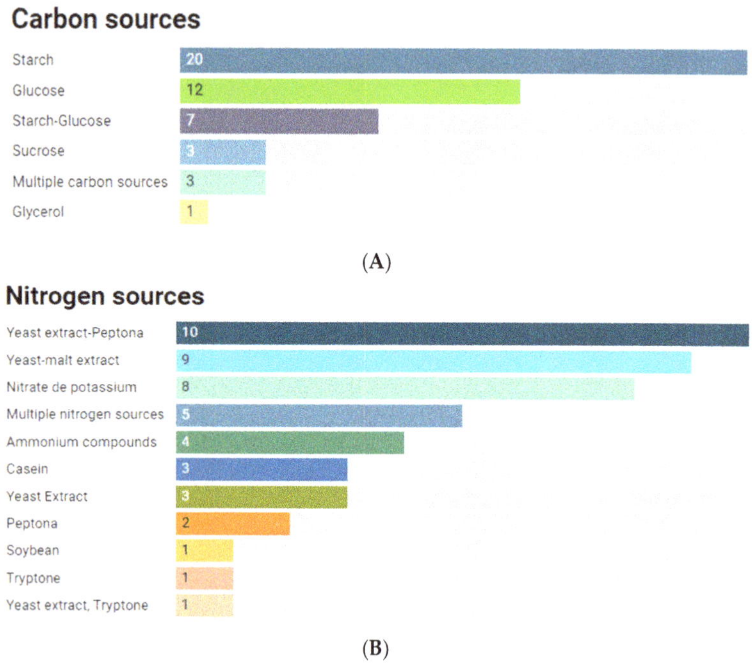

Figure 6. (**A**). Most used carbon sources to maximize the anti-infective compound. (**B**). Most used nitrogen sources to maximize the anti-infective compounds. Only 48 of 177 papers reported culture conditions.

Other carbon sources reported to increase the production of compounds with anti-infective potential were glucose [18,27,33], sucrose [37,132,144] starch–glucose [23,31,94] among others.

Regarding the nitrogen source, the most common were yeast extract–peptone [81,94,102], yeast–malt extract [20,38,145], potassium nitrate [10,40,146], ammonium compounds as ammonium sulfate [147], ammonium chloride [79], ammonium nitrate [114], and casein [22,43,148] as shown in Figure 6B. The quenching potential also has been subjected to optimization processes to maximize its performance, finding that soybean meal and sodium chloride were two crucial factors in the culture medium that significantly increased both the bioactivity and metabolite production (302 and 241%, respectively) when compared to the original condition [29].

Also, some studies highlighted the need for seawater not only for the cultivation of the strains, but also to produce antibiotics [43,105,149], making it clear that this depends on the concentration of salt [142,149]. Figure 6 presents the carbon and nitrogen sources most used in the rise of anti-infective compound production.

On the other hand, Xu et al., reported that the supplementation of the rare earth salt Lanthanum chloride (LaCl3) during fermentation of HB-J378 significantly increased the yield of these angucyclines [102]. This similarly occurred with the strains N816 and S355 isolated from marine sponge actinomycetes, which showed potent anti-MRSA activity elicited due to the addition of LaCl3 that was significantly enhanced in the J378 strain, which shows LaCl3 to be an effective elicitor [102].

2.6.2. Co-Culture Combination as Strategies to Maximize Anti-Infective Metabolites in Marine Actinobacteria

The co-culture of microbial strains can activate the production of compounds that in monoculture are not obtained or the accumulation of metabolites is less. In addition, it has been considered that this strategy also contributes to activating silent biosynthetic gene clusters, leading to the improved production of natural compounds that do not occur under laboratory conditions [86]. In the marine environments, bacterial secondary metabolites production usually depends on their interactions with other microbes or is regulated by environmental or stressing conditions such as competition for nutrients or space [27,86,150].

There are diverse ways to have a microbial strain co-culture; one of the most common is between fungus and bacteria as was reported in the microbial co-culture combination of a sponge-derived actinomycete *Streptomyces rochei* MB037 and a gorgonian-derived fungus *Rhinocladiella similis,* which induced the production of related polyketides and exhibited significant antibacterial activity against methicillin-resistant *S. aureus* with a MIC value of 0.195 mg/mL [28]. Furthermore, another way of co-culture is the co-cultivation between bacteria of different or the same genus, such as the co-culture of two red marine sponge-associated actinomycetes *Micromonospora* sp. UR56 and *Actinokinespora* sp. EG49, which induced the accumulation of metabolites with antibacterial and antibiofilm activity, that were not traced in their axenic cultures [27]. The compounds belong to the phenazine class and have been isolated and characterized previously. In total, authors obtained five compounds; from them, Compounds 1 (dimethyl phenazine-1,6-dicarboxylate), 2 (phencomycin), and 9 (*N*-(2-hydroxyphenyl)-acetamide) showed considerable antibacterial activity against *S. aureus* with growth inhibitions of 47, 69, and 53% respectively. In addition, Compounds 3 (tubermycin) and 10 (p-anisamide) displayed potent antibacterial activity against *P. aeruginosa* with growth inhibition of 94 and 70% respectively [27]. Also, the co-culture between marine-derived actinobacteria and human pathogens in this systematic review has been reported, which resulted in increased production of three antibiotics: granaticin, granatomycin D, and dihydrogranaticin B, and it also strongly enhanced biological activity against the Gram-positive human pathogens such as MRSA [25].

2.7. Main Families of Compounds Found in Marine Actinobacteria with Antibacterial Activity

An enormous variety of compounds were reported in the papers included in this systematic review; these have been arranged considering the type of activity that they exhibited and grouped in families.

Among families, polyketides were the most reported; these types of compounds are a vast variety of constituents and represent a highly diverse structural class of products, demonstrating varied biological functions [72]. Polyketides are secondary metabolites produced from bacteria, fungi, plants, and animals, and bacteria from the *Streptomyces* genus, which are thought one of the polyketides producers [28]. Polyketides are made up of many compounds, including macrolides, reported in 7 of 177 papers, aromatic polyketides in 9 of 177 (including angucyclines), and so on.

Table 6 displays the family compounds, their constituents, and the frequency that were presented.

Table 6. Family compounds with antibacterial activity.

Compound	Frequency	Constituents	Ref.
Polyketide	19	Naphthoquinone-based meroterpenoids	[37]
		Naphthoquinone Derivatives	[25]
		Chlorinated Meroterpenoids (Merochlorins G–J)	[78]
		Angucycline	[23,42,50,54,100,102]
		Aromatic Polyketides	[61,68,151]
		Polyketide [1]	[72]
		Compound 1 [2]	[43]
		Macrolides [3]	[52,57,67,96,107,132,152]
Phenolic compound	1	Bromophenol derivative	[101]
Phthalate	1	Bis (2-ethylhexyl)	[101]
Acetamide	2	4-methoxyacetanilide	[18]
		2-ethylhexyl 1H-imidazole-4-carboxylate	[97]
Alkaloids	3	butyl 1Himidazole-4-carboxylate	[97]
		Chlorinated bis-indole alkaloids	[45]
		Indolizinium alkaloid	[58]
Pyrrole	3	Chlorinated Bisindole Pyrrole	[106]
		Pyrrole-derivative	[41,60]
Chromopeptides	6	Actinomycins ($X_{0\beta}$, X_2, D, D1–D4, A)	[56]
		Neo-actinomycin A, B, actinomycins D and C4, X_2,)	[64,77,80,153]
Cyclo peptides	3	Desotamides A–D	[154]
		cyclo(L-Val-L-Pro),	[79]
		cyclo-(L-Pro-4-OH-L-Leu)	[55]
Antracycline	1	Bisanhydroaklavi-none 1-Hydroxybisanhydroaklavinone	[19]
Marinopyrroles	1	(−)-marinopyrroles A (−)-marinopyrroles B	[70]
Phenazines	5	phenazine-1,6-dicarboxylate, phencomycin, tubermycin	[27]
		Streptophenazines G	[63]
		1,6-Dihydroxy phenazine, dimethoxy phenazine	[155]
		Actinomycins D1 and D2	[41]
Spirotetronate antibiotics	2	Lobophorins L and M Lobophorins E	[62]
Proteins	2	Enzyme PA720 (Thermophilic Hemoglobin-degrading Protease)	[156]
		β-lactamase inhibitory protein	[157]
Pyranonaphthoquinones	3	Medermycin-type naphthoquinones	[158]
		Medermycin derivative	[51]
		Lactoquinomycin A (LQM-A)	[53]
Quinomycin family antibiotics	1	Quinomycin G	[55]
Quinona	1	1- hydroxy-1-norresistomycin	[38]
Siderophore native	3	S1, S2, S3 [4]	[57,144,159]
Thiazolyl Peptide Antibiotic Family	1	Kocurin	[108]
Pigment	1	Melanin pigment	[137]
Aminofuran natural products	1	Proximicin F and G	[94]
Type I lasso peptide natural products	1	Aborycin	[48]
Natural product class diazaanthraquinone	1	Diazaanthraquinone	[160]
Benzoic acid	1	2,4-dichloro-5-sulfamoyl benzoic acid	[44]
4-oxazolidinone antibiotics	1	Lipoxazolidinone A, B and C.	[105]
Cyslabdan-like compound	1	Cyslabdan-like compound	[93]
Benzene Derivative	1	1,3-Benzodioxole	[81]
Flavonoids	3	Citreamicin θ A Citreamicin θ B Citreaglycon A Dehydrocitreaglycon A	[84]

[1] Polyketide: Compound name no identified. [2] Compound 1: [2-hydroxy-5-((6-hydroxy-4-oxo-4Hpyran-2-yl) methyl)-2-propylchroman-4-one]. [3] Polyketide: Elaiophylin Derivatives, Nargeninas, Desertomycin G, Kendomycin analogues, N-Arylpyrazinone Derivative. [4] S1: 5,6-dihydro-1,8-dihydroxy-3-methylbenz[a]anthracene-7,12-quinone; S2: 1,4-dihidroxy-2-(3-hydroxybutyl)-9, 10-antraquinone; S3: Desferrioxamine B and the New Desferrioxamine B2.

Macrolides are a class of antibiotics derived from *Saccharopolyspora erythraea* (originally called *Streptomyces erythreus*), a type of soil-borne bacteria. They are bacteriostatic antibiotics in that they suppress or inhibit bacterial growth rather than killing bacteria completely and possess a macrocyclic lactone ring containing eight or more atoms, and polyketide [161]. They act by inhibiting the protein synthesis of bacteria by binding to the 50S ribosomal element [162].

Macrolides, especially erythromycin together with clindamycin, which is a lincosamide (isolated from an actinobacterium, *Streptomyces lincolnensis* obtained from the soil in the region of Lincoln, Nebraska, United States), are the main antibiotics recommended as the first-line therapy in the acute inflammatory phase of acne [163].

Both have similar mechanisms of action, and lincosamides have even been integrated with macrolides in a group called "macrolides and similar" [164].

The angucycline group of antibiotics and aromatic polyketide natural products belong to a specific group of polycyclic aromatic polyketides, which exhibit anticancer and antimicrobial activities [165]. This type of antibiotic was first discovered as a tetrangomycin isolated from *Streptomyces rimosus* in 1965. Members of angucyclines are characterized by an angular tetracyclic (benz[α]anthracene) structure with a hydrolyzable sugar moiety and they are biosynthesized by type II polyketide synthases (PKSs) via decarboxylative condensations of a short acyl-CoA starter and nine extender units [146,165]. *Streptomyces* sp. is known as the major producer of angucyclines [54].

Aromatic polyketides, representative substances of type II polyketides, have significant therapeutic properties, including tetracycline and anthracycline-type doxorubicin, which are typical of aromatic polyketides with pharmacological applications [53].

Flavonoid structures are characterized by a 15-carbon skeleton, in two aromatic ring systems (A, and B rings) and a heterocyclic ring C, the ring containing embedded oxygen [166]. This carbon structure can be abbreviated as C6–C3–C6 rings and with different substitution patterns to produce a series of subclass compounds, such as flavones and flavonols, as the quercetin, isoflavones, etc. [166]. Nevertheless, there are other flavonoids without a C6–C3–C6 skeleton, for instance, biflavones, furan chromones, and xanthones [166].

Another family of compounds reported to have antibacterial activity are phenazines; these are heterocyclic nitrogenous compounds that consist of two benzene rings attached through two nitrogen atoms and substituted at different sites of the core ring system. They have been isolated in substantial amounts from terrestrial bacteria such as *Pseudomonas*, *Streptomyces*, and other genera from marine habitats [27]. Based on earlier reports on the biological activities of this class of compounds, it was suggested both DNA gyrase B (Gyr-B) and pyruvate kinase (PK) were the possible molecular targets of their antibacterial activity [27].

Chromopeptide lactone antibiotics is another family of compounds among which actinomycins are one of their constituents; actinomycin D is one of the older anticancer drugs and has been studied extensively and widely used clinically for the treatment of several types of malignant tumors. Despite their initial discovery more than 70 years ago, actinomycins continue to be a focus of many research areas, especially in their biological activity and medicinal use [116].

2.8. Main Family Compounds Found in Marine Actinobacteria with QQ Activity

The inhibition of quorum sensing is a therapeutic target for the treatment of diseases generated by bacteria that has gradually been gaining interest, since to date, there have been no reports of the development of resistance by bacteria against this mechanism. Few studies to date have reported compounds isolated from marine actinobacteria with the ability to inhibit quorum sensing; however, some families of compounds that have exhibited this activity have already been identified. Among these, fatty acyl compounds, phenoxazines, lactones, and similar brominated furanones have been reported, the latter being potent antibiofilm agents whose mechanism of action has been attributed to their capability to inhibit QS processes in bacteria [138]. Interestingly, a melanin pigment was informed with QQ activity. Table 7 shows these compounds' families.

Table 7. Family compounds with QQ activity.

Compound	Frequency	Constituents	Ref.
Fatty acyl compounds	1	13Z-Octadecenal.	[140]
Phenoxazines	1	Questiomycin A	[29]
		2-hydroxyacetate-3-hydroxyacetamido-phenoxazine (HHP)	[29]
Lactones	1	Butenolide	[138]
Pigment	1	Melanin	[137]
Strain IM20 [1]	1	NA [2]	[141]

[1] Compound not identified. [2] Information not reported.

Some of the compounds reported with biological activities such as antibacterial, antimicrobial, antibiofilm, and QQ effects have been extensively studied and their structure–activity relationships (SAR) have been described; some of them are the following:

Phenazines, which are compounds with both antibacterial and antibiofilm activity, which is related to the presence of carboxylic acids on both C1 and C6 of the phenazine ring system, decreased the antibiofilm effect towards Gram-negative strains, but made these derivatives active against Gram-positive ones, particularly, *S. aureus*. Regarding that antibacterial activity, an analogous situation occurs in which the addition of another carboxylic acid or carboxyl ester at C-6 significantly decreased the inhibitory activity against Gram-negative bacteria and converts these phenazine derivates to be active against Gram-positive strains [27], as shown in Figure 7.

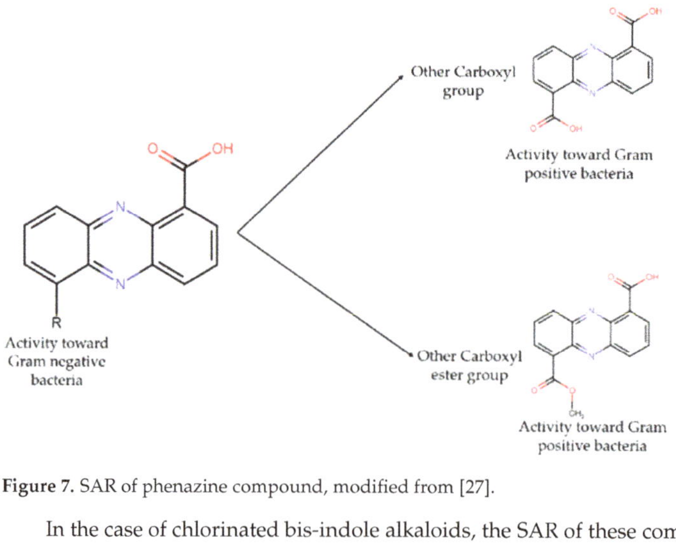

Figure 7. SAR of phenazine compound, modified from [27].

In the case of chlorinated bis-indole alkaloids, the SAR of these compounds, which showed antibacterial activity, reveals that the chlorine atom at C-6″ could be pivotal for conferring their bioactivity, thus providing hints on chemical modifications on bis-indole alkaloid scaffold in drug design [45].

Also, niphimycin is a type of macrolide with antibacterial activity against methicillin-resistant *S. epidermidis* (MRSE) and *S. aureus* (MRSA) [167]. Another type of macrolide is glycosidic antibiotics: similar to other macrolides, these compounds have antibacterial activity against Gram positive organisms and are inactive against Gram-negative bacteria. This compound activity is related to the presence of hemiketal groups at C-11 and C-11′ in the structure. This is concluded because compounds that did not have this group showed an approximately two-fold decrease in activity against most strains [52]. Likewise, borrelidins J and K are macrolides that showed activity against MRSA, and their activity could enhance the cleavage of the ester bond. The cleavage of the ester bond in borrelidins makes them long-chain unsaturated fatty acids and it has been reported by previous studies that long-chain unsaturated fatty acids could exhibit strong activity against *S. aureus* by inhibiting the enoyl–acyl carrier protein reductase (FabI), which is the essential component in bacterial fatty acid synthesis [28].

Nocardiopsistins are angucycline compounds that belong to the polyketides family. These compounds presented antibacterial activity toward MRSA, and their activity is related to the presence of a hydroxyl group (-OH) at C3 in this structure [102].

Napyradiomycin is a large class of unique meroterpenoids with different halogenation patterns that present significant growth-inhibitory activity against MRSA. The specific mechanism of action for this family of meroterpenoids is not clear, however, studies about its SAR have shown that structural variations among the napyradiomycin metabolites, such as the different halogenation patterns or the presence or absence of the methyl group at C-7 among others, can attenuate or enhance their biological activities [113].

Lobophorin analogs are spirotetronate antibiotics with antibacterial activity against Gram-positive bacteria such as *Bacillus subtilis* and *S. aureus*. Their activity was related to compounds such as Lobophorin B, F H, I, and Lobophorin L, which has been related to the increase of the number of monosaccharide units in its structure, increasing inhibitory activity and indicating that monosaccharides might play a significant role in the antimicrobial activity of lobophorins [62,115]. In the same way, the antimicrobial activity showed by Lobophorins E and F is related to the absence of the hydroxyl group in C-32, which seems to enhance the bioactivity at a 416-fold improvement. On the contrary, the presence of the terminal sugar moiety is disadvantageous for the antimicrobial property [59].

Another compound that has reported SAR is Citreamicin, which is a xanthone commonly found in plants. It showed antibacterial activity against *S. aureus*; this may be due to the five-member nitrogen heterocycle in their structure. This five-member nitrogen heterocycle is similar to that in oxazolidinones, which are an approved class of antibiotics [84].

2.9. Biosynthetic Gene Clusters, BGCs

The capability of actinobacterial strains to produce bioactive secondary metabolites is considered to rely on their genomic potential, which typically contains many biosynthetic gene clusters (BGCs), including genes encoding for polyketide synthases (PKS) and non-ribosomal peptide synthetases (NRPS) [168]. However, nowadays, other biosynthetic gene clusters have been found, especially in marine actinobacteria, which, due to environmental conditions, are targets for the search for compounds with anti-infective activity that could provide alternative treatments for acne vulgaris. In addition to the PKS/NRPS clusters, in this study, other biosynthetic gene clusters have been reported such as the phenazine cluster, (this has been related to QQ and antibiofilm activity), which is directly involved in the production of phenazine compounds, the DSA cluster, related to the production of desotamides, the nes gene cluster, involved in the production of nenestatin A (Benzofluorene angucyclines), the abo cluster related to the aborycin compound, among others. Table 8 presents the details of the biosynthetic cluster genes reported in this study [168].

Table 8. Biosynthetic gene clusters identified in marine actinobacteria reported in this study.

Genus	BGS	Genes	Metabolites Production	Ref.
Streptomyces sp.	PKS gene cluster	PKS-I and PKS-II Genes	Polyketide Angucycline	[20,30] [23]
		PKS-II Genes	Angucyclinone derivatives	[146]
		PKS-KS	NA [1]	[169]

Table 8. *Cont.*

Genus	BGS	Genes	Metabolites Production	Ref.
		PKS	Niphimycins	[117]
		PKS II	Analogue of paulomenol	[103]
	PKS/NRPS	NA [1]	Antimycin A analogues	[77]
		NA [1]	NA [1]	[168,170]
		NRPS, PKS Type I, II, and III	Naphthoquinone antibiotics	[25]
	NRPS gene cluster	NRPS-A	NA [1]	[169]
		NA [1]	NA [1]	[49]
	Aborycin biosynthetic gene cluster (abo)	NA [1]	Aborycin	[48]
	Lassopeptide cluster	NA [1]	Lasso peptide family	[48]
	Phenazine cluster	phzE and phzF	Streptophenazines (Phenazines)	[83]
	dsa cluster	DsaA y DsaN, dsaB y dsaJ	Desotamides	[154]
	PKS/terpenoid biosynthetic pathways	NA [1]	Napyradiomycin derivatives (Terpenoids)	[71,113,171]
Micromonospora sp.	nes gene cluster	NA [1]	nenestatin A (Benzofluorene angucyclines)	[100]
Co-culture of *Actinokineospora* sp. and *Micromonospora* sp.	NA	NA [1]	Phenazine	[27]
	PKS/NRPS	NA [1]	Polyketide	[170]
Nocardiopsis sp.		PKS-II	α-pyrone compound	[103]
	PKS gene cluster	ACP synthase α-subunit (KSα), β-subunit (KSβ) and acyl carrier protein (ACP)	Angucyclines	[102]
		PKS-II	Angucycline	[102]
		phzE	Phenazines	[155]
Rhodococcus sp.	NRPS/NRPS	NA [1]	NA [1]	[22]
	PKS/NRPS	NA [1]	Polyketide	[170]
Salinispora sp.	PKS gene cluster	PKS I, II	Rifamycin B	[103]
	PKS/NRPS	NA [1]	Polyketide	[170]
Verrucosispora sp.	PKS gene cluster	PKSI (pks1 and pks2), two PKSII (pks3 and pks4), PKSIII (pks5);	New salicylic derivative, brevianamide F, abyssomicin B	[95]
	NRPS gene cluster	NA [1]		
	Terpene clusters	terp1, terp2, terp3 and terp4		
	NRPS-PKSI hybrid clusters	np1 and np2		
	Lanthipeptide clusters	lant1 and lant2		
	Siderophore cluster	sid		
Brachybacterium paraconglomeratum	NRPS/PKS	NRPS genes, PKS type I genes, and PKS type II gene	NA [1]	[172]

[1] Information no reported.

Natural products derived from these biosynthetic pathways have been extensively described for cultured and uncultured marine strains. Metabolites derived from marine actinobacteria include, among others, the polyketide synthase-derived abyssomicin C, a unique polycyclic polyketide from a marine *Verrucosispora* [97,130], salinisporamide A, from *Salinispora tropica* [108] that is currently in clinical trials as one of the most potent anticancer agents isolated until today [173], all isolated from the phylum of Actinomycetales.

BGCs sequences have been reported in marine actinobacteria isolated from a wide variety of environments and with a high occurrence variability. Of the articles included in this systematic review, only 21 reported the presence of biosynthetic gene clusters related to the biological activity of the promising strains. Of these, five articles reported the complete genomes and four reported the BGC sequences. Among the BGCs, the most common were type I and type II polyketide synthases (PKS-I, PKS-II), and nonribosomal synthetase (NRPS), mostly identified in *Streptomyces* sp., followed by *Salinispora* sp. isolated from marine sediments, as well as *Nocardipsis* sp., isolated from a sponge. This type of BGS has been the most studied; nevertheless, other BGCs have been reported in *Streptomyces* sp. such as the *abo* cluster, which is related to the synthesis of a compound with anti-infective activity, aborycin; the *dsa* cluster that is directly involved in the biosynthesis of the antibacterial compound desotamide, which has activity against *S. aureus* ATCC 29213, and methicillin-resistant *S. epidermidis* (MRSE) shhs-E1; phenazine cluster (*phe*), which has also been described in the genera *Nocardipsis* and *Salinispora*. Likewise, other BGCs have been found in genera such as Micromonospora, such as the *nes* cluster, involved in the biosynthesis of homo-dehydrorabelomycin E, which had antibacterial activity against *S. aureus* ATCC 29213, as presented in Figure 8. Despite this fact, it is important to note that the detection of genes associated with these biosynthetic clusters does not guarantee the expression of the genes involved in the production of secondary metabolites; notwithstanding, the detection of secondary metabolite biosynthetic pathways can be used as an indicator of metabolic potential, and suitable culture conditions are generally needed to express most of these pathways as well as the use of the appropriate targets to reveal the biological activity of the compounds [108].

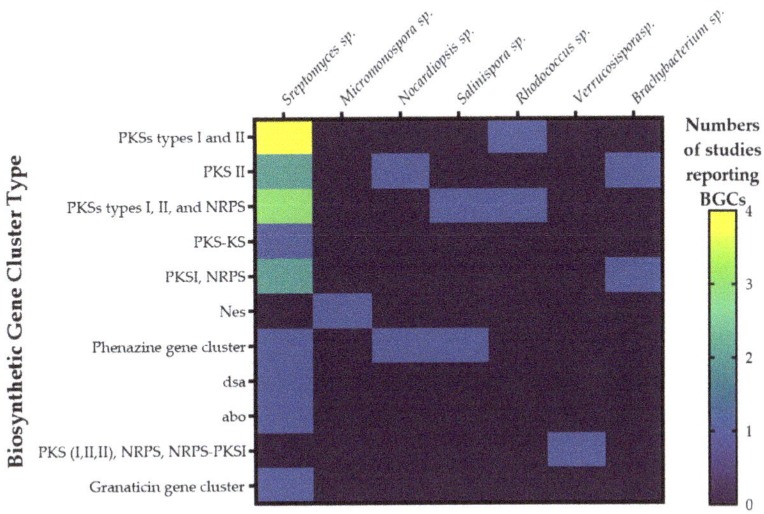

Figure 8. Heatmap of the number and type of biosynthetic gene clusters (BGCs) in the genomes of bioactive strains belongs to different genera collected in this study. Clusters are arranged top to bottom, beginning with the greatest number of BGC types in the top left. Strains are shown left to right by the highest number of BGCs. The most abundant BGCs were Type I and II PKSs followed by NRPS clusters for the *Streptomyces* genus. The color bar represents the number of studies that reported a type of BGCs, Purple to blue (minor values), blue to green (middle values), and green to yellow (high values).

3. Discussion

Microbial secondary metabolites are prevalent sources of natural products and they have been known as immense reservoirs of chemical classes of compounds with strong biological activities such as promising therapeutic potential [37].

Among the microorganisms, the actinobacteria phylum is one of the most known groups, being biologically active secondary metabolite producers, and it continues to represent an exciting source for the identification of novel natural products; due to this, it is considered the most economical and biotechnological important prokaryote source [101].

Out of these Actinobacteria, *Streptomyces* is the genera known as the most prolific, with many natural products with antibacterial, antifungal, antioxidant, antitumor activity, etc., from which products have been developed with a wide range of pharmaceutical applications contributing to a high number of antibiotics with current pharmaceutical applications, potentially useful to treat acne vulgaris [101]. Nevertheless, in the last years, other actinomycetes genera have received more attention as producers of commercially important secondary metabolites due to the probability of the rediscovery of novel compounds with new chemical structures from *Streptomyces* being increased [101], especially if they are obtained from terrestrial environments. Whereby, environments less explored as oceans, which cover about two-thirds of the Earth's surface, have become important because they are considered a source in which microorganisms are submitted to extreme conditions and they are more challenging to culture compared to their terrestrial relatives. Therefore, the sea offers an enormous resource for novel compounds. The field of marine drug discovery has been growing over the past 20 years, with currently almost 35,000 research articles on natural products of marine origin [22].

The present review showed a significant increase of studies from 2002 to 2022, which demonstrates the interest in the marine environment to search for new bioactive compounds in addition to the need for the discovery of new compounds with anti-infective activity, finding that the majority of molecules reported are derived from *Streptomyces,* with a rising potential of finding new active compounds from rare actinobacteria genera such as *Nocardiopsis,* producing compounds with antibacterial, antimicrobial, anti-biofilm and QQ activity [29,137].

As expected, the antibacterial activity is the most reported biological activity and with the higher number of molecules discovered. These have very varied modes of action, such as affecting the membrane of the target bacteria and interrupting protein synthesis, among others. Likewise, in this systematic review, molecules, extracts, and fractions were reported as being highly active with MICs ranging from 0.01 to >1000 μg/mL. This shows that reported MICs are variable and that there is no consensus on the minimum value of the MIC to consider whether the compounds, fractions, or extracts are active and whether they have true pharmaceutical potential to produce commercial alternative treatments for acne vulgaris.

In addition, although there is a wide variety in the MICs reported, compounds with extremely low MICs are ideal, as this would allow the use of the compound in low proportions, this being more favorable than compounds that require a large amount to achieve the desired activity.

Likewise, some specific isolation sources have been prevalent, such as the marine sediment being the most frequent, becoming a reference hotspot for the bioprospecting of marine actinobacteria with antibiotic activities in the last decades [19]. The sea floor has been reported as a unique system with many forms of actinomycetes [174] and this is attributable to marine sediments, which are mixtures of complex organic and inorganic particles that have accumulated due to the accretion and erosion of the continents, oceanic biological activities, volcanic eruptions, and chemical processes within the ocean. Given their vast coverage, marine sediments harbor remarkably diverse microbial communities accounting for 12–45% of the total microbial biomass [23]. Proof of this is the fact that in compounds with antibacterial activity, the most predominant isolation source was marine sediment, followed by sponges, and ascidians, which are sessile marine invertebrates, making them vulnerable to predation and therefore are hypothesized to use host-associated bacteria that produce biologically active secondary metabolites for chemical defense [25].

Moreover, compounds with antibiofilm activity and metabolites with antibacterial and antibiofilm activity also have been isolated from sponges. It is well known that the sponges are of great biotechnological interest because these are well known for hosting a complex microbial consortium with the potential of producing biologically active secondary metabolites. Three-fourths of all discovered new bioactive microbial products from the oceans have originated from bacteria associated with marine invertebrates [175]. Two articles included in this systematic review with antibacterial, antibiofilm, and antimicrobial activity were reported by Joseph et al. and Sing et al., respectively, in which these bioactive compounds were isolated from a marine sponge symbiont, *Streptomyces pharmamarensis*, and marine-sponge-derived *Salinispora* sp., showing the enormous potential of marine sponge-associated actinomycetes that represent an exciting resource for the identification of new and novel natural products [110,134]. In the same way, another paper was reported by Hifnawy et al., in which two rare actinomycetes (*Micromonospora* sp. UR56 and *Actinokineospora* sp. EG49) were co-cultures and this led to the isolation of antibacterial metabolites of the phenazine class with antibiofilm, and cytotoxic properties [27].

Similarly, some compounds isolated from marine sponges, including angucyclines, antibacterial metabolites generating cell wall disruption in MRSA, have been reported previously [10,132]. Furthermore, one of the bacteria of interest in this paper is *S. epidermidis*, however, there are few articles reporting the action of actinobacterial compounds against this bacterium. Nevertheless, one article reported its growth and biofilm inhibition by *Streptomyces* sp. SBT348 extract [133] isolated from the marine sponge *Petrosia ficiformis*.

Concerning compounds with QQ activity, the sources from which the bacteria that produce them have been isolated are very varied: these are the intestines of marine fish, marine sediments, sponges, and water [29,137,138,140,141]. This may be due to the few studies that have so far been reported or have had their activity evaluated in extracts or isolated compounds of marine actinobacteria, indicating that there is no specific marine source for the isolation of marine actinobacteria with such activity.

Regarding places of isolation, two sites where more actinobacterial strains with anti-infective activity were isolated were the South China Sea and the Bay of Bengal in India. The former has emerged as a potentially abundant source of new species or genera of marine actinomycetes. Some new bioactive compounds, lobophorins E and F, were reported from marine actinomycetes isolated from the South China Sea [59]. The second is a well-known potential source for marine-derived bacteria rich in bioactive compounds [148] and is a point of access for diverse sets of marine fauna and flora, in particular sponges, sea anemones, sea cucumbers, sea urchins, soft corals, and many marine algae that, due to being little explored, have given rise to their bioprospecting as reported by Gandhimathi et al. [176].

The compounds most commonly produced by marine actinobacteria that have been recovered in this study are compounds with antibacterial activity against *S. aureus* and methicillin-resistant *S. aureus* (MRSA) (that could be present in skin diseases, but are also related), which cause a wide range of infections such as furuncles, pneumonia, osteomyelitis, endocarditis, bacteremia, etc. [171]. These same compounds in some cases have shown antibiotic activity against other Gram-positive bacteria such as *S. epidermidis*, *Bacillus* sp., vancomycin-resistant *Enterococcus faecalis* (VRE), among others, and to a lesser extent, against Gram-negative bacteria such as *E. coli*, *P. aeruginosa*, among others [133]. This phenomenon may be due to the morphological differences between Gram-positive and Gram-negative microorganisms. Whilst Gram-negative bacteria have an outer lipopolysaccharide membrane that makes the cell wall impermeable to lipophilic solutes, Gram-positive bacteria are more susceptible as they have a more permeable outer peptidoglycan layer [30]. However, this demonstrated the potential of compounds from marine actinobacteria to contribute to infectious disease control. This indicates a great possibility of using these compounds to treat acne vulgaris and the bacteria commonly associated with it, which are mainly Gram-positive bacteria.

In this same sense, it is noteworthy that few studies with activity against *S. epidermidis* [83] were retrieved, and there are none with activity against *Propionibacterium acnes*, currently renamed *C. acnes*, which is also an actinobacterium, but to date, there is no study on the action of compounds isolated from marine actinobacteria against this bacterium, which can become pathogenic due to unknown effects and participate in the development of the pathology of acne vulgaris. The fact that *C. acnes* is an anaerobe could increase the technical requirements to carry out the antibacterial activity screening; nevertheless, it is highly expected that the antibacterial compounds here described also have antibacterial activity against this bacterium.

It is important to point out that two of the antibiotics most currently used to treat acne were obtained from actinobacteria of the *Streptomyces* genus from soil samples. These compounds are erythromycin, belonging to the macrolide class, which has also been isolated from marine actinobacteria as reported in this systematic review, and clindamycin, a semi-synthetic derivative of lincosamide with a mechanism of action similar to macrolides, which binds to the 50S ribosomal subunit of bacteria, inhibiting protein synthesis [164]. This demonstrates the potential of actinobacteria as a source of new compounds for the treatment of acne vulgaris and the opportunity to study, search and develop compounds with antibacterial activity or QQ activity against this bacterium, the latter activity being a therapeutic target since in most cases it does not affect bacterial growth, which would be positive for *C. acnes* since it is a bacterium that, in a normal environment of the skin, protects from the invasion of pathogenic bacteria, contributing to its homeostasis.

Regarding antibacterial metabolites, various studies have reported bioactive metabolites that belong to the polyketide family, it being one the most isolated families from marine actinobacteria. Among these were found aromatic polyketides as described by Dong et al., Ahamad et al., and Govindarajan et al. [61,68,151], angucyclines described by Akhter et al. [54], polyketide–terpenoid as Naphthoquinone, reported by Shen et al. [37], macrolides, described by Braña, Zhang and Wu [52,67,107,152], etc. Likewise, some chlorinated compounds were frequent, such as chlorinated bis-indole alkaloids and chlorinated 3-phenylpropanoic acid described by Song et al. [45] and Shaala et al. [124]; this may be due to the concentrations of chloride and bromide ions in the ocean [177]. Interestingly, marine-derived bis-indole compounds typically contain halogen atoms in their structures. Such halogenated bis-indole alkaloids display potent cytotoxic or antibacterial activities or both, and they are thus considered promising anti-cancer or antibacterial leads. In the same way, a series of marine-derived chlorinated bis-indoles were shown to inhibit methicillin-resistant *S. aureus* (MRSA) pyruvate kinase significantly, with their halogenated indole ring being implicated as a critical pharmacophore [45]; this has been reported by Wang et al. [177] and is well known that marine actinomycetes produce a variety of halogenated compounds with diverse structures and a range of biological activities owing to their unique metabolic pathways [31,177].

Similarly, compounds in the bis-indole family are ubiquitously distributed in plants and microorganisms, similar to phenolic compounds, which can be defined as plant substances, are the most widely distributed in the plant kingdom, and are the most abundant secondary metabolites of plants [178]. However, some of them have been isolated from marine actinobacteria as described by Siddharth and Rai [101], specifically, from rare actinomycetes *Nocardiopsis* sp. This metabolite (4-bromophenol, a bromophenol derivative) exhibited a significant antioxidant activity through DPPH and ABTS assays, as was expected due to the antioxidant capacity that has been described in these compounds; in addition, it showed broad-spectrum inhibitory activity against MRSA, *Klebsiella pneumonia* ATCC 13883, *B. subtilis* ATCC 6633 [101]. Likewise, other plant-derived compounds have been isolated from marine actinobacteria as Cinnamaldehyde, produced by *Streptomyces chartreusis*, which showed antibacterial activity, and other studies reported its effect on the swarming motility of *P. aeruginosa*, which is related to quorum sensing in this bacterium, which shows the possible ability of Cinnamaldehyde to inhibit quorum sensing [174].

As for the compounds, there is a wide diversity, finding polyketides, macrolides, quinolones, terpenes, phenazines, naphthoquinones, and phenolic compounds that displayed antibacterial, antimicrobial, antibiofilm activity, and QQ; within these are some compounds that mainly have been discovered in plants, but nowadays have been discovered in marine actinobacteria, such as cinnamaldehyde, flavonoids, and xanthone natural products, which exhibit a wide array of bioactivities including antioxidant, antibacterial, antimalarial, antituberculosis and cytotoxic activities as reported earlier [179].

Most of the compounds obtained from actinobacteria have been isolated using organic solvents. Among the articles collected in this systematic review, the most reported was ethyl acetate, which has a medium to high polarity. This solvent is described as the ideal solvent for obtaining metabolites with antibacterial activity. This may be because it is possible that actinobacteria, especially streptomyces, produce semipolar antibacterial compounds so that they can be extracted by solvents with the same polarity as ethyl acetate, as mentioned in the study by Kurnianto et al. [180]. Likewise, Satish et al. evaluated the activity of extracts obtained from different solvents such as chloroform, butanol, and ethyl acetate against MRSA, finding that only the extracts obtained with the latter exhibited antibacterial activity [87].

On the other hand, traditionally, marine invertebrates are considered a prolific source of exceptional natural products, with a diverse range of biological activities. However, current studies on invertebrate-associated microbial communities are revealing microorganisms as the real producers of many of these compounds. In this study, one article with *Streptomyces* strains was reported with QQ and antimicrobial activity isolated from the gut of marine fish *Rastrelliger kanagurta* [141]. This compound was not identified, however, in this study the findings revealed that there is a wide variety of compounds of the family, with polyketides being the most frequent, as expected, because they have been the most studied and are synthesized by the enzyme polyketide synthase, encoded by PKS genes against which genetic mobilization through horizontal gene transfer (HGT) has been reported with a high frequency, and this could be due to multiple factors. Some PKSs are encoded on plasmids or located within pathogenic islands, which facilitate gene transfer through conjugation, transposition, or transduction, as was reported by Nivina et al. [181].

In this same sense, the PKS gene cluster was the most reported, together with NRPS and the *phe* gene cluster, however, the detection of genes associated with these biosynthetic clusters does not guarantee the expression of genes involved in the production of secondary metabolites due to recent studies have demonstrated that the abundance of biosynthesis gene clusters in actinobacteria genomes do not appear to be expressed under standard laboratory culture conditions. Activation of these gene clusters would considerably enhance the ability to discover novel natural products. Studies by Xu et al. have shown that LaCl3 induced antifungal or antibacterial activities in strains that did not show such activities under normal cultivation conditions [167]. In addition, the culture condition such as agitation speed, temperature, pH, etc., apart from helping improve the performance of compounds, could also be related to the expression of the biosynthetic gene cluster. Furthermore, carbon and nitrogen sources have been reported with a profound influence on secondary metabolite production; regarding carbon sources, glucose favors a high growth rate, nevertheless, this represses secondary metabolite production through carbon catabolite repression [182,183]. Due to this, other sources have been used as starch; for this reason, glucose was reported in this review with less frequency compared to starch. Regarding the nitrogen source, ammonium is reported as the preferred nitrogen source for most actinobacteria; its presence in high concentrations is positively related to the growth rate, however, it delays the onset of secondary metabolite production. On the contrary, nitrate can be assimilated by actinobacteria as an alternative nitrogen source. Interestingly, the nitrate excess enhances secondary metabolite production in actinobacteria [182], which explains why nitrate has been reported in twice as many articles as ammonium in this systematic revision.

Respecting the antibiofilm activity, there are about 5027 anti-biofilm agents against Gram-positive and -negative bacteria, and fungi have been reported between 1988 and 2017 [133]. However, up to date, few have been successfully translated to the market for clinical and medical applications or against whom bacteria have developed action mechanisms, because of this is required to continue in the search for new options and despite the huge expectations on synthetic molecules with effective antimicrobial properties, natural products are still worthy of promise as reported by Newman and Cragg [30,184].

Although the compounds with QQ activity were few, the present investigation confirms the ability of actinobacteria to produce secondary metabolites with this effect, being one of the novel approaches to counter the drug-resistant bacteria and target therapeutic that could be inhibited the virulence factors of some bacteria such as *C. acnes* and generate new treatment options to acne vulgaris disease.

Seeing these results in an integrated manner, it is possible to guide research towards the isolation of marine actinobacteria obtained from sediments and marine invertebrates, paying more attention to the *Streptomyces* genus, and looking for families of compounds such as polyketides, macrolides, phenazines, among others. In the same way, the variation of the culture condition may promote the production of bioactive metabolites, especially carbon and nitrogen sources.

In short, our results reinforce the need to further explore marine actinomycetes and their enormous potential of them as a rich source of novel metabolites relevant for biotechnological applications.

4. Materials and Methods

4.1. Search Strategy

A systematic search was conducted in PubMed, Scopus, and Web of Science (WOS) without limits of timeframe (The first search was in May 2021 and the last updated in April 2022). The search strategy for all databases included the descriptors: "streptomyces", "actino", "acne", "antibacterial", "quorum quenching" and other terms combined with Boolean operators AND and OR and it defined as follows.

((streptomyces OR action *) AND (acne * OR "staphylococcus epidermidis" OR "staphylococcus aureus" OR "cutibacterium acnes" OR "propionibacterium acnes") AND (antibacterial OR quorum OR "quorum quenching")).

"Acne" was used instead of "acne vulgaris" as it is more general and commonly used and the term "quorum" was included for researchers that used quorum-sensing inhibitors instead of "Quorum Quenching".

In addition, for the synthesis, the papers were grouped by the type of biological activity reported.

4.2. Eligibility Criteria

Studies were included in this systematic review to see if they met all the following eligibility criteria:

Original research articles, studies on extraction of compounds or extracts or metabolites derived from marine actinobacteria strains, and studies evaluating the activities of antibacterial, antimicrobial, anti-biofilm, and quorum quenching.

The following were considered to be exclusion criteria: compounds or extracts isolated from soil actinobacteria or another environment different from marine, compounds or extracts obtained from microorganisms other than actinobacteria, compounds were not identified, reviews, communication, and letters to the editor were not considered and articles whose language was not English.

Three researchers performed all the literature selection steps individually and then discussed the differences within the research team. An article was eligible to be included in the review when at least two authors indicated that it met the inclusion/exclusion criteria. Eligible articles were read at a full-text level and those who met the inclusion/exclusion criteria were selected to carry out the data extraction.

4.3. Data Extraction

Data were extracted and sorted by the title of studies, author, year, the number of strains, isolation country, isolation source (sediment, sponge, seawater, mangrove, coral, marine invertebrates, and so on), genus of actinobacteria (*Streptomyces* sp., *Nocardiopsis* sp., *Micromonospora* sp., *Verrucosispora* sp., *Salinispora* sp., among others), type of activity (antibacterial, antimicrobial, antibiofilm, quorum-sensing inhibition), extracts or compounds used, the organic solvent used to get the extracts or compounds (EtOAc, MeOH, Butanone, Butanol, Methanol, Acetone, Chloroform, Dichloromethane, or the combination of them), the family of compounds, genes associated with the compounds' production, the biosynthetic gene clusters (BGCs) and the structure of the compounds if reported.

5. Conclusions

The marine ecosystems are one of the most dynamic, under-explored environments and are a natural reservoir of metabolites with a wide spectrum of biological activities. *Streptomyces* sp. remains the most prolific genus of actinobacteria in the phylum, however, the so-called rare actinobacteria have gained interest due to the variety of compounds they can produce, such as those that show antibiofilm activity and quorum quenching. In the same way, marine sediments and sponges were the most outstanding source to isolate bioactive actinobacteria. Regarding compounds with antibacterial activity, polyketides were most frequently comprised of angucyclines, aromatics polyketides, and naphthoquinones, among others, followed by phenazines which displayed antibacterial, antimicrobial, and quorum-sensing inhibition, finding an exciting potential in this type of secondary metabolite. Likewise, compounds originally found in plants were reported to be isolated from marine actinobacteria, evidence that the bioactivity of some plants or animals like fishes is due to microorganisms and not to the host organism. Furthermore, it was evident that there are few studies of the compounds obtained from marine actinobacteria with antibacterial, antibiofilm, or QQ activity against *C. acnes*, giving us a wonderful opportunity to investigate future studies in this interesting area. Finally, biosynthetic gene clusters in the production of secondary metabolites in actinobacteria play an important role, and although the presence of this in the genome of actinobacteria does not imply that they will be expressed, they are indicators of the potential of strains to produce compounds and it was clear that in most cases that they must be activated through some strategies such as co-culture, stress-generated external factors such as pH, temperature, agitation speed, variation of co-culture conditions and so on. This makes evident the need to sequence the genomes, since these allow us to know the bacteria in-depth and put into practice different strategies, establish the relationship between gene clusters of genes and functions, postulating this methodology as an alternative for the extraction of metabolites, its performance and use. In short, the findings in this research support the evidence of the potential of marine-derived actinobacteria to produce anti-infective compounds and suggested the search for this microorganism of compounds with novel approaches as QQ.

Supplementary Materials: The following supporting information can be downloaded at: https://www.mdpi.com/article/10.3390/antibiotics11070965/s1, Table S1: PRISMA checklist, Table S2: Source of isolation and type of activity of actinobacterial strains reported in this systematic review. Table S3: Antibacterial activity of actinobacterial crude extracts or compounds presented through inhibition zone (mm). Table S4: Antimicrobial activity of actinobacterial crude extracts or compounds presented through inhibition zone (mm). References [185–215] are cited in the supplementary tables.

Author Contributions: Conceptualization, M.C.D.L.H.-R., L.V. and L.D.; methodology, M.C.D.L.H.-R., L.V. and L.D.; software, M.C.D.L.H.-R.; validation, L.V. and L.D.; formal analysis, M.C.D.L.H.-R. and L.V.; investigation, M.C.D.L.H.-R.; resources, L.V. and L.D.; data curation, M.C.D.L.H.-R. and L.V.; writing—original draft preparation, M.C.D.L.H.-R.; writing—review and editing, M.C.D.L.H.-R., L.V. and L.D.; visualization, M.C.D.L.H.-R.; supervision, L.V. and L.D.; funding acquisition, L.V. and L.D. All authors have read and agreed to the published version of the manuscript.

Funding: This research was funded by Minciencias (Ministerio de Ciencia, Tecnología e Innovación—Colombia-,and Fondo Francisco José De Caldas project code 2105-905-87457 contract 80740-458-2021) and by Universidad de La Sabana and Clinica Universidad de La Sabana—the Biomedical Campus Call (General Research Directorate, project ING-PHD-42-2021).

Institutional Review Board Statement: Not applicable.

Informed Consent Statement: Not applicable.

Data Availability Statement: Data supporting reported results can be found in this document and the Supplementary Materials. If they become required, please request them by mail at luisa.villamil@unisabana.edu.co.

Acknowledgments: To Universidad de La Sabana for the Carlos Jordana PhD scholarship. To GIBP and Actinos Group for their support, especially to Aixa Sarmiento.

Conflicts of Interest: The authors declare no conflict of interest.

References

1. Tuchayi, S.M.; Makrantonaki, E.; Ganceviciene, R.; Dessinioti, C.; Feldman, S.R.; Zouboulis, C. Acne vulgaris. *Nat. Rev. Dis. Prim.* **2015**, *1*, 15029. [CrossRef] [PubMed]
2. Heng, A.H.S.; Chew, F.T.; Heng, A.H.S.; Chew, F.T. Systematic review of the epidemiology of acne vulgaris. *Sci. Rep.* **2020**, *10*, 5754. [CrossRef] [PubMed]
3. Dréno, B.; Araviiskaia, E.; Kerob, D.; Andriessen, A.; Anfilova, M.; Arenbergerova, M.; Barrios, O.L.F.; Mokos, Z.B.; Haedersdal, M.; Hofmann, M.A.; et al. Nonprescription acne vulgaris treatments: Their role in our treatment armamentarium—An international panel discussion. *J. Cosmet. Dermatol.* **2020**, *19*, 2201–2211. [CrossRef] [PubMed]
4. Gannesen, A.V.; Zdorovenko, E.L.; Botchkova, E.A.; Hardouin, J.; Massier, S.; Kopitsyn, D.S.; Gorbachevskii, M.V.; Kadykova, A.A.; Shashkov, A.S.; Zhurina, M.V.; et al. Composition of the Biofilm Matrix of *Cutibacterium acnes* Acneic Strain RT5. *Front. Microbiol.* **2019**, *10*, 1284. [CrossRef]
5. Platsidaki, E.; Dessinioti, C. Recent advances in understanding Propionibacterium acnes (*Cutibacterium acnes*) in acne. *F1000 Res.* **2018**, *7*, 1953. [CrossRef]
6. Fournière, M.; Latire, T.; Souak, D.; Feuilloley, M.G.J.; Bedoux, G. Staphylococcus epidermidis and *Cutibacterium acnes*: Two Major Sentinels of Skin Microbiota and the Influence of Cosmetics. *Microorganisms* **2020**, *8*, 1752. [CrossRef]
7. de Sousa, I.C.V.D. Evaluating FMX-101 as a promising therapeutic for the treatment of acne. *Expert Opin. Pharmacother.* **2020**, *21*, 741–746. [CrossRef]
8. Farrah, G.; Tan, E. The use of oral antibiotics in treating acne vulgaris: A new approach. *Dermatol. Ther.* **2016**, *29*, 377–384. [CrossRef]
9. Abdelmohsen, U.R.; Pimentel-Elardo, S.M.; Hanora, A.; Radwan, M.; Abou-El-Ela, S.H.; Ahmed, S.; Hentschel, U. Isolation, Phylogenetic Analysis and Anti-infective Activity Screening of Marine Sponge-Associated Actinomycetes. *Mar. Drugs* **2010**, *8*, 399–412. [CrossRef]
10. Mary, T.R.J.; Kannan, R.R.; Iniyan, A.M.; Ramachandran, D.; Vincent, S.G.P. Cell wall distraction and biofilm inhibition of marine Streptomyces derived angucycline in methicillin resistant *Staphylococcus aureus*. *Microb. Pathog.* **2020**, *150*, 104712. [CrossRef]
11. O'Neill, J. The Review on Antimicrobial Resistance. *Rev. Laryngol. Otol. Rhinol.* **2016**, *136*, 29–31.
12. World Health Organization (WHO). La Escasez Mundial de Antibióticos Innovadores Favorece La Aparición y Propagación de La Farmacorresistencia. Available online: https://www.who.int/es/news/item/15-04-2021-global-shortage-of-innovative-antibiotics-fuels-emergence-and-spread-of-drug-resistance (accessed on 28 January 2022).
13. World Health Organization. Lack of New Antibiotics Threatens Global Efforts to Contain Drug-Resistant Infections. Available online: https://www.who.int/news-room/detail/17-01-2020-lack-of-new-antibiotics-threatens-global-efforts-to-contain-drug-resistant-infections (accessed on 29 April 2020).
14. Szántó, M.; Dózsa, A.; Antal, D.; Szabó, K.; Kemény, L.; Bai, P. Targeting the gut-skin axis—Probiotics as new tools for skin disorder management? *Exp. Dermatol.* **2019**, *28*, 1210–1218. [CrossRef] [PubMed]
15. Wang, Y.; Kuo, S.; Shu, M.; Yu, J.; Huang, S.; Dai, A.; Two, A.; Gallo, R.L.; Huang, C.-M. *Staphylococcus epidermidis* in the human skin microbiome mediates fermentation to inhibit the growth of *Propionibacterium acnes*: Implications of probiotics in acne vulgaris. *Appl. Microbiol. Biotechnol.* **2014**, *98*, 411–424. [CrossRef] [PubMed]
16. Brandwein, M.; Steinberg, D.; Meshner, S. Microbial biofilms and the human skin microbiome. *NPJ Biofilms Microb.* **2016**, *2*, 3. [CrossRef]
17. Pham, J.V.; Yilma, M.A.; Feliz, A.; Majid, M.T.; Maffetone, N.; Walker, J.R.; Kim, E.; Cho, H.J.; Reynolds, J.M.; Song, M.C.; et al. A Review of the Microbial Production of Bioactive Natural Products and Biologics. *Front. Microbiol.* **2019**, *10*, 1404. [CrossRef]
18. Siddharth, S.; Vittal, R.R. Isolation, Characterization, and Structural Elucidation of 4-Methoxyacetanilide from Marine Actinobacteria *Streptomyces* sp. SCA29 and Evaluation of Its Enzyme Inhibitory, Antibacterial, and Cytotoxic Potential. *Arch. Microbiol.* **2019**, *201*, 737–746. [CrossRef]
19. Paderog, M.J.V.; Suarez, A.F.L.; Sabido, E.M.; Low, Z.J.; Saludes, J.P.; Dalisay, D.S. Anthracycline Shunt Metabolites From Philippine Marine Sediment-Derived Streptomyces Destroy Cell Membrane Integrity of Multidrug-Resistant *Staphylococcus aureus*. *Front. Microbiol.* **2020**, *11*, 743. [CrossRef]
20. Dholakiya, R.N.; Kumar, R.; Mishra, A.; Mody, K.H.; Jha, B. Antibacterial and Antioxidant Activities of Novel Actinobacteria Strain Isolated from Gulf of Khambhat, Gujarat. *Front. Microbiol.* **2017**, *8*, 2420. [CrossRef]
21. Aljelawi, R.O.; Kadhem, M.F. Production, Purification, and Characterization of Bioactive Metabolites Produced from Rare Actinobacteria Pseudonocardia Alni. *Asian J. Pharm. Clin. Res.* **2016**, *9*, 264. [CrossRef]
22. Gavriilidou, A.; Mackenzie, T.; Sánchez, P.; Tormo, J.; Ingham, C.; Smidt, H.; Sipkema, D. Bioactivity Screening and Gene-Trait Matching across Marine Sponge-Associated Bacteria. *Mar. Drugs* **2021**, *19*, 75. [CrossRef]

23. Sabido, E.M.; Tenebro, C.P.; Suarez, A.F.L.; Ong, S.D.C.; Trono, D.J.V.L.; Amago, D.S.; Evangelista, J.J.E.; Reynoso, A.M.Q.; Villalobos, I.G.M.; Alit, L.D.D.; et al. Marine Sediment-Derived Streptomyces Strain Produces Angucycline Antibiotics against Multidrug-Resistant *Staphylococcus aureus* Harboring SCCmec Type 1 Gene. *J. Mar. Sci. Eng.* **2020**, *8*, 734. [CrossRef]
24. Moher, D.; Liberati, A.; Tetzlaff, J.; Altman, D.G.; PRISMA Group. Preferred reporting items for systematic reviews and meta-analyses: The PRISMA statement. *PLoS Med.* **2009**, *6*, e1000097. [CrossRef] [PubMed]
25. Sung, A.A.; Gromek, S.M.; Balunas, M.J. Upregulation and Identification of Antibiotic Activity of a Marine-Derived Streptomyces sp. via Co-Cultures with Human Pathogens. *Mar. Drugs* **2017**, *15*, 250. [CrossRef] [PubMed]
26. Lim, H.J.; An, J.S.; Bae, E.S.; Cho, E.; Hwang, S.; Nam, S.J.; Oh, D.C. Ligiamycins A and B, Decalin-Amino-Maleimides from the Co-Culture of *Streptomyces* sp. and *Achromobacter* sp. Isolated from the Marine Wharf Roach, Ligia Exotica. *Mar. Drugs* **2022**, *20*, 93. [CrossRef] [PubMed]
27. Hifnawy, M.S.; Hassan, H.M.; Mohammed, R.; Fouda, M.M.; Sayed, A.M.; Hamed, A.A.; Abouzid, S.F.; Rateb, M.E.; Alhadrami, H.A.; Abdelmohsen, U.R. Induction of Antibacterial Metabolites by Co-Cultivation of Two Red-Sea-Sponge-Associated Actinomycetes *Micromonospora* sp. UR56 and *Actinokinespora* sp. EG49. *Mar. Drugs* **2020**, *18*, 243. [CrossRef]
28. Yu, M.; Li, Y.; Banakar, S.P.; Liu, L.; Shao, C.; Li, Z.; Wang, C. New Metabolites from the Co-culture of Marine-Derived Actinomycete Streptomyces rochei MB037 and Fungus *Rhinocladiella similis* 35. *Front. Microbiol.* **2019**, *10*, 915. [CrossRef]
29. Miao, L.; Qian, S.; Qi, S.; Jiang, W.; Dong, K. Culture Medium Optimization and Active Compounds Investigation of an Anti-Quorum Sensing Marine Actinobacterium Nocardiopsis Dassonvillei JS106. *Microbiology* **2021**, *90*, 112–123. [CrossRef]
30. Arasu, M.V.; Duraipandiyan, V.; Ignacimuthu, S. Antibacterial and antifungal activities of polyketide metabolite from marine *Streptomyces* sp. AP-123 and its cytotoxic effect. *Chemosphere* **2013**, *90*, 479–487. [CrossRef]
31. Zhang, Z.; Sibero, M.T.; Kai, A.; Fukaya, K.; Urabe, D.; Igarashi, Y. TMKS8A, an Antibacterial and Cytotoxic Chlorinated α-Lapachone, from a Sea Slug-Derived Actinomycete of the Genus Streptomyces. *J. Antibiot.* **2021**, *74*, 464–469. [CrossRef]
32. Newaz, A.W.; Yong, K.; Lian, X.Y.; Zhang, Z. Streptoindoles A–D, Novel Antimicrobial Indole Alkaloids from the Marine-Associated Actinomycete Streptomyces Sp. ZZ1118. *Tetrahedron* **2022**, *104*, 132598. [CrossRef]
33. Santos, J.D.; Vitorino, I.; De la Cruz, M.; Díaz, C.; Cautain, B.; Annang, F.; Lage, O.M.; Pérez-Moreno, G.; Martinez, I.G.; Tormo, J.R.; et al. Bioactivities and Extract Dereplication of Actinomycetales Isolated from Marine Sponges. *Front. Microbiol.* **2019**, *10*, 727. [CrossRef] [PubMed]
34. Elsayed, Y.; Refaat, J.; Abdelmohsen, U.R.; Othman, E.M.; Stopper, H.; Fouad, M.A. Metabolomic profiling and biological investigation of the marine sponge-derived bacterium *Rhodococcus* sp. UA13. *Phytochem. Anal.* **2018**, *29*, 543–548. [CrossRef] [PubMed]
35. Cheng, Y.-B.; Jensen, P.R.; Fenical, W. Cytotoxic and Antimicrobial Napyradiomycins from Two Marine-Derived *Streptomyces* Strains. *Eur. J. Org. Chem.* **2013**, *2013*, 3751–3757. [CrossRef]
36. Hughes, C.C.; Prieto-Davo, A.; Jensen, P.R.; Fenical, W. ChemInform Abstract: The Marinopyrroles, Antibiotics of an Unprecedented Structure Class from a Marine *Streptomyces* sp. *ChemInform* **2008**, *10*, 629–631. [CrossRef]
37. Shen, X.; Wang, X.; Huang, T.; Deng, Z.; Lin, S. Naphthoquinone-Based Meroterpenoids from Marine-Derived *Streptomyces* sp. B9173. *Biomolecules* **2020**, *10*, 1187. [CrossRef]
38. Kock, I.; Maskey, R.P.; Biabani, M.A.F.; Helmke, E.; Laatsch, H. 1-Hydroxy-1-norresistomycin and Resistoflavin Methyl Ether: New Antibiotics from Marine-derived *Streptomycetes*. *J. Antibiot.* **2005**, *58*, 530–534. [CrossRef]
39. Yi, W.; Li, Q.; Song, T.; Chen, L.; Li, X.-C.; Zhang, Z.; Lian, X.-Y. Isolation, structure elucidation, and antibacterial evaluation of the metabolites produced by the marine-sourced *Streptomyces* sp. ZZ820. *Tetrahedron* **2019**, *75*, 1186–1193. [CrossRef]
40. Nandhagopal, S.; Iniyan, A.M.; Kannan, R.R.; Vincent, S.G.P. In Vivo Evaluation of Anti-MRSA Compound from *Streptomyces* Collinus ICN1 in Zebrafish Embryos. *Indian J. Geo-Marine Sci.* **2017**, *46*, 1155–1161.
41. Jiao, W.-H.; Yuan, W.; Li, Z.-Y.; Li, J.; Li, L.; Sun, J.-B.; Gui, Y.-H.; Wang, J.; Ye, B.-P.; Lin, H.-W. Anti-MRSA actinomycins D1-D4 from the marine sponge-associated *Streptomyces* sp. LHW52447. *Tetrahedron* **2018**, *74*, 5914–5919. [CrossRef]
42. Yang, L.; Hou, L.; Li, H.; Li, W. Antibiotic angucycline derivatives from the deepsea-derived *Streptomyces lusitanus*. *Nat. Prod. Res.* **2019**, *34*, 3444–3450. [CrossRef]
43. Djinni, I.; Defant, A.; Kecha, M.; Mancini, I. Metabolite Profile of Marine-Derived Endophytic *Streptomyces* Sundarbansensis WR1L1S8 by Liquid Chromatography-Mass Spectrometry and Evaluation of Culture Conditions on Antibacterial Activity and Mycelial Growth. *J. Appl. Microbiol.* **2014**, *116*, 39–50. [CrossRef] [PubMed]
44. Rajan, B.M.; Kannabiran, K. Antibiotic Potency of 2,4-Dichloro-5-Sulfamoyl Benzoic Acid Extracted from Marine Bacterium *Streptomyces* sp. VITBRK3 against Methicillin Resistant Staphylococcus aureus. *Pharm. Lett.* **2015**, *7*, 244–252.
45. Song, Y.; Yang, J.; Yu, J.; Li, J.; Yuan, J.; Wong, N.-K.; Ju, J. Chlorinated bis-indole alkaloids from deep-sea derived *Streptomyces* sp. SCSIO 11791 with antibacterial and cytotoxic activities. *J. Antibiot.* **2020**, *73*, 542–547. [CrossRef] [PubMed]
46. León, J.; Aponte, J.J.; Rojas, R.; Cuadra, D.; Ayala, N.; Tomás, G.; Guerrero, M. Estudio de actinomicetos marinos aislados de la costa central del Perú y su actividad antibacteriana frente a Staphylococcus aureus Meticilina Resistentes y Enterococcus faecalis Vancomicina Resistentes. *Rev. Peru. Med. Exp. Salud Publ.* **2011**, *28*, 237–246. [CrossRef]
47. Al-Dhabi, N.A.; Ghilan, A.-K.M.; Esmail, G.A.; Arasu, M.V.; Duraipandiyan, V.; Ponmurugan, K. Bioactivity assessment of the Saudi Arabian Marine Streptomyces sp. Al-Dhabi-90, metabolic profiling and its in vitro inhibitory property against multidrug resistant and extended-spectrum beta-lactamase clinical bacterial pathogens. *J. Infect. Public Health* **2019**, *12*, 549–556. [CrossRef]

48. Shao, M.; Ma, J.; Li, Q.; Ju, J. Identification of the Anti-Infective Aborycin Biosynthetic Gene Cluster from Deep-Sea-Derived Streptomyces sp. SCSIO ZS0098 Enables Production in a Heterologous Host. *Mar. Drugs* **2019**, *17*, 127. [CrossRef]
49. Padmanaban, V.P.; Verma, P.; Venkatabaskaran, S.; Keppayan, T.; Gopal, D.; Sekar, A.K.; Ramalingam, K. Antimicrobial potential and taxonomic investigation of piezotolerant Streptomyces sp. NIOT-Ch-40 isolated from deep-sea sediment. *World J. Microbiol. Biotechnol.* **2017**, *33*, 27. [CrossRef]
50. Song, Y.; Liu, G.; Li, J.; Huang, H.; Zhang, X.; Zhang, H.; Ju, J. Cytotoxic and Antibacterial Angucycline- and Prodigiosin-Analogues from the Deep-Sea Derived Streptomyces sp. SCSIO 11594. *Mar. Drugs* **2015**, *13*, 1304–1316. [CrossRef]
51. Lacret, R.; Oves-Costales, D.; Pérez-Victoria, I.; de la Cruz, M.; Díaz, C.; Vicente, F.; Genilloud, O.; Reyes, F. MDN-0171, a new medermycin analogue from Streptomyces albolongus CA-186053. *Nat. Prod. Res.* **2018**, *33*, 66–73. [CrossRef]
52. Wu, C.; Tan, Y.; Gan, M.; Wang, Y.; Guan, Y.; Hu, X.; Zhou, H.; Shang, X.; You, X.; Yang, Z.; et al. Identification of Elaiophylin Derivatives from the Marine-Derived Actinomycete Streptomyces sp. 7-145 Using PCR-Based Screening. *J. Nat. Prod.* **2013**, *76*, 2153–2157. [CrossRef]
53. Chung, B.; Kwon, O.-S.; Shin, J.; Oh, K.-B. Antibacterial Activity and Mode of Action of Lactoquinomycin A from Streptomyces bacillaris. *Mar. Drugs* **2020**, *19*, 7. [CrossRef] [PubMed]
54. Akhter, N.; Liu, Y.; Auckloo, B.N.; Shi, Y.; Wang, K.; Chen, J.; Wu, X.; Wu, B. Stress-Driven Discovery of New Angucycline-Type Antibiotics from a Marine Streptomyces pratensis NA-ZhouS1. *Mar. Drugs* **2018**, *16*, 331. [CrossRef] [PubMed]
55. Zhen, X.; Gong, T.; Liu, F.; Zhang, P.C.; Zhou, W.Q.; Li, Y.; Zhu, P. A New Analogue of Echinomycin and a New Cyclic Dipeptide from a Marine-Derived Streptomyces Sp. LS298. *Mar. Drugs* **2015**, *13*, 6947–6961. [CrossRef] [PubMed]
56. Wang, D.; Wang, C.; Gui, P.; Liu, H.; Khalaf, S.M.H.; Elsayed, E.A.; Wadaan, M.A.M.; Hozzein, W.N.; Zhu, W. Identification, Bioactivity, and Productivity of Actinomycins from the Marine-Derived Streptomyces heliomycini. *Front. Microbiol.* **2017**, *8*, 1147. [CrossRef]
57. Ravikumar, S.; Gnanadesigan, M.; Saravanan, A.; Monisha, N.; Brindha, V.; Muthumari, S. Antagonistic properties of seagrass associated Streptomyces sp. RAUACT-1: A source for anthraquinone rich compound. *Asian Pac. J. Trop. Med.* **2012**, *5*, 887–890. [CrossRef]
58. Zhang, X.; Chen, L.; Chai, W.; Lian, X.-Y.; Zhang, Z. A unique indolizinium alkaloid streptopertusacin A and bioactive bafilomycins from marine-derived Streptomyces sp. HZP-2216E. *Phytochemistry* **2017**, *144*, 119–126. [CrossRef]
59. Niu, S.; Li, S.; Chen, Y.; Tian, X.; Zhang, H.; Zhang, G.; Zhang, W.; Yang, X.; Zhang, S.; Ju, J.; et al. Lobophorins E and F, new spirotetronate antibiotics from a South China Sea-derived Streptomyces sp. SCSIO 01127. *J. Antibiot.* **2011**, *64*, 711–716. [CrossRef]
60. Norouzi, H.; Khorasgani, M.R.; Danesh, A. Anti-MRSA activity of a bioactive compound produced by a marine Streptomyces and its optimization using statistical experimental design. *Iran. J. Basic Med. Sci.* **2019**, *22*, 1073–1084. [CrossRef]
61. Dong, Y.; Ding, W.; Sun, C.; Ji, X.; Ling, C.; Zhou, Z.; Chen, Z.; Chen, X.; Ju, J. Julichrome Monomers from Marine Gastropod Mollusk-Associated *Streptomyces* and Stereochemical Revision of Julichromes Q(3.5) and Q(3.3). *Chem. Biodivers.* **2020**, *17*, e2000057. [CrossRef]
62. Luo, M.; Tang, L.; Dong, Y.; Huang, H.; Deng, Z.; Sun, Y. Antibacterial natural products lobophorin L and M from the marine-derived Streptomyces sp. 4506. *Nat. Prod. Res.* **2020**, *35*, 5581–5587. [CrossRef]
63. Liang, Y.; Chen, L.; Ye, X.; Anjum, K.; Lian, X.Y.; Zhang, Z. New streptophenazines from marine Streptomyces sp. 182SMLY. *Nat. Prod. Res.* **2017**, *31*, 411–417. [CrossRef]
64. Wang, Q.; Zhang, Y.; Wang, M.; Tan, Y.; Chunling, X.; He, H.; Xiao, C.; You, X.; Wang, Y.; Gan, M. Neo-actinomycins A and B, natural actinomycins bearing the 5H-oxazolo[4,5-b]phenoxazine chromophore, from the marine-derived Streptomyces sp. IMB094. *Sci. Rep.* **2017**, *7*, 3591. [CrossRef] [PubMed]
65. Song, Y.; Huang, H.; Chen, Y.; Ding, J.; Zhang, Y.; Sun, A.; Zhang, W.; Ju, J. Cytotoxic and Antibacterial Marfuraquinocins from the Deep South China Sea-Derived Streptomyces niveus SCSIO 3406. *J. Nat. Prod.* **2013**, *76*, 2263–2268. [CrossRef] [PubMed]
66. Miller, B.W.; Torres, J.P.; Tun, J.O.; Flores, M.S.; Forteza, I.; Rosenberg, G.; Haygood, M.G.; Schmidt, E.W.; Concepcion, G.P. Synergistic anti-methicillin-resistant Staphylococcus aureus (MRSA) activity and absolute stereochemistry of 7,8-dideoxygriseorhodin C. *J. Antibiot.* **2020**, *73*, 290–298. [CrossRef] [PubMed]
67. Braña, A.F.; Sarmiento-Vizcaíno, A.; Pérez-Victoria, I.; Martín, J.; Otero, L.; Palacios-Gutiérrez, J.J.; Fernández, J.; Mohamedi, Y.; Fontanil, T.; Salmón, M.; et al. Desertomycin G, a New Antibiotic with Activity against Mycobacterium tuberculosis and Human Breast Tumor Cell Lines Produced by Streptomyces althioticus MSM3, Isolated from the Cantabrian Sea Intertidal Macroalgae Ulva sp. *Mar. Drugs* **2019**, *17*, 114. [CrossRef]
68. Govindarajan, G.; Santhi, V.S.; Jebakumar, S.R.D. Antimicrobial potential of phylogenetically unique actinomycete, Streptomyces sp. JRG-04 from marine origin. *Biologicals* **2014**, *42*, 305–311. [CrossRef]
69. Wu, Z.; Li, S.; Li, J.; Chen, Y.; Saurav, K.; Zhang, Q.; Zhang, H.; Zhang, W.; Zhang, W.; Zhang, S.; et al. Antibacterial and Cytotoxic New Napyradiomycins from the Marine-Derived Streptomyces sp. SCSIO 10428. *Mar. Drugs* **2013**, *11*, 2113–2125. [CrossRef]
70. Hughes, C.C.; Kauffman, C.; Jensen, P.; Fenical, W. Structures, Reactivities, and Antibiotic Properties of the Marinopyrroles A−F. *J. Org. Chem.* **2010**, *75*, 3240–3250. [CrossRef]
71. Haste, N.M.; Farnaes, L.; Perera, V.R.; Fenical, W.; Nizet, V.; Hensler, M.E. Bactericidal Kinetics of Marine-Derived Napyradiomycins against Contemporary Methicillin-Resistant Staphylococcus aureus. *Mar. Drugs* **2011**, *9*, 680–689. [CrossRef]
72. Djinni, I.; Defant, A.; Kecha, M.; Mancini, I. Antibacterial Polyketides from the Marine Alga-Derived Endophitic Streptomyces sundarbansensis: A Study on Hydroxypyrone Tautomerism. *Mar. Drugs* **2013**, *11*, 124–135. [CrossRef]

73. Sun, P.; Maloney, K.N.; Nam, S.-J.; Haste, N.M.; Raju, R.; Aalbersberg, W.; Jensen, P.R.; Nizet, V.; Hensler, M.E.; Fenical, W. Fijimycins A–C, three antibacterial etamycin-class depsipeptides from a marine-derived Streptomyces sp. *Bioorganic Med. Chem.* **2011**, *19*, 6557–6562. [CrossRef] [PubMed]
74. Haste, N.M.; Perera, V.R.; Maloney, K.N.; Tran, D.N.; Jensen, P.; Fenical, W.; Nizet, V.; Hensler, E.M. Activity of the streptogramin antibiotic etamycin against methicillin-resistant Staphylococcus aureus. *J. Antibiot.* **2010**, *63*, 219–224. [CrossRef] [PubMed]
75. Furumai, T.; Eto, K.; Sasaki, T.; Higuchi, H.; Onaka, H.; Saito, N.; Fujita, T.; Naoki, H.; Igarashi, Y. TPU-0037-A, B, C and D, Novel Lydicamycin Congeners with Anti-MRSA Activity from Streptomyces platensis TP-A0598. *J. Antibiot.* **2002**, *55*, 873–880. [CrossRef] [PubMed]
76. Hassan, H.M.; Degen, D.; Jang, K.H.; Ebright, R.H.; Fenical, W. Salinamide F, new depsipeptide antibiotic and inhibitor of bacterial RNA polymerase from a marine-derived streptomyces sp. *J. Antibiot.* **2015**, *68*, 206–209. [CrossRef]
77. Han, Z.; Xu, Y.; McConnell, O.; Liu, L.; Li, Y.; Qi, S.; Huang, X.; Qian, P. Two Antimycin A Analogues from Marine-Derived Actinomycete Streptomyces lusitanus. *Mar. Drugs* **2012**, *10*, 668–676. [CrossRef] [PubMed]
78. Ryu, M.-J.; Hillman, P.F.; Lee, J.; Hwang, S.; Lee, E.-Y.; Cha, S.-S.; Yang, I.; Oh, D.-C.; Nam, S.-J.; Fenical, W. Antibacterial Meroterpenoids, Merochlorins G-J from the Marine Bacterium Streptomyces sp. *Mar. Drugs* **2021**, *19*, 618. [CrossRef]
79. Yang, W.Z.; Liang, G.J.; Sun, Y.; Gong, Z.J. Bioactive Secondary Metabolites from Marine Streptomyces Griseorubens F8: Isolation, Identification and Biological Activity Assay. *J. Mar. Sci. Eng.* **2021**, *9*, 978. [CrossRef]
80. Qureshi, K.A.; Bholay, A.D.; Rai, P.K.; Mohammed, H.A.; Khan, R.A.; Azam, F.; Jaremko, M.; Emwas, A.-H.; Stefanowicz, P.; Waliczek, M.; et al. Isolation, Characterization, Anti-MRSA Evaluation, and in-Silico Multi-Target Anti-Microbial Validations of Actinomycin X(2) and Actinomycin D Produced by Novel Streptomyces Smyrnaeus UKAQ_23. *Sci. Rep.* **2021**, *11*, 14539. [CrossRef]
81. Thi, D.P.; Mai, H.D.T.; Cao, D.D.; Thi, Q.V.; Nguyen, M.A.; Le Thi, H.M.; Tran, D.T.; Chau, V.M.; Pham, V.C. Novel 1,3-Benzodioxole From Marine-Derived Actinomycete in East Vietnam Sea. *Nat. Prod. Commun.* **2020**, *15*, 1934578X20920042. [CrossRef]
82. Song, Y.; Li, Q.; Liu, X.; Chen, Y.; Zhang, Y.; Sun, A.; Zhang, W.; Zhang, J.; Ju, J. Cyclic Hexapeptides from the Deep South China Sea-Derived Streptomyces scopuliridis SCSIO ZJ46 Active Against Pathogenic Gram-Positive Bacteria. *J. Nat. Prod.* **2014**, *77*, 1937–1941. [CrossRef]
83. Kunz, A.L.; Labes, A.; Wiese, J.; Bruhn, T.; Bringmann, G.; Imhoff, J.F. Nature's Lab for Derivatization: New and Revised Structures of a Variety of Streptophenazines Produced by a Sponge-Derived Streptomyces Strain. *Mar. Drugs* **2014**, *12*, 1699–1714. [CrossRef] [PubMed]
84. Liu, L.-L.; Xu, Y.; Han, Z.; Li, Y.-X.; Lu, L.; Lai, P.-Y.; Zhong, J.-L.; Guo, X.-R.; Zhang, X.-X.; Qian, P.-Y. Four New Antibacterial Xanthones from the Marine-Derived Actinomycetes Streptomyces caelestis. *Mar. Drugs* **2012**, *10*, 2571–2583. [CrossRef] [PubMed]
85. Iniyan, A.M.; Sudarman, E.; Wink, J.; Kannan, R.R.; Vincent, S.G.P. Ala-geninthiocin, a new broad spectrum thiopeptide antibiotic, produced by a marine Streptomyces sp. ICN19. *J. Antibiot.* **2018**, *72*, 99–105. [CrossRef] [PubMed]
86. Moghaddam, H.S.; Shahnavaz, B.; Makhdoumi, A.; Iranshahy, M. Evaluating the effect of various bacterial consortia on antibacterial activity of marine Streptomyces sp. AC117. *Biocontrol Sci. Technol.* **2021**, *31*, 1248–1266. [CrossRef]
87. Kumar, S.S.; Rao, K.V.B. In–vitro antimicrobial activity of marine actinobacteria against multidrug resistance Staphylococcus aureus. *Asian Pac. J. Trop. Biomed.* **2012**, *2*, S1802–S1807. [CrossRef]
88. Sabido, E.; Tenebro, C.; Trono, D.; Vicera, C.; Leonida, S.; Maybay, J.; Reyes-Salarda, R.; Amago, D.; Aguadera, A.; Octaviano, M.; et al. Insights into the Variation in Bioactivities of Closely Related Streptomyces Strains from Marine Sediments of the Visayan Sea against ESKAPE and Ovarian Cancer. *Mar. Drugs* **2021**, *19*, 441. [CrossRef]
89. Ramesh, C.; Vinithkumar, N.V.; Kirubagaran, R.; Venil, C.K.; Dufossé, L. Applications of Prodigiosin Extracted from Marine Red Pigmented Bacteria Zooshikella sp. and Actinomycete Streptomyces sp. *Microorganisms* **2020**, *8*, 556. [CrossRef]
90. Kurnianto, M.A.; Kusumaningrum, H.D.; Lioe, H.N.; Chasanah, E. Antibacterial and Antioxidant Potential of Ethyl Acetate Extract from Streptomyces AIA12 and AIA17 Isolated from Gut of Chanos chanos. *J. Biol. Divers.* **2021**, *22*, 813. [CrossRef]
91. Shin, H.J.; Lee, H.-S.; Lee, D.-S. The synergistic antibacterial activity of 1-acetyl-beta-carboline and beta-lactams against methicillin-resistant Staphylococcus aureus (MRSA). *J. Microbiol. Biotechnol.* **2010**, *20*, 501–505.
92. Kim, M.C.; Li, Z.; Cullum, R.; Molinski, T.F.; Eid, M.A.G.; Hebishy, A.M.S.; Faraag, A.H.I.; Abdel Moneim, A.E.; Abdelfattah, M.S.; Fenical, W. Chlororesistoflavins A and B, Chlorinated Benzopyrene Antibiotics Produced by the Marine-Derived Actinomycete Streptomyces Sp. Strain EG32. *J. Nat. Prod.* **2022**, *85*, 270–275. [CrossRef]
93. Shanthi, J.; Senthil, A.; Gopikrishnan, V.; Balagurunathan, R. Characterization of a Potential β-Lactamase Inhibitory Metabolite from a Marine Streptomyces sp. PM49 Active Against Multidrug-Resistant Pathogens. *Appl. Biochem. Biotechnol.* **2015**, *175*, 3696–3708. [CrossRef] [PubMed]
94. Fang, C.; Zhang, Q.; Zhu, Y.; Zhang, L.; Zhang, W.; Ma, L.; Zhang, H.; Zhang, C. Proximicins F and G and Diproximicin A: Aminofurans from the Marine-Derived Verrucosispora sp. SCSIO 40062 by Overexpression of PPtase Genes. *J. Nat. Prod.* **2020**, *83*, 1152–1156. [CrossRef] [PubMed]
95. Huang, P.; Xie, F.; Ren, B.; Wang, Q.; Wang, J.; Wang, Q.; Abdel-Mageed, W.M.; Liu, M.; Han, J.; Oyeleye, A.; et al. Anti-MRSA and anti-TB metabolites from marine-derived Verrucosispora sp. MS100047. *Appl. Microbiol. Biotechnol.* **2016**, *100*, 7437–7447. [CrossRef] [PubMed]
96. Zhang, S.; Xie, Q.; Sun, C.; Tian, X.P.; Gui, C.; Qin, X.; Ju, J. Cytotoxic Kendomycins Containing the Carbacylic Ansa Scaffold from the Marine-Derived Verrucosispora Sp. SCSIO 07399. *Nat. Prod.* **2019**, *82*, 3366–3371. [CrossRef] [PubMed]

97. Chen, M.-H.; Lian, Y.-Y.; Fang, D.-S.; Chen, L.; Jia, J.; Zhang, W.-J.; Lin, R.; Xie, Y.; Bi, H.-K.; Jiang, H. Identification and antimicrobial properties of a new alkaloid produced by marine-derived Verrucosispora sp. FIM06-0036. *Nat. Prod. Res.* **2019**, *35*, 4211–4217. [CrossRef]
98. Zhang, Y.; Adnani, N.; Braun, D.R.; Ellis, G.A.; Barns, K.J.; Parker-Nance, S.; Guzei, I.A.; Bugni, T.S. Micromonohalimanes A and B: Antibacterial Halimane-Type Diterpenoids from a Marine Micromonospora Species. *J. Nat. Prod.* **2016**, *79*, 2968–2972. [CrossRef]
99. Zhang, W.; Liu, Z.; Li, S.; Lu, Y.; Chen, Y.; Zhang, H.; Zhang, G.; Zhu, Y.; Zhang, G.; Zhang, W.; et al. Fluostatins I–K from the South China Sea-Derived Micromonospora rosaria SCSIO N160. *J. Nat. Prod.* **2012**, *75*, 1937–1943. [CrossRef]
100. Jiang, X.; Zhang, Q.; Zhu, Y.; Nie, F.; Wu, Z.; Yang, C.; Zhang, L.; Tian, X.; Zhang, C. Isolation, structure elucidation and biosynthesis of benzo[b]fluorene nenestatin A from deep-sea derived Micromonospora echinospora SCSIO 04089. *Tetrahedron* **2017**, *73*, 3585–3590. [CrossRef]
101. Siddharth, S.; Rai, V.R. Isolation and characterization of bioactive compounds with antibacterial, antioxidant and enzyme inhibitory activities from marine-derived rare actinobacteria, Nocardiopsis sp. SCA21. *Microb. Pathog.* **2019**, *137*, 103775. [CrossRef]
102. Xu, D.; Nepal, K.K.; Chen, J.; Harmody, D.; Zhu, H.; McCarthy, P.J.; Wright, A.E.; Wang, G. Nocardiopsistins A-C: New angucyclines with anti-MRSA activity isolated from a marine sponge-derived Nocardiopsis sp. HB-J378. *Synth. Syst. Biotechnol.* **2018**, *3*, 246–251. [CrossRef]
103. Yang, N.; Song, F. Bioprospecting of Novel and Bioactive Compounds from Marine Actinomycetes Isolated from South China Sea Sediments. *Curr. Microbiol.* **2017**, *75*, 142–149. [CrossRef] [PubMed]
104. Rajivgandhi, G.; Vijayan, R.; Kannan, M.; Santhanakrishnan, M.; Manoharan, N. Molecular characterization and antibacterial effect of endophytic actinomycetes Nocardiopsis sp. GRG1 (KT235640) from brown algae against MDR strains of uropathogens. *Bioact. Mater.* **2016**, *1*, 140–150. [CrossRef] [PubMed]
105. Sunga, M.J.; Teisan, S.; Tsueng, G.; Macherla, V.R.; Lam, K.S. Seawater requirement for the production of lipoxazolidinones by marine actinomycete strain NPS8920. *J. Ind. Microbiol. Biotechnol.* **2008**, *35*, 761–765. [CrossRef] [PubMed]
106. McArthur, K.A.; Mitchell, S.S.; Tsueng, G.; Rheingold, A.; White, D.J.; Grodberg, J.; Lam, K.S.; Potts, B.C.M. Lynamicins A–E, Chlorinated Bisindole Pyrrole Antibiotics from a Novel Marine Actinomycete. *J. Nat. Prod.* **2008**, *71*, 1732–1737. [CrossRef]
107. Braña, A.F.; Sarmiento-Vizcaíno, A.; Pérez-Victoria, I.; Otero, L.; Fernández, J.; Palacios, J.J.; Martín, J.; de la Cruz, M.; Díaz, C.; Vicente, F.; et al. Branimycins B and C, Antibiotics Produced by the Abyssal Actinobacterium Pseudonocardia Carboxydivorans M-227. *J. Nat. Prod.* **2017**, *80*, 569–573. [CrossRef]
108. Palomo, S.; González, I.; de la Cruz, M.; Martín, J.; Tormo, J.R.; Anderson, M.; Hill, R.T.; Vicente, F.; Reyes, F.; Genilloud, O. Sponge-Derived Kocuria and Micrococcus spp. as Sources of the New Thiazolyl Peptide Antibiotic Kocurin. *Mar. Drugs* **2013**, *11*, 1071–1086. [CrossRef]
109. Ellis, G.; Wyche, T.P.; Fry, C.G.; Braun, D.R.; Bugni, T.S. Solwaric Acids A and B, Antibacterial Aromatic Acids from a Marine Solwaraspora sp. *Mar. Drugs* **2014**, *12*, 1013–1022. [CrossRef]
110. Singh, S.; Prasad, P.; Subramani, R.; Aalbersberg, W. Production and Purification of a Bioactive Substance against Multi-Drug Resistant Human Pathogens from the Marine-Sponge-Derived Salinispora Sp. *Asian Pac. J. Trop. Biomed.* **2014**, *4*, 825–831. [CrossRef]
111. Zhang, D.; Yi, W.; Ge, H.; Zhang, Z.; Wu, B. Bioactive Streptoglutarimides A–J from the Marine-Derived Streptomyces Sp. ZZ741. *J. Nat. Prod.* **2019**, *82*, 2800–2808. [CrossRef]
112. Zhou, B.; Huang, Y.; Zhang, H.-J.; Li, J.-Q.; Ding, W.-J. Nitricquinomycins A-C, uncommon naphthopyrrolediones from the Streptomyces sp. ZS-A45. *Tetrahedron* **2019**, *75*, 3958–3961. [CrossRef]
113. Carretero-Molina, D.; Ortiz-López, F.J.; Martín, J.; Oves-Costales, D.; Díaz, C.; de la Cruz, M.; Cautain, B.; Vicente, F.; Genilloud, O.; Reyes, F. New Napyradiomycin Analogues from Streptomyces sp. Strain CA-271078. *Mar. Drugs* **2020**, *18*, 22–41. [CrossRef] [PubMed]
114. Sujatha, P.; Raju, K.B.; Ramana, T. Studies on a new marine streptomycete BT-408 producing polyketide antibiotic SBR-22 effective against methicillin resistant Staphylococcus aureus. *Microbiol. Res.* **2005**, *160*, 119–126. [CrossRef] [PubMed]
115. Pan, H.-Q.; Zhang, S.-Y.; Wang, N.; Li, Z.-L.; Hua, H.-M.; Hu, J.-C.; Wang, S.-J. New Spirotetronate Antibiotics, Lobophorins H and I, from a South China Sea-Derived Streptomyces sp. 12A35. *Mar. Drugs* **2013**, *11*, 3891–3901. [CrossRef] [PubMed]
116. Zhang, X.; Ye, X.; Chai, W.; Lian, X.-Y.; Zhang, Z. New Metabolites and Bioactive Actinomycins from Marine-Derived Streptomyces sp. ZZ338. *Mar. Drugs* **2016**, *14*, 181. [CrossRef] [PubMed]
117. Hu, Y.; Wang, M.; Wu, C.; Tan, Y.; Li, J.; Hao, X.; Duan, Y.; Guan, Y.; Shang, X.; Wang, Y.; et al. Identification and Proposed Relative and Absolute Configurations of Niphimycins C–E from the Marine-Derived *Streptomyces* sp. IMB7-145 by Genomic Analysis. *J. Nat. Prod.* **2018**, *81*, 178–187. [CrossRef] [PubMed]
118. Cao, D.D.; Do, T.Q.; Doan Thi Mai, H.; Vu Thi, Q.; Nguyen, M.A.; Le Thi, H.M.; Tran, D.T.; Chau, V.M.; Cong Thung, D.; Pham, V.C. Antimicrobial lavandulylated flavonoids from a sponge-derived actinomycete. *Nat. Prod. Res.* **2020**, *34*, 413–420. [CrossRef]
119. Cong, Z.; Huang, X.; Liu, Y.; Liu, Y.; Wang, P.; Liao, S.; Wang, J. Cytotoxic Anthracycline and Antibacterial Tirandamycin Analogues from a Marine-Derived Streptomyces Sp. SCSIO 41399. *J. Antibiot.* **2019**, *72*, 45–49. [CrossRef]
120. Saurav, K.; Kannabiran, K. In vitro activity of 5-(2,4-dimethylbenzyl) pyrrolidin-2-one extracted from marine Streptomyces VITSVK5 spp. against fungal and bacterial human pathogens. *Rev. Iberoam. Micol.* **2012**, *29*, 29–33. [CrossRef] [PubMed]

121. Cho, E.; Kwon, O.-S.; Chung, B.; Lee, J.; Sun, J.; Shin, J.; Oh, K.-B. Antibacterial Activity of Chromomycins from a Marine-Derived Streptomyces microflavus. *Mar. Drugs* **2020**, *18*, 522. [CrossRef] [PubMed]
122. Chen, M.; Chai, W.; Zhu, R.; Song, T.; Zhang, Z.; Lian, X.-Y. Streptopyrazinones A−D, rare metabolites from marine-derived Streptomyces sp. ZZ446. *Tetrahedron* **2018**, *74*, 2100–2106. [CrossRef]
123. Lacret, R.; Pérez-Victoria, I.; Oves-Costales, D.; de la Cruz, M.; Domingo, E.; Martín, J.; Díaz, C.; Vicente, F.; Genilloud, O.; Reyes, F. MDN-0170, a New Napyradiomycin from Streptomyces sp. Strain CA-271078. *Mar. Drugs* **2016**, *14*, 188. [CrossRef] [PubMed]
124. Shaala, L.A.; Youssef, D.T.A.; Alzughaibi, T.A.; Elhady, S.S. Antimicrobial Chlorinated 3-Phenylpropanoic Acid Derivatives from the Red Sea Marine Actinomycete Streptomyces coelicolor LY001. *Mar. Drugs* **2020**, *18*, 450. [CrossRef]
125. Jiang, L.; Huang, P.; Ren, B.; Song, Z.; Zhu, G.; He, W.; Zhang, J.; Oyeleye, A.; Dai, H.; Zhang, L.; et al. Antibacterial polyene-polyol macrolides and cyclic peptides from the marine-derived Streptomyces sp. MS110128. *Appl. Microbiol. Biotechnol.* **2021**, *105*, 4975–4986. [CrossRef] [PubMed]
126. Setiawati, S.; Nuryastuti, T.; Sholikhah, E.N.; Lisdiyanti, P.; Pratiwi, S.U.T.; Sulistiyani, T.R.; Mustofa, M. The Potency of Actinomycetes Extracts Isolated from Pramuka Island, Jakarta, Indonesia as Antimicrobial Agents. *Biodivers. J. Biol. Divers.* **2021**, *22*, 150933. [CrossRef]
127. Dalisay, D.; Williams, D.E.; Wang, X.L.; Centko, R.; Chen, J.; Andersen, R.J. Marine Sediment-Derived Streptomyces Bacteria from British Columbia, Canada Are a Promising Microbiota Resource for the Discovery of Antimicrobial Natural Products. *PLoS ONE* **2013**, *8*, e77078. [CrossRef]
128. Uzair, B.; Menaa, F.; Khan, B.A.; Mohammad, F.V.; Ahmad, V.U.; Djeribi, R.; Menaa, B. Isolation, purification, structural elucidation and antimicrobial activities of kocumarin, a novel antibiotic isolated from actinobacterium Kocuria marina CMG S2 associated with the brown seaweed Pelvetia canaliculata. *Microbiol. Res.* **2018**, *206*, 186–197. [CrossRef]
129. Kwon, H.C.; Kauffman, C.A.; Jensen, P.R.; Fenical, W. Marinomycins A-D, Antitumor-Antibiotics of a New Structure Class from a Marine Actinomycete of the Recently Discovered Genus "Marinispora". *J. Am. Chem. Soc.* **2006**, *128*, 1622–1632. [CrossRef]
130. Chen, M.H.; Zhang, W.L.; Chen, L.; Lin, R.; Xie, Y.; Fang, D.S.; Jiang, H.; Lian, Y.-Y. Isolation, Purification and Identification of Two New Alkaloids Metabolites from Marine-Derived Verrucosispora Sp. FIM06025. *Nat. Prod. Res.* **2019**, *33*, 2897–2903. [CrossRef]
131. Panche, A.N.; Diwan, A.D.; Chandra, S.R. Flavonoids: An overview. *J. Nutr. Sci.* **2016**, *5*, e47. [CrossRef]
132. Iniyan, A.M.; Mary, T.R.J.; Joseph, F.-J.R.S.; Kannan, R.R.; Vincent, S.G.P. Cell wall distracting anti-Methicillin-resistant Staphylococcus aureus compound PVI331 from a marine sponge associated Streptomyces. *J. Appl. Biomed.* **2016**, *14*, 273–283. [CrossRef]
133. Balasubramanian, S.; Skaf, J.; Holzgrabe, U.; Bharti, R.; Förstner, K.U.; Ziebuhr, W.; Humeida, U.H.; Abdelmohsen, U.R.; Oelschlaeger, T.A. A New Bioactive Compound From the Marine Sponge-Derived Streptomyces sp. SBT348 Inhibits Staphylococcal Growth and Biofilm Formation. *Front. Microbiol.* **2018**, *9*, 1473. [CrossRef] [PubMed]
134. Joseph, F.-J.R.S.; Iniyan, A.M.; Vincent, S.G.P. HR-LC-MS based analysis of two antibacterial metabolites from a marine sponge symbiont Streptomyces pharmamarensis ICN40. *Microb. Pathog.* **2017**, *111*, 450–457. [CrossRef] [PubMed]
135. Devescovi, G.; Kojic, M.; Covaceuszach, S.; Camara, M.; Williams, P.; Bertani, I.; Subramoni, S.; Venturi, V. Negative Regulation of Violacein Biosynthesis in Chromobacterium violaceum. *Front. Microbiol.* **2017**, *8*, 349. [CrossRef]
136. Lu, L.; Li, M.; Yi, G.; Liao, L.; Cheng, Q.; Zhu, J.; Zhang, B.; Wang, Y.; Chen, Y.; Zeng, M. Screening strategies for quorum sensing inhibitors in combating bacterial infections. *J. Pharm. Anal.* **2021**, *12*, 1–14. [CrossRef] [PubMed]
137. Kamarudheen, N.; Naushad, T.; Rao, K.V.B. Biosynthesis, Characterization and Antagonistic Applications of Extracellular Melanin Pigment from Marine Nocardiopsis Sps. *Indian J. Pharm. Educ. Res.* **2019**, *53*, s112–s120. [CrossRef]
138. Yin, Q.; Liang, J.; Zhang, W.; Zhang, L.; Hu, Z.-L.; Zhang, Y.; Xu, Y. Butenolide, a Marine-Derived Broad-Spectrum Antibiofilm Agent Against Both Gram-Positive and Gram-Negative Pathogenic Bacteria. *Mar. Biotechnol.* **2019**, *21*, 88–98. [CrossRef]
139. Le, K.Y.; Otto, M. Quorum-Sensing Regulation in Staphylococci-an Overview. *Front. Microbiol.* **2015**, *6*, 1174. [CrossRef]
140. Kamarudheen, N.; Rao, K.B. Fatty acyl compounds from marine Streptomyces griseoincarnatus strain HK12 against two major bio-film forming nosocomial pathogens; an in vitro and in silico approach. *Microb. Pathog.* **2018**, *127*, 121–130. [CrossRef]
141. Vignesh, A.; Ayswarya, S.; Gopikrishnan, V.; Radhakrishnan, M. Bioactive Potential of Actinobacteria Isolated from the Gut of Marine Fishes. *Indian J. Geo-Marine Sci.* **2019**, *48*, 1280–1285.
142. Abd-Elnaby, H.; Abo-Elala, G.; Abdel-Raouf, U.; Abd-Elwahab, A.; Hamed, M. Antibacterial and anticancer activity of marine Streptomyces parvus: Optimization and application. *Biotechnol. Biotechnol. Equip.* **2015**, *30*, 180–191. [CrossRef]
143. Mohamedin, A.H.; El-Naggar, N.E.-A.; Sherief, A.E.-D.A.; Hussien, S.M. Optimization of Bioactive Metabolites production by a Newly Isolated Marine Streptomyces sp. Using Statistical Approach. *Biotechnology* **2015**, *14*, 211–224. [CrossRef]
144. Katif, C.; Chilczuk, T.; Sabour, B.; Belattmania, Z.; Hilmi, A.; Niedermeyer, T.H.J.; Barakate, M. Isolation and Structure Elucidation of Desferrioxamine B and the New Desferrioxamine B2 Antibiotics from a Brown Marine Macroalga Carpodesmia Tamariscifolia Associated Streptomyces Isolate. *Biointerface Res. Appl. Chem.* **2022**, *12*, 5647–5662.
145. Contreras-Castro, L.; Martínez-García, S.; Cancino-Diaz, J.C.; Maldonado, L.A.; Hernández-Guerrero, C.J.; Martínez-Díaz, S.F.; González-Acosta, B.; Quintana, E.T. Marine Sediment Recovered Salinispora sp. Inhibits the Growth of Emerging Bacterial Pathogens and other Multi-Drug-Resistant Bacteria. *Pol. J. Microbiol.* **2020**, *69*, 321–330. [CrossRef] [PubMed]
146. Zhou, B.; Ji, Y.-Y.; Zhang, H.-J.; Shen, L. Gephyyamycin and cysrabelomycin, two new angucyclinone derivatives from the Streptomyces sp. HN-A124. *Nat. Prod. Res.* **2019**, *35*, 2117–2122. [CrossRef] [PubMed]
147. Iniyan, A.M.; Joseph, F.R.S.; Kannan, R.R.; Vincent, S.G.P. Anti-MRSA Potential of Phenolic Compound Isolated from a Marine Derived Actinomycete Micromonospora Sp. ICN36. *Indian J. Geo-Marine Sci.* **2016**, *45*, 1279–1287.

148. Manikandan, M.; Gowdaman, V.; Duraimurugan, K.; Prabagaran, S.R. Taxonomic characterization and antimicrobial compound production from Streptomyces chumphonensis BDK01 isolated from marine sediment. *3 Biotech* **2019**, *9*, 167. [CrossRef]
149. Suthindhiran, K.; Kannabiran, K. Diversity and exploration of bioactive marine actinomycetes in the Bay of Bengal of the Puducherry coast of India. *Indian J. Microbiol.* **2010**, *50*, 76–82. [CrossRef]
150. Ibrahimi, M.; Korichi, W.; Hafidi, M.; Lemee, L.; Ouhdouch, Y.; Loqman, S. Marine Actinobacteria: Screening for Predation Leads to the Discovery of Potential New Drugs against Multidrug-Resistant Bacteria. *Antibiotics* **2020**, *9*, 91. [CrossRef]
151. Ahmad, S.; Nazir, M.; Tousif, M.I.; Saleem, M.; Mustafa, R.; Khatoon, T. A New Polyketide Antibiotic from the Marine Bacterium Streptomyces sp. PGC 32. *Chem. Nat. Compd.* **2019**, *55*, 1–4. [CrossRef]
152. Zhang, Z.; Chen, L.; Zhang, X.; Liang, Y.; Anjum, K.; Chen, L.; Lian, X.-Y. Bioactive Bafilomycins and a New N-Arylpyrazinone Derivative from Marine-derived Streptomyces sp. HZP-2216E. *Planta Med.* **2017**, *83*, 1405–1411. [CrossRef]
153. Huang, H.; Song, Y.; Li, X.; Wang, X.; Ling, C.; Qin, X.; Zhou, Z.; Li, Q.; Wei, X.; Ju, J. Abyssomicin Monomers and Dimers from the Marine-Derived Streptomyces koyangensis SCSIO 5802. *J. Nat. Prod.* **2018**, *81*, 1892–1898. [CrossRef] [PubMed]
154. Ding, W.; Dong, Y.; Ju, J.; Li, Q. The roles of genes associated with regulation, transportation, and macrocyclization in desotamide biosynthesis in Streptomyces scopuliridis SCSIO ZJ46. *Appl. Microbiol. Biotechnol.* **2020**, *104*, 2603–2610. [CrossRef] [PubMed]
155. Karuppiah, V.; Li, Y.; Sun, W.; Feng, G.; Li, Z. Functional gene-based discovery of phenazines from the actinobacteria associated with marine sponges in the South China Sea. *Appl. Microbiol. Biotechnol.* **2015**, *99*, 5939–5950. [CrossRef] [PubMed]
156. Yang, J.; Li, J.; Hu, Y.; Li, L.; Long, L.; Wang, F.; Zhang, S. Characterization of a thermophilic hemoglobin-degrading protease from Streptomyces rutgersensis SCSIO 11720 and its application in antibacterial peptides production. *Biotechnol. Bioprocess Eng.* **2015**, *20*, 79–90. [CrossRef]
157. Abdulkhair, W.M.; Alghuthaymi, M.A. Double Inhibitory Effect of Extracellular Protein of Marine Streptomyces Tendae against Different Strains of MRSA. *Der Pharm. Lett.* **2016**, *8*, 11–20.
158. Jiang, Y.-J.; Zhang, D.-S.; Zhang, H.-J.; Li, J.-Q.; Ding, W.-J.; Xu, C.-D.; Ma, Z.-J. Medermycin-Type Naphthoquinones from the Marine-Derived Streptomyces sp. XMA39. *J. Nat. Prod.* **2018**, *81*, 2120–2124. [CrossRef]
159. Kurata, A.; Sugiura, M.; Kokoda, K.; Tsujimoto, H.; Numata, T.; Kato, C.; Nakasone, K.; Kishimoto, N. Taxonomy of actinomycetes in the deep-sea Calyptogena communities and characterization of the antibacterial compound produced by Actinomadura sp. DS-MS-114. *Biotechnol. Biotechnol. Equip.* **2017**, *31*, 1000–1006. [CrossRef]
160. Li, S.; Tian, X.; Niu, S.; Zhang, W.; Chen, Y.; Zhang, H.; Yang, X.; Zhang, W.; Li, W.; Zhang, S.; et al. Pseudonocardians A–C, New Diazaanthraquinone Derivatives from a Deap-Sea Actinomycete Pseudonocardia sp. SCSIO 01299. *Mar. Drugs* **2011**, *9*, 1428–1439. [CrossRef]
161. Arslan, I. Trends in Antimicrobial Resistance in Healthcare-Associated Infections: A Global Concern. *Ref. Modul. Biomed. Sci.* **2022**, *4*, 652–661. [CrossRef]
162. Kirst, H.A. Macrolide Antibiotics. In *Antimicrobials*; Springer: Berlin/Heidelberg, Germany, 2013; pp. 211–230. ISBN 978-3-642-39968-8. [CrossRef]
163. Xu, H.; Li, H. Acne, the Skin Microbiome, and Antibiotic Treatment. *Am. J. Clin. Dermatol.* **2019**, *20*, 335–344. [CrossRef] [PubMed]
164. Stahl, J.-P. Lincosamidas. *EMC Tratado Med.* **2017**, *21*, 1–4. [CrossRef]
165. Ma, M.; Rateb, M.E.; Teng, Q.; Yang, D.; Rudolf, J.D.; Zhu, X.; Huang, Y.; Zhao, L.-X.; Jiang, Y.; Li, X.; et al. Angucyclines and Angucyclinones from Streptomyces sp. CB01913 Featuring C-Ring Cleavage and Expansion. *J. Nat. Prod.* **2015**, *78*, 2471–2480. [CrossRef] [PubMed]
166. Wang, T.-Y.; Li, Q.; Bi, K.-S. Bioactive flavonoids in medicinal plants: Structure, activity and biological fate. *Asian J. Pharm. Sci.* **2017**, *13*, 12–23. [CrossRef] [PubMed]
167. Xu, D.; Han, L.; Li, C.; Cao, Q.; Zhu, D.; Barrett, N.H.; Harmody, D.; Chen, J.; Zhu, H.; McCarthy, P.J.; et al. Bioprospecting Deep-Sea Actinobacteria for Novel Anti-infective Natural Products. *Front. Microbiol.* **2018**, *9*, 787. [CrossRef]
168. Cumsille, A.; Undabarrena, A.; González, V.; Claverías, F.; Rojas, C.; Cámara, B. Biodiversity of Actinobacteria from the South Pacific and the Assessment of Streptomyces Chemical Diversity with Metabolic Profiling. *Mar. Drugs* **2017**, *15*, 286. [CrossRef]
169. Su, P.; Wang, D.-X.; Ding, S.-X.; Zhao, J. Isolation and diversity of natural product biosynthetic genes of cultivable bacteria associated with marine sponge Mycale sp. from the coast of Fujian, China. *Can. J. Microbiol.* **2014**, *60*, 217–225. [CrossRef]
170. Meena, B.; Anburajan, L.; Vinithkumar, N.V.; Kirubagaran, R.; Dharani, G. Biodiversity and antibacterial potential of cultivable halophilic actinobacteria from the deep sea sediments of active volcanic Barren Island. *Microb. Pathog.* **2019**, *132*, 129–136. [CrossRef]
171. Bauermeister, A.; Pereira, F.; Grilo, I.R.; Godinho, C.C.; Paulino, M.; Almeida, V.; Gobbo-Neto, L.; Prieto-Davó, A.; Sobral, R.G.; Lopes, N.P.; et al. Intra-clade metabolomic profiling of MAR4 Streptomyces from the Macaronesia Atlantic region reveals a source of anti-biofilm metabolites. *Environ. Microbiol.* **2019**, *21*, 1099–1112. [CrossRef]
172. Anggelina, A.C.; Pringgenies, D.; Setyati, W.A. Presence of Biosynthetic Gene Clusters (NRPS/PKS) in Actinomycetes of Mangrove Sediment in Semarang and Karimunjawa, Indonesia. *Environ. Nat. Res. J.* **2021**, *19*, 391–401. [CrossRef]
173. Íñiguez-Martínez, A.M.; Cardoso-Martínez, F.; De La Rosa, J.; Cueto, M.; Díaz-Marrero, A.; Darias, J.; Becerril-Espinosa, A.; Rosas, L.J.P.; Soria-Mercado, E.I. Compounds isolated from Salinispora arenicola of the Gulf of California, México. *Rev. Biol. Mar. Oceanogr.* **2016**, *51*, 161–170. [CrossRef]
174. El-Naggar, M.Y.; Barakat, K.M.; Aly, N.S. Physiological Response, Antibacterial Activity and Cinnamaldehyde Production by a Marine Streptomyces Chartreusis. *J. Pure Appl. Microbiol.* **2016**, *10*, 1797–1808.

175. Rajasabapathy, R.; Ghadi, S.C.; Manikandan, B.; Mohandass, C.; Surendran, A.; Dastager, S.G.; Meena, R.M.; James, R.A. Antimicrobial profiling of coral reef and sponge associated bacteria from southeast coast of India. *Microb. Pathog.* **2020**, *141*, 103972. [CrossRef] [PubMed]
176. Gandhimathi, R.; Arunkumar, M.; Selvin, J.; Thangavelu, T.; Sivaramakrishnan, S.; Kiran, G.; Shanmughapriya, S.; Natarajaseenivasan, K. Antimicrobial potential of sponge associated marine actinomycetes. *J. Mycol. Med.* **2008**, *18*, 16–22. [CrossRef]
177. Wang, C.; Du, W.; Lu, H.; Lan, J.; Liang, K.; Cao, S. A Review: Halogenated Compounds from Marine Actinomycetes. *Molecules* **2021**, *26*, 2754. [CrossRef] [PubMed]
178. Machmudah, S.; Kanda, H.; Goto, M. *Hydrolysis of Biopolymers in Near-Critical and Subcritical Water*; Elsevier Inc.: Amsterdam, The Netherlands, 2017; ISBN 9780128093801.
179. Eltamany, E.E.; Abdelmohsen, U.R.; Ibrahim, A.K.; Hassanean, H.A.; Hentschel, U.; Ahmed, S.A. New antibacterial xanthone from the marine sponge-derived Micrococcus sp. EG45. *Bioorganic Med. Chem. Lett.* **2014**, *24*, 4939–4942. [CrossRef]
180. Kurnianto, M.A.; Kusumaningrum, H.D.; Lioe, H.N.; Chasanah, E. Partial Purification and Characterization of Bacteriocin-Like Inhibitory Substances Produced by Streptomyces Sp. Isolated from the Gut of Chanos chanos. *Biomed Res. Int.* **2021**, *2021*, 7190152. [CrossRef]
181. Nivina, A.; Yuet, K.P.; Hsu, J.; Khosla, C. Evolution and Diversity of Assembly-Line Polyketide Synthases. *Chem. Rev.* **2019**, *119*, 12524–12547. [CrossRef]
182. Wink, J.; Mohammadipanah, F.; Hamedi, J. *Biology and Biotechnology of Actinobacteria*; Springer International Publishing: Berlin/Heidelberg, Germany, 2017; ISBN 9783319603391.
183. Luti, K.J.K. Mixture Design of Experiments for the Optimization of Carbon Source for Promoting Undecylprodigiosin and Actinorhodin Production. *J. Pure Appl. Microbiol.* **2018**, *12*, 1783–1793. [CrossRef]
184. Newman, D.J.; Cragg, G.M. Natural Products As Sources of New Drugs over the 30 Years from 1981 to 2010. *J. Nat. Prod.* **2012**, *75*, 311–335. [CrossRef]
185. Ballav, S.; Kerkar, S.; Thomas, S.; Augustine, N. Halophilic and halotolerant actinomycetes from a marine saltern of Goa, India producing anti-bacterial metabolites. *J. Biosci. Bioeng.* **2015**, *119*, 323–330. [CrossRef] [PubMed]
186. Priyanka, S.; Jayashree, M.; Shivani, R.; Anwesha, S.; Rao, K.B. Characterisation and identification of antibacterial compound from marine actinobacteria: In vitro and in silico analysis. *J. Infect. Public Health* **2018**, *12*, 83–89. [CrossRef] [PubMed]
187. Cheng, C.; MacIntyre, L.; Abdelmohsen, U.R.; Horn, H.; Polymenakou, P.N.; Edrada-Ebel, R.; Hentschel, U. Biodiversity, Anti-Trypanosomal Activity Screening, and Metabolomic Profiling of Actinomycetes Isolated from Mediterranean Sponges. *PLoS ONE* **2015**, *10*, e0138528. [CrossRef] [PubMed]
188. Rajan, B.M.; Kannabiran, K. Extraction and Identification of Antibacterial Secondary Metabolites from Marine Streptomyces sp. VITBRK2. *Int. J. Mol. Cell. Med.* **2014**, *3*, 130–137.
189. Thosar, A.; Satpathy, P.; Devi, C.S. Marine Streptomyces sp. VITASP as a Source of New Bioactive Secondary Metabolites. *Curr. Bioact. Compd.* **2020**, *16*, 611–617. [CrossRef]
190. Mercy, R.B.; Kannabiran, K. Identification of Antibacterial Secondary Metabolite from Marine Streptomyces Sp. VITBRK4 and Its Activity against Drug Resistant Gram Positive Bacteria. *Int. J. Drug Dev. Res.* **2013**, *5*, 224.
191. Mohseni, M.; Norouzi, H.; Hamedi, J.; Roohi, A. Screening of Antibacterial Producing Actinomycetes from Sediments of the Caspian Sea. *Int. J. Mol. Cell. Med.* **2013**, *2*, 64–71.
192. Yuan, X.W.; Yang, R.L.; Cao, X.; Gao, J.J. Taxonomic identification of a novel strain of Streptomyces cavourensis subsp. washingtonensis, ACMA006, exhibiting antitumor and antibacteria activity. *Drug Discov. Ther.* **2010**, *4*, 405–411.
193. Wahaab, F.; Subramaniam, K. Bioprospecting marine actinomycetes for multidrug-resistant pathogen control from Rameswaram coastal area, Tamil Nadu, India. *Arch. Microbiol.* **2017**, *200*, 57–71. [CrossRef]
194. Rajan, B.M.; Kannabiran, K. Antagonistic Activity of Marine Streptomyces Sp. VITBRK1 on Drug Resistant Gram Positive Cocci. *Der Pharm. Lett.* **2013**, *5*, 185–191.
195. Kokare, C.R.; Mahadik, K.R.; Kadam, S.S.; Chopade, B.A. Isolation, Characterization and Antimicrobial Activity of Marine Halophilic Actinopolyspora Species AH1 from the West Coast of India. *Curr. Sci.* **2004**, *86*, 593–597.
196. Tenebro, C.P.; Trono, D.J.V.L.; Vicera, C.V.B.; Sabido, E.M.; Ysulat, J.J.A.; Macaspac, A.J.M.; Tampus, K.A.; Fabrigar, T.A.P.; Saludes, J.P.; Dalisay, D.S. Multiple Strain Analysis of Streptomyces Species from Philippine Marine Sediments Reveals Intraspecies Heterogeneity in Antibiotic Activities. *Sci. Rep.* **2021**, *11*, 17544. [CrossRef] [PubMed]
197. Mane, M.; Mahadik, K.; Kokare, C. Purification, Characterization and Applications of Thermostable Alkaline Protease from Marine Streptomyces Sp. D1. *Int. J. Pharma Bio. Sci.* **2013**, *4*, 572–582.
198. Ouchene, R.; Intertaglia, L.; Zaatout, N.; Kecha, M.; Suzuki, M.T. Selective isolation, antimicrobial screening and phylogenetic diversity of marine actinomycetes derived from the Coast of Bejaia City (Algeria), a polluted and microbiologically unexplored environment. *J. Appl. Microbiol.* **2021**, *132*, 2870–2882. [CrossRef]
199. Mani, A.; Ravi, L.; Krishnan, K. Antibacterial and antifungal potential of marine Streptomyces sp. VITAK1 derived novel compound Pyrrolidinyl-Hexadeca-Heptaenone by in Silico docking analysis. *Res. J. Pharm. Technol.* **2018**, *11*, 1901. [CrossRef]
200. Undabarrena, A.; Beltrametti, F.; Claverías, F.P.; González, M.; Moore, E.R.B.; Seeger, M.; Cámara, B. Exploring the Diversity and Antimicrobial Potential of Marine Actinobacteria from the Comau Fjord in Northern Patagonia, Chile. *Front. Microbiol.* **2016**, *7*, 1135. [CrossRef]

201. Liang, Y.; Xie, X.; Chen, L.; Yan, S.; Ye, X.; Anjum, K.; Huang, H.; Lian, X.; Zhang, Z. Bioactive Polycyclic Quinones from Marine Streptomyces sp. 182SMLY. *Mar. Drugs* **2016**, *14*, 10. [CrossRef]
202. Attimarad, S.L.; Gaviraj, E.N.; Nagesh, C.; Kugaji, M.S.; Sutar, R.S. Screening, Isolation and Purification of Antibiotic(s) from Marine Actinomycetes. *Int. J. Res. Ayurveda Pharm.* **2012**, *3*, 447–453.
203. León, J.; Liza, L.; Soto, I.; Cuadra, D.; Patiño, L.; Zerpa, R. Bioactives Actinomycetes of Marine Sediment from the Central Coast of Peru [Actinomycetes Bioactivos de Sedimento Marino de La Costa Central Del Perú]. *Rev. Peru. Biol.* **2007**, *14*, 259–270.
204. Jagan Mohan, Y.S.Y.V.; Sirisha, B.; Haritha, R.; Ramana, T. Selective Screening, Isolation and Characterization of Antimicrobial Agents from Marine Actinomycetes. *Int. J. Pharm. Pharm. Sci.* **2013**, *5*, 443–449.
205. Song, Y.; Li, Q.; Qin, F.; Sun, C.; Liang, H.; Wei, X.; Wong, N.-K.; Ye, L.; Zhang, Y.; Ju, J.; et al. Neoabyssomicins A–C, Polycyclic Macrolactones from the Deep-Sea Derived Streptomyces Koyangensis SCSIO 5802. *Tetrahedron* **2017**, *73*, 5366–5372. [CrossRef]
206. Eliwa, E.M.; Abdel-Razek, A.S.; Frese, M.; Halawa, A.H.; El-Agrody, A.M.; Bedair, A.H.; Sewald, N.; Shaaban, M. New naturally occurring phenolic derivatives from marine Nocardiopsis sp. AS23C: Structural elucidation and in silico computational studies. *Vietnam J. Chem.* **2019**, *57*, 164–174. [CrossRef]
207. Tangjitjaroenkun, J.; Pluempanupat, W.; Tangchitcharoenkhul, R.; Yahayo, W.; Supabphol, R. Antibacterial, antioxidant, cytotoxic effects and GC-MS analysis of mangrove-derived Streptomyces achromogenes TCH4 extract. *Arch. Biol. Sci.* **2021**, *73*, 223–235. [CrossRef]
208. Asnani, A.; Purwanti, A.; Bakrudin, W.A.; Anjarwati, D.U. The Production of Streptomyces W-5B Extract for Antibiofilm against Methicillin-resistant Staphylococcus aureus. *J. Pure Appl. Microbiol.* **2022**, *16*, 337–346. [CrossRef]
209. Devi, N.A. Isolation and Identification of Marine Actinomycetes and their Potential in Antimicrobial Activity. *Pak. J. Biol. Sci.* **2006**, *9*, 470–472. [CrossRef]
210. Cao, D.T.; Tran, V.H.; Vu, V.N.; Mai, H.D.T.; Le, T.H.M.; Vu, T.Q.; Nguyen, H.H.; Chau, V.M.; Pham, V.C. Antimicrobial metabolites from a marine-derived Actinomycete Streptomyces sp. G278. *Nat. Prod. Res.* **2018**, *33*, 3223–3230. [CrossRef]
211. Asolkar, R.N.; Kirkland, T.N.; Jensen, P.; Fenical, W. Arenimycin, an antibiotic effective against rifampin- and methicillin-resistant Staphylococcus aureus from the marine actinomycete Salinispora arenicola. *J. Antibiot.* **2009**, *63*, 37–39. [CrossRef] [PubMed]
212. Sandoval-Powers, M.; Králová, S.; Nguyen, G.-S.; Fawwal, D.V.; Degnes, K.; Lewin, A.S.; Klinkenberg, G.; Wentzel, A.; Liles, M.R. Streptomyces Poriferorum Sp. Nov., a Novel Marine Sponge-Derived Actinobacteria Species Expressing Anti-MRSA Activity. *Syst. Appl. Microbiol.* **2021**, *44*, 126244. [CrossRef]
213. Dharmaraj, S.; Sumantha, A. Bioactive Potential of Streptomyces Associated with Marine Sponges. *World J. Microbiol. Biotechnol.* **2009**, *25*, 1971–1979. [CrossRef]
214. Cristianawati, O.; Sibero, M.T.; Ayuningrum, D.; Nuryadi, H.; Syafitri, E.; Radjasa, O.K.; Riniarsih, I. Screening of Antibacterial Activity of Seagrass-Associated Bacteria from the North Java Sea, Indonesia against Multidrug-Resistant Bacteria. *AACL Bioflux* **2019**, *12*, 1054–1064.
215. Antoniraj, A.; Anandan, V.; Ganesan, T.; Gunasingh, A. Isolation of Marine Actinomycetes Associated with the Carangid Fish Alepes Melanoptera, (Swainson, 1839) and an Evaluation of Their Antimicrobial Activity. *J. Microbiol.* **2018**, *20*, 235–247.

antibiotics

MDPI

Brief Report

Bactericidal Activity of Sodium Bituminosulfonate against *Staphylococcus aureus*

Elisa Heuser [1], Karsten Becker [1] and Evgeny A. Idelevich [1,2,*]

1 Friedrich Loeffler-Institute of Medical Microbiology, University Medicine Greifswald, Ferdinand-Sauerbruch-Straße 1, 17489 Greifswald, Germany; elisa.heuser@med.uni-greifswald.de (E.H.); karsten.becker@med.uni-greifswald.de (K.B.)

2 Institute of Medical Microbiology, University Hospital Münster, Domagkstraße 10, 48149 Münster, Germany

* Correspondence: evgeny.idelevich@med.uni-greifswald.de; Tel.: +49-3834-86-5563; Fax: +49-3834-86-5561

Abstract: Antibiotic resistance is increasing worldwide making it necessary to search for alternative antimicrobials. Sodium bituminosulfonate is a long-known substance, whose antimicrobial inhibitory activity has recently been re-evaluated. However, to the best of our knowledge, the bactericidal mode of action of this substance has not been systematically characterized. The aim of this study was to investigate the in vitro bactericidal activity of sodium bituminosulfonate by determining the minimal bactericidal concentrations (MBC), as well as the rapidity of bactericidal effect by time-kill curves. Clinical isolates of methicillin-susceptible (MSSA, $n = 20$) and methicillin-resistant (*mecA/mecC*-MRSA, $n = 20$) *Staphylococcus aureus* were used to determine MBC by a broth microdilution method. Sodium bituminosulfonate (Ichthyol® light) was tested in double-dilution concentration steps ranging from 0.03 g/L to 256 g/L. For time-kill analysis, two reference and two clinical *S. aureus* strains were tested with different concentrations of sodium bituminosulfonate (1 × minimal inhibitory concentration (MIC), 2 × MIC, 4 × MIC, 16 × MIC and 256 × MIC). For MSSA isolates, MBC_{50}, MBC_{90} and the MBC range were 0.5 g/L, 1.0 g/L and 0.125–1.0 g/L; $(MBC/MIC\ ratio)_{50}$, $(MBC/MIC\ ratio)_{90}$ and the range of the MBC/MIC ratio were 4, 4 and 1–8, respectively. Among MRSA isolates, MBC_{50}, MBC_{90} and the MBC range amounted to 0.5 g/L, 1.0 g/L and 0.06–1.0 g/L; $(MBC/MIC\ ratio)_{50}$, $(MBC/MIC\ ratio)_{90}$ and the range of the MBC/MIC ratio were 2, 4 and 1–8, respectively. Time-kill kinetics revealed a bactericidal effect after 30 min for sodium bituminosulfonate concentrations of 16 × MIC and 256 × MIC. The bactericidal activity against MSSA and MRSA was demonstrated for sodium bituminosulfonate. The killing was very rapid with the initial population reduced by 99.9% after only short incubation with concentrations of 16 × MIC and higher.

Keywords: sodium bituminosulfonate; *Staphylococcus aureus*; bactericidal activity

Citation: Heuser, E.; Becker, K.; Idelevich, E.A. Bactericidal Activity of Sodium Bituminosulfonate against *Staphylococcus aureus*. *Antibiotics* **2022**, *11*, 896. https://doi.org/10.3390/antibiotics11070896

Academic Editors: Valério Monteiro-Neto, Elizabeth S. Fernandes and Marc Maresca

Received: 23 May 2022
Accepted: 30 June 2022
Published: 5 July 2022

1. Introduction

The burden of bacterial resistance against antibiotics is steadily increasing [1,2]. The development of antimicrobial resistance (AMR) is driven by use of antibiotics in humans, animals and the environment, as well as the worldwide spread of resistant bacteria [3]. Although considerable improvements in the control of methicillin-resistant *Staphylococcus aureus* (MRSA) have been made in certain countries, in the WHO priority list of antibiotic-resistant bacteria, MRSA is still ranked as a high priority pathogen [4]. In 2019, MRSA caused more than 100,000 deaths and 3.5 million disability-adjusted life-years attributable to AMR [5]. Despite some progress achieved in the antibacterial pipeline in recent years, the pipeline outlook remains unfavorable [6,7]. In the context of a scarcity of new antimicrobial treatments, the re-examination of old substances with antimicrobial activity has potential value [8]. Such re-evaluation includes the collection of in vitro susceptibility data as well as consideration of clinical evidence, including pharmacokinetic/pharmacodynamic analyses [9,10]. Sodium bituminosulfonate, a long-known substance derived from sulfur-rich oil shale and commonly known as "Ichthyol®, light" [11], has been used to treat various conditions, particularly in dermatology, including skin infections, for nearly a century [12–14]. The in vitro antimicrobial activity of sodium bituminosulfonate has recently been re-evaluated according to current international guidelines on antimicrobial susceptibility testing (AST) [15]. To the best of our knowledge the first time this has been undertaken, we performed a systematic investigation and characterization of the bactericidal activity of sodium bituminosulfonate against *S. aureus*.

2. Results

2.1. Determination of MIC and MBC

Of 20 clinical MRSA strains, the *mecA* gene was detected in 19 strains, whereas one strain possessed the *mecC* gene. All MSSA isolates tested *mecA*/*mecC*-negative. For MSSA as well as for MRSA clinical isolates, the MIC_{50} and MIC_{90} values were 0.125 g/L and 0.25 g/L, respectively. The MIC range was 0.06–0.5 g/L for MSSA and 0.06–0.25 g/L for MRSA. The MBC_{50} and MBC_{90} values were equal for MSSA and MRSA and amounted to 0.5 g/L and 1.0 g/L, respectively (Table 1). MBC ranges were 0.125–1.0 g/L and 0.06–1.0 g/L for MSSA and MRSA, respectively. The $(MBC/MIC\ ratio)_{50}$, $(MBC/MIC\ ratio)_{90}$ and the range of the MBC/MIC ratio are shown in Table 1.

The MICs of vancomycin determined for *S. aureus* ATCC 29213 by general QC testing were within the ranges recommended by EUCAST and CLSI [15,16].

2.2. Time-Kill Curves

Time-kill kinetic investigations demonstrated a rapid bactericidal effect of sodium bituminosulfonate against MSSA and MRSA. Confirmation of the initial inoculum revealed a mean real bacterial density of 1.1×10^6 CFU/mL and 1.3×10^6 CFU/mL for the reference strains MSSA ATCC 29213 and MRSA ATCC 43300, respectively, 1×10^6 for clinical MSSA and 1.5×10^6 for clinical MRSA strains. At concentrations of 16 × MIC and 256 × MIC, a bactericidal efficacy of 99.9% inoculum reduction by sodium bituminosulfonate was already achieved after 30 min of incubation. Regrowth was observed at 1 × MIC, 2 × MIC and 4 × MIC within 4 to 24 h, whereas killing without regrowth was documented at 16 × MIC and 256 × MIC (Figure 1).

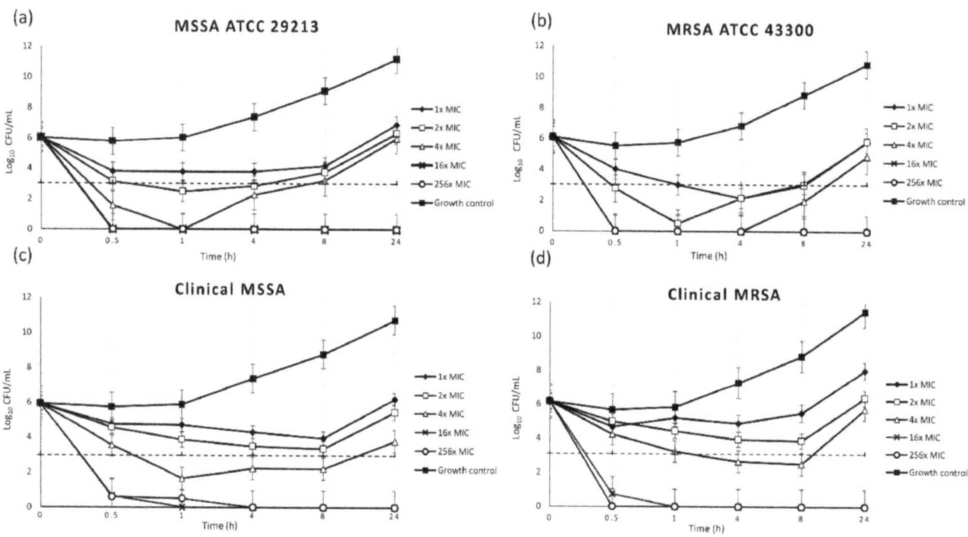

Figure 1. Time-kill curves for sodium bituminosulfonate against two reference strains (**a**,**b**) and two clinical strains (**c**,**d**) of *Staphylococcus aureus*. The threshold (dashed line) denotes 99.9% decreases in CFU/mL. MSSA, methicillin-susceptible *S. aureus*; MRSA, methicillin-resistant *S. aureus*; MIC, minimal inhibitory concentration.

Table 1. Bactericidal activity of sodium bituminosulfonate against methicillin-susceptible (MSSA, $n = 20$) and methicillin-resistant (MRSA, $n = 20$) *Staphylococcus aureus* isolates.

Strains	MIC_{50} (g/L)	MIC_{90} (g/L)	MIC Range (g/L)	MBC_{50} (g/L)	MBC_{90} (g/L)	MBC Range (g/L)	(MBC/MIC Ratio)$_{50}$	(MBC/MIC Ratio)$_{90}$	(MBC/MIC Ratio) Range
MSSA	0.125	0.25	0.06–0.5	0.5	1.0	0.125–1.0	4	4	1–8
MRSA	0.125	0.25	0.06–0.25	0.5	1.0	0.06–1.0	2	4	1–8

3. Discussion

S. aureus is one of the most common and important pathogens that can cause infections of skin and soft tissues, which are particularly difficult to treat when originating from methicillin-resistant strains [17]. Methicillin resistance in *S. aureus* is mediated by the *mecA, mecB* or *mecC* genes and leads to resistance to almost all β-lactam antibiotics [18–21]. Alternative antimicrobial substances may be important for the successful treatment of such infections or for the eradication of cutaneous colonization by *S. aureus*. Bituminosulfonate compounds have been known for over 100 years [22,23]. Data obtained from older studies—performed with various derivatives and formulations—showed good antimicrobial effects against Gram-positive bacteria, such as staphylococci and streptococci [24–26]. The activity of sodium bituminosulfonate against Gram-positive bacteria has recently been confirmed in an in vitro study [27]. Notably, another very recent study found only a low potential for resistance development in *S. aureus* after exposure to this derivate [28]. For comparison purposes, the antimicrobial inhibitory effect of sodium bituminosulfonate against *S. aureus* was also demonstrated in our study (Table 1). Its MIC values against 40 clinical and two reference *S. aureus* strains were found to be below the concentrations contained in commercially available preparations and the results were comparable to previous results [25–28]. The bactericidal efficacy of sodium bituminosulfonate was demonstrated by MBC determination. In agreement with international guidelines, a 99.9% killing of bacteria was considered as the criterion for a bactericidal effect [29]. To the best of our knowledge, no data are available on the rapidity of the bactericidal effect of sodium bituminosulfonate. Therefore, we characterized the bactericidal effect of sodium bituminosulfonate using a time-kill methodology [29,30]. In this first study of sodium bituminosulfonate's bactericidity, a very rapid bactericidal effect against MSSA and MRSA was observed, with 99.9% of the inoculum already killed after a short incubation time of 30 min by concentrations corresponding to 16 × MIC and 256 × MIC (Figure 1). Regrowth—a phenomenon occasionally observed with the time-kill methodology [31,32]—was noted at lower concentrations of 1 × MIC, 2 × MIC and 4 × MIC, while the higher concentrations of 16 × MIC and 256 × MIC caused irreversible killing. In conclusion, a bactericidal effect was demonstrated for sodium bituminosulfonate by MBC determination on a clinical collection of MSSA and MRSA strains. Detailed analyses of time-kill kinetics revealed very rapid bactericidal activity. This study contributes to the deeper in vitro characterization of the antimicrobial effects of sodium bituminosulfonate by investigating MBCs and time-kill curves under standardized conditions and has confirmed sodium bituminosulfonate as a promising alternative to commonly used topical antibiotics.

4. Materials and Methods

4.1. Bacterial Strains and Antimicrobial Substance

For the determination of minimal inhibitory concentrations (MICs) and minimal bactericidal concentrations (MBCs), 40 clinical isolates of *S. aureus* from routine diagnostics of the Friedrich Loeffler-Institute of Medical Microbiology, University Medicine Greifswald, in 2019–2021 were used, including 20 consecutive MSSA isolates and 20 consecutive MRSA isolates. The isolates were recovered from blood, urine, wounds, aspirates and respiratory samples. Only one isolate per patient was eligible. For time-kill experiments, two of these clinical strains (one MSSA and one MRSA) were used, as well as two reference strains (MSSA ATCC 29213 and MRSA ATCC 43300). The presence of *mec* genes in MRSA strains was confirmed using a loop-mediated isothermal amplification assay (eazyplex® MRSAplus, Amplex Diagnostics, Gars-Bahnhof, Germany) according to the user's manual [33].

Sodium bituminosulfonate (Ichthyol® light) was provided by the Ichthyol-Gesellschaft Cordes, Hermanni & Co. (Hamburg, Germany).

4.2. Determination of MIC and MBC

For MIC determination, a broth microdilution method was used according to the recommendations of the Clinical and Laboratory Standards Institute (CLSI) and the International Organization for Standardization (ISO) [15,34,35]. The strains were cultivated overnight on Columbia blood agar plates at 35 °C. Afterwards, colony material from the overnight culture was adjusted to the turbidity standard of McFarland 0.5 in 0.9% saline. The cultures were diluted in cation-adjusted Mueller–Hinton broth (CA-MHB, BD Diagnostics, Heidelberg, Germany) to obtain a final test inoculum of 5×10^5 CFU/mL. The testing was performed in sterile U-bottom microtiter plates (Brand, Wertheim, Germany) with an incubation time of 18 ± 2 h at 35 °C in ambient air. Sodium bituminosulfonate was tested in double-dilution concentration steps ranging from 0.03 g/L to 256 g/L. As no quality control (QC) ranges exist for the susceptibility testing of sodium bituminosulfonate, MIC determination was additionally performed for *S. aureus* ATCC 29213 with vancomycin to control for the overall performance of the testing procedures.

For the determination of MBCs, 10 µL samples from the clear wells were sub-cultured on tryptic soy agar plates (TSA, BD Diagnostics, Heidelberg, Germany) and colonies were counted after overnight incubation at 35 °C. The minimal concentration needed to kill at least 99.9% of the initial inoculum was considered as the MBC, according to CLSI [29]. All MIC and MBC determinations were performed in triplicate and median values were calculated for analysis.

4.3. Time-Kill Curves

To determine the rapidity of the bactericidal effect of sodium bituminosulfonate against *S. aureus*, a time-kill kinetic methodology was used. The initial liquid culture prepared in 10 mL tryptic soy broth (TSB) from overnight growth on Columbia blood agar was incubated for 3 h at 35 °C and 160 rpm. After adjusting to the 0.5 McFarland turbidity standard and dilution, 5 mL of a suspension containing approximately 10^6 CFU/mL was transferred into glass flasks with different concentrations of sodium bituminosulfonate (1 × MIC, 2 × MIC, 4 × MIC, 16 × MIC and 256 × MIC for the respective isolate). The suspensions were incubated at 35 °C and 160 rpm for 0.5 h, 1 h, 4 h, 8 h and 24 h. After incubation, 200-µL samples were collected and serially diluted, followed by plating of 10 µL in triplicate on TSA plates. After overnight incubation at 35 °C, the colonies were counted, and average values were calculated. A growth control without antimicrobial substance, as well as a sterile control, were used in each experiment. The time-kill curves were performed in triplicate.

Author Contributions: Conceptualization, E.A.I. and K.B.; methodology, E.A.I. and E.H.; investigation, E.H.; data curation, E.H.; writing—original draft preparation, E.H.; writing—review and editing, E.A.I. and K.B.; project administration, E.A.I. and K.B.; funding acquisition, K.B. All authors have read and agreed to the published version of the manuscript.

Funding: This research was funded by the Ichthyol-Gesellschaft Cordes, Hermanni & Co. (GmbH & Co. KG); Hamburg, Germany.

Institutional Review Board Statement: Not applicable.

Informed Consent Statement: Not applicable.

Data Availability Statement: The data for the study are available in the article.

Acknowledgments: The authors are thankful to Katrin Darm, Betty Nedow, Annemarie Mawick and Claudia Wiede for their technical assistance. Parts of the study were presented as a poster at the European Congress of Clinical Microbiology and Infectious Diseases (ECCMID) in 2021, online, (2977), and as an oral presentation at the Annual Meeting of the German Society for Hygiene and Microbiology (DGHM) in 2021, online (069/DKMV).

Conflicts of Interest: K.B. received a speaker fee from Ichthyol-Gesellschaft. E.H. and E.A.I. declare no conflict of interest.

References

1. Roca, I.; Akova, M.; Baquero, F.; Carlet, J.; Cavaleri, M.; Coenen, S.; Cohen, J.; Findlay, D.; Gyssens, I.; Heure, O.E.; et al. The global threat of antimicrobial resistance: Science for intervention. *New Microbes New Infect.* **2015**, *6*, 22–29. [CrossRef] [PubMed]
2. Cassini, A.; Högberg, L.D.; Plachouras, D.; Quattrocchi, A.; Hoxha, A.; Simonsen, G.S.; Colomb-Cotinat, M.; Kretzschmar, M.E.; Devleesschauwer, B.; Cecchini, M.; et al. Attributable deaths and disability-adjusted life-years caused by infections with antibiotic-resistant bacteria in the EU and the European Economic Area in 2015: A population-level modelling analysis. *Lancet Infect. Dis.* **2019**, *19*, 56–66. [CrossRef]
3. McEwen, S.A.; Collignon, P.J. Antimicrobial Resistance: A One Health Perspective. *Microbiol. Spectr.* **2018**, *6*. [CrossRef] [PubMed]
4. Tacconelli, E.; Carrara, E.; Savoldi, A.; Harbarth, S.; Mendelson, M.; Monnet, D.L.; Pulcini, C.; Kahlmeter, G.; Kluytmans, J.; Carmeli, Y.; et al. Discovery, research, and development of new antibiotics: The WHO priority list of antibiotic-resistant bacteria and tuberculosis. *Lancet Infect. Dis.* **2018**, *18*, 318–327. [CrossRef]
5. Murray, C.J.; Ikuta, K.S.; Sharara, F.; Swetschinski, L.; Aguilar, G.R.; Gray, A.; Han, C.; Bisignano, C.; Rao, P.; Wool, E.; et al. Global burden of bacterial antimicrobial resistance in 2019: A systematic analysis. *Lancet* **2022**, *399*, 629–655. [CrossRef]
6. Butler, M.S.; Gigante, V.; Sati, H.; Paulin, S.; Al-Sulaiman, L.; Rex, J.H.; Fernandes, P.; Arias, C.A.; Paul, M.; Thwaites, G.E.; et al. Analysis of the clinical pipeline of treatments for drug resistant bacterial infections: Despite progress, more action is needed. *Antimicrob. Agents Chemother.* **2022**, *63*, e01991-21. [CrossRef]
7. World Health Organization (Ed.) *2020 Antibacterial Agents in Clinical and Preclinical Development: An Overview and Analysis*; WHO: Geneva, Switzerland, 2020; ISBN 978-92-4-002130-3.
8. Zayyad, H.; Eliakim-Raz, N.; Leibovici, L.; Paul, M. Revival of old antibiotics: Needs, the state of evidence and expectations. *Int. J. Antimicrob. Agents* **2017**, *49*, 536–541. [CrossRef]
9. Theuretzbacher, U.; Paul, M. Revival of old antibiotics: Structuring the re-development process to optimize usage. *Clin. Microbiol. Infect.* **2015**, *21*, 878–880. [CrossRef]
10. Muller, A.E.; Theuretzbacher, U.; Mouton, J.W. Use of old antibiotics now and in the future from a pharmacokinetic/pharmacodynamic perspective. *Clin. Microbiol. Infect.* **2015**, *21*, 881–885. [CrossRef]
11. Wernicke, E.A. Schieferöl-Präparate. *Münchener Med. Wochenzeitschr.* **1936**, *13*, 522. (In German)
12. Korting, H.C.; Schöllmann, C.; Cholcha, W.; Wolff, L. Efficacy and tolerability of pale sulfonated shale oil cream 4% in the treatment of mild to moderate atopic eczema in children: A multicentre, randomized vehicle-controlled trial. *J. Eur. Acad. Dermatol. Venereol.* **2010**, *24*, 1176–1182. [CrossRef]
13. Beckert, S.; Warnecke, J.; Zelenkova, H.; Kovnerystyy, O.; Stege, H.; Cholcha, W.; Königsrainer, A.; Coerper, S. Efficacy of topical pale sulfonated shale oil in the treatment of venous leg ulcers: A randomized, controlled, multicenter study. *J. Vasc. Surg.* **2006**, *43*, 94–100. [CrossRef] [PubMed]
14. Weber, G. Klinische Erfahrungen mit Ichtholan T; therapeutische Mitteilung. *Arztl. Wochensch.* **1954**, *9*, 521–522. [PubMed]
15. Clinical and Laboratory Standards Institute (CLSI). *Performance Standards for Antimicrobial Susceptibility Testing*, 31st ed.; M100-S31; CLSI: Wayne, PA, USA, 2021.
16. The European Committee on Antimicrobial Susceptibility Testing. *Routine and Extended Internal Quality Control for MIC Determination and Disk Diffusion as Recommended by EUCAST*; Version 11.0; EUCAST: Växjö, Sweden, 2021; Available online: http://www.eucast.org (accessed on 1 September 2021).
17. Sunderkötter, C.; Becker, K. Frequent bacterial skin and soft tissue infections: Diagnostic signs and treatment. *J. Dtsch. Dermatol. Ges.* **2015**, *13*, 501–524. [CrossRef]
18. Becker, K.; van Alen, S.; Idelevich, E.A.; Schleimer, N.; Seggewiß, J.; Mellmann, A.; Kaspar, U.; Peters, G. Plasmid-Encoded Transferable *mecB*-Mediated Methicillin Resistance in Staphylococcus aureus. *Emerg. Infect. Dis.* **2018**, *24*, 242–248. [CrossRef]
19. Ubukata, K.; Nonoguchi, R.; Matsuhashi, M.; Konno, M. Expression and inducibility in *Staphylococcus aureus* of the *mecA* gene, which encodes a methicillin-resistant *S. aureus*-specific penicillin-binding protein. *J. Bacteriol.* **1989**, *171*, 2882–2885. [CrossRef] [PubMed]
20. Garcí a-Álvarez, L.; Holden, M.T.; Lindsay, H.; Webb, C.R.; Brown, D.F.; Curran, M.D.; Walpole, E.; Brooks, K.; Pickard, D.J.; Teale, C.; et al. Meticillin-resistant *Staphylococcus aureus* with a novel *mecA* homologue in human and bovine populations in the UK and Denmark: A descriptive study. *Lancet Infect. Dis.* **2011**, *11*, 595–603. [CrossRef]
21. Shore, A.C.; Deasy, E.C.; Slickers, P.; Brennan, G.; O'Connell, B.; Monecke, S.; Ehricht, R.; Coleman, D.C. Detection of staphylococcal cassette chromosome *mec* type XI carrying highly divergent *mecA*, *mecI*, *mecR1*, *blaZ*, and *ccr* genes in human clinical isolates

of clonal complex 130 methicillin-resistant *Staphylococcus aureus*. *Antimicrob. Agents Chemother.* **2011**, *55*, 3765–3773. [CrossRef] [PubMed]
22. Baumann, E.; Schotten, C. Über das "Ichthyol". *Mon. Prakt. Dermatol.* **1883**, *2*, 257–262. (In German)
23. Unna, P.G. Über Ichthyol. *Mon. Prakt. Dermatol.* **1897**, *25*, 533–539. (In German)
24. Latteux, P. Bakteriologische Untersuchungen, die antiseptischen Eigenschaften des Ichthyols betreffend. *Mon. Prakt. Dermatol.* **1892**, *14*, 389–397.
25. Pantke, R. Zur Kenntnis der Wirkung von Seefelder Schieferölprodukten auf Staphylococcen und Streptococcen (Action of Seefeld slate oil products on staphylococci and streptococci). *Arzneim. Forsch. Drug Res.* **1951**, *1*, 415–416. (In German)
26. Pantke, R. Bakteriologische Untersuchung von Arzneimitteln aus Schieferöl (Bacteriological studies of drugs from shale oil). *Arzneim. Forsch. Drug. Res.* **1965**, *15*, 570–573.
27. Idelevich, E.A.; Becker, K. In vitro activity of sodium bituminosulfonate: Susceptibility data for the revival of an old antimicrobial. *Microb. Drug Resist.* **2020**, *26*, 1405–1409. [CrossRef]
28. Blisse, M.; Idelevich, E.; Becker, K. Investigation of In-Vitro Adaptation toward Sodium Bituminosulfonate in *Staphylococcus aureus*. *Microorganisms* **2020**, *8*, 1962. [CrossRef]
29. Clinical and Laboratory Standards Institute (CLSI). *Methods for Determining Bactericidal Activity of Antimicrobial Agents*; Approved Guideline, CLSI Document M26-A; CLSI: Wayne, PA, USA, 1999.
30. Schleimer, N.; Kaspar, U.; Knaack, D.; von Eiff, C.; Molinaro, S.; Grallert, H.; Idelevich, E.A.; Becker, K. In Vitro Activity of the Bacteriophage Endolysin HY-133 against *Staphylococcus aureus* Small-Colony Variants and Their Corresponding Wild Types. *Int. J. Mol. Sci.* **2019**, *20*, 716. [CrossRef]
31. Tängdén, T.; Lundberg, C.V.; Friberg, L.E.; Huttner, A. How preclinical infection models help define antibiotic doses in the clinic. *Int. J. Antimicrob. Agents* **2020**, *56*, 106008. [CrossRef]
32. Kaspar, U.; Schleimer, N.; Idelevich, E.A.; Molinaro, S.; Becker, K. Exploration of Bacterial Re-Growth as In Vitro Phenomenon Affecting Methods for Analysis of the Antimicrobial Activity of Chimeric Bacteriophage Endolysins. *Microorganisms* **2022**, *10*, 445. [CrossRef]
33. Leikeim, R.S.M.; Kesselmeier, M.; Löffler, B.; Rödel, J.; Höring, S. Diagnostic accuracy and clinical impact of loop-mediated isothermal amplification for rapid detection of *Staphylococcus aureus* bacteremia: A retrospective observational study. *Eur. J. Clin. Microbiol. Infect. Dis.* **2020**, *39*, 679–688. [CrossRef]
34. Clinical and Laboratory Standards Institute (CLSI). *Methods for Dilution Antimicrobial Susceptibility Tests for Bacteria that Grow Aerobically, Approved Standard*, 11th ed.; M07-A11; CLSI: Wayne, PA, USA, 2018.
35. *ISO 20776-1:2006(E)*; Clinical Laboratory Testing and In Vitro Diagnostic Test Systems—Susceptibility Testing of Infectious Agents and Evaluation of Performance of Antimicrobial Susceptibility Test Devices—Part 1: Reference Method for Testing the In Vitro Activity of Antimicrobial Agents Against Rapidly Growing Aerobic Bacteria Involved in Infectious Diseases. ISO International Standard: Geneva, Switzerland, 2006.

antibiotics

Article

The Array of Antibacterial Action of Protocatechuic Acid Ethyl Ester and Erythromycin on Staphylococcal Strains

Maria Miklasińska-Majdanik *, Małgorzata Kępa, Monika Kulczak, Maciej Ochwat and Tomasz J. Wąsik

Department of Microbiology and Virology, Faculty of Pharmaceutical Sciences in Sosnowiec, Medical University of Silesia in Katowice, Jagiellońska 4, 41-200 Sosnowiec, Poland; mkepa@sum.edu.pl (M.K.); monikakulczak7@gmail.com (M.K.); maciekochwat@wp.pl (M.O.); twasik@sum.edu.pl (T.J.W.)

* Correspondence: mmiklasinska@sum.edu.pl; Tel.: +48-32-364-1621

Abstract: The spread of antibiotic resistance among bacteria has become one of the major health problems worldwide. Methicillin-resistant staphylococcal strains are especially dangerous because they are often resistant to other antibiotics. The increasing insensitivity to macrolides, lincosamides and streptogramin B antibiotics of methicillin-resistant staphylococcal isolates has limited the use of these drugs in therapy. The combination of natural compounds and antibiotics can be considered as an alternative tool to fight multi-drug-resistant pathogen infections. The aim of the presented study was to examine the antibacterial activity of protocatechuic acid ethyl ester–erythromycin combination towards *Staphylococcus aureus* and *Staphylococcus epidermidis* strains with various resistance profiles to methicillin and macrolides, lincosamides and streptogramin B (MLS_B) antibiotics. The in-vitro antibacterial potential of the above combination was investigated by minimum inhibitory concentration assays and checkerboard testing. The observed effects were strain dependent, with 8 of 12 tested staphylococcal strains showing an indifferent effect on the natural compound and erythromycin; for 2 strains, the tested combination had an additive effect, while for another 2, the effect was synergistic. Interestingly, the multi-drug-resistant strains were more sensitive to the cooperative action of the protocatechuic acid ethyl ester and the antibiotic.

Keywords: protocatechuic acid ethyl ester; erythromycin; fractional inhibitory concentration; *Staphylococcus* spp.

Citation: Miklasińska-Majdanik, M.; Kępa, M.; Kulczak, M.; Ochwat, M.; Wąsik, T.J. The Array of Antibacterial Action of Protocatechuic Acid Ethyl Ester and Erythromycin on Staphylococcal Strains. *Antibiotics* **2022**, *11*, 848. https://doi.org/10.3390/antibiotics11070848

Academic Editors: Valério Monteiro-Neto and Elizabeth S. Fernandes

Received: 26 May 2022
Accepted: 22 June 2022
Published: 24 June 2022

1. Introduction

New resistance mechanisms developed by bacteria have greatly reduced the number of available therapeutic agents effective against bacterial infections. Frequently, microorganisms isolated from hospitalized patients show resistance to more than one group of antibiotics. In recent years, multi-drug-resistant bacteria have become the major concern for global health, so the search for new antimicrobial agents is now a priority for researchers [1]. To date, the beneficial interactions of antibiotics with compounds of plant origin were suggested by many studies [2–4]. *Staphylococcus aureus* is one of the World Health Organization (WHO) priority pathogens for research on new antibiotics. This mainly commensal microorganism can induce diseases of the skin and soft tissues, respiratory, urinary or digestive system. What is more, it could form biofilms on medical devices or surgical sites. In turn, *Staphylococcus epidermidis*, which colonizes mucous membranes and skin, is one of the main etiological factors of nosocomial infections [2,5], and is a major element in the human body's microbiota [6]. Colonization with coagulase-negative staphylococci (CoNS) takes place at birth and accompanies us throughout all our life [7]. It is currently believed that this opportunistic pathogen is responsible for infections associated with medical accessories, such as central venous ports, catheters, hip prostheses, knee prostheses and various procedures, such as coronary artery bypass surgery, cholecystectomy, laminectomy, colon surgery and cesarean section [8,9]. Due to the ability of *S. epidermidis* to form biofilms on the surfaces of medical devices, they are one of the reservoirs of infection [8]. Because multi-drug-resistant staphylococci strains are common in both hospital and community environments, it is very important to search for alternative treatments in staphylococcal infections. Many studies proved the synergistic effect of plant-derived compounds in combination with antibiotics [2,7,10–12]. New drug combinations may lead to the development of new, cheaper, more accessible, more effective, safer-for-the-environment and, above all, for patients, antimicrobial therapies [1]. Protocatechuic acid (PCA) is one of the plant-derived compounds shown to have antimicrobial activity against Gram-positive, Gram-negative bacteria and fungi [13–16]. Its antimicrobial activity is related to the ability to inhibit bacterial growth and enhance the action of antibiotics and, thus, reduce resistance development [15]. What is more, PCA is non-toxic to humans at an oral dose of 100 mg/kg [15,17]. A usage limitation of phenolic acids, such as PCA as medicinal substances, is their low bioavailability. Moreover, phenolic acids are rapidly metabolized and excreted in the urine. Chemical modifications of phenolic acids may increase their biological activity, e.g., esterification increases the lipophilicity of PCA [18,19]. Figure 1 shows the differences in the chemical structure of PCA and protocatechuic acid ethyl ester (EDHB, ethyl 3–4 dihydroxybenzoate). EDHB has not yet been extensively studied for its antibacterial activity. This compound is found in the leaves and roots of many plant species, peanut seed casings and also in tea and wine [20]. EDHB is widely used as a food stabilizer [21] and its antioxidant, neuroprotective, myo- and cardioprotective properties were indicated by many authors [21–26]. Our previous study proved the antibacterial properties of EDHB against reference and clinical strains of *S. aureus* [2]. What is more, a decrease in the minimal inhibitory concentrations (MIC) of erythromycin in the presence of this ester was observed. Therefore, it seems important to thoroughly evaluate the interactions of EDHB with erythromycin and to define its potential clinical usefulness. Since erythromycin and PCA have different targets in the bacterial cell, the choice of this combination seems to be justified. Erythromycin inhibits bacterial protein synthesis, while PCA causes membrane lysis of bacteria [16,27,28]. PCA can, thus, enhance the antibacterial effects of erythromycin without fear of interference. Since the esterification of phenolic acid may increase it bioavailability, research on the antibacterial properties of EDHB is more promising than on PCA; therefore, the authors decided to focus on EDHB. The aim of this study was to determine the direction of the influence of EDHB and erythromycin on the reference and clinical strains of *S. aureus* and *S. epidermidis*.

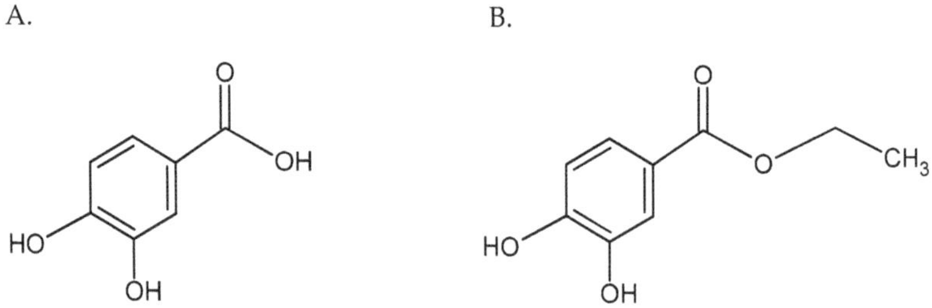

Figure 1. The chemical structure of protocatechuic acid (**A**) and protocatechuic acid ethyl ester (**B**).

2. Results

2.1. Antibiotic Resistance Profile of Tested Strains

The resistance profile of the tested strains is presented in Table 1. Among the tested *S. aureus* strains, three were resistant to methicillin (*S. aureus* ATCC 43300, *S. aureus* 3 and 4) and three strains of *S. epidermidis* were resistant to this antibiotic (*S. epidermidis* ATCC 35984, *S. epidermidis* 1 and 2). The remaining isolates showed sensitivity to methicillin (*S. aureus* ATCC 25923, *S. epidermidis* ATCC 12228, *S. aureus* 1 and 2, *S. epidermidis* 3 and 4). *S. aureus* ATCC 43300, *S. epidermidis* ATCC 35984, *S. aureus* 3, *S. aureus* 4, *S. epidermidis* 1 and *S. epidermidis* 2 strains demonstrated the constitutive phenotype of resistance to MLS_B antibiotics ($cMLS_B$). The remaining isolates did not show the MLS_B resistance phenotype.

Table 1. The methicillin and MLS_B resistance profiles of examined strains.

Strain	Methicillin Resistance Profile	MLS_B Resistance Profile
S. aureus ATCC 25923	MSSA	-
S. aureus ATCC 43300	MRSA	$cMLS_B$
S. epidermidis ATCC 12228	MSSE	-
S. epidermidis ATCC 35984	MRSE	$cMLS_B$
S. aureus 1	MSSA	-
S. aureus 2	MSSA	-
S. aureus 3	MRSA	$cMLS_B$
S. aureus 4	MRSA	$cMLS_B$
S. epidermidis 1	MRSE	$cMLS_B$
S. epidermidis 2	MRSE	$cMLS_B$
S. epidermidis 3	MSSE	-
S. epidermidis 4	MSSE	-

MSSA—methicillin-sensitive *Staphylococcus aureus*, MRSA—methicillin-resistant *Staphylococcus aureus*, MRSE—methicillin-resistant *Staphylococcus epidermidis*, MSSE—methicillin-resistant *Staphylococcus epidermidis*, $cMLS_B$—constitutive macrolide, lincosamide and streptogramin B mechanism of resistance.

2.2. The Fractional Inhibitory Concentration (FIC) Values for Protocatechuic Acid Ethyl Ester and Erythromycin against Staphylococcal Strains

The EDHB inhibited the growth of all the tested *S. aureus* strains, with MIC values ranging from 16 to 1024 μg/mL. The growth of the *S. aureus* reference strains was inhibited at a concentration of 512 μg/mL. *S. aureus* 1 and 2 strains proved to be sensitive to EDHB with MIC 16 μg/mL, while the growth of the *S. aureus* 3 and 4 strains was inhibited at a concentration of 1024 μg/mL. The MIC values for erythromycin against *S. aureus* isolates ranged from 0.25 to 2048 μg/mL. *S. aureus* ATCC 25925, *S. aureus* 1 and *S. aureus* 2 strains showed the lowest MIC values, 0.25 μg/mL, while *S. aureus* strains ATCC 25923, 3 and 4 demonstrated MICs at 2048 μg/mL.

EDHB at the concentrations used in this study inhibited the growth of all tested *S. epidermidis* strains. The MICs of EDHB ranged from 512 to 1024 μg/mL. Five *S. epidermidis* strains showed identical susceptibility to EDHB with MIC at 512 μg/mL. The highest MIC values at the level of 1024 μg/mL were characteristic for the *S. epidermidis* ATCC 12228 strain. The MIC values for erythromycin against *S. epidermidis* strains ranged from 0.125 to 2048 μg/mL. *S. epidermidis* ATCC 12228, 3 and 4 strains showed the lowest MIC values (0.125, 0.125 and 2 μg/mL, respectively), while *S. epidermidis* ATCC 35984, 1 and 2 strains demonstrated MICs at 2048 μg/mL. The MIC values for both EDHB and erythromycin are presented in Table 2.

Based on the checkerboard assay, MIC values were determined for EDHB in combination with erythromycin and for erythromycin in combination with EDHB.

Erythromycin and EDHB exerted an indifferent effect against five *S. aureus* strains. The synergistic effect of the compounds was noted only against *S. aureus* 3. The erythromycin–EDHB combination turned out to be more active against *S. epidermidis* strains, where in two cases, an additive effect was found (*S. epidermidis* 12228 and 3) and in one, it was synergistic (*S. epidermidis* 4). For the remaining strains, the effect of combining the compounds was indifferent.

Table 2. The MIC values for staphylococcal strains in the second stage of the study.

Staphylococcal Strain	MIC Values	
	EDHB	Erythromycin
S. aureus ATCC 25923	512	0.25
S. aureus ATCC 43300	512	2048
S. aureus 1	16	0.25
S. aureus 2	16	0.25
S. aureus 3	1024	2048
S. aureus 4	1024	2048
S. epidermidis ATCC 12228	1024	0.125
S. epidermidis ATCC 35984	512	2048
S. epidermidis 1	512	2048
S. epidermidis 2	512	2048
S. epidermidis 3	512	0.125
S. epidermidis 4	512	2

MIC—minimum inhibitory concentration, EDHB—protocatechuic acid ethyl ester.

The results of the checkerboard assay for each strain are shown in Figure S1 (Supplementary Materials), while the FIC index and their interpretation are presented in Table 3.

Table 3. FIC index and their interpretation.

Strain	FIC Index	Interacion
S. aureus ATCC 25923	1.031	indifference
S. aureus ATCC 43300	1.016	indifference
S. epidermidis ATCC 12228	0.628	additive
S. epidermidis ATCC 35984	1.063	indifference
S. aureus 1	2	indifference
S. aureus 2	1.125	indifference
S. aureus 3	0.078	synergism
S. aureus 4	1.016	indifference
S. epidermidis 1	1.015	indifference
S. epidermidis 2	1.015	indifference
S. epidermidis 3	0.750	additive
S. epidermidis 4	0.281	synergism

FIC index—fractional inhibitory concentration index.

Table 4 shows the MIC value changes of erythromycin in the presence of EDHB at different concentrations, together with MICs of EDHB. The EDHB concentrations were determined for each strain on the basis of the MIC values obtained in the first stage of the study [29,30]. The changes in the MIC value of erythromycin after the addition of EDHB were statistically significant ($p = 0.005$). Statistical analysis also revealed significant differences between MIC changes for resistant versus susceptible strains ($p = 0.002$). Interestingly, the strains resistant to MLS_B antibiotics and methicillin were more sensitive to the erythromycin–EDHB combination (Table 4).

Table 4. The MIC values of erythromycin alone and in combination with EDHB towards staphylococcal strains.

Strain	MIC of EDHB	EDHB Concentration in Well with FIC Index	MIC of Erythromycin	MIC of Erythromycin with EDHB	Decrease of the MIC Value after the EDHB Addition [%]
S. aureus ATCC 25923	512	16	0.25	0.25	0
S. aureus ATCC 43300	512	512	2048	32	98.44
S. epidermidis ATCC 12228	1024	512	0.125	0.016	12.8
S. epidermidis ATCC 35984	512	32	2048	1	99.95
S. aureus 1	16	16	0.25	0.25	0
S. aureus 2	16	16	0.25	0.03125	87.5
S. aureus 3	1024	64	2048	32	98.44
S. aureus 4	1024	1024	2048	32	98.44
S. epidermidis 1	512	512	2048	1	99.95
S. epidermidis 2	512	512	2048	32	99.95
S. epidermidis 3	512	128	0.125	0.063	50.4
S. epidermidis 4	512	32	2	0.5	75

EDHB—protocatechuic acid ethyl ester, MIC—minimum inhibitory concentration.

3. Discussion

Staphylococcal infections have become one of the most important public health problems, as multi-drug-resistant strains of this microbe are spreading rapidly. This fact stimulates scientists to search for new antimicrobial compounds and therapeutic strategies for staphylococcal disease treatment. A very promising direction of research is the implementation of substances of natural origin to augment routine antimicrobial therapies. The enhancement of antibiotic action by such compounds is due to sensitizing bacterial strains to drugs and enhancing their activity by increasing the bioavailability or simultaneously affecting a different site in the bacterial cell [4].

In this study, the antibacterial effect of the EDHB and erythromycin combination on reference and clinical *Staphylococcus* spp. strains was assessed. EDHB in combination with erythromycin showed an indifferent effect against five *S. aureus* isolates, while a synergistic interaction was found for one strain. Interestingly, a synergistic effect was noted against a multi-drug-resistant clinical strain, which, due to its character, should be less susceptible to the combined action of the compounds. In turn, among *S. epidermidis* strains, three indifferent, two additive and one synergistic effect were noted. A synergistic interaction was found against the sensitive *S. epidermidis* 4. It should be noted that if one substance is much more active, sometimes it is difficult to distinguish an indifferent effect from an additive effect, especially when using dilutions of the antibiotics [31]. The statistical analysis showed that the resistant strains of the tested staphylococci were more sensitive to the EDHB–erythromycin combination then sensitive isolates. Significant differences were observed in the decrease in erythromycin MIC values after the addition of EDHB (Table 4). Therefore, it is likely that EDHB blocks the mechanisms of bacterial resistance, thus, increasing its sensitivity.

What is more, the combination of "erythromycin–EDHB" turned out to be more effective against coagulase-negative staphylococci. Because *S. epidermidis* strains produce biofilm, which is the CoNSs' main virulence factor, evaluation of the effective combination of EDHB–erythromycin is of importance. The antibiofilm properties of PCA have been described in many studies [32–35]. The mechanism of PCA antibiofilm action is attributed to the changes in the properties of bacterial surfaces and inhibition of quorum sensing [32,34,35]. The reference *S. epidermidis* ATCC 35983 strain has a biofilm-forming ability and possesses *icaADBC* operon. The direction of the interaction of EDHB–erythromycin for this strain was indifferent. All tested clinical *S. epidermidis* isolates also have biofilm formation ability (data not shown). Among these strains, two indifferent, one synergistic and one additive interaction were noted. Since the EDHB–erythromycin combination has been shown to be more effective against the CoNS strains, it is possible that EDHB also affects the biofilm formation process. It is believed that natural compounds showing synergism with antibiotics could have potential use in the effective treatment of nosocomial infections caused by the CoNS, especially in cases requiring a non-standard pharmacological approach [3].

Previously, Miklasińska et al. [2] investigated the antibacterial effect of EDHB alone and in combination with four antibiotics: erythromycin, clindamycin, vancomycin and cefoxitin. Twenty clinical strains and three reference strains of *S. aureus* were tested. The MIC values for EDHB ranged from 64 to 1024 µg/mL, with median 512 µg/mL, while the MIC values for EDHB in this study ranged from 16 µg/mL to 1024 µg/mL (median 512 µg/mL) and were strain dependent. In our previous study, we found that the differences in the EDHB MIC values did not depend on the mechanism of resistance to MLS_B antibiotics [2]. What is more, we found out that the presence of EDHB increased the sensitivity of the studied strains to erythromycin, as well as to clindamycin and vancomycin. Among examined strains were two reference isolates (*S. aureus* ATCC 43300 and *S. aureus* ATCC 25923), which were also included in this work and showed an indifferent effect to the EDHB–erythromycin combination. For the *S. aureus* ATCC 43300 strain, both studies showed the same MIC value of 512 µg/mL, while for *S. aureus* 25923, the obtained MICs were different. In our previous study, it was 256 µg/mL, while in the present work, the MIC was 512 µg/mL. As both experiments were carried out on the same strains, stored in our department and with the use of a compounds purchased from the same company, the probable cause of the difference is a laboratory error. However, it should be noted that the above MIC values do not differ significantly, and these results do not affect the assessment of the antibacterial activity of EDHB against the tested strains. The results of both studies failed to unambiguously demonstrate whether the combination of erythromycin and EDHB would bring a noticeable therapeutic benefit but showed a tendency to decrease erythromycin resistance under the influence of the EDHB.

To the best of our knowledge, there are no studies, except ours, on the antibacterial activity of EDHB, but works on the antimicrobial potential of the PCA are worth discussing. Chai et al. [36] studied the action of PCA and chlorogenic acid in combination with antibiotics against *Escherichia coli*, *S. aureus*, *Streptococcus iniae* and *Proteus mirabilis*. The authors found that the growth of Gram-negative bacteria was inhibited to a lesser extent by PCA and chlorogenic acid than the growth of Gram-positive pathogens, and that the effect of PCA was more prominent than that of chlorogenic acid. Both of these acids showed the strongest antibacterial activity against *S. aureus*. Similarly, Stojković et al. [28], in their study, noted the highest PCA activity against the *S. aureus* strain compared with other Gram-positive and also Gram-negative bacteria. The MIC value of PCA against *S. aureus* in their study was 300 µg/mL, so it was close to the median obtained in our work for EDHB (512 µg/mL). Chai et al. [36] also studied an interaction between PCA and antibiotics. A synergistic effect against *S. aureus* was noted for the PCA with clinafloxacin and gatifloxacin, an additive effect for the combination with ciprofloxacin, and an indifferent effect for the combination with sulfamonomethoxine. Sanhueza et al. [37] also assessed the interactions between PCA and antibiotics with a different mechanism of action (oxacillin, ampicillin, nalidixic acid, ciprofloxacin, norfloxacin, levofloxacin, tetracycline and chloramphenicol) against five clinical and one reference strain of *S. aureus* (ATCC 6538). The analysis of the FIC index by the checkerboard method showed values below 0.5 for all tested combinations of antibiotics, phenolic compounds and bacterial strains, which pointed to a synergistic effect. The results of the above studies may suggest that the direction of EDHB's effect does not depend on the mechanism of action of the antibiotic. Erythromycin, used in our work, inhibits the synthesis of bacterial proteins by binding to the 50S ribosome subunit, while clinafloxacin, gatifloxacin, ciprofloxacin, nalidixic acid, norfloxacin and levofloxacin interfere with the synthesis of bacterial DNA. On the other hand, oxacillin and ampicillin interfere with the synthesis of the cell wall and tetracycline inhibits bacterial proteins synthesis by binding to the 30S ribosome subunit. Since PCA causes membrane lysis of bacteria [16,27,28], it is more probable that the synergistic effect is related to the stimulation of other sites in the bacterial cell. Because the influence of PCA–antibiotics combinations is strain dependent, it may be related to other bacterial resistance mechanisms. Mandalari et al. [38] evaluated the FIC values for catechin, PCA and epicatechin combinations against Gram-negative and Gram-positive bacteria. Further, in their study, the *S. aureus* strain turned out to be most susceptible to the examined compounds. An indifference towards synergism was observed between naringenin and PCA against *S. aureus*. Since naringenin affects the bacterial cell membrane [39], and PCA causes membrane lysis [16,27,28], the direction of the interaction may be justified by reinforcing one other's action.

The above studies and the results of the presented work indicate that natural compounds, including EDHB, can modify the antibacterial action of antibiotics against staphylococcal strains. The experiments showed a better effect from PCA on Gram-positive bacteria, in particular, *S. aureus*, and this should be the focus of future studies. As numerous works imply the antibacterial properties of protocatechuic acid and its chemical derivatives, such studies should be carried out with a greater number of strains to precisely evaluate the antimicrobial potential of phenolic compounds, with associations with antibiotics with different mechanisms of action. Efforts should also be directed to precisely determine the mechanism of action of EDHB on bacterial cells and its potential applications in the treatment of staphylococcal infections. This line of research may, in the future, provide a new, effective method of antibacterial therapy.

4. Materials and Methods

4.1. Bacterial Strains

The antibacterial activity of EDHB was assessed against four clinical *S. aureus* strains isolated (*S. aureus* 1, 2, 3 and 4) from clinical wound samples and four clinical *S. epidermids* strains (*S. epidermidis* 1, 2, 3 and 4) isolated from blood together with four reference strains: *S. aureus* ATCC 25923, *S. aureus* ATCC 43300, *S. epidermidis* ATCC 12228 and *S. epidermidis* ATCC 35984. The species of the clinical strains was confirmed by assessing their phenotypic features, such as: morphology of colonies grown on blood agar, type of hemolysis, growth on Chapman medium and production of coagulase and catalase. The Oxoid Staphytect Plus test and the API Staph test (bioMerieux, Marcy-l'Étoile, France) were also performed. To ensure that the clinical strains were identified correctly the PCR-RFLP reaction was carried out. GeneMATRIX Tissue & Bacterial DNA Purification KIT (EuRx Ltd., Gdańsk, Poland) was used for bacterial genomic DNA isolation [40]. The PCR reaction was performed using 10× PCR RED master mix kit (BLIRT SA, Gdańsk, Poland) in a MJ Mini Personal Thermal Cycler (Bio-Rad, Hercules, CA, USA). To confirm the classification of species, the specific restriction profiles after cleaving of PCR products with 10 U of restriction enzymes XapI and Bsp143I (Fermentas, Vilnius, Lithuania) were analyzed. All the tested strains were stored in the Trypticase Soy Broth medium with 20% glycerol at −80 °C. EDHB used in this study was received from Sigma Chemical Co. (St. Louis, MO, USA) and dissolved in DMSO immediately prior to use.

4.2. Antibiotic Resistance Profile

The disc-diffusion method was used to assess the resistance profile of examined strains to methicillin, macrolides and lincosamides with use of antibiotic discs (EMAPOL) of cefoxitin (30 μg), clindamycin (2 μg) and erythromycin (15 μg) and Mueller–Hinton Agar (BTL) [31].

4.3. Susceptibility Testing of Staphylococcal Strains to Erythromycin and EDHB Using the Microdilution Method

The standard microdilution method in sterile 96-well polystyrene plates (FL Medical, Torreglia, Italy) was used to determine the minimum inhibitory concentrations of EDHB and erythromycin towards the staphylococcal strains [41,42]. Serial dilutions were made as follows: 11 wells of 96-well polystyrene plates were filled with Mueller–Hinton; in the next step, 100 µL of EDHB or erythromycin stock solution was added to the first well and mixed thoroughly, and subsequently 100 µL was transferred to the next and remaining wells in the same manner, and finally, from the last well 100 µL was removed. In the next step, 100 µL of the 0.5 McFarland bacterial suspension was added to the wells containing the EDHB and erythromycin dilutions. The organization of the titration plate is shown in Figure 2. The absorbance was assessed in wavelength λ = 595 nm by spectrophotometry (Thermo Electron Corp., Vantoa, Finland). The MIC is defined as the lowest compound concentration that yields no visible microorganism growth, and it indicates the resistance of bacteria to an antimicrobial agent and determines the potency of new antimicrobial agents [31,41]. All experiments were carried out in triplicate. The obtained MIC values were used to design a "checkerboard" to determine the FIC value.

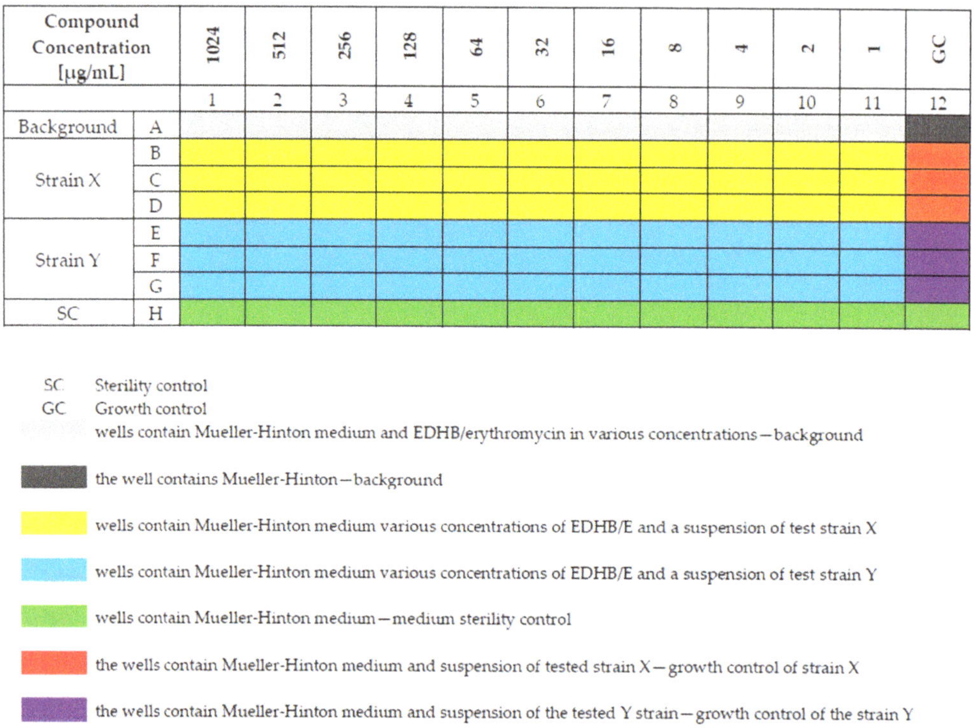

Figure 2. The organization of the titration plate.

4.4. Determination of the Susceptibility of Staphylococcal Strains to the Erythromycin and EDHB Combination

The susceptibility of staphylococcal strains to the combination of erythromycin and EDHB was assessed by determining the fractional inhibitory concentration (FIC) value for each strain. The checkerboard microdilution method with modifications was used to determine the total susceptibility effect of the tested strains [35,36]. Briefly, the erythromycin and EDHB solutions corresponding to an MIC value of 8 were prepared. Then, a series of 1/8 MIC dilutions was made. The MIC values of the substances were determined in the previous step. As such, 95 µL of dual-concentrated Mueller–Hinton medium was added to each well of the titration plate. Then, 50 µL of EDHB and appropriate concentration of erythromycin were added. Finally, 5 µL of a 0.5 McFarland staphylococcal bacterial suspension was added. The volume of each well was 200 µL. Therefore, the solutions corresponding to the concentrations of 8 MIC, 4 MIC, 2 MIC, MIC, 1/2 MIC, 1/4 MIC and 1/8 MIC were prepared for erythromycin and EDHB to take into account the dilution of

50 μL of these substances in 200 μL of the solution and obtained as follows: 2 MIC, MIC, 1/2 MIC, 1/4 MIC, 1/8 MIC, 1/16 MIC and 1/32 MIC in each of the wells. To the last column and row, respectively, instead of EDHB and erythromycin, 50 μL of medium was added. The resulting "checkerboard" is shown in Figure 3.

	1	2	3	4	5	6	7	8	9	10	11	12
A	E 2 MIC	E 1 MIC	E 1/2 MIC	E 1/4 MIC	E 1/8 MIC	E 1/16 MIC	E 1/32 MIC	E 0	B	SC		
	EDHB 2 MIC	EDHB 2 MIC	EDHB 2 MIC	EDHB 2 MIC	EDHB 2 MIC	EDHB 2 MIC	EDHB 2 MIC	EDHB 2 MIC	B	SC		
B	E 2 MIC	E 1 MIC	E 1/2 MIC	E 1/4 MIC	E 1/8 MIC	E 1/16 MIC	E 1/32 MIC	E 0	B	SC		
	EDHB 1 MIC	EDHB 1 MIC	EDHB 1 MIC	EDHB 1 MIC	EDHB 1 MIC	EDHB 1 MIC	EDHB 1 MIC	EDHB 1 MIC	B	SC		
C	E 2 MIC	E 1 MIC	E 1/2 MIC	E 1/4 MIC	E 1/8 MIC	E 1/16 MIC	E 1/32 MIC	E 0	B	SC		
	EDHB 1/2 MIC	EDHB 1/2 MIC	EDHB 1/2 MIC	EDHB 1/2 MIC	EDHB 1/2 MIC	EDHB 1/2 MIC	EDHB 1/2 MIC	EDHB 1/2 MIC	B	SC		
D	E 2 MIC	E 1 MIC	E 1/2 MIC	E 1/4 MIC	E 1/8 MIC	E 1/16 MIC	E 1/32 MIC	E 0	B	SC		
	EDHB 1/4 MIC	EDHB 1/4 MIC	EDHB 1/4 MIC	EDHB 1/4 MIC	EDHB 1/4 MIC	EDHB 1/4 MIC	EDHB 1/4 MIC	EDHB 1/4 MIC	B	SC		
E	E 2 MIC	E 1 MIC	E 1/2 MIC	E 1/4 MIC	E 1/8 MIC	E 1/16 MIC	E 1/32 MIC	E 0	B	SC		
	EDHB 1/8 MIC	EDHB 1/8 MIC	EDHB 1/8 MIC	EDHB 1/8 MIC	EDHB 1/8 MIC	EDHB 1/8 MIC	EDHB 1/8 MIC	EDHB 1/8 MIC	B	SC		
F	E 2 MIC	E 1 MIC	E 1/2 MIC	E 1/4 MIC	E 1/8 MIC	E 1/16 MIC	E 1/32 MIC	E 0	B	SC		
	EDHB 1/16 MIC	EDHB 1/16 MIC	EDHB 1/16 MIC	EDHB 1/16 MIC	EDHB 1/16 MIC	EDHB 1/16 MIC	EDHB 1/16 MIC	EDHB 1/16 MIC	B	SC		
G	E 2 MIC	E 1 MIC	E 1/2 MIC	E 1/4 MIC	E 1/8 MIC	E 1/16 MIC	E 1/32 MIC	E 0	B	SC		
	EDHB 1/32 MIC	EDHB 1/32 MIC	EDHB 1/32 MIC	EDHB 1/32 MIC	EDHB 1/32 MIC	EDHB 1/32 MIC	EDHB 1/32 MIC	EDHB 1/32 MIC	B	SC		
H	E 2 MIC	E 1 MIC	E 1/2 MIC	E 1/4 MIC	E 1/8 MIC	E 1/16 MIC	E 1/32 MIC	E 0 (GC)	B	SC		
	EDHB 0	EDHB 0	EDHB 0	EDHB 0	EDHB 0	EDHB 0	EDHB 0	EDHB 0 (GC)	B	SC		

Figure 3. The scheme of the checkerboard assay to evaluate FIC index for tested strains. B—medium + solvent (background). SC—sterility control of the medium (sterility control). The first line of each row (A–H)—erythromycin (E) dilutions, the darkest color shows the lowest dilution (A1–H1), and the lightest shows the highest dilution (A8–H8). The second line of each row (A–H)—EDHB (protocatechuic acid ethyl ester) concentrations, the darkest color indicates the lowest dilution (A1–A8), while the lightest the highest (H1–H9).

The prepared plates were incubated at 37 °C for 24 h. The absorbances were then read at 595 nm. The percent increase in individual wells in relation to the growth control was calculated using the formula:

$$\text{GROWTH} = (\text{A well} - \text{A background})/(\text{A growth control} - \text{A background}) \times 100\% \quad (1)$$

where A is the absorbance. The MIC was re established for EDHB and erythromycin for each strain. Then the FIC [35,36] was calculated for EDHB and erythromycin and their sum for each well. The following formula was used:

$$\text{FIC index} = \text{FICA} + \text{FICB} = (\text{MICA} + \text{B})/\text{MICA} + (\text{MICB} + \text{A})/\text{MICB} \quad (2)$$

where:

MICA + B—MIC of an antibiotic in the presence of a polyphenolic compound.

MICB + A—MIC of a polyphenolic compound in the presence of an antibiotic.

MICA—MIC of the antibiotic alone.

MICB—MIC of a polyphenolic compound alone.

Many different combinations are observed in the checkerboard test; therefore, only the FIC values of the most efficient combination of compounds are used to calculate the FICI [35,36]. Based on the FIC index value for each strain, the relationship between EDHB and erythromycin was assessed according to the following scale:

FIC $\leq$ 0.5—means synergism;

0.5 < FIC $\leq$ 1—means additive effect;

1 < FIC $\leq$ 4—means indifference;

FIC > 4—means antagonism [1].

A synergistic effect can be found when the joint effect of the substances is greater than the sum of the individual effects. An additive effect can be observed when the sum of the effects of the substances themselves is equal to the joint effect. A neutral effect is characterized by a lack of interaction between the compounds [29]. An antagonistic effect is defined as a decreased collective interaction of the compounds compared to the interaction of the compounds themselves [31].

4.5. *Statistical Analysis*

Wilcoxon signed-rank test was used to determine whether the differences in the MIC values of erythromycin after the addition of EDHB were statistically significant. To assess the relationship between the presence of resistance mechanisms and the change in erythromycin MIC values following the addition of EDHB for a given strain the t-student test was used. For all used tests $p \leq 0.05$ was considered as statistically significant. The data were analyzed with the use of STATISTICA v 13.0 software (StatSoft, Krakow, Poland).

5. Conclusions

The results of this study demonstrate that multi-drug-resistant strains turned out to be more sensitive to the combination of antibiotics and EDHB than sensitive isolates. The combination of "erythromycin–EDHB" was more effective against coagulase-negative staphylococci. The in vitro additive effect and synergy of EDHB and erythromycin can indicate that EDHB is capable of augmenting the antimicrobial potential of antibiotics in vivo, but since this effect is strain dependent, further studies are necessary to evaluate the exact mechanisms of action of protocatechuic acid and EDHB.

Supplementary Materials: The following supporting information can be downloaded at: https://www.mdpi.com/article/10.3390/antibiotics11070848/s1, Figure S1: The checkerboard assay results for staphylococcal strains. The dark grey represents greater than 50% bacterial growth in a well, while the light grey shows smaller than 50% bacterial growth in a well compared to the growth control. The blue represents an indifferent, pink additive, while green synergistic interaction. The red framed box represents FIC index for each strain. The values highlighted in yellow show the MICs for EDHB and erythromycin.

Author Contributions: Conceptualization: T.J.W., M.M.-M., M.K. (Małgorzata Kępa); Methodology: M.M.-M., M.K. (Małgorzata Kępa), M.K. (Monika Kulczak), M.O.; Validation: T.J.W., M.M.-M., M.K. (Małgorzata Kępa); Formal analysis: T.J.W., M.M.-M., M.K. (Małgorzata Kępa); Investigation: M.K. (Monika Kulczak), M.O.; Data curation: M.M.-M., M.K. (Małgorzata Kępa), M.K. (Monika Kulczak), M.O.; Writing—original draft preparation: M.M.-M., M.K. (Małgorzata Kępa), T.J.W.; Writing—review and editing: T.J.W.; Supervision: T.J.W.; Funding acquisition: M.M.-M., T.J.W. All authors have read and agreed to the published version of the manuscript.

Funding: This research was funded by the Medical University of Silesia in Katowice, Poland, grant number PCN-2-087/N/0/I.

Institutional Review Board Statement: Not applicable.

Informed Consent Statement: Not applicable.

Data Availability Statement: Not applicable.

Conflicts of Interest: All authors declare no competing interests.

References

1. Mazur, P.; Skiba-Kurek, I.; Mrowiec, P.; Karczewska, E.; Drożdż, R. Synergistic Ros-Associated Antimicrobial Activity of Silver Nanoparticles and Gentamicin against *Staphylococcus epidermidis*. *Int. J. Nanomed.* **2020**, *15*, 3551–3562. [CrossRef] [PubMed]
2. Miklasińska, M.; Kepa, M.; Wojtyczka, R.D.; Idzik, D.; Zdebik, A.; Orlewska, K.; Wasik, T.J. Antibacterial Activity of Protocatechuic Acid Ethyl Ester on *Staphylococcus aureus* Clinical Strains Alone and in Combination with Antistaphylococcal Drugs. *Molecules* **2015**, *20*, 13536–13549. [CrossRef] [PubMed]
3. Kępa, M.; Miklasińska-Majdanik, M.; Wojtyczka, R.D.; Idzik, D.; Korzeniowski, K.; Smoleń-Dzirba, J.; Wasik, T.J. Antimicrobial Potential of Caffeic Acid against *Staphylococcus aureus* Clinical Strains. *Biomed Res. Int.* **2018**, *2018*, 7413504. [CrossRef] [PubMed]
4. El Atki, Y.; Aouam, I.; El Kamari, F.; Taroq, A.; Nayme, K.; Timinouni, M.; Lyoussi, B.; Abdellaoui, A. Antibacterial Activity of Cinnamon Essential Oils and Their Synergistic Potential with Antibiotics. *J. Adv. Pharm. Technol. Res.* **2019**, *10*, 63–67. [CrossRef]
5. Ghassemi, A.; Farhangi, H.; Badiee, Z.; Banihashem, A.; Mosaddegh, M.R. Evaluation of Nosocomial Infection in Patients at Hematology-Oncology Ward of Dr. Sheikh Children's Hospital. *Iran. J. Pediatr. Hematol. Oncol.* **2015**, *5*, 179–185.
6. Vuong, C.; Otto, M. *Staphylococcus epidermidis* infections. *Microbes Infect.* **2002**, *4*, 481–489. [CrossRef]
7. Wojtyczka, R.D.; Kępa, M.; Idzik, D.; Kubina, R.; Kabała-Dzik, A.; Dziedzic, A.; Wąsik, T.J. In Vitro Antimicrobial Activity of Ethanolic Extract of Polish Propolis against Biofilm Forming *Staphylococcus epidermidis* Strains. *Evid. Based Complement. Altern. Med.* **2013**, *2013*, 590703. [CrossRef]
8. European Centre for Disease Prevention and Control. *Multidrug-Resistant Staphylococcus Epidermidis*; ECDC: Stockholm, Sweden, 2018; pp. 1–6.
9. European Centre for Disease Prevention and Control. *Healthcare-Associated Infections: Surgical Site Infections—Annual Epidemiological Report for 2016*; ECDC: Stockholm, Sweden, 2018; pp. 1–15.
10. Gadisa, E.; Weldearegay, G.; Desta, K.; Tsegaye, G.; Hailu, S.; Jote, K.; Takele, A. Combined Antibacterial Effect of Essential Oils from Three Most Commonly Used Ethiopian Traditional Medicinal Plants on Multidrug Resistant Bacteria. *BMC Complement. Altern. Med.* **2019**, *19*, 24. [CrossRef]
11. Wojtyczka, R.D.; Dziedzic, A.; Kępa, M.; Kubina, R.; Kabała-Dzik, A.; Mularz, T.; Idzik, D. Berberine enhances the antibacterial activity of selected antibiotics against coagulase-negative staphylococcus strains in vitro. *Molecules* **2014**, *19*, 6583–6596. [CrossRef]
12. Kyaw, B.M.; Arora, S.; Lim, C.S. Bactericidal Antibiotic-Phytochemical Combinations against Methicillin Resistant *Staphylococcus aureus*. *Braz. J. Microbiol.* **2012**, *43*, 938–945. [CrossRef]
13. Institute of Medicine (US) Forum on Microbial Threats. *Antibiotic Resistance: Implications for Global Health and Novel Intervention Strategies: Workshop Summary*; National Academies Press: Washington, DC, USA, 2010; pp. 3–17.
14. Semaming, Y.; Pannengpetch, P.; Chattipakorn, S.C.; Chattipakorn, N. Pharmacological Properties of Protocatechuic Acid and Its Potential Roles as Complementary Medicine. *Evid. Based Complement. Altern. Med.* **2015**, *2015*, 593902. [CrossRef] [PubMed]
15. Kakkar, S.; Bais, S. A Review on Protocatechuic Acid and Its Pharmacological Potential. *ISRN Pharmacol.* **2014**, *2014*, 952943. [CrossRef] [PubMed]
16. Khan, A.K.; Rashid, R.; Fatima, N.; Mahmood, S.; Mir, S.; Khan, S.; Jabeen, N.; Murtaza, G. Pharmacological Activities of Protocatechuic Acid. *Acta Pol. Pharm.* **2015**, *72*, 643–650. [PubMed]
17. Abouaitah, K.; Piotrowska, U.; Wojnarowicz, J.; Swiderska-Sroda, A.; El-Desoky, A.H.H.; Lojkowski, W. Enhanced Activity and Sustained Release of Protocatechuic Acid, a Natural Antibacterial Agent, from Hybrid Nanoformulations with Zinc Oxide Nanoparticles. *Int. J. Mol. Sci.* **2021**, *22*, 5287. [CrossRef]
18. Torrisi, C.; Malfa, G.A.; Acquaviva, R.; Castelli, F.; Sarpietro, M.G. Effect of Protocatechuic Acid Ethyl Ester on Biomembrane Models: Multilamellar Vesicles and Monolayers. *Membranes* **2022**, *12*, 283. [CrossRef]

19. Reis, B.; Martins, M.; Barreto, B.; Millhazes, N.; Garrido, E.M.; Silva, P.; Garrido, J.; Borges, F. Structure—Property—Activity Relationship of Phenolic Acids and Derivatives. Protocatechuic Acid Alkyl Esters. *J. Agric. Food Chem.* **2010**, *58*, 6986–6993. [CrossRef]
20. Li, W.; Hou, G.; Zhou, D.; Lou, X.; Xu, Y.; Liu, S.; Zhao, X. The Roles of AKR1C1 and AKR1C2 in Ethyl-3,4-Dihydroxybenzoateinduced Esophageal Squamous Cell Carcinoma Cell Death. *Oncotarget* **2016**, *7*, 21542–21555. [CrossRef]
21. Han, B.; Li, W.; Sun, Y.; Zhou, L.; Xu, Y.; Zhao, X. A Prolyl-Hydroxylase Inhibitor, Ethyl-3,4-Dihydroxybenzoate, Induces Cell Autophagy and Apoptosis in Esophageal Squamous Cell Carcinoma Cellsvia up-Regulation of BNIP3 and N-Myc Downstream-Regulated Gene-1. *PLoS ONE* **2014**, *9*, e107204. [CrossRef]
22. Nimker, C.; Kaur, G.; Revo, A.; Chaudhary, P.; Bansal, A. Ethyl 3,4-Dihydroxy Benzoate, a Unique Preconditioning Agent for Alleviating Hypoxia-Mediated Oxidative Damage in L6 Myoblasts Cells. *J. Physiol. Sci.* **2015**, *65*, 77–87. [CrossRef]
23. Merkl, R.; Hrádková, I.; Filip, V.; Šmidrkal, J. Antimicrobial and Antioxidant Properties of Phenolic Acids Alkyl Esters. *Czech. J. Food Sci.* **2010**, *28*, 275–279. [CrossRef]
24. Philipp, S.; Cui, L.; Ludolph, B.; Kelm, M.; Schulz, R.; Cohen, M.V.; Downey, J.M. Desferoxamine and Ethyl-3,4-Dihydroxybenzoate Protect Myocardium by Activating NOS and Generating Mitochondrial ROS. *Am. J. Physiol. Hear. Circ. Physiol.* **2006**, *290*, 450–457. [CrossRef] [PubMed]
25. Nandan, D.; Clarke, E.P.; Ball, E.H.; Sanwal, B.D. Ethyl-3,4-Dihydroxybenzoate Inhibits Myoblast Differentiation: Evidence for an Essential Role of Collagen. *J. Cell Biol.* **1990**, *110*, 1673–1679. [CrossRef] [PubMed]
26. Wang, H.; Huo, X.; Chen, H.; Li, B.; Liu, J.; Ma, W.; Wang, X.; Xie, K.; Yu, Y.; Shi, K. Hydrogen-Rich Saline Activated Autophagy via HIF-1 α Pathways in Neuropathic Pain Model. *Biomed. Res. Int.* **2018**, *2018*, 4670834. [CrossRef] [PubMed]
27. Chao, C.Y.; Yin, M.C. Antibacterial effects of roselle calyx extracts and protocatechuic acid in ground beef and apple juice. *Foodborne Pathog. Dis.* **2009**, *6*, 201–206. [CrossRef]
28. Stojković, D.S.; Zivković, J.; Soković, M.; Glamočlija, J.; Ferreira, I.C.; Janković, T.; Maksimović, Z. Antibacterial activity of *Veronica montana* L. extract and of protocatechuic acid incorporated in a food system. *Food Chem. Toxicol.* **2013**, *55*, 209–213. [CrossRef] [PubMed]
29. Bassolé, I.H.N.; Juliani, H.R. Essential oils in combination and their antimicrobial properties. *Molecules* **2012**, *17*, 3989–4006. [CrossRef]
30. Fratini, F.; Mancini, S.; Turchi, B.; Friscia, E.; Pistelli, L.; Giusti, G.; Cerri, D. A novel interpretation of the Fractional Inhibitory Concentration Index: The case *Origanum vulgare* L. and *Leptospermum scoparium* JR et G. Forst essential oils against Staphylococcus aureus strains. *Microbiol. Res.* **2017**, *195*, 11–17. [CrossRef]
31. European Committee for Antimicrobial Susceptibility Testing (EUCAST) of the European Society of Clinical Microbiology and Infectious Diseases (ESCMID). Terminology relating to methods for the determination of susceptibility of bacteria to antimicrobial agents. *Clin. Microbiol. Infect.* **2000**, *6*, 503–508. [CrossRef]
32. Mostafa, I.; Abbas, H.A.; Ashour, M.L.; Yasri, A.; El-Shazly, A.M.; Wink, M.; Sobeh, M. Polyphenols from Salix Tetrasperma Impair Virulence and Inhibit Quorum sensing of *Pseudomonas Aeruginosa*. *Molecules* **2020**, *25*, 1341. [CrossRef]
33. Bernal-Mercado, A.T.; Vazquez-Armenta, F.J.; Tapia-Rodriguez, M.R.; Islas-Osuna, M.A.; Mata-Haro, V.; Gonzalez-Aguilar, G.A.; Lopez-Zavala, A.A.; Ayala-Zavala, J.F. Comparison of Single and Combined Use of Catechin, Protocatechuic, and Vanillic Acids as Antioxidant and Antibacterial Agents against Uropathogenic *Escherichia Coli* at Planktonic and Biofilm Levels. *Molecules* **2018**, *23*, 2813. [CrossRef]
34. Bernal-Mercado, A.T.; Gutierrez-Pacheco, M.M.; Encinas-Basurto, D.; Mata-Haro, V.; Lopez-Zavala, A.A.; Islas-Osuna, M.A.; Gonzalez-Aguilar, G.A.; Ayala-Zavala, J.F. Synergistic mode of action of catechin, vanillic and protocatechuic acids to inhibit the adhesion of uropathogenic *Escherichia coli* on silicone surfaces. *J. Appl. Microbiol.* **2019**, *128*, 387–400. [CrossRef] [PubMed]
35. Srivastava, N.; Tiwari, S.; Bhandari, K.; Biswal, A.K.; Rawat, A.K.S. Novel derivatives of plant monomeric phenolics: Act as inhibitors of bacterial cell-to-cell communication. *Microb. Pathog.* **2020**, *141*, 103856. [CrossRef] [PubMed]
36. Chai, B.; Jiang, W.; Hu, M.; Wu, Y.; Si, H. In vitro synergistic interactions of Protocatechuic acid and Chlorogenic acid in combination with antibiotics against animal pathogens. *Synergy* **2019**, *9*, 6–11. [CrossRef]
37. Sanhueza, L.; Melo, R.; Montero, R.; Maisey, K.; Mendoza, L.; Wilkens, M. Synergistic interactions between phenolic compounds identified in grape pomace extract with antibiotics of different classes against *Staphylococcus aureus* and *Escherichia coli*. *PLoS ONE* **2017**, *12*, e0172273. [CrossRef] [PubMed]
38. Mandalari, G.; Bisignano, C.; D'Arrigo, M.; Ginestra, G.; Arena, A.; Tomaino, A.; Wickham, M.S. Antimicrobial potential of polyphenols extracted from almond skins. *Lett. Appl. Microbiol.* **2010**, *51*, 83–89. [CrossRef]
39. Wang, L.H.; Zeng, X.A.; Wang, M.S.; Brennan, C.S.; Gong, D. Modification of Membrane Properties and Fatty Acids Biosynthesis-Related Genes in Escherichia Coli and Staphylococcus Aureus: Implications for the Antibacterial Mechanism of Naringenin. *Biochim. Biophys. Acta Biomembr.* **2018**, *1860*, 481–490. [CrossRef]
40. Murakami, K.; Minamide, W.; Wada, K.; Nakamura, E.; Teraoka, H.; Watanabe, S. Identification of methicillin-resistant strains of Staphylococci by polymerase chain reaction. *J. Clin. Microbiol.* **1991**, *29*, 2240–2244. [CrossRef]
41. Amsterdam, D. Susceptibility Testing of Antimicrobials in Liquid Media. In *Antibiotics in Laboratory Medicine*, 5th ed.; Loman, V., Ed.; Williams and Wilkins: Philadelphia, PA, USA, 2005; pp. 61–143.
42. Devienne, K.F.; Raddi, M.S.G. Screening for antimicrobial activity of natural Products using a microplate photometer. *Braz. J. Microbiol.* **2002**, *33*, 166–168. [CrossRef]

antibiotics

Article

Ethyl Acetate Fraction of *Bixa orellana* and Its Component Ellagic Acid Exert Antibacterial and Anti-Inflammatory Properties against *Mycobacterium abscessus* subsp. *massiliense*

Roberval Nascimento Moraes-Neto [1], Gabrielle Guedes Coutinho [1], Ana Caroline Santos Ataíde [1], Aline de Oliveira Rezende [1], Camila Evangelista Carnib Nascimento [1], Rafaela Pontes de Albuquerque [1], Cláudia Quintino da Rocha [2], Adriana Sousa Rêgo [3], Maria do Socorro de Sousa Cartágenes [1], Ana Lúcia Abreu-Silva [1,4], Igor Victor Ferreira dos Santos [5], Cleydson Breno Rodrigues dos Santos [5], Rosane Nassar Meireles Guerra [1], Rachel Melo Ribeiro [1], Valério Monteiro-Neto [1], Eduardo Martins de Sousa [1,6] and Rafael Cardoso Carvalho [1,*]

1 Graduate Program in Health Sciences, Federal University of Maranhão—UFMA, São Luís 65080-805, MA, Brazil; roberval.moraes@discente.ufma.br (R.N.M.-N.); gabrielle.guedes@discente.ufma.br (G.G.C.); ana.ataide@discente.ufma.br (A.C.S.A.); aline.rezende@discente.ufma.br (A.d.O.R.); camila.carnib@ufma.br (C.E.C.N.); rafaela.pontes@hotmail.com (R.P.d.A.); cartagenes.maria@ufma.br (M.d.S.d.S.C.); anasilva1@professor.uema.br (A.L.A.-S.); rosane.guerra@ufma.br (R.N.M.G.); melo.rachel@ufma.br (R.M.R.); valerio.monteiro@ufma.br (V.M.-N.); eduardo.martins@ceuma.br (E.M.d.S.)

2 Graduate Program in Chemistry, Federal University of Maranhão—UFMA, São Luís 65080-805, MA, Brazil; rocha.claudia@ufma.br

3 Graduate Program in Health and Services Management, CEUMA University—UniCEUMA, São Luís 65075-120, MA, Brazil; adriana.dsousa@ceuma.br

4 Graduate Program in Animal Science, University of Maranhão State—UEMA, São Luís 65055-310, MA, Brazil

5 Graduate Program in Biotechnology and Biodiversity Network BIODSNORTE, Federal University of Amapá—UFAP, Macapá 68903-419, AP, Brazil; igorsantosvictor@gmail.com (I.V.F.d.S.); breno@unifap.br (C.B.R.d.S.)

6 Graduate Program in Microbial Biology, CEUMA University—UniCEUMA, São Luís 65075-120, MA, Brazil

* Correspondence: carvalho.rafael@ufma.br

Citation: Moraes-Neto, R.N.; Coutinho, G.G.; Ataíde, A.C.S.; de Oliveira Rezende, A.; Nascimento, C.E.C.; de Albuquerque, R.P.; da Rocha, C.Q.; Rêgo, A.S.; de Sousa Cartágenes, M.d.S.; Abreu-Silva, A.L.; et al. Ethyl Acetate Fraction of *Bixa orellana* and Its Component Ellagic Acid Exert Antibacterial and Anti-Inflammatory Properties against *Mycobacterium abscessus* subsp. *massiliense*. *Antibiotics* **2022**, *11*, 817. https://doi.org/10.3390/antibiotics11060817

Academic Editor: Carlos M. Franco

Received: 10 May 2022
Accepted: 15 June 2022
Published: 17 June 2022

Abstract: *Mycobacterium abscessus* subsp. *massiliense* (*Mabs*) causes chronic infections, which has led to the need for new antimycobacterial agents. In this study, we investigated the antimycobacterial and anti-inflammatory activities of the ethyl acetate fraction of *Bixa orellana* leaves (BoEA) and ellagic acid (ElAc). In silico analysis predicted that ElAc had low toxicity, was not mutagenic or carcinogenic, and had antimicrobial and anti-inflammatory activities. Apparently, ElAc can interact with COX2 and Dihydrofolate reductase (DHFR) enzymes, which could explain both activities. In vitro analysis showed that BoEA and ElAc exerted antimicrobial activity against *Mabs* (minimum inhibitory concentration of 1.56, 1.56 mg/mL and bactericidal concentration of 6.25, 3.12 mg/mL, respectively. Clarithromycin showed MIC and MBC of 1 and 6 μg/mL). Treatment with BoEA or ElAc increased survival of *Tenebrio molitor* larvae after lethal infection with *Mabs* and reduced carrageenan-induced paw edema in mice, around 40% of edema volume after the fourth hour, similarly to diclofenac. In conclusion, BoEA and ElAc exert antimicrobial effects against *Mabs* and have anti-inflammatory effects, making them potential sources of antimycobacterial drugs. The biological activities of ElAc may be due to its high binding affinities predicted for COX2 and DHFR enzymes.

Keywords: *Bixa orellana*; bioprospecting; in silico; *Tenebrio molitor*; anti-mycobacteria activity; anti-inflammatory activity

1. Introduction

Mycobacterium abscessus (*Mabs*) is a nontuberculous mycobacterium that has recently emerged as the causative agent for a wide spectrum of clinical manifestations, including pulmonary infections, mainly in patients with cystic fibrosis (95% of cases) [1]. This fast-growing microorganism is one of the most important drug-resistant mycobacterial species, owing to its intrinsic resistance to several classes of antibiotics [2]. The low permeability of mycobacterial cell wall plays a major role in this intrinsic antibiotic resistance [3,4]. Dormancy and latency causing increased tolerance to antimicrobial agents that are lethal to replicating bacilli, a phenomenon referred to as phenotypic drug resistance, and efflux pumps that actively transport many antibiotics out of the cell and are major contributors to the intrinsic resistance of mycobacteria to many drugs. Furthermore, the intrinsic resistance to several important antibiotics in mycobacteria can also be observed by modification of antibiotics (enzymatic degradation of antibiotics) and modification of target sites, for example, erythromycin resistance methylase (*erm*) gene [5].

Patterns of genetic susceptibility can be used as predictors of antibiotic efficacy, thereby guiding the choice of antimicrobials to be used in *Mabs* infections [6]. The molecular mechanism associated with macrolide resistance is associated to the expression of the erythromycin ribosome methylase (*erm*) gene. Other resistance mechanisms include the *rrs* gene, which is associated with resistance to aminoglycosides, class A beta-lactamase associated with resistance to majority of the beta-lactams (except cefoxitin and imipenem), enzymatic inactivation of majority of the tetracyclines (except tigecycline) [7], and *gyrA* and *gyrB* genes, which are associated with quinolone resistance [6].

In this context, the use of natural products (phytochemical compounds) to treat conditions caused by resistant microorganisms can be advantageous because of their wide range of biological activities, which make them potential sources for the development of new drugs, including antimycobacterial agents [8–11]. *Bixa orellana* is a plant that possesses anti-inflammatory, antioxidant, antinociceptive, anticonvulsant, cardioprotective, and antidiabetic properties. This plant species has also been considered a promising source for antimicrobial agents against several microorganisms, given its various biological properties, ethnopharmacological relevance, and widespread distribution [12,13].

Our group previously characterized the ethyl acetate fraction obtained from *B. orellana* leaf hydroalcoholic extract (BoEA) [14,15] and identified ellagic acid (ElAc) as one of the active compounds. ElAc exerts several pharmacological activities that can potentially be used for the treatment and prevention of cancer, diabetes, and cardiovascular and neurodegenerative diseases. Additionally, it also has nephroprotective, hepatoprotective, antiviral, and antiparasitic properties [16,17]. Previous studies from our group demonstrated the antimicrobial activity of BoEA against Gram-positive bacteria, *Staphylococcus aureus* and *Streptococcus pyogenes* [18] and its anti-inflammatory activity was confirmed in vivo by reduction in total leukocytes migration to the peritoneum in mice [14].

In this research, specifically, it is important to highlight that the use of ellagic acid was performed in a bioguided study. After *B. orellana* leaf collection and processing with the preparation of the crude extract, a lyophilized compound called BOHE (Hydroalcoholic Extract of *B. orellana*) was prepared. Then, the BOHE was subjected to liquid–liquid fractionation with hexane, chloroform, and ethyl acetate, producing hexane (BoHex), chloroform (BoCl), and ethyl acetate (BoEA) fractions. The BoHex and BoCl fractions were stored to be used in further studies and, the ethyl acetate fraction (BoEa) was used in this study. The BoEA was subjected to column chromatography, which identified 10 groups or subfractions, and for each subfraction in vitro assays against *Mabs* were performed. The best results were shown by subfraction or group 10. Then, using HPLC-PDA and FIA-ESI-IT/MS it was characterized that the group/subfraction 10, is ellagic acid, the compound that we directed our study.

Considering the biological properties and antimicrobial potential of *B. orellana* as well as the clinical relevance of *Mabs*, this study aimed to evaluate the anti-inflammatory and antimicrobial activities of the ethyl acetate fraction of *B. orellana* and ElAc against the microorganism in silico, in vitro, and in vivo. In addition to providing a strategy for alternative therapies for mycobacterial infections, this study aimed to contribute to the development of new drug formulations.

2. Results

2.1. Identification of Phytochemical Compounds–Profile of Fraction Groups

A total of 10 fraction groups (G1–G10) of different compositions were obtained using classical column chromatography. The results of each fraction group, as well as the representative spectra for each peak obtained in the ultraviolet (UV) region, are presented in Table 1.

Table 1. Data taken from high-performance liquid chromatography with photodiode array detector analyses of fractions obtained from *Bixa orellana* fractions.

Group	UV Max (nm)
G1 (mixture)	300; 325; 278
G2 (mixture)	300; 325; 278
G3 (mixture)	300; 325; 278
G4 (mixture)	300; 325; 278
G5 (mixture)	255; 267; 349
G6 (mixture)	253; 267; 345
G7 (mixture)	253; 267; 345
G8 (mixture)	253; 267; 345
G9 (mixture)	300; 325; 278
G10 (pure compound)	254; 300; 366

Group G10 was a pure compound, and showed the same spectrum in the UV region and retention time when compared with the ElAc standard by high-performance liquid chromatography with photodiode array detector (HPLC-PDA). Flow injection analysis-electrospray ionization-ion trap mass spectrometry (FIA-ESI-IT/MS) further confirmed that the isolated substance was ElAc (Figure 1 and Table 2).

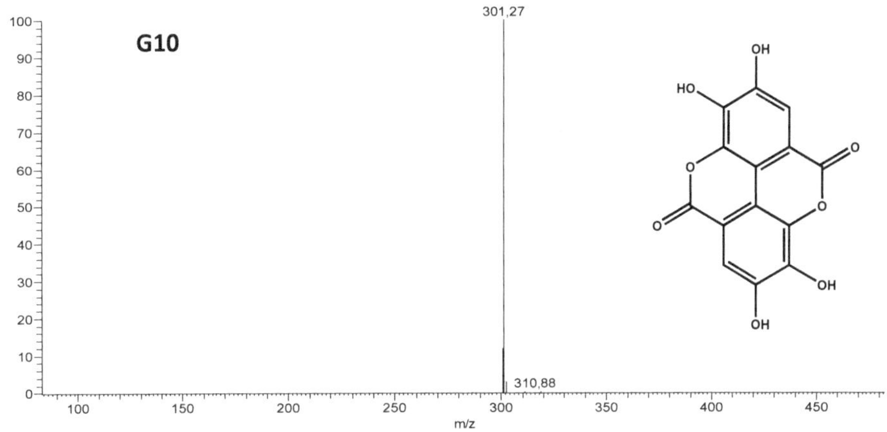

Figure 1. Mass spectrum of Group 10, obtained by flow injection analysis-electrospray ionization-ion trap mass spectrometry (FIA-ESI-IT/MS), ionization by negative mode and structure of the isolated compound (Ellagic acid).

Table 2. Chemical characterization of G10 by FIA-ESI-MS.

[M-H]-	MS^n ions	Identification
301	257 [M-44-H]- 229 [M-44-28-H]-	Ellagic Acid

Abbreviations: [M-H]-, deprotonated molecule; MS^n, multi-stage mass.

From the HPLC–DAD analysis and the construction of the quantification curve, it was possible to measure that the concentration of ElAc in BoEA was 5.93%.

2.2. In Silico Analysis of the Biological Activities of ElAc and Prediction of Its Toxic Effects, Including Hepatotoxicity

The biological activity spectra of ElAc were determined using the prediction of activity spectra for substances (PASS) software. Table 3 shows the probable activity (Pa) and probable inactivity (Pi) values. Several activities of ElAc were predicted, including anti-inflammatory, antioxidant, antibacterial, antimycobacterial, and hepatoprotective activities. The highest Pa value was obtained for anti-inflammatory activity (0.749).

Table 3. In silico analysis of the biological activities of Ellagic Acid (ElAc).

Activities	PASS Predictions of Ellagic Acid	
	Pa	Pi
Anti-inflammatory	0.749	0.010
Antioxidant	0.699	0.004
Antibacterial	0.380	0.035
Antimycobacterial	0.301	0.080
Hepatoprotective	0.599	0.012
DNA ligase (ATP) inhibitor	0.466	0.009
Superoxide dismutase inhibitor	0.423	0.058
DNA-3-methyladenine glycosylase I inhibitor	0.379	0.029
RNA directed DNA polymerase inhibitor	0.323	0.026
DNA polymerase I inhibitor	0.330	0.041
Nitric oxide scavenger	0.282	0.011
DNA synthesis inhibitor	0.261	0.074
Antibiotic	0.158	0.044
Antibacterial Ophthalmic	0.154	0.051
Cell wall synthesis inhibitor	0.094	0.072
tRNA-pseudouridine synthase I inhibitor	0.617	0.014

Abbreviations: Pa, probable activity; Pi, probable inactivity; PASS, prediction of activity spectra for substances.

The predictive analysis of ADME-Tox properties was performed for the compounds clarithromycin, diclofenac and ellagic acid. Table 4 shows the results of the ADME parameters of PPB, hepatotoxic, solubility, BBBP, HIA and violations of Lipinski's rule of five [19].

Table 4. ADME prediction of controls and Ellagic Acid.

Compound	PPB	Hepatotoxic	Solubility	BBBP	HIA	Lipinski Violations (Max 4)
Clarithromycin	false	True	2	4	3	3
Diclofenac	true	True	2	1	0	0
Ellagic Acid	true	True	3	4	1	0

Abbreviations: PPB = plasma–protein binding (false: does not bind to plasma proteins, true: binds to plasma proteins); Hepatotoxic (true: hepatotoxic effect, false: no hepatotoxic effect); Aqueous solubility: (acceptable range: range is 0–3, where 3 is a good solubility) [20,21]; BBBP = 0 (very high penetrant), 1 (high), 2 (medium), 3 (low), 4 (undefined/very low), HIA: 0 (good); 1 (moderate), 2 (poor), 3 (very poor) [22].

The prediction of toxicity through models of carcinogenicity in rodents (female mouse and rat female), Ames mutagenicity, skin irritancy, skin sensitization and ocular irritancy is shown in Table 5. The risk of toxicity and the prediction of carcinogenic potential are described in Tables 6 and 7.

Table 5. Topkat models (Mouse Female, Rat Female, Ames Mutagenicity, Skin Irritancy, Skin Sensitization and Ocular Irritancy) of clarithromycin, diclofenac and Ellagic Acid.

Compound	Mouse Female	Rat Female	Ames Mutagenicity	Skin Irritancy	Skin Sensitization	Ocular Irritancy
Clarithromycin	NC	NC	NM	Mild	None	Mild
Diclofenac	NC	NC	NM	None	Strong	Mild
Ellagic Acid	NC	NC	NM	None	Strong	Mild

Abbreviations: NC: non-carcinogen; NM: non-mutagen.

Table 6. Toxicity risk prediction via TOPKAT (Rate oral LD50, Daphnia EC50, rat chronic LOAEL, fathead minnow LC50).

Compound	Rat Oral LD50 (g/kg Body Weight)	Rat Chronic LOAEL (g/kg Body Weight)	Fathead Minnow LC50 (g/L)	*Daphnia* EC50 (mg/L)
Clarithromycin	0.787	0.001	0.004	0.278
Diclofenac	0.564	0.034	0.002	9.665
Ellagic Acid	0.428	0.302	0.049	16.407

Abbreviations: Daphnia EC50 = the effect concentration of a substance that causes adverse effects on 50% of the test population (*Daphnia magna*); LOAEL—lowest observed adverse effect level; Fathead minnow—short-term toxicity to fish.

Table 7. Carcinogenic potential via TOPKAT (Rat TD50 and RMTD).

Compound	Rat TD50 (mg/kg Body Weight/Day)	RMTD—Feed (mg/kg Body Weight/Day)
Clarithromycin	0.002	9.995
Diclofenac	51.448	742.520
Ellagic Acid	5.838	721.906

Abbreviations: RMTD: rat maximum tolerated dose; TD50: tumorigenic dose rate.

2.3. Molecular Docking

For docking studies, in an attempt to identify potential targets that could explain anti-inflammatory mechanisms of action and also the antimicrobial activity against *Mabs*, some targets were selected from the Protein Data Bank. Table 8 lists their data, including native ligands and grid center coordinates.

Table 8. Description of receptors related to antibacterial and anti-inflammatory activity used in the study of molecular docking.

Enzyme	Ligands Ligand ID/(Synonyms)	Coordinates of Grid Center
COX2 (Prostaglandin G/H synthase 2) (*Mus musculus*) PDB ID: 1PXX Resolution: 2.90 Å Reference: [23]	2-[2-[(2,6-dichlorophenyl)amino]phenyl]ethanoic acid DIF Diclofenac	X = 27,115 Y = 24,090 Z = 14,936
COX2 Prostaglandin G/H synthase 2 (*Homo sapiens*) PDB ID: 5IKQ Resolution: 2.41 Å Reference: [24]	2-[(2,6-dichloro-3-methyl-phenyl)amino]benzoic acid 2- / JMS Meclofenamic Acid	X = 21,597 Y = 51,876 Z = 17,696
Dihydrofolate reductase (*Mycobacteroides abscessus* ATCC 19977) PDB ID: 7K6C Resolution: 2.00 Å Reference: [25]	3-(2-{3-[(2,4-diamino-6-ethylpyrimidin-5-yl)oxy]propoxy}phenyl)propanoic acid/ MMV	X = -33,882 Y = −7502 Z = 56,281

Table 8. *Cont.*

Enzyme	Ligands Ligand ID/(Synonyms)	Coordinates of Grid Center
Phosphoribosylaminoimidazole-succinocarboxamide synthase (*Mycobacteroides abscessus* ATCC 19977) PDB ID: 6YYB Resolution: 1.51 Å Reference: [26]	4-azanyl-6-[1-[(1~{R})-1-phenylethyl]pyrazol-4-yl]pyrimidine-5-carbonitrile Q0Q	X = 21,1853 Y = 14,726 Z = 34,7921
tRNA (guanine-N(1)-)-methyltransferase (*Mycobacteroides abscessus*) PDB ID: 6QR4 Resolution: 1.52 Å Reference: [27]	zanyl-3-[1-[[(2~{R})-1-methylpiperidin-2-yl]methyl]indol-6-yl]-1~{H}-pyrazole-4-carbonitrile JD8	X = -12,718 Y = 7688 Z = −27,062

The comparison of the crystallographic ligands diclofenac, meclofenamic acid, MMV, Q0Q and JD8 (yellow color) and the best conformation predicted by molecular docking via DockThor [28] (green color), as well as the respective RMSD obtained in each analysis, can be seen in Figure 2.

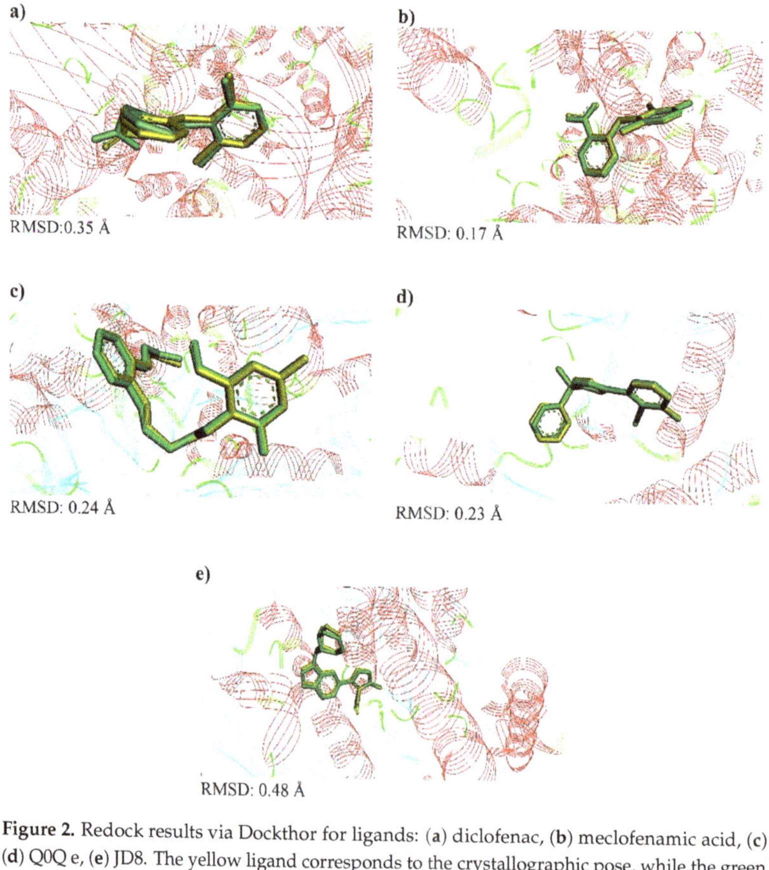

Figure 2. Redock results via Dockthor for ligands: (**a**) diclofenac, (**b**) meclofenamic acid, (**c**) MMV, (**d**) Q0Q e, (**e**) JD8. The yellow ligand corresponds to the crystallographic pose, while the green ligand corresponds to the computationally generated pose. The RMSD values for each analysis are shown below each image.

The validation of molecular docking was accepted through the calculation of RMSD, where values in the range of 0.17–0.48 Å were observed (Figure 2). Then, a study of the interactions between native ligands/controls and selected targets, in their respective binding pockets, was performed. The types of interactions as well as their distances are shown in Figures S1 and S2.

After validation of molecular docking protocols, predictions of binding affinity (ΔG) and modes of interaction of native ligands, diclofenac (control used experimentally) and ellagic acid, were performed (Figure 3 and Figure S3). The murine COX2 enzyme presented binding affinities (ΔG) of −9.1 and −8.2 (kcal/mol), with diclofenac and ellagic acid, respectively (Figure 3a). Predictions with the human COX2 also showed comparable binding affinities of −9.4, −9.3 and −9.1 (kcal/mol) with meclofenanic acid, diclofenac and ellagic acid, respectively (Figure 3b). Predictions by Dockthor showed that ellagic acid interacts with SER498 of the murine COX2 and with TYR322 and SER497 of the human COX2 enzyme (Figure S3).

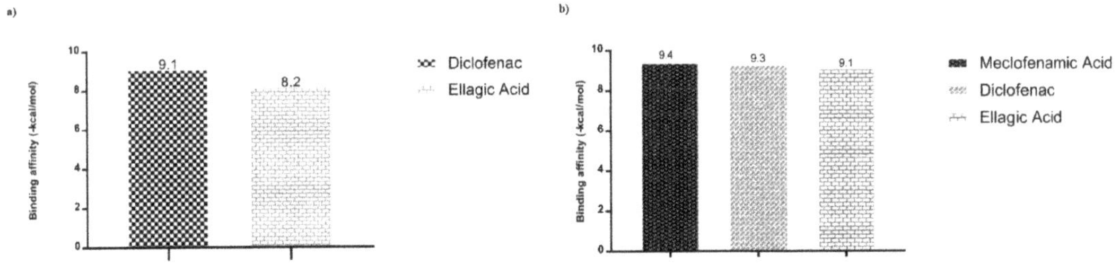

Figure 3. Binding affinity values of the ligands with the inflammation targets obtained by molecular docking via Dockthor: (**a**) with the murine COX2 (PDB ID 1PXX); and (**b**) with the human COX2 (PDB ID 5KIQ).

Antimicobacterial activity was investigated using specific *Mabs* targets that were available in the Protein Data Bank, which included Dihydrofolate reductase (PDB ID 7K6C, DHFR), Phosphoribosylaminoimidazole-succinocarboxamide synthase (PDB ID 6YYB, SAICAR Synthetase), and tRNA (guanine-N(1)-)-methyltransferase (PDB ID 6QR4). Predictions of binding affinities showed that ellagic acid had binding affinity value that is similar (−8.3 kcal/mol) to that of the specific inhibitor (MMV) for the DHFR enzyme (Figure 4a).

Predictions of the modes of interaction of ellagic acid showed interactions with the DHFR enzyme through 5 conventional hydrogen bonds with ALA8, ASP28 and THR47 (Figure S4). Its interaction with the SAICAR Synthetase and methyltransferase (TrmD) targets also occurred through conventional hydrogen bonds (Figure S4); however, they resulted in lower binding affinities than that observed with DHFR (Figure 4).

The diagram of the interaction modes of the control compounds (diclofenac and clarithromycin) and the referred targets are presented in Figure S5.

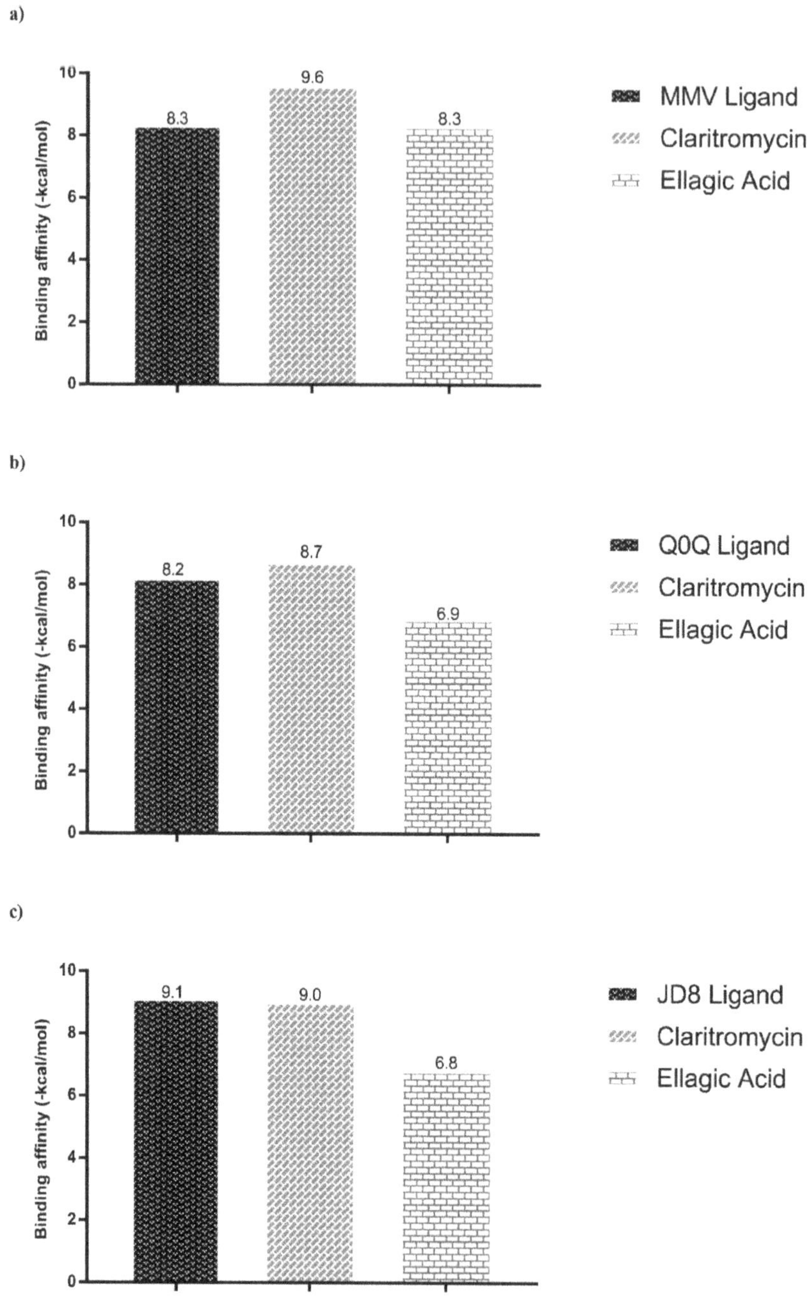

Figure 4. Binding affinity values of the native ligands, clarithromycin, and ellagic acid with targets of antimicrobial activity obtained by molecular docking via Dockthor. The selected targets were: (**a**) Dihydrofolate reductase (PDB ID 7K6C); (**b**) Phosphoribosylaminoimidazole-succinocarboxamide synthase (PDB ID 6YYB), and (**c**) tRNA (guanine-N(1)-)-methyltransferase (PDB ID 6QR4).

2.4. Antibacterial Activity of BoEA and ElAc

BoEA showed a MIC of 1.56 mg/mL and MBC of 6.25 mg/mL, while ElAc showed a MIC of 1.56 mg/mL and MBC of 3.12 mg/mL. The Clarithromycin showed an MIC of 1 µg/mL and MBC of 6 µg/mL.

2.5. Time–Kill Assay

To evaluate the antimicrobial activity of BoEA and ElAc at different time intervals, a time–kill curve assay was performed over a period of 12–120 h with Mabs in the presence of various concentrations of BoEA and ElAc (MIC, 2 × MIC, and 4 × MIC). Table 9 shows that BoEA at 2 × MIC and 4 × MIC inhibited the growth of Mabs when compared to MIC and the control. On the other hand, ElAc considerably reduced the growth of Mabs compared to the control.

Table 9. Time–kill evaluation of Mabs with BoEA and ElAc. Clarithromycin concentration was used at MIC. Growth control: no compound was added to the cell suspension.

Time (h)	Control Group	Antibacterial Agents						
		CLA	BoEA MIC	BoEA 2 × MIC	BoEA 4 × MIC	ElAc MIC	ElAc 2 × MIC	ElAc 4 × MIC
0	5.9 ± 0.08	5.9 ± 0.08	5.9 ± 0.08 **	5.9 ± 0.08 **	5.9 ± 0.08 **	5.9 ± 0.08 **	5.9 ± 0.08 **	5.9 ± 0.08 **
12	7.03 ± 0.12	6.76 ± 0.16	6.83 ± 0.12 **	6.83 ± 0.12 **	6.9 ± 0.08 **	6.83 ± 0.12 **	6.66 ± 0.23 **	5.66 ± 0.12 **
24	8.9 ± 0.08	6.73 ± 0.16	7.23 ± 0.16 **	7.23 ± 0.16 **	6.53 ± 0.12 **	6.3 ± 0.08 **	6.06 ± 0.09 **	5.76 ± 0.26 **
36	9.1 ± 0.16	6.9 ± 0.08	7.4 ± 0.37 **	7.1 ± 0.08 **	6.33 ± 0.12 **	6.53 ± 0.12 **	5.83 ± 0.16 **	5.73 ± 0.20 **
48	9.76 ± 0.04	7.03 ± 0.12	8.5 ± 0.40 *	7.86 ± 0.04 *	6.66 ± 0.12 *	6.8 ± 0.08 **	5.76 ± 0.04 **	5.30 ± 0.08 **
72	10.26 ± 0.44	7.5 ± 0.21	8.5 ± 0.37 *	7.23 ± 0.30 *	6.66 ± 0.12 *	7.0 ± 0.08 **	6.23 ± 0.16 **	5.36 ± 0.26 **
96	10.7 ± 0.16	8.1 ± 0.14	9.7 ± 0.24 *	8.46 ± 0.41 *	7.16 ± 0,12 *	7.36 ± 0.09 **	6.23 ± 0.30 **	5.33 ± 0.20 **
120	9.7 ± 0.16	8.33 ± 0.12	9.9 ± 0.08 *	9.1 ± 0.08 *	8.33 ± 0.23 *	6.86 ± 0.12 **	7.03 ± 0.12 **	6.26 ± 0.18 **

Data are presented as the mean ± standard deviation (SD) of Colony-Forming Unit (CFU) and representing of one of three independent experiments performed in triplicate. Statistical analysis was performed using analysis of variance and post hoc t-test. * $p < 0.05$, ** $p < 0.005$ when compared to the growth control. Abbreviations: BoEA, ethyl acetate fraction of *B. orellana*; E1Ac, ellagic acid; MIC, minimal inhibitory concentration.

2.6. BoEA and ElAc Increase Survival of *Tenebrio molitor* Infected with Mabs

Tenebrio molitor larvae inoculated with *Mabs* were treated with BoEA and ElAc at doses corresponding to the MIC, 2 × MIC, and 4 × MIC. Treatment with BoEA or ElAc increased the survival of infected *T. molitor* compared to that of the untreated group (phosphate-buffered saline (PBS)-treated) (Figure 5). Both treatments showed a dose-dependent effect, as treatment with the highest concentration (4 × MIC) resulted in larval survival until day 8. In addition, the lowest dose of ElAc (MIC) was able to achieve 50% larval survival on day 10.

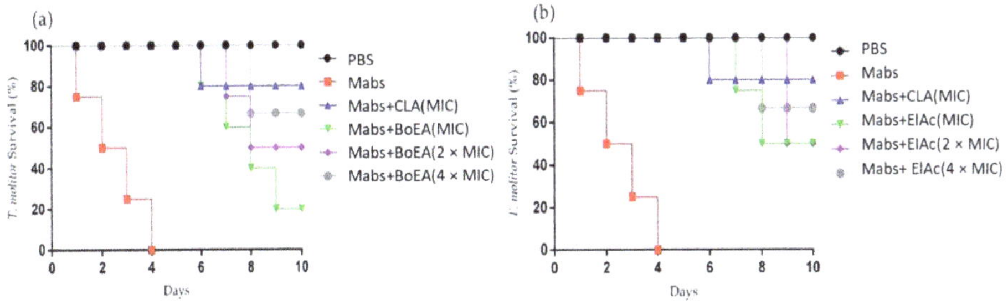

Figure 5. Effects of BoEA and ElAc on the *T. molitor* survival after lethal infection with *Mabs*. Effect of MIC, 2 × MIC, and 4 × MIC of (**a**) BoEA and (**b**) ElAc on larval survival. Ten larvae were included in each group and were monitored daily to assess mortality. PBS-treated group represents the negative control, while the clarithromycin-treated group represent the positive control. The results are representative of one of three independent experiments. Abbreviations: PBS, phosphate-buffered saline; CLA, clarithromycin; MIC, minimum inhibitory concentration; BoEA, ethyl acetate fraction of *B. orellana*; ElAc, ellagic acid.

2.7. Anti-Inflammatory Activity of BoEA and ElAc

BoEA and ElAc showed anti-inflammatory effects on carrageenan-induced paw edema in mice (Figure 6). ElAc (50 mg/kg) acted similarly to diclofenac (15 mg/kg) and showed a long-lasting effect when compared to BoEA, as the inhibition could be observed until the third time.

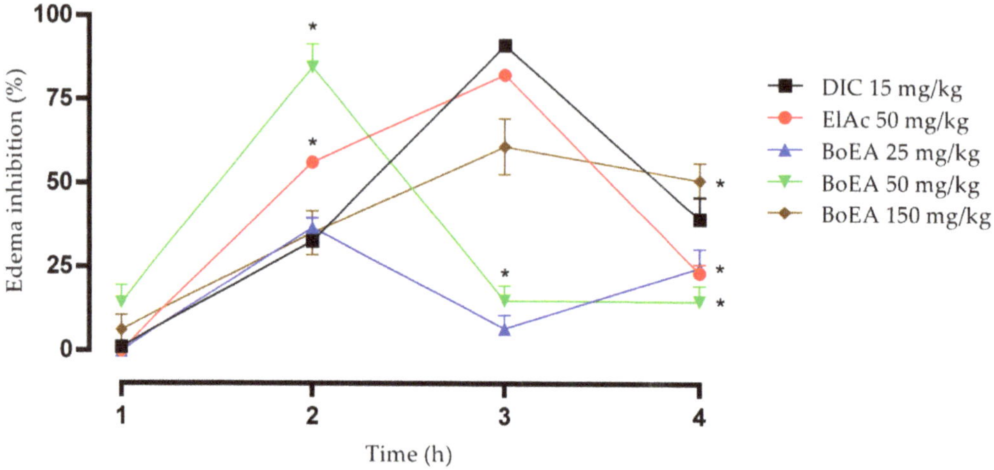

Figure 6. Anti-inflammatory effect of BoEA and ElAc in carrageenan-induced paw edema in mice. The thickness of the paw was measured 1, 2, 3, and 4 h after carrageenan injection. Data are presented as the mean ± SD. * $p < 0.05$ compared with the diclofenac-treated group. The results are representative of one of three independent experiments. Abbreviations: BoEA, ethyl acetate fraction of *B. orellana*; ElAc, ellagic acid; DIC, diclofenac.

3. Discussion

In the present study, we demonstrated through in silico, in vitro and in vivo analyses that the ethyl acetate fraction of *B. orellana* leaves (BoEA) and ellagic acid (ElAc) have both antimicrobial against *Mabs* and anti-inflammatory activity. ElAc was chosen for further studies after HPLC revealed its presence as one of the components of BoEA and because the literature suggests that the compound has many biological activities [16–18].

In silico prediction of the biological effects of ElAc revealed antioxidant, hepatoprotective, anti-inflammatory, and antibacterial activities. Zhao et al. [20] previously reported the hepatoprotective properties of ElAc, which reinforced the findings of our in silico prediction. They reported that mice with alcohol-induced liver disease treated orally with ElAc showed increased activity of alanine aminotransferase and serum aspartate aminotransferase, increased levels of triglyceride, low-density lipoprotein, free fatty acid, and total cholesterol, and decreased high-density lipoprotein levels. ElAc also showed antioxidant activities in the hepatic milieu, based on the levels of glutathione peroxidase, catalase, malondialdehyde, superoxide dismutase, and glutathione. Additionally, treatment with ElAc reduced proinflammatory cytokines, such as interleukin (IL)-6, IL-1β, and tumor necrosis factor-alpha (TNF-α), and the expression of several genes associated with the inflammatory immune response, including *TLR4*, *Myd88*, *CD14*, cyclooxygenase-2 [*COX2*], nitric oxide synthase 2 (*NOS2*), and nuclear factor kappa β (*NF-κβ1*) [20]. The authors concluded that ElAc decreased oxidative stress, inflammatory response, and steatosis in mice with alcohol-induced liver disease.

Oxidative stress impairs antioxidant defenses (enzymatic and non-enzymatic), and the consequence of this imbalance is damage to major cellular components, which further leads to a gradual loss of tissue and organ function [29]. Consistent with our findings, Aslan et al. [30] demonstrated the antioxidant potential of ElAc in lung damage induced by carbon tetrachloride in rats. Rats with carbon tetrachloride-induced lung damage treated with ElAc showed reduced levels of inflammatory proteins NF-κβ and COX2, and proinflammatory cytokine TNF-α, establishing the protective role of ElAc in lung damage.

The parameters used to predict bioavailability of clarithromycin were used to predict bioavailability of ElAc in order to compare the results with previous reports. With regard to intestinal absorption, ElAc was found to be superior to clarithromycin, which suggests that in an in vivo study, a larger amount of clarithromycin would be needed to obtain the same serum concentration of ElAc [31]. Additionally, in silico analysis revealed that ElAc does not have mutagenic, tumorigenic, and irritating effects, or harmful effects on the reproductive system [32,33].

To validate the molecular docking methodology in our study, native ligands were submitted to redocking simulations, using the DockThor receptor–ligand docking program [28]. Gowthaman et al. [34] and Hevener et al. [35] reported that the molecular docking methodology will be validated when the RMSD calculated through redocking analysis—i.e., between the crystallographic pose and the computationally predicted pose is less than 2.00 Å. Then, based on the results shown in this analysis, the molecular docking analysis was validated for the five crystal structures used.

To evaluate the in silico anti-inflammatory activity of ElAc via COX_2 inhibition, two crystal structures were selected: the crystal structure PDB ID 1PXX deposited by Rowlinson et al. [23] and PDB ID 5IKQ, deposited by Orlando and Malkowski [24]. The interactions predicted by Dockthor with TYR353 and SER498 are the same as those observed in the crystallographic pose, since during the protein preparation, the sequence is altered by the software itself. However, ellagic acid interacts with COX_2 through its phenolic group linked to SER498, and it is possible to notice several interactions such as Pi-Sigma, Pi-Alkyl, Carbon Hydrogen Bond, and Amide Pi-Stacked, between amino acid residues and the atoms of its aromatic rings.

Docking studies carried out to identify possible targets related to inhibitory activity against *Mabs* revealed that ElAc has a binding affinity value (-8.3 kcal/mol) equal to that presented by the specific inhibitor (MMV) for DHFR, suggesting that its antibacterial effect may be linked to the inhibition of this enzyme than to the other two targets analyzed (SAICAR Synthetase and tRNA methyltransferase), since the predictions revealed that ElAc has lower binding affinities to these latter structures, when compared to their native ligands. DHFR is an enzyme with an important role in the folic acid pathway, participating in the synthesis of nucleic acid precursors, among other functions related to an improvement in DNA translation, RNA transcription and protein replication and, thus, controlling bacterial multiplication [36]. Therefore, DHFR inhibitor compounds can lead to bacterial death. We cannot yet be assured of whether ellagic acid can be considered an inhibitor of this enzyme, but our in silico data show considerable binding affinity indicating this enzyme as a possible target that needs further investigation.

Regarding the antibacterial effect, we observed that BoEA and ElAc exerted bactericidal effects. The antimicrobial activity of BoEA has been reported previously against various microorganisms, such as Gram-positive bacteria (*S. aureus* and *S. pyogenes*) [18], Gram-negative bacteria *(Escherichia coli, Salmonella typhi, Shigella dysenteriae, Klebsiella pneumoniae, Proteus vulgaris*) [37], yeast (*Candida albicans*) [38], antileishmanial [39], and antimalarial activities [40]. Our group previously demonstrated that BoEA has antimycobacterial activity against *Mabs* [14]. In addition, in the present study, we provide evidence for the antimycobacterial activity of ElAc, which was isolated from BoEA.

The antimycobacterial activity of BoEA can be explained by the chemical characteristics of its constituent polar compounds. Compounds with intermediate to high polarity inhibit the growth of mycobacterial species [41]. Although we could not confirm the exact phytochemicals responsible for the observed antimycobacterial activity against *Mabs*, it is highly possible that the observed activity was due to the presence of ElAc in BoEA. We used various approaches including in silico, in vitro, and in vivo analyses to confirm that the phytochemical ElAc from BoEA demonstrated antimicrobial activity against *Mabs*.

Other studies have evaluated the antimycobacterial activity of ElAc isolated from other plants. Dey et al. [42] reported that ElAc from pomegranate fruit exerts antimycobacterial activity against *Mycobacterium tuberculosis* (MIC 64–512 μg/mL). Sridevi et al. [43] reported that ElAc from pomegranate peels inhibited *M. tuberculosis* with an MIC of 0.3–3.5 mg/mL. Salih et al. [44] reported that ElAc from *Combretum hartmannianum* exerts antimycobacterial activity against *Mycobacterium smegmatis* (MIC 500–1000 μg/mL) and in another study, the same group [45] reported that ElAc from *Anogeissus leiocarpa* exhibited growth inhibitory activity against *M. smegmatis* with an MIC value of 500 μg/mL. However, to the best of our knowledge, we are the first to report the biological activity of ElAc from *B. orellana* leaves against *Mabs*. The MIC and MBC values of ElAc against *Mabs* in the present study were 1.56–3.12 mg/mL. These values are higher than those reported for E1Ac in previous studies. This can be explained by the intrinsic resistance of *Mabs* compared with that of other mycobacterial species. Treatment of *M. abscessus* infection is challenging due to the high level of innate resistance of the bacteria and is often associated with lengthy, costly, and non-standardized administration of antimicrobial agents. Moreover, adverse effects, drug toxicities, and high relapse rates result in poor treatment outcomes [46].

In the present study, the antimicrobial effects of BoEA and ElAc were tested in *Mabs*-infected *T. molitor* larvae. We observed that BoEA and ElAc at MIC, 2× MIC, and 4× MC inhibited *Mabs* infection in *T. molitor* larvae [47]. Larvae treated with BoEA and ElAc exhibited greater survival than untreated larvae, emphasizing the protective effect of these compounds against *Mabs* infection. The ability of *T. molitor* to produce reactive species in response to deleterious stimuli makes it a potential model for studying antimicrobial substances. Despite these advantages, to the best of our knowledge, no study has used *T. molitor* larvae for screening plant-derived antioxidant compounds against *Mabs*. This approach could provide more information than models traditionally used for studying antimicrobial agents, based on the chemical interaction of compounds and without biological activity relevance [48].

To evaluate the anti-inflammatory activity of BoEA and E1Ac, we measured the inhibition rate of carrageenan-induced paw edema in mice. Inflammation limits the damage to cells after invasion by foreign organisms or mechanical injury. Histamine is a common inflammatory mediator [49]. Inflammatory responses occur rapidly and eventually lead to vasodilatation and plasma exudation, which induces recruitment of inflammatory mediators and causes edema [50]. Reduction in carrageenan-induced acute inflammation and the resulting paw edema is a useful marker for investigating the anti-inflammatory potential of drugs and plant extracts [51].

Development of carrageenan-induced edema occurs in two phases: the initial phase involves histamine and serotonin, and the later phase is characterized by a marked increase in prostaglandin production [52]. The results obtained in the present study suggest that the anti-inflammatory effect of BoEA on carrageenan-induced edema in mice affects both the early and late phases, observed at doses of 50 mg/kg and 150 mg/kg, respectively, where its bioactive components, including ElAc, possibly act by suppressing the inflammatory response mediated by histamine, serotonin, and/or prostaglandins, presenting a potential source for cyclooxygenase inhibitors [53]. Previously, it was reported that the anti-inflammatory mechanisms of ElAc were related to a decrease in the level of COX_2 via the suppression of proinflammatory cytokines (TNF-α, IL-1β), NO, and Prostaglandin E2 overproduction [54].

Some limitations can be highlighted in our study, including that we evaluated the antimicrobial activity against only one isolate of *M. abscessus* subsp. *massaliense*. Although we did not include a reference strain in the study, the *Mabs* used in our study (strain G01) is a well-characterized and relevant clinical isolate, since it was cultured from 18 patients in an outbreak of post-surgical infections in a hospital in the city of Goiânia (Brazil) [55]. A second limitation concerns the small number of potential targets for antimicrobial activity that we analyzed. However, at the time of writing, the Protein Data Bank only has these three crystallographic structures obtained from *Mabs* with their respective inhibitors, which restricted our analysis.

4. Materials and Methods

4.1. Drugs, Chemicals, and Reagents

Hexane, ethanol, methanol, chloroform, acetic acid, dimethyl sulfoxide (DMSO), PBS, resazurin, clarithromycin, carrageenan, and ElAc were purchased from Sigma–Aldrich (St. Louis, MO, USA). Mueller-Hinton broth was purchased from Merck KGaA (Darmstadt, Germany). Middlebrook 7H11 agar base was purchased from HiMedia (Mumbai, India). Diclofenac sodium was purchased from Novartis Biociências S.A. (São Paulo, Brazil).

4.2. Preparation of BoEA and Isolation/Identification of ElAc

4.2.1. Collection of *B. orellana* Leaves and Separation of Fractions

B. orellana leaves were collected in the municipality of São José de Ribamar, Maranhão, Brazil (April–May 2019) and identified at the herbarium of the Federal University of Maranhão (São Luís, Brazil, specimen voucher 1147, SISGen AE80277). The leaves were dried in an oven at 40 °C for three days and then dried at room temperature (24–26 °C) for another four days. The leaves were then ground in a mill and their crude extract was extracted by maceration in 70% ethanol for 24 h.

The sample was filtered, and the resulting filtrate was concentrated on a rotary evaporator under low pressure at 50 °C. The concentrate was lyophilized and labeled BoHE. BoHE was then subjected to liquid–liquid fractionation using hexane, chloroform, and ethyl acetate with a series of increasing polarities to produce hexane (BoHex), chloroform (BoCl), and ethyl acetate (BoEA) fractions. BoEA was used in the present study and the other fractions were stored for future use.

4.2.2. Isolation and Characterization of Compounds in BoEA

The phytochemical compounds in BoEA were purified using a chromatography column (230–400 mesh; 8 × 100 cm) and eluted with increasing polarities of mixtures of n-hexane/ethyl acetate and ethyl acetate/methanol to obtain subfractions. The subfractions (395) were then grouped using thin-layer chromatography and HPLC-PDA. Ten fractions were obtained and analyzed using HPLC-PDA. Group 10 showed a single peak in the chromatogram, suggesting that it may have been a single compound.

The structure of group 10 was determined using HPLC-PDA and FIA-ESI-IT/MS. Mass spectrometry was performed using an LCQ Fleet Equipment (Thermo Fisher Scientific, Waltham, MA, USA) equipped with a device for direct sample insertion via FIA. The studied matrix was analyzed by ESI and multi-stage fragmentation (MS2, MS3, and MSn) was performed in an IT interface. The negative and positive modes were selected for the generation and analysis of mass spectra for the first order (MS) and the other multi-stage experiments were performed under the following conditions: capillary voltage of 25 V, spray voltage of 5 kV, and capillary temperature of 275 °C. A carrier gas (N_2) with a flux of eight arbitrary units (was used and the collision gas was helium (He). The range of acquisition was m/z 100–1000. Xcalibur software version 1.3 (Thermo Fisher Scientific) was used to acquire and process the data.

4.3. In Silico Analysis

4.3.1. Prediction of Biological Activities of ElAc

The biological activities of ElAc and clarithromycin (standard drug) were evaluated using the PASS Online platform (version 2.0, Way2Drug.com©2011–2022, Moscow, Russia) (www.way2drug.com/passonline/, accessed on 15 October 2021), which can predict several biological characteristics of a substance. The PASS program describes biological activity as "active" (Pa) or "inactive" (Pi), where the estimated probability varies from 0 to 1. The chances of finding a particular activity increase when the Pa values are higher and Pi values are lower. The results of PASS prediction were interpreted as follows: (i) only biological activities with Pa > Pi were considered probable for a particular compound; (ii) if Pa > 0.7, the substance is likely to exhibit the said biological activity and the probability of the compound being an analog of a known pharmaceutical drug is also high; (iii) if 0.5 < Pa < 0.7, the compound is likely to exhibit the said biological activity, but the compound is not similar to known drugs; (iv) if Pa < 0.5, the compound likely does not exhibit the said biological activity and is perhaps a structurally new compound [56].

4.3.2. Prediction of Pharmacokinetic Characteristics and Toxic Effects of ElAc

Predictive values of absorption, distribution, metabolism, excretion, and toxicity (ADME-Tox) properties were calculated using Discovery Studio Client v16 software 1.0.15350 [21], following the methodological proposal described by Shukla et al. [57] and Ramos et al. [21]. The ADMET Descriptors protocol, run by the aforementioned software, uses a variety of QSAR models to estimate pharmacokinetic/toxicological properties for small molecules [22]. Thus, the properties analyzed were: binding to plasma proteins (PBB), hepatotoxicity, penetration of the blood brain barrier (BBBP), solubility and human intestinal absorption (HIA). Another important step for the ADME-Tox analysis is to evaluate possible violations of Lipinski's Rule of Fives [19]. To carry out this methodological step, the "calculate molecular properties" module present in the Discovery Studio Client v16 software was used 1.0.15350 [22].

Toxicity predictions were performed using TOPKAT (Toxicity Prediction function by Komputer Assisted Technology) [19]. Therefore, such a module can predict the toxicity of chemicals based solely on their 2D molecular structure, using a variety of robust, cross-validated quantitative structure-toxicity relationship (QSTR) models to assess specific toxicological parameters. Therefore, the toxicological properties analyzed were: carcinogenicity in rodents (female mouse and rat female), Ames mutagenicity, skin irritancy, and skin sensitization. Toxicity risk prediction calculations were performed via TOPKAT and measured the following parameters: rate oral LD50 (g/kg BodyWeight), Daphnia EC50 (mg/L), rat chronic LOAEL (g/kg BodyWeight), fathead minnow LC50 (g/L). In addition, the carcinogenic potential was also predicted through the parameters: TD50 (mg/kg body weight/day-mouse/rat) and RMTD (Rat Maximum Tolerated Dose-mg/kg body weight).

4.3.3. Molecular Docking Study

To investigate the anti-inflammatory activity by the inhibition of COX2 (Prostaglandin G/H synthase 2) two crystal structures were selected according to Rowlinson et al. [23] and Orlando et al. [24]. Furthermore, to evaluate the multi-target antibacterial effect, three crystal structures of the following enzymes were selected: dihydrofolate reductase (DHFR), SAICAR Synthetase (PurC), and tRNA (m 1 G37) methyltransferase (TrmD) [25–27]. To carry out the molecular docking study, it was necessary to preprocess the structures with the help of Discovery Studio Client software v16. 1. 0. 15350 [22], which allowed the exclusion of water molecules, metals, cofactors, separation of ligands and protein structures. Then, the DockThor software, a receiver–ligand docking program, was used (https://dockthor.lncc.br/v2/, accessed on 30 May 2022) [28]. The parameters used were: number of evaluations "1,000,000", population size "750", initial seed "−1985" and number of runs "24". In addition, hydrogen was added (pH 7.0) by the software's own module. The grid center was defined using Discovery Studio Client software [22], and the grid size and discretization were kept with their default values presented by the software, i.e., (X = 20, Y = 20 and Z = 20) and 0.25, respectively. The validation of each molecular docking analysis was performed through redocking simulations. This procedure was performed with the submission of the crystallized native ligand itself to DockThor [28]. Then, the RMSD (root-mean-square deviation) was calculated using the Discovery Studio Client software, using the crystallographic ligand pose and the computationally generated pose, using all atoms. In addition, comparative poses were generated. Finally, the analysis of interactions between the respective crystallized ligand and the enzymes in each binding pocket was performed, generating 2D diagrams using the Discovery Studio Client software.

The control compounds (diclofenac and clarithromycin) and the studied compound ellagic acid were submitted to molecular docking analysis with the parameters already described (Table S1). In addition, the binding affinity (ΔG) results predicted by DockThor as a score (−kcal/mol) were plotted using the GraphPadPrism 7.0 software (GraphPad Software Inc., San Diego, CA, USA). Finally, to qualitatively analyze the interactions between the compounds and the aforementioned enzymes, 2D diagrams were generated in the respective binding pocket, using the Discovery Studio Client software.

4.4. *In Vitro Analysis*

4.4.1. Strain

In the present study, we used the *Mabs* GO01 strain, which was cryopreserved (−80 °C) in the Laboratory of Immunology and Microbiology of Respiratory Infections of the CEUMA University, São Luís, Brazil. This isolate was collected from a patient with infections caused by contaminated instruments, following invasive procedures [55].

The use of the clinical isolate was authorized through a written informed consent form approved by the Certificate of Ethics Appraisal Presentation (CAAE:21357413.4.0000.5084).

4.4.2. Determining MIC and MBC

MIC determination was performed based on the standard recommendations [58], using the Mueller Hinton broth microdilution technique in a 96-well plate, with different concentrations of clarithromycin (positive control; 0.001–1 mg/mL), BoEA (0.1–12.5 mg/mL), and ElAc (0.1–12.5 mg/mL). *Mabs* culture without any additive was used as the negative culture and was set as 100% growth (i.e., 0% inhibition). The experiment was performed in triplicate. On the third day of incubation, 30 µL of 0.01% resazurin was added to each well, and the plate was further incubated overnight. A change in color from blue (oxidized state) to pink (reduced state) indicates bacterial growth [59]; MIC value was defined as the lowest drug concentration that prevented bacterial growth.

The concentrations from the MIC determination assay that did not show visible growth were inoculated on a 7H11 agar plate for MBC determination. MBC was defined as the lowest concentration of the BoEA and E1Ac that inhibited mycobacterial growth (≥99.9% cell death) [60].

4.4.3. Time–Kill Assay

Individual bottles with 20 mL of Middlebrook 7H9 with OADC growth supplement and 0.05% Tween 80 were inoculated with *Mabs* (10^5–10^6 CFU/mL) and BoEA or ElAc at MIC, 2× MIC, and 4× MIC was added to the bottles, which were then cultured at 37 °C. All bottles were shaken (100 rpm) and ventilated through a bacterial filter (FP 30/0.2 Ca/S, Whatman GmbH, Dassel, Germany). To perform CFU counting, 200 µL was taken from each bottle and serial 10-fold dilutions in 0.85% sterile saline solution were plated on Middlebrook 7H11 plates (BD Bioscience, Franklin Lakes, NJ, USA) at different time intervals (12, 24, 36, 48, 72, 96, and 120 h). Experiments were performed thrice in triplicate. The results are expressed as Log_{10} CFU.

4.5. *In Vivo Analysis*

4.5.1. *T. molitor* Survival Assay

T. molitor larvae were fed ad libitum with bran flour and water supplemented with apples. For experiments, 10th–12th instar larvae without evident body injuries or melanization were used. The larvae were infected with 10 µL of *Mabs* (1×10^5 CFU) using a Hamilton syringe equipped with a 26-gauge needle. They were injected intrahemocele, at the second or third sternite, above the legs, in the ventral portion. After inoculation, the larvae were kept in petri dishes at 37 °C, which contained chopped apples to avoid dehydration. Silk on the larval surface was removed as soon as possible to delay transition to the pupal stage.

Each experimental group was composed of 10 larvae, and they were distributed as follows: PBS group, negative control, not infected or treated, inoculated with 10 μL of sterile PBS; positive control group, infected with *Mabs* (1×10^5 CFU) and treated with clarithromycin (10 μL/mL); *Mabs* group, infected with *Mabs* (1×10^5 CFU), not treated; BoEA group, infected with *Mabs* (1×10^5 CFU) and treated with BoEA at MIC, $2 \times$ MIC, and $4 \times$ MIC; and ElAc group, infected with *Mabs* (1×10^5 CFU) and treated with ElAc at MIC, $2 \times$ MIC, and $4 \times$ MIC. The larvae were treated with clarithromycin, BoEA, or ElAc 1 h after inoculation with *Mabs* were monitored daily for 10 days. Dead animals were defined as those showing no signs of irritability, extensive body melanization, or shrinking [61].

4.5.2. Carrageenan-Induced Paw Edema Model

C57BL/6 females, 6–8 weeks of age and weighing between 18 and 22 g, were procured from the Central Vivarium of CEUMA University. Mice were housed in an upper cage and allowed ad libitum food and water. The protocol for animal experiments was approved by the Ethics Committee on the Use of Animals, CEUA/UniCEUMA (protocol number 147/18). All animal experiments were performed according to the guidelines of the National Council for the Control of Animal Experimentation (CONCEA), Ministry of Science, Technology and Innovation (MCTI), Brazil, and the Brazilian Society of Science in Laboratory Animals (SBCAL).

Paw edema was induced by injecting 0.05 mL of 1% (*w*/*v*) carrageenan suspended in saline into the subplantar tissue of the right hind paw of each mouse [62]. A digital caliper (ZAAS Amatools, São Paulo, Brazil) was used to measure the paw thickness (Ct). An increase in paw thickness was considered an indicator of inflammation. Paw thickness was measured at various time points: at 0 h (C0), i.e., immediately after edema induction, and at 1, 2, 3, and 4 h after paw edema induction. All animals were pretreated by gavage 1 h before the induction of paw edema.

Mice were divided into six groups (n = 4 each): control group, paw edema was induced, and animals were treated with vehicle (10 mg/kg); BoEA groups, paw edema was induced and animals were treated with 25, 50, or 150 mg/kg of BoEA; ElAc group, paw edema was induced and animals were treated with 50 mg/kg of ElAc; and diclofenac group, paw edema was induced and animals were treated with 15 mg/kg.

The following equation was used to calculate percentage inhibition of inflammation [63]:

$$\%\text{Inhibition} = \frac{(\text{Ct} - \text{C0})\text{Control} - (\text{Ct} - \text{C0})\text{treated}}{(\text{Ct} - \text{C0})\text{Control}} \times 100$$

Legend: Ct, paw thickness after treatment with carrageenan at time "t"; C0, initial (basal) paw thickness.

4.6. Statistical Analysis

Data are presented as the mean ± standard variation or percentages. The normality of distributions was determined using the Shapiro–Wilk test, and differences between groups were evaluated by analysis of variance followed by Tukey's multiple comparison test using GraphPad Prism software (version 6.0, GraphPad Prism software Inc., San Diego, CA, USA). Statistical significance was set at $p < 0.05$. Larval survival assays were analyzed using the Kaplan–Meier method to calculate survival fractions, and the log-rank test was used to compare survival curves.

5. Conclusions

Our in silico, in vitro, and in vivo data indicate that BoEA and ElAc have anti-inflammatory and antimicrobial activity against *Mabs*. Apparently, the mechanisms of action are related to the binding affinity of ellagic acid to both COX_2 and DHFR, respectively, which warranted further studies. Our results suggest that ElAc can be optimized to develop new lead compounds against antimicrobial resistant pathogens, such *Mabs* and/or anti-inflammatory drugs.

Supplementary Materials: The following supporting information can be downloaded at: https://www.mdpi.com/article/10.3390/antibiotics11060817/s1, Figure S1: Diagram 2D of Interactions between controls / native ligands and targets: (a) diclofenac and COX2 (PDB ID: 1PXX); (b) meclofenamic acid and COX2 (PDB ID: 5IKQ); (c) MMV ligand and dihydrofolate reductase (PDB ID: 7K6C); (d) Q0Q ligand and phosphoribosylaminoimidazole-succinocarboxamide synthase (PDB ID: 6YYB), and (e) JD8 ligand and tRNA (guanine-N(1)-)-methyltransferase (PDB ID: 6QR4). Interactions: Conventional Hydrogen Bond: ■, Pi-Sigma: ■, Alkyl/ Pi-Alkyl: ■, Van der Walls: ■, Salt Bridge: ■ (present in c), Pi-Pi stacked: ■ (present in c), Unfavorable Aceptor-Aceptor: ■, Pi-cation: ■ (present in d), Carbon Hydrogen Bond: . Figure S2: Diagram 2D of Interactions between controls and targets. (a) Interactions between diclofenac and COX2 (PDB ID: 5IKQ); (b) Interactions between clarithromycin and Dihydrofolate reductase (PDB ID 7K6C); (c) Interactions between clarithromycin and Phosphoribosylaminoimidazole-succinocarboxamide synthase (PDB ID 6YYB); (d) Interactions between clarithromycin and tRNA (guanine-N(1)-)-methyltransferase (PDB ID 6QR4). Figure S3: Diagram 2D of Interactions between ellagic acid and COX2 targets: (a) with the murine COX2 (PDB ID: 1PXX) and (b) with the human COX2 (PDB ID: 5IKQ). Interactions: ■—Conventional Hydrogen Bond, ■—Pi-Sigma, ■—Alkyl/ Pi-Alkyl, —Carbon Hydrogen Bond, ■—Amide Pi-Stacked. Figure S4: Diagram 2D of Interactions between ellagic acid and targets for antimicrobial activity: (a) Dihydrofolate reductase (PDB ID 7K6C); (b) Phosphoribosylaminoimidazole-succinocarboxamide synthase (PDB ID 6YYB), and (c) tRNA (guanine-N(1)-)-methyltransferase (PDB ID 6QR4). Interactions: ■—Conventional Hydrogen Bond, ■—Alkyl, —Carbon Hydrogen Bond. Figure S5: Diagram 2D of Interactions between: (a) diclofenac and COX2 (PDB ID: 5IKQ); (b) clarithromycin and Dihydrofolate reductase (PDB ID 7K6C); (c) clarithromycin and Phosphoribosylaminoimidazole-succinocarboxamide synthase (PDB ID 6YYB), and (d) clarithromycin and tRNA (guanine-N(1)-)-methyltransferase (PDB ID 6QR4). Table S1: Predicting molecular properties related to Lipinski's rule.

Author Contributions: Conceptualization, E.M.d.S. and R.C.C.; methodology, R.N.M.-N., G.G.C., A.C.S.A., A.d.O.R., C.E.C.N. and R.P.d.A.; validation, C.Q.d.R., A.S.R., M.d.S.d.S.C., A.L.A.-S., R.N.M.G., V.M.-N., R.M.R., E.M.d.S. and R.C.C.; formal analysis, C.Q.d.R., I.V.F.d.S.; C.B.R.d.S.; R.N.M.G., V.M.-N., R.M.R., E.M.d.S. and R.C.C.; investigation, R.N.M.-N., G.G.C., A.C.S.A., A.d.O.R., C.E.C.N., R.P.d.A., A.S.R., I.V.F.d.S. and C.B.R.d.S.; resources, M.d.S.d.S.C., R.M.R., E.M.d.S. and R.C.C.; data curation, E.M.d.S. and R.C.C.; writing—original draft preparation, R.N.M.-N., C.Q.d.R., M.d.S.d.S.C., A.L.A.-S., I.V.F.d.S.; C.B.R.d.S.; R.N.M.G., V.M.-N., R.M.R., E.M.d.S. and R.C.C.; writing—review and editing, C.Q.d.R., R.N.M.G., V.M.-N., R.M.R., E.M.d.S. and R.C.C.; supervision, E.M.d.S. and R.C.C.; project administration, E.M.d.S. and R.C.C.; funding acquisition, R.C.C. All authors have read and agreed to the published version of the manuscript.

Funding: This research was funded by FAPEMA—Fundação de Amparo à Pesquisa e ao Desenvolvimento Científico e Tecnológico do Maranhão—UNIVERSAL-01032/19; UNIVERSAL-0965/19; BEPP-01667/21, BM-01729/19; FAPEMA/CAPES ACT- 05691/21; CAPES- Coordenação de Aperfeiçoamento de Pessoal de Nível Superior—Finance Code 001, and Conselho Nacional de Desenvolvimento Científico e Tecnológico (CNPQ) grant number 315072/2020-2.

Institutional Review Board Statement: This study was performed following protocols approved by the Ethics Committee on the Use of Animals—CEUA/UniCEUMA (Protocol number 147/18) of CEUMA University, São Luís, Maranhão, Brazil.

Informed Consent Statement: Not applicable.

Data Availability Statement: Not applicable.

Conflicts of Interest: The authors declare no conflict of interest.

References

1. Degiacomi, G.; Sammartino, J.C.; Chiarelli, L.R.; Riabova, O.; Makarov, V.; Pasca, M.R. Mycobacterium Abscessus, an Emerging and Worrisome Pathogen among Cystic Fibrosis Patients. *Int. J. Mol. Sci.* **2019**, *20*, 5868. [CrossRef] [PubMed]
2. Boudehen, Y.M.; Kremer, L. Mycobacterium Abscessus. *Trends Microbiol.* **2021**, *29*, 951–952. [CrossRef] [PubMed]
3. Johansen, M.D.; Herrmann, J.L.; Kremer, L. Non-Tuberculous Mycobacteria and the Rise of Mycobacterium Abscessus. *Nat. Rev. Microbiol.* **2020**, *18*, 392–407. [CrossRef] [PubMed]
4. Strnad, L.; Winthrop, K.L. Treatment of Mycobacterium Abscessus Complex. *Semin. Respir. Crit. Care Med.* **2018**, *39*, 362–376. [CrossRef]

5. Nasiri, M.J.; Haeili, M.; Ghazi, M.; Goudarzi, H.; Pormohammad, A.; Fooladi, A.A.I.; Feizabadi, M.M. New Insights in to the Intrinsic and Acquired Drug Resistance Mechanisms in Mycobacteria. *Front. Microbiol.* **2017**, *8*, 681. [CrossRef]
6. Meir, M.; Barkan, D. Alternative and Experimental Therapies of Mycobacterium Abscessus Infections. *Int. J. Mol. Sci.* **2020**, *21*, 6793. [CrossRef]
7. Victoria, L.; Gupta, A.; Gómez, J.L.; Robledo, J. Mycobacterium Abscessus Complex: A Review of Recent Developments in an Emerging Pathogen. *Front. Cell. Infect. Microbiol.* **2021**, *11*, 659997. [CrossRef]
8. Liu, Q.; Meng, X.; Li, Y.; Zhao, C.N.; Tang, G.Y.; Li, H. Bin Antibacterial and Antifungal Activities of Spices. *Int. J. Mol. Sci.* **2017**, *18*, 1283. [CrossRef]
9. Tran, N.; Pham, B.; Le, L. Bioactive Compounds in Anti-Diabetic Plants: From Herbal Medicine to Modern Drug Discovery. *Biology* **2020**, *9*, 252. [CrossRef]
10. Santos, D.C.; Barboza, A.S.; Ribeiro, J.S.; Rodrigues Junior, S.A.; Campos, Â.D.; Lund, R.G. Bixa Orellana L. (Achiote, Annatto) as an Antimicrobial Agent: A Scoping Review of Its Efficiency and Technological Prospecting. *J. Ethnopharmacol.* **2022**, *287*, 114961. [CrossRef]
11. Sousa-Andrade, L.M.; Oliveira, A.B.M.; Leal, A.L.A.B.; Oliveira, F.A.A.; Portela, A.L.; Lima-Neto, J.S.; Siqueira-Júnior, J.P.; Kaatz, G.W.; Rocha, C.Q.; Barreto, H.M. Antimicrobial Activity and Inhibition of the NorA Efflux Pump of Staphylococcus Aureus by Extract and Isolated Compounds from Arrabidaea Brachypoda. *Microb. Pathog.* **2020**, *140*, 103935. [CrossRef] [PubMed]
12. Vilar, D.D.A.; Vilar, M.S.D.A.; Moura, T.F.A.D.L.E.; Raffin, F.N.; Oliveira, M.R.D.; Franco, C.F.D.O.; De Athayde-Filho, P.F.; Diniz, M.D.F.F.M.; Barbosa-Filho, J.M. Traditional Uses, Chemical Constituents, and Biological Activities of Bixa Orellana L.: A Review. *Sci. World J.* **2014**, *2014*, 857292. [CrossRef] [PubMed]
13. Medina-Flores, D.; Ulloa-Urizar, G.; Camere-Colarossi, R.; Caballero-García, S.; Mayta-Tovalino, F.; del Valle-Mendoza, J. Antibacterial Activity of Bixa Orellana L. (Achiote) against Streptococcus Mutans and Streptococcus Sanguinis. *Asian Pac. J. Trop. Biomed.* **2016**, *6*, 400–403. [CrossRef]
14. Lima-Viana, J.; Zagmignan, A.; Lobato, L.F.L.; Abreu, A.G.; Nascimento da Silva, L.C.; Cortez-Sá, J.; Monteiro, C.A.; Lago, J.H.G.; Gonçalves, L.M.; Carvalho, R.C.; et al. Hydroalcoholic Extract and Ethyl Acetate Fraction of Bixa Orellana Leaves Decrease the Inflammatory Response to Mycobacterium Abscessus Subsp. Massiliense. *Evid. -Based Complement. Altern. Med.* **2018**, *2018*, 6091934. [CrossRef]
15. Moraes-Neto, R.N.; Coutinho, G.G.; Oliveira Rezende, A.; Pontes, D.B.; Ferreira, R.L.P.S.; Morais, D.A.; Albuquerque, R.P.; Lima-Neto, L.G.; Nascimento-Silva, L.C.; Rocha, C.Q.; et al. Compounds Isolated from Bixa Orellana: Evidence-Based Advances to Treat Infectious Diseases. *Rev. Colomb. Cienc. Quím. Farm* **2020**, *49*, 581–601. [CrossRef]
16. Alfei, S.; Turrini, F.; Catena, S.; Zunin, P.; Grilli, M.; Pittaluga, A.M.; Boggia, R. Ellagic Acid a Multi-Target Bioactive Compound for Drug Discovery in CNS? A Narrative Review. *Eur. J. Med. Chem.* **2019**, *183*, 111724. [CrossRef]
17. Sharifi-Rad, J.; Quispe, C.; Castillo, C.M.S.; Caroca, R.; Lazo-Vélez, M.A.; Antonyak, H.; Polishchuk, A.; Lysiuk, R.; Oliinyk, P.; De Masi, L.; et al. Ellagic Acid: A Review on Its Natural Sources, Chemical Stability, and Therapeutic Potential. *Oxid. Med. Cell. Longev.* **2022**, *2022*, 1–24. [CrossRef]
18. Zagminan, A.; Viana, J.; Dutra, R.; Pinheiro, A.; Carvalho, L.; Paschoal, M.A.; Falcai, A.; Lima-Neto, L.; Gonçalves, L.; Sousa, E. Effect of Ethyl Acetate Fraction of Urucum Extract on Cell Viability of Peripheral Blood Mononuclear Cells and Antimicrobial Activity in Human Pathogenic Bacteria. In Proceedings of the 11th Congress of the Latin American Association of Immunology, Medellin, Colombia, 13–16 October 2015; Volume 6. [CrossRef]
19. Lipinski, C.A.; Lombardo, F.; Dominy, B.W.; Feeney, P.J. Experimental and Computational Approaches to Estimate Solubility and Permeability in Drug Discovery and Development Settings. *Adv. Drug Deliv. Rev.* **2001**, *46*, 3–26. [CrossRef]
20. Zhao, L.; Mehmood, A.; Soliman, M.M.; Iftikhar, A.; Iftikhar, M.; Aboelenin, S.M.; Wang, C. Protective Effects of Ellagic Acid Against Alcoholic Liver Disease in Mice. *Front. Nutr.* **2021**, *8*, 744520. [CrossRef]
21. Ramos, R.S.; Borges, R.S.; de Souza, J.S.N.; Araujo, I.F.; Chaves, M.H.; Santos, C.B.R. Identification of Potential Antiviral Inhibitors from Hydroxychloroquine and 1,2,4,5-Tetraoxanes Analogues and Investigation of the Mechanism of Action in SARS-CoV-2. *Int. J. Mol. Sci.* **2022**, *23*, 1781. [CrossRef]
22. BIOVIA Discovery Studio-BIOVIA-Dassault Systèmes®. Available online: https://www.3ds.com/products-services/biovia/products/molecular-modeling-simulation/biovia-discovery-studio/ (accessed on 11 June 2022).
23. Rowlinson, S.W.; Kiefer, J.R.; Prusakiewicz, J.J.; Pawlitz, J.L.; Kozak, K.R.; Kalgutkar, A.S.; Stallings, W.C.; Kurumbail, R.G.; Marnett, L.J. A Novel Mechanism of Cyclooxygenase-2 Inhibition Involving Interactions with Ser-530 and Tyr-385. *J. Biol. Chem.* **2003**, *278*, 45763–45769. [CrossRef] [PubMed]
24. Orlando, B.J.; Malkowski, M.G. Substrate-Selective Inhibition of Cyclooxygeanse-2 by Fenamic Acid Derivatives Is Dependent on Peroxide Tone. *J. Biol. Chem.* **2016**, *291*, 15069–15081. [CrossRef] [PubMed]
25. RCSB PDB—7K6C: Crystal Structure of Dihydrofolate Reductase (DHFR) from Mycobacterium Abscessus ATCC 19977/DSM 44196 with NADP and Inhibitor P218. Available online: https://www.rcsb.org/structure/7K6C (accessed on 11 June 2022).
26. Charoensutthivarakul, S.; Thomas, S.E.; Curran, A.; Brown, K.P.; Belardinelli, J.M.; Whitehouse, A.J.; Acebrón-García-de-Eulate, M.; Sangan, J.; Gramani, S.G.; Jackson, M.; et al. Development of Inhibitors of SAICAR Synthetase (PurC) from *Mycobacterium Abscessus* Using a Fragment-Based Approach. *ACS Infect. Dis.* **2022**, *8*, 296–309. [CrossRef] [PubMed]

27. Whitehouse, A.J.; Thomas, S.E.; Brown, K.P.; Fanourakis, A.; Chan, D.S.-H.; Libardo, M.D.J.; Mendes, V.; Boshoff, H.I.M.; Floto, R.A.; Abell, C.; et al. Development of Inhibitors against *Mycobacterium Abscessus* TRNA (m [1] G37) Methyltransferase (TrmD) Using Fragment-Based Approaches. *J. Med. Chem.* **2019**, *62*, 7210–7232. [CrossRef] [PubMed]
28. Guedes, I.A.; Barreto, A.M.S.; Marinho, D.; Krempser, E.; Kuenemann, M.A.; Sperandio, O.; Dardenne, L.E.; Miteva, M.A. New Machine Learning and Physics-Based Scoring Functions for Drug Discovery. *Sci. Rep.* **2021**, *11*, 3198. [CrossRef]
29. Djedjibegovic, J.; Marjanovic, A.; Panieri, E.; Saso, L. Ellagic Acid-Derived Urolithins as Modulators of Oxidative Stress. *Oxid. Med. Cell. Longev.* **2020**, *2020*, 5194508. [CrossRef]
30. Aslan, A.; Hussein, Y.T.; Gok, O.; Beyaz, S.; Erman, O.; Baspinar, S. Ellagic Acid Ameliorates Lung Damage in Rats via Modulating Antioxidant Activities, Inhibitory Effects on Inflammatory Mediators and Apoptosis-Inducing Activities. *Environ. Sci. Pollut. Res.* **2020**, *27*, 7526–7537. [CrossRef]
31. Kiela, P.R.; Ghishan, F.K. Physiology of Intestinal Absorption and Secretion. *Best Pract. Res. Clin. Gastroenterol.* **2016**, *30*, 145–159. [CrossRef]
32. Baby, J.; Devan, A.R.; Kumar, A.R.; Gorantla, J.N.; Nair, B.; Aishwarya, T.S.; Nath, L.R. Cogent Role of Flavonoids as Key Orchestrators of Chemoprevention of Hepatocellular Carcinoma: A Review. *J. Food Biochem.* **2021**, *45*, e13761. [CrossRef]
33. Montenegro, C.; Gonçalves, G.; Oliveira Filho, A.; Lira, A.; Cassiano, T.; Lima, N.; Barbosa-Filho, J.; Diniz, M.; Pessôa, H. In Silico Study and Bioprospection of the Antibacterial and Antioxidant Effects of Flavone and Its Hydroxylated Derivatives. *Molecules* **2017**, *22*, 869. [CrossRef]
34. Gowthaman, U.; Jayakanthan, M.; Sundar, D. Molecular Docking Studies of Dithionitrobenzoic Acid and Its Related Compounds to Protein Disulfide Isomerase: Computational Screening of Inhibitors to HIV-1 Entry. *BMC Bioinform.* **2008**, *9*, S14. [CrossRef] [PubMed]
35. Hevener, K.E.; Zhao, W.; Ball, D.M.; Babaoglu, K.; Qi, J.; White, S.W.; Lee, R.E. Validation of Molecular Docking Programs for Virtual Screening against Dihydropteroate Synthase. *J. Chem. Inf. Model.* **2009**, *49*, 444–460. [CrossRef] [PubMed]
36. Raju, A.; Kulkarni, S.; Ray, M.K.; Rajan, M.G.R.; Degani, M.S. E84G Mutation in Dihydrofolate Reductase from Drug Resistant Strains of Mycobacterium Tuberculosis (Mumbai, India) Leads to Increased Interaction with Trimethoprim. *Int. J. Mycobacteriol.* **2015**, *4*, 97–103. [CrossRef] [PubMed]
37. Alim, S.; Bairagi, N.; Shahriyar, S.; Masnoon Kabir, M.; Habibur Rahman, M.; Sayeed Shahriyar, C. In Vitro Antibacterial Potential of Bixa Orellana L. against Some Pathogenic Bacteria and Comparative Investigation on Some Standard Antibiotics. *J. Pharmacogn. Phytochem.* **2016**, *5*, 178–181.
38. Poma-Castillo, L.; Espinoza-Poma, M.; Mauricio, F.; Mauricio-Vilchez, C.; Alvítez-Temoche, D.; Mayta-Tovalino, F. Antifungal Activity of Ethanol-Extracted Bixa Orellana (L) (Achiote) on Candida Albicans, at Six Different Concentrations. *J. Contemp. Dent. Pract.* **2019**, *20*, 1159–1163. [CrossRef] [PubMed]
39. Machín, L.; Tamargo, B.; Piñón, A.; Atíes, R.C.; Scull, R.; Setzer, W.N.; Monzote, L. *Bixa Orellana* L. (Bixaceae) and *Dysphania Ambrosioides* (L.) Mosyakin & Clemants (Amaranthaceae) Essential Oils Formulated in Nanocochleates against *Leishmania Amazonensis*. *Molecules* **2019**, *24*, 4222. [CrossRef]
40. Fernández-Calienes Valdés, A.; Mendiola Martínez, J.; Acuña Rodríguez, D.; Scull Lizama, R.; Gutiérrez Gaitén, Y. Antimalarial Activity of Hydroalcoholic Extract from Bixa *Orellana* L. *Rev. Cubana Med. Trop.* **2011**, *63*, 181–185.
41. Obakiro, S.B.; Kiprop, A.; K'owino, I.; Andima, M.; Owor, R.O.; Chacha, R.; Kigondu, E. Phytochemical, Cytotoxicity, and Antimycobacterial Activity Evaluation of Extracts and Compounds from the Stem Bark of Albizia Coriaria Welw Ex. Oliver. *Evid.-Based Complement. Altern. Med.* **2022**, *2022*, 7148511. [CrossRef]
42. Dey, D.; Ray, R.; Hazra, B. Antimicrobial Activity of Pomegranate Fruit Constituents against Drug-Resistant *Mycobacterium Tuberculosis* and β-Lactamase Producing *Klebsiella Pneumoniae*. *Pharm. Biol.* **2015**, *53*, 1474–1480. [CrossRef]
43. Sridevi, D.; Sudhakar, K.U.; Ananthathatmula, R.; Nankar, R.P.; Doble, M. Mutation at G103 of MtbFtsZ Altered Their Sensitivity to Coumarins. *Front. Microbiol.* **2017**, *8*, 578. [CrossRef]
44. Salih, E.Y.A.; Julkunen-Tiitto, R.; Luukkanen, O.; Fahmi, M.K.M.; Fyhrquist, P. Hydrolyzable Tannins (Ellagitannins), Flavonoids, Pentacyclic Triterpenes and Their Glycosides in Antimycobacterial Extracts of the Ethnopharmacologically Selected Sudanese Medicinal Plant Combretum Hartmannianum Schweinf. *Biomed. Pharmacother.* **2021**, *144*, 112264. [CrossRef] [PubMed]
45. Salih, E.Y.A.; Julkunen-Tiitto, R.; Luukkanen, O.; Sipi, M.; Fahmi, M.K.M.; Fyhrquist, P.J. Potential Anti-Tuberculosis Activity of the Extracts and Their Active Components of Anogeissus Leiocarpa (DC.) Guill. and Perr. with Special Emphasis on Polyphenols. *Antibiotics* **2020**, *9*, 364. [CrossRef] [PubMed]
46. Luthra, S.; Rominski, A.; Sander, P. The Role of Antibiotic-Target-Modifying and Antibiotic-Modifying Enzymes in Mycobacterium Abscessus Drug Resistance. *Front. Microbiol.* **2018**, *9*, 364. [CrossRef]
47. Da Fonseca Amorim, E.F.; Castro, E.J.M.; da Silva Souza, V.; Alves, M.S.; Dias, L.R.L.; Melo, M.H.F.; da Silva, I.M.A.; Villis, P.C.M.; Bonfim, M.R.Q.; Falcai, A.; et al. Antimicrobial Potential of Streptomyces Ansochromogenes (PB3) Isolated From a Plant Native to the Amazon Against Pseudomonas Aeruginosa. *Front. Microbiol.* **2020**, *11*, 574693. [CrossRef]
48. Silva, T.F.; Filho, J.R.N.C.; Fonsêca, M.M.L.B.; Dos Santos, N.M.; da Silva, A.C.B.; Zagmignan, A.; Abreu, A.G.; da Silva, A.P.S.; de Menezes Lima, V.L.; da Silva, N.H.; et al. Products Derived from Buchenavia Tetraphylla Leaves Have in Vitro Antioxidant Activity and Protect Tenebrio Molitor Larvae against Escherichia Coli-induced Injury. *Pharmaceuticals* **2020**, *13*, 46. [CrossRef] [PubMed]

49. Yong, Y.K.; Sulaiman, N.; Hakim, M.N.; Lian, G.E.C.; Zakaria, Z.A.; Othman, F.; Ahmad, Z. Suppressions of Serotonin-Induced Increased Vascular Permeability and Leukocyte Infiltration by Bixa Orellana Leaf Extract. *Biomed Res. Int.* **2013**, *2013*, 463145. [CrossRef]
50. Zouari Bouassida, K.; Makni, S.; Tounsi, A.; Jlaiel, L.; Trigui, M.; Tounsi, S. Effects of *Juniperus Phoenicea* Hydroalcoholic Extract on Inflammatory Mediators and Oxidative Stress Markers in Carrageenan-Induced Paw Oedema in Mice. *Biomed Res. Int.* **2018**, *2018*, 3785487. [CrossRef]
51. Mayouf, N.; Charef, N.; Saoudi, S.; Baghiani, A.; Khennouf, S.; Arrar, L. Antioxidant and Anti-Inflammatory Effect of Asphodelus Microcarpus Methanolic Extracts. *J. Ethnopharmacol.* **2019**, *239*, 111914. [CrossRef]
52. Yilmazer, N.; Coskun, C.; Gurel-Gurevin, E.; Yaylim, I.; Eraltan, E.H.; Ikitimur-Armutak, E.I. Antioxidant and Anti-Inflammatory Activities of a Commercial Noni Juice Revealed by Carrageenan-Induced Paw Edema. *Pol. J. Vet. Sci.* **2016**, *19*, 589–595. [CrossRef]
53. Laavola, M.; Leppänen, T.; Eräsalo, H.; Hämäläinen, M.; Nieminen, R.; Moilanen, E. Anti-Inflammatory Effects of Nortrachelogenin in Murine J774 Macrophages and in Carrageenan-Induced Paw Edema Model in the Mouse. *Planta Med.* **2017**, *83*, 519–526. [CrossRef]
54. Ghorbanzadeh, B.; Mansouri, M.; Hemmati, A.; Naghizadeh, B.; Mard, S.; Rezaie, A. A Study of the Mechanisms Underlying the Anti-Inflammatory Effect of Ellagic Acid in Carrageenan-Induced Paw Edema in Rats. *Indian J. Pharmacol.* **2015**, *47*, 292. [CrossRef] [PubMed]
55. Cardoso, A.M.; Martins de Sousa, E.; Viana-Niero, C.; Bonfim de Bortoli, F.; Pereira das Neves, Z.C.; Leão, S.C.; Junqueira-Kipnis, A.P.; Kipnis, A. Emergence of Nosocomial Mycobacterium Massiliense Infection in Goiás, Brazil. *Microbes Infect.* **2008**, *10*, 1552–1557. [CrossRef] [PubMed]
56. Mendonça, A.M.S.; de Andrade Monteiro, C.; Moraes-Neto, R.N.; Monteiro, A.S.; Mondego-Oliveira, R.; Nascimento, C.E.C.; da Silva, L.C.N.; Lima-Neto, L.G.; Carvalho, R.C.; de Sousa, E.M. Ethyl Acetate Fraction of Punica Granatum and Its Galloyl-HHDP-Glucose Compound, Alone or in Combination with Fluconazole, Have Antifungal and Antivirulence Properties against Candida Spp. *Antibiotics* **2022**, *11*, 265. [CrossRef]
57. Shukla, A.; Sharma, P.; Prakash, O.; Singh, M.; Kalani, K.; Khan, F.; Bawankule, D.U.; Luqman, S.; Srivastava, S.K. QSAR and Docking Studies on Capsazepine Derivatives for Immunomodulatory and Anti-Inflammatory Activity. *PLoS ONE* **2014**, *9*, e100797. [CrossRef]
58. Woods, G.L. *Susceptibility Testing of Mycobacteria, Nocardiae and Other Aerobic Actinomycetes: Approved Standard*; Clinical and Laboratory Standards Institute: Wayne, PA, USA, 2011; ISBN 1562387464.
59. Garcia de Carvalho, N.F.; Sato, D.N.; Pavan, F.R.; Ferrazoli, L.; Chimara, E. Resazurin Microtiter Assay for Clarithromycin Susceptibility Testing of Clinical Isolates of Mycobacterium Abscessus Group. *J. Clin. Lab. Anal.* **2016**, *30*, 751–755. [CrossRef]
60. Neves, R.C.; Andrade, R.G.; Carollo, C.A.; Boaretto, A.G.; Kipnis, A.; Junqueira-Kipnis, A.P. Hyptis Brevipes and Paullinia Pinnata Extracts and Their Fractions Presenting Activity against Mycobacterium Abscessus SubSP. Massiliense. *J. Trop. Pathol.* **2019**, *48*, 148–160. [CrossRef]
61. Lozoya-Pérez, N.E.; García-Carnero, L.C.; Martínez-Álvarez, J.A.; Martínez-Duncker, I.; Mora-Montes, H.M. Tenebrio Molitor as an Alternative Model to Analyze the Sporothrix Species Virulence. *Infect. Drug Resist.* **2021**, *14*, 2059–2072. [CrossRef]
62. Santos, A.L.E.; Júnior, C.P.S.; Neto, R.N.M.; Santos, M.H.C.; Santos, V.F.; Rocha, B.A.M.; Sousa, E.M.; Carvalho, R.C.; Menezes, I.R.A.; Oliveira, M.R.C.; et al. Machaerium Acutifolium Lectin Inhibits Inflammatory Responses through Cytokine Modulation. *Process Biochem.* **2020**, *97*, 149–157. [CrossRef]
63. Meshram, G.G.; Kumar, A.; Rizvi, W.; Tripathi, C.D.; Khan, R.A. Evaluation of the Anti-Inflammatory Activity of the Aqueous and Ethanolic Extracts of the Leaves of Albizzia Lebbeck in Rats. *J. Tradit. Complement. Med.* **2016**, *6*, 172–175. [CrossRef]

antibiotics

Article

Cinnamaldehyde Increases the Survival of Mice Submitted to Sepsis Induced by Extraintestinal Pathogenic *Escherichia coli*

Isabella F. S. Figueiredo [1,2], Lorena G. Araújo [1,3], Raissa G. Assunção [1,2], Itaynara L. Dutra [1,2], Johnny R. Nascimento [2,4], Fabrícia S. Rego [1], Carolina S. Rolim [1], Leylane S. R. Alves [1], Mariana A. Frazão [1], Samilly F. Cadete [1], Luís Cláudio N. da Silva [1,3], Joicy C. de Sá [1,3], Eduardo M. de Sousa [2,3], Waldir P. Elias [5], Flávia R. F. Nascimento [2,4] and Afonso G. Abreu [1,2,3,*]

1 Laboratório de Patogenicidade Microbiana, Programa de Pós-Graduação em Biologia Microbiana, Universidade Ceuma, São Luís 65075-120, Brazil; bellaafigueiredo@hotmail.com (I.F.S.F.); araujo.lorenabio@gmail.com (L.G.A.); raissa_guara@hotmail.com (R.G.A.); itaynaradutra@hotmail.com (I.L.D.); fabricia_sr@hotmail.com.br (F.S.R.); carolina_silvaa04@outlook.com (C.S.R.); leylanesusy@hotmail.com (L.S.R.A.); frazaoamariana@gmail.com (M.A.F.); samillyfrancad@gmail.com (S.F.C.); luiscn.silva@ceuma.br (L.C.N.d.S.); joicyvet@hotmail.com (J.C.d.S.)

2 Programa de Pós-Graduação em Ciências da Saúde, Universidade Federal do Maranhão, São Luís 65080-805, Brazil; john_nyramos@yahoo.com.br (J.R.N.); edmsousa@hotmail.com (E.M.d.S.); flavia.nascimento@ufma.br (F.R.F.N.)

3 Programa de Pós-Graduação em Biologia Microbiana, Universidade Ceuma, São Luís 65075-120, Brazil

4 Laboratório de Imunofisiologia, Departamento de Patologia, Universidade Federal do Maranhão, São Luís 65080-805, Brazil

5 Laboratório de Bacteriologia, Instituto Butantan, São Paulo 05503-900, Brazil; waldir.elias@butantan.gov.br

* Correspondence: afonso.abreu@ceuma.br

Citation: Figueiredo, I.F.S.; Araújo, L.G.; Assunção, R.G.; Dutra, I.L.; Nascimento, J.R.; Rego, F.S.; Rolim, C.S.; Alves, L.S.R.; Frazão, M.A.; Cadete, S.F.; et al. Cinnamaldehyde Increases the Survival of Mice Submitted to Sepsis Induced by Extraintestinal Pathogenic *Escherichia coli*. *Antibiotics* **2022**, *11*, 364. https://doi.org/10.3390/antibiotics11030364

Academic Editor: Carlos M. Franco

Received: 28 January 2022
Accepted: 4 March 2022
Published: 9 March 2022

Abstract: Several natural products have been investigated for their bactericidal potential, among these, cinnamaldehyde. In this study, we aimed to evaluate the activity of cinnamaldehyde in the treatment of animals with sepsis induced by extraintestinal pathogenic *E. coli*. Initially, the *E. coli* F5 was incubated with cinnamaldehyde to evaluate the minimum inhibitory and minimum bactericidal concentration. Animal survival was monitored for five days, and a subset of mice were euthanized after 10 h to evaluate histological, hematological, and immunological parameters, as well as the presence of bacteria in the organs. On the one hand, inoculation of bacterium caused the death of 100% of the animals within 24 h after infection. On the other hand, cinnamaldehyde (60 mg/kg) was able to keep 40% of mice alive after infection. The treatment significantly reduced the levels of cytokines in serum and peritoneum and increased the production of cells in both bone marrow and spleen, as well as lymphocytes at the infection site. Cinnamaldehyde was able to reduce tissue damage by decreasing the deleterious effects for the organism and contributed to the control of the sepsis and survival of animals; therefore, it is a promising candidate for the development of new drugs.

Keywords: sepsis; inflammation; *Escherichia coli*; cinnamaldehyde

1. Introduction

Sepsis is characterized by a life-threatening organ dysfunction resulting from an unregulated immune response to an infection [1]. It represents a serious global public health problem due to its elevated rate of morbidity and mortality in intensive care units; it is also a difficult and costly treatment [2,3].

Escherichia coli is a frequent bacterial agent of sepsis [4]. It is widely distributed and found in the large intestine of humans, as well as warm-blooded animals as part of the microbiota [5]. Virulence factors are encoded by genes restricted to pathogenic *E. coli*, being absent in commensal bacteria, making these microorganism etiologic agents of intestinal and extraintestinal infections, including bacteremia and sepsis [6].

Recently, when studying the function of protein involved in colonization (Pic) in the pathogenesis of sepsis caused by *E. coli*, our group demonstrated that this serine protease was responsible for inducing lethal sepsis in Swiss mice, causing their death in up to 12 h, with a significant increase in cytokine levels and other inflammatory mediators [7]. Pic is produced by *E. coli*, *Shigella flexneri*, and *Citrobacter rodentium* [8–10] and has several functions such as hemagglutination, coagulation cascade factor V degradation, mucinolytic activity [8,11], leukocyte surface glycoprotein cleavage [12], intestinal colonization of mice and rabbits [8,13], and cleavage of several molecules belonging to the complement system [14].

According to the World Health Organization, antibacterial drugs have become less effective, resulting in an accelerating worldwide health security emergency [15]. Therefore, due to the difficult treatment of several microbial infections, it is important to identify and characterize natural products that can be used in the treatment of microbial diseases. As an example, cinnamon bark has been widely studied for therapeutic purposes and has an important role for both research and popular use, since it contains a large amount of essential oil. Cinnamaldehyde is the main active constituent of cinnamon essential oil, and the major bioactive compound isolated from the leaves of *Cinnamomum* spp. [16].

Several functions have been described for the compound, such as: anticancer/antitumor activity [17]; cardioprotective effect [18]; anti-inflammatory properties [19,20]; and antimicrobial activity against several pathogens, such as fungi [21] and bacteria [16,22].

Therefore, due to several properties of cinnamaldehyde, especially the antimicrobial and anti-inflammatory activity, the objective of this work was to evaluate the activity of cinnamaldehyde in the treatment of animals with sepsis induced by extraintestinal pathogenic *E. coli*.

2. Results

2.1. Cinnamaldehyde Increases the Survival Rate of Septic Animals

On the one hand, cinnamaldehyde at a concentration of 30 mg/kg showed no effect in the survival rate in prophylactic or therapeutic treatments. On the other hand, when animals were treated with cinnamaldehyde at a concentration of 60 mg/kg, the survival rate increased by 16.6% in the prophylactic treatment and by 40% in the therapeutic treatments as compared with non-treated infected mice (Figure 1).

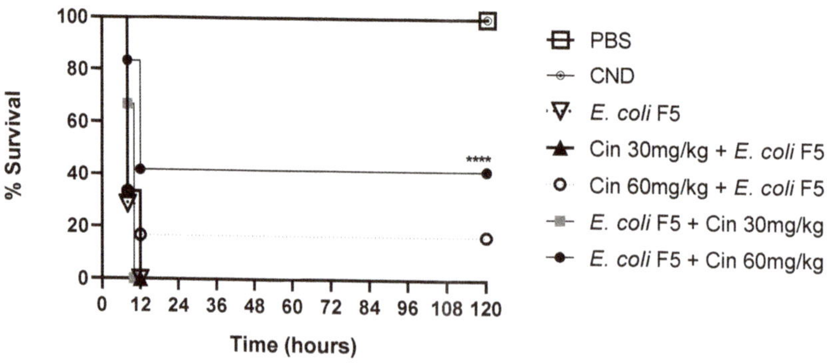

Figure 1. Survival curve of animals infected with *E. coli* F5 and cinnamaldehyde treatment. The mice used in the experiment were distributed into seven groups ($n = 6$). The animals in the non-infected control groups received PBS or cinnamaldehyde (CIN 60 mg/kg). The remaining 3 groups were infected with *E. coli* F5. Prophylactic cinnamaldehyde groups received 30 (prophylactic CIN 30) and 60 mg/kg (prophylactic CIN 60) of cinnamaldehyde, 5 days before the infection, and therapeutic cinnamaldehyde groups were treated with 30 (therapeutic CIN 30) and 60 mg/kg (therapeutic CIN 60) of cinnamaldehyde, 2 h after *E. coli* F5 infection. Then, the survival was evaluated every 12 h until the 5th day after infection. **** $p < 0.0001$ as compared with the *E. coli* F5 group.

After these results, the therapeutic treatment with the dose of 60 mg/Kg was chosen to follow the study and the mice were euthanized ten hours post infection to evaluate the immunological parameters.

2.2. *Cinnamaldehyde Promotes Bactericidal Effect In Vitro, but Not In Vivo*

Using the MIC method, it was shown that cinnamaldehyde inhibited *E. coli* F5 growth (6 mg/mL). In addition, cinnamaldehyde was bactericidal at the concentration of 12 mg/mL.

E. coli F5 was present in both bacterial groups (F5 and treated) on the peritoneal lavage, blood, liver, spleen, and lungs, indicating that the mice had systemic bacteremia. In the PBS and cinnamaldehyde controls, as expected, there was no bacterial growth (data not shown). Nevertheless, no significant difference was observed among the infected groups (Figure 2), suggesting that the increase in survival was not related to bactericidal activity of cinnamaldehyde.

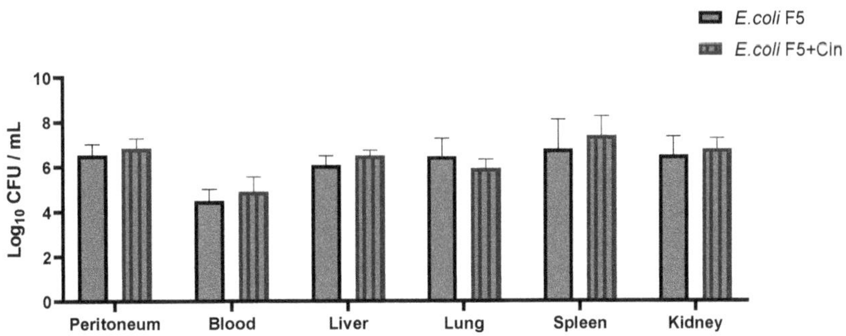

Figure 2. CFU counting from blood, peritoneal lavage, lungs, liver, spleen, and kidneys from animals infected with *E. coli* F5 and cinnamaldehyde treatment. The animals used in the experiment were distributed into 4 groups (n = 6). The groups received *E. coli* F5 or *E. coli* F5 followed by cinnamaldehyde treatment 60 mg/kg after two hours. After 10 h, the mice were euthanized. Aliquots of peritoneal fluid and blood were diluted in sterile PBS and serial dilutions were cultivated on MacConkey for evaluation of CFU. To quantify bacteria in the organs, part of them were macerated, and dilutions were cultivated on MacConkey. After overnight incubation at 37 °C, the CFUs were counted, and the results expressed as Log of CFU/mL.

2.3. *Cinnamaldehyde Decreases Tissue Damage in Animals Promoted by E. coli F5*

The animals of infected groups presented histopathological alterations typical of inflammation in the lungs, spleen, and liver as compared with the non-infected control groups (PBS and CIN groups), with the presence of edema, hemorrhage, and cellular infiltrate. However, it was possible to observe a moderate reduction in the hemorrhagic process, as well as a reduction in the inflammatory infiltrate in animals infected and treated with cinnamaldehyde in relation to the non-treated infected group (Table 1 and Figure 3).

The presence of congestion in the groups analyzed was due to the euthanasia procedure. Necrosis was not observed in any analyzed organs; however, the presence of mild periportal edema was detected in the liver of most animals in the non-treated infected group.

In summary, the cinnamaldehyde treatment was able to reduce tissue damage, especially in the lungs, contributing to animal survival.

Table 1. Histopathological evaluation of kidneys, lungs, and liver from animals submitted to intraperitoneal inoculation of bacteria and cinnamaldehyde treatment.

		Hemorrhage	Infiltrated	Edema	Necrosis
Kidney	PBS	0	0.7 ± 0.5	0	0
	Cin	0	1.0 ± 0.8	0	0
	E. coli F5	1.3 ± 0.5 [a]	1.7 ± 0.5	0	0
	E. coli F5+Cin	0.7 ± 0.7	1.7 ± 0.5	0	0
Lung	PBS	0	0.7 ± 0.5	0	0
	Cin	0	1.0 ± 0.0	0	0
	E. coli F5	1.3 ± 0.7	2.2 ± 0.4	0	0
	E. coli F5+Cin	1.2 ± 0.4	1.8 ± 0.4	0	0
Liver	PBS	0	1.0 ± 0.0	0	0
	Cin	0	1.0 ± 0.0	0	0
	E. coli F5	2.8 ± 0.7	2.2 ± 0.4	0.7 ± 0.5	0
	E. coli F5+Cin	1.2 ± 0.4	2.0 ± 0.0	0.3 ± 0.5	0

[a] The results are expressed as mean ± SD of the scores: 0, absent; 1, mild; 2, moderate; 3, intense.

In Figure 3, arrows show inflammatory infiltrated area and asterisks show hemorrhagic area.

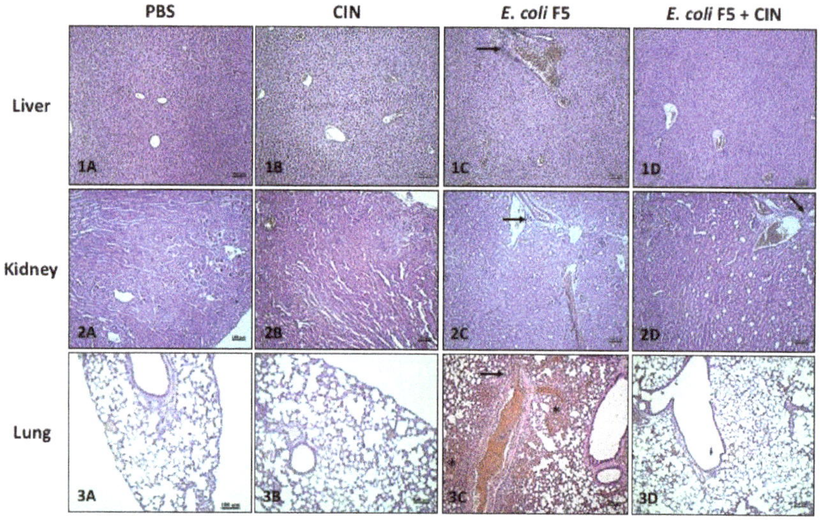

Figure 3. Histopathological analyzes of lungs, liver, and kidneys from animals infected with *E. coli* F5 and cinnamaldehyde treatment. Ten hours after peritoneal inoculation of *E. coli* F5, the organs of the animals were removed for histopathological analysis. Liver: (**1A**) PBS; (**1B**) CIN; (**1C**) *E. coli* F5; (**1D**) *E. coli* F5+CIN. Kidney: (**2A**) PBS; (**2B**) CIN; (**2C**) *E. coli* F5; (**2D**) *E. coli* F5+CIN. Lung: (**3A**) PBS; (**3B**) CIN; (**3C**) *E. coli* F5; (**3D**) *E. coli* F5+CIN. The scale bar is 100 μm.

2.4. Cinnamaldehyde Treatment Promotes an Increase in Leukocytes in the Bone Marrow, Peritoneal Cavity, and Spleen

The infection induced a reduction in the total blood cell count as compared with PBS and CIN non-infected groups, which was related to a decreasing in lymphocytes number. The neutrophils and monocytes were not reduced after infection. The treatment with cinnamaldehyde did not alter the cell profile in the blood in the infected mice (Figure 4).

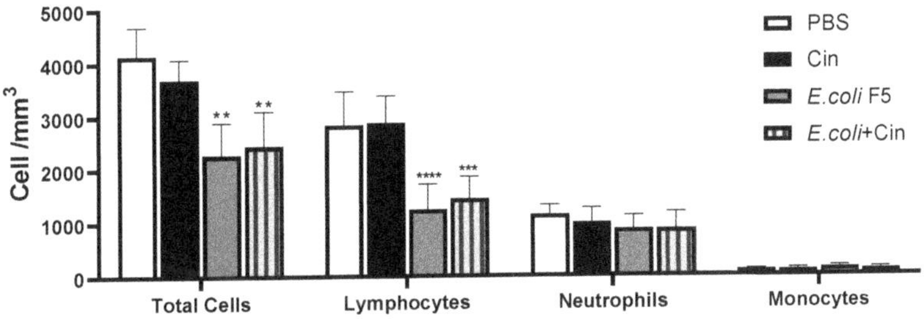

Figure 4. Total and differential blood cell counts from animals infected with *E. coli* F5 and cinnamaldehyde treatment. Ten hours after peritoneal inoculation of *E. coli* F5, blood samples were collected for total and differential cell counting. The results represent the mean ± S.D. of 6 animals/group. ** $p < 0.01$ as compared with the PBS group, *** $p < 0.001$ as compared with the PBS group, and **** $p < 0.0001$ as compared with the PBS group.

In the same way, the number of peritoneal cells was decreased in the infected animals as compared with the PBS and CIN non-infected groups. The treatment with CIN induced an increase in the number of peritoneal cells as compared with the non-treated infected group (Figure 5).

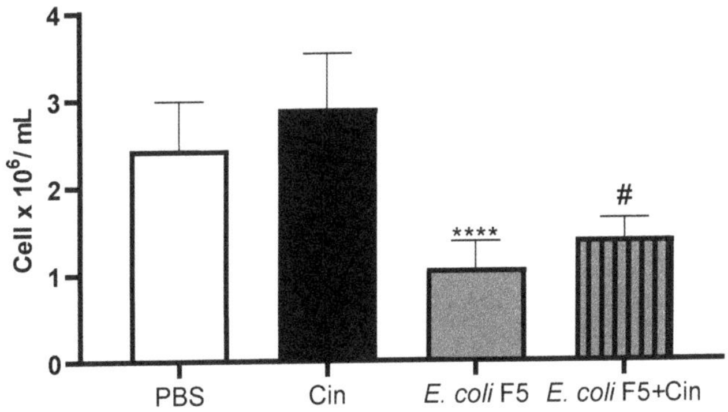

Figure 5. Peritoneal cell counts of animals infected with *E. coli* F5 and cinnamaldehyde treatment. Ten hours after peritoneal inoculation of *E. coli* F5, the peritoneal cells of the mice were collected from the lavage of the peritoneal cavity. The cells were stained with 0.05% violet crystal in 30% acetic acid and counted under a light microscope. The results represent the mean ± S.D. of 6 animals/group. **** $p < 0.0001$ as compared with the CIN group and # $p < 0.05$ as compared with the *E. coli* F5 group.

The decrease in blood cell and peritoneal cell numbers in the infected mice was not related to a decrease in cell production by bone marrow, considering that no differences among infected and PBS groups were observed in bone marrow cell counts (Figure 6).

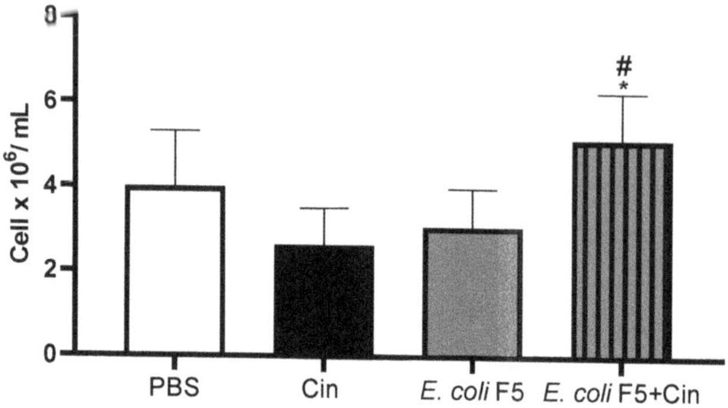

Figure 6. Total bone marrow cell counts from animals infected with *E. coli* F5 and cinnamaldehyde treatment. Ten hours after peritoneal inoculation of *E. coli* F5, bone marrow cells were obtained. The cells were stained with 0.05% violet crystal in 30% acetic acid and counted under a light microscope. The results represent the mean ± S.D. of 6 animals/group. * $p < 0.05$ as compared with the *E. coli* F5 group and # $p < 0.01$ as compared to the CIN group.

It is possible to observe a small reduction in the number of cells in the non-infected group. However, this reduction is not statistically significant. On the other hand, the administration of cinnamaldehyde to infected animals leads to a significant increase in the number of total cells in both bone marrow and spleen (Figure 7) as compared with the non-treated infected group.

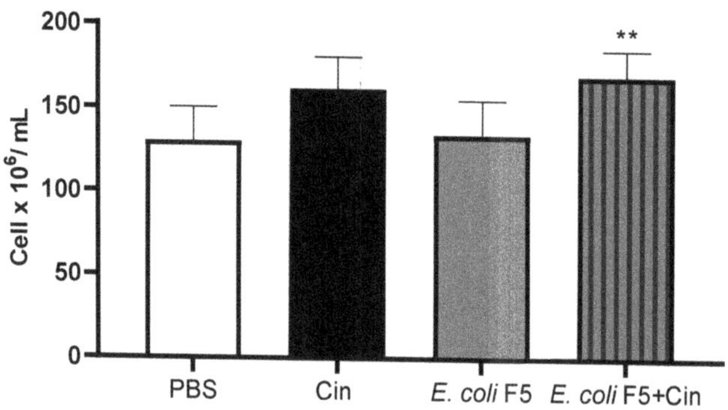

Figure 7. Total spleen cell counts from animals infected with *E. coli* F5 and cinnamaldehyde treatment. Ten hours after peritoneal inoculation of *E. coli* F5, spleen cells were obtained by removing and macerating the organ using PBS. The cells were stained with 0.05% violet crystal in 30% acetic acid and counted under a light microscope. The results represent the mean ± S.D. of 6 animals/group. ** $p < 0.05$ as compared with the *E. coli* F5 group.

2.5. Cinnamaldehyde Promotes a Decrease in Cytokine Levels in Serum and Peritoneum

Proinflammatory cytokines (IL-6, IL-12, IFN-γ, and TNF-α), anti-inflammatory cytokine (IL-10), and chemokine (MCP-1) were detected in serum (Figure 8) and peritoneal lavage (Figure 9) from the *E. coli* F5 infected group. The cinnamaldehyde treatment significantly reduced the levels of all these mediators, with the exception of TNF-α in the serum.

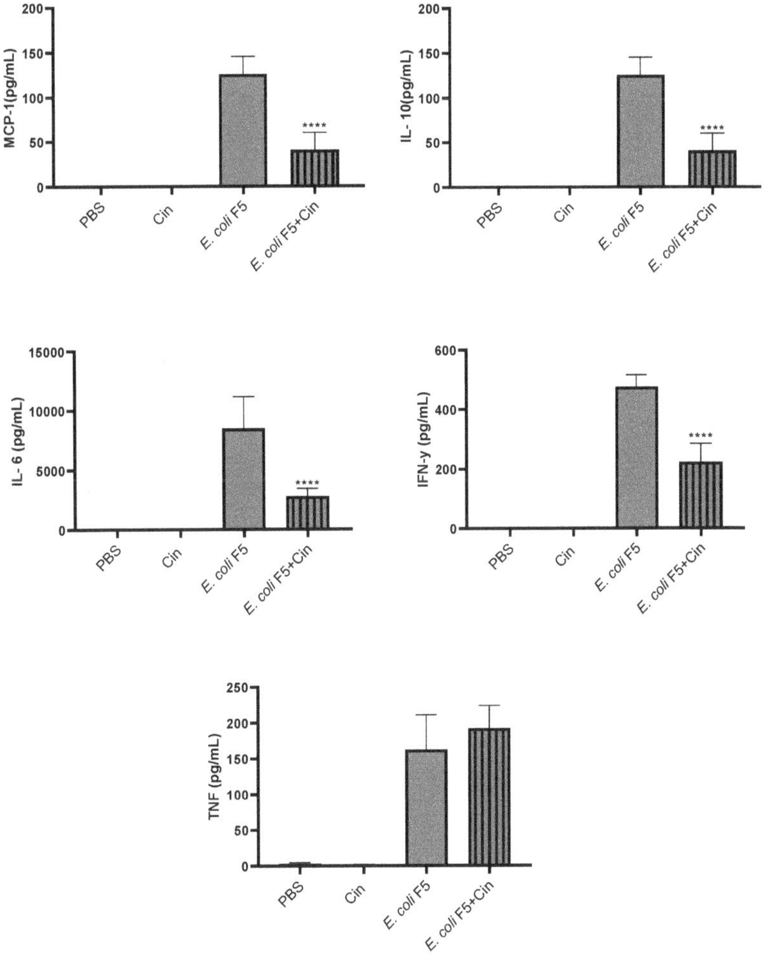

Figure 8. Detection of cytokines in the sera of animals infected with *E. coli* F5 and cinnamaldehyde treatment. Ten hours after peritoneal inoculation of *E. coli* F5, animal sera were used for cytokines dosage by flow cytometry. The results represent the mean ± S.D. of 6 animals/group. **** $p < 0.0001$ as compared with the *E. coli* F5 group.

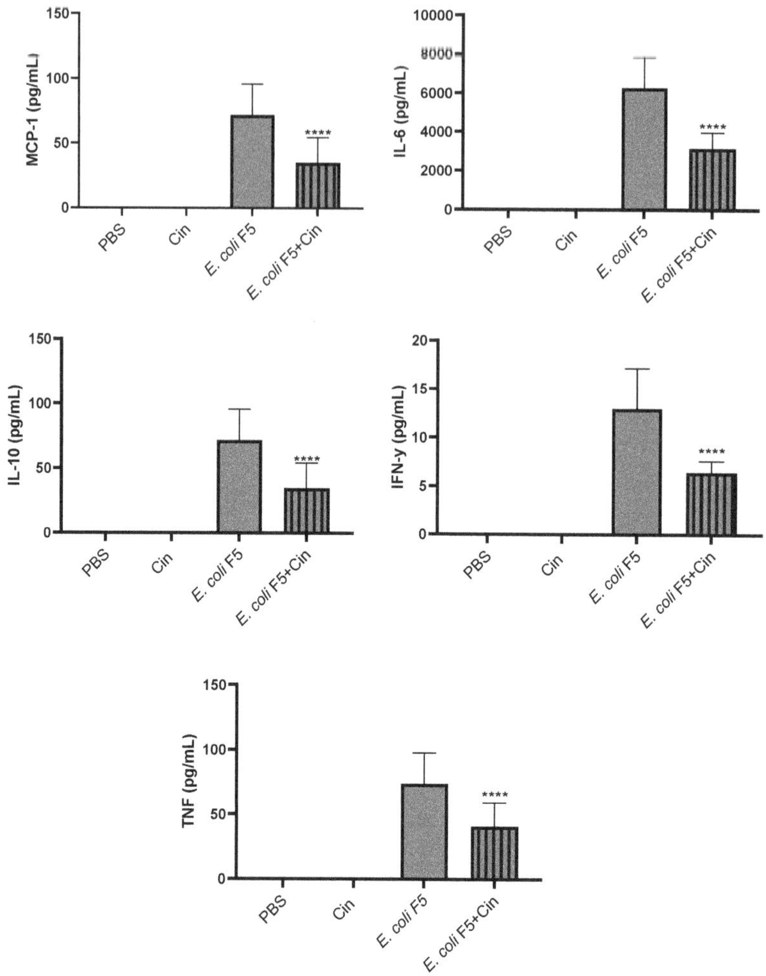

Figure 9. Detection of cytokines in the peritoneal lavage of animals infected with *E. coli* F5 and cinnamaldehyde treatment. Ten hours after peritoneal inoculation of *E. coli* F5, animal peritoneal lavages were used for cytokines dosage by flow cytometry. The results represent the mean ± S.D. of 6 animals/group. **** $p < 0.0001$ as compared with the *E. coli* F5 group.

3. Discussion

We have recently shown that the intraperitoneal inoculation of *E. coli* strain F5 caused the death of 100% of the animals within 24 h after infection [7]. In this work, we demonstrated that a treatment with cinnamaldehyde (60 mg/kg) improved the survival of animals infected with *E. coli* F5 using the same murine sepsis model. This effect was not due to a bactericidal activity, since there was no reduction in bacteria in the blood and organs of animals after cinnamaldehyde treatment, but likely, due to an anti-inflammatory effect indicated by a reduction in systemic inflammatory mediators and decreased lung damage.

Histological analyses from liver, lungs, and kidneys of the animals infected with *E. coli* F5 presented cellular infiltrate, hemorrhage, and edema. However, the treatment with cinnamaldehyde (60 mg/kg) reduced tissue damage. Probably, the animals survived longer due to the decrease in lung inflammation induced by the compound.

As observed here, several studies have also demonstrated protective effects of cinnamaldehyde in controlling infections. Recently, we demonstrated that cinnamaldehyde treatment reduced the intestinal colonization of mice infected by *E. coli* [22]. Yang et al. [23] showed that trans-cinnamaldehyde attenuated the intestinal histological damage in *Cronobacter sakazakii*-infected newborn mice. Tung et al. [24] showed that bioactive phytochemicals of leaf essential oils of *Cinnamomum osmophloeum* significantly reduced the incidence of liver lesions due to lipopolysaccharide/D-galactosamine-induced acute hepatitis. In addition, cinnamaldehyde was able to protect against rat intestinal ischemia/reperfusion injuries [25], reduced the infarct area in cerebral ischemia mouse model [26], as well as exhibited cardioprotective effects against acute ischemic myocardial injury induced by isoproterenol in rats [18].

In this study, the blood cell counts showed a decrease in leucocytes from animals infected with *E. coli* F5 as compared with the other groups not infected, which may have occurred due to a reduction in lymphocytes. Previously, we described that Pic reduced the number of T and B lymphocytes and compromised the expression of several cell surface molecules in leukocytes, such as CD80 and CD86 [7]. Probably, this reduction in blood cell and peritoneal cell numbers was related to a failure in the recruitment caused by Pic, since in the bone marrow the production was not compromised.

Other studies have also shown that lymphocyte apoptosis was increased in septic patients, leading to a persistent and profound lymphopenia [27,28]. Here, we showed that cinnamaldehyde induced an increase in leukocytes in the bone marrow, and spleen, as well as in the peritoneum, the focus of infection. However, treatment with cinnamaldehyde did not alter the cell profile in the blood in the infected mice. The same was observed by Mendes et al. [29] who investigated the effects of cinnamaldehyde in the inflammatory response triggered by LPS injection in mice. The study showed that cinnamaldehyde treatment did not affect the number of total peripheral blood leukocytes or the number of circulating polymorphonuclear cells in LPS-injected mice. However, an evident reduction in the number of circulating mononuclear cells was observed in the same animals.

It is important to mention that neutrophils are important during the initial immune response to eliminate the microorganisms in the sepsis [30]. As previously demonstrated by Dutra et al. [7], the production of neutrophils is affected by Pic. Thus, the reduction in neutrophils in the *E. coli* F5 group may also have contributed to the high mortality of animals. Moreover, cinnamaldehyde was also not able to induce an increase in the number of neutrophils in serum. In fact, even with an increase in the number of cells in the peritoneum, the treatment with cinnamaldehyde promoted a reduction in the level of cytokines in the peritoneal lavage.

We demonstrated that Pic induced increased production of cytokines, since mice infected with *E. coli* F5 significantly produced more cytokines than its defective Pic mutant [7]. Other studies have also described increased production of cytokine in murine model of sepsis, such as IL-6, as well as the chemokine MCP-1 [29,31–33]. Here, when animals with sepsis were treated with cinnamaldehyde, it was possible to observe a reduction in the levels of all cytokines in both serum and peritoneum, with the exception of TNF-α in the serum. Interestingly, there was a reduction in the level of anti-inflammatory cytokine IL-10, suggesting that cinnamaldehyde induces an overall suppression in the immune response.

According to our data, other studies have described the anti-inflammatory effect of cinnamaldehyde in reducing the release of several cytokines. Pannee et al. [34] showed that, in a concentration-dependent manner, cinnamaldehyde significantly inhibited the expression of IL-6, IL-1β, and TNF-α in activated J774A.1 cells. Liao et al. [20] reported that cinnamic aldehyde reduced TNF in LPS-induced murine macrophages, as well as in a carrageenan-induced paw edema model, demonstrating that the compound had excellent anti-inflammatory activities. Mendes et al. [29] demonstrated that cinnamaldehyde also significantly diminished TNFα in plasma.

It is worth noting that the reduction in the level of inflammatory cytokines is important, since exacerbated increased levels of cytokines cause a systemic disorder, which can lead to death. It has already been reported that high concentrations of MCP-1 have also been linked to mortality in sepsis in humans. Zhu et al. [35] reported that plasma levels of MCP-1 were significantly higher in a non-survivor as compared with a survivor group. Mera et al. [36] also showed that the expression levels of MCP-1 soon after intensive care unit admission were higher in non-survivors as compared with survivors.

4. Materials and Methods

4.1. Bacterial Strain

E. coli strain F5, an extraintestinal pathogenic *E. coli* (ExPEC) isolated from the bloodstream [7], was used in this study as the sepsis inducing bacterium. The strain was aerobically grown at 37 °C for 18 h in Luria-Bertani (LB) broth, MacConkey, or LB agar. The strain was kept in LB broth supplemented with 15% glycerol at −80 °C until experimental procedures.

4.2. Minimum Inhibitory (MIC) and Minimum Bactericidal Concentration (MBC)

E. coli F5 was cultivated in Mueller Hinton broth and grown until it reached an optical density of 0.1, determined at 600 nm. A total of 10 µL of this suspension were added to wells containing cinnamaldehyde at concentrations ranging from 200 to 2500 µg/mL. After 24 h incubation at 37 °C, resazurin 0.03% was used to determine the MIC.

Aliquots from all the wells which showed no visible bacterial growth were seeded on MacConkey agar plates and incubated for 24 h at 37 °C to determine the MBC [37]. All assays were performed in triplicate with at least two repetitions.

Cinnamaldehyde (trans-cinnamaldehyde 99%) used in this study was obtained commercially from Sigma-Aldrich (Darmstadt, Germany). To carry out the experiments, 20% dimethyl sulfoxide (DMSO) was used as solvent.

4.3. Animals

Six-to-eight-week-old female Swiss mice, weighing ~25 g, were used in the study. The animals were obtained from the Central Animal House of the Ceuma University (São Luís, Brazil) and maintained at 26 + 2 °C, 44% to 56% relative humidity, under 12 h light-dark cycles and maintained with free access to sterile food and water.

4.3.1. Experimental Design

A total of 42 animals were distributed into seven experimental groups ($n = 6$): two control groups, without infection, which received only phosphate buffered solution (PBS) or cinnamaldehyde (CIN 60 mg/kg); a third group infected with *E. coli* F5 and did not receive any treatment (*E. coli* F5); two prophylactic groups that received cinnamaldehyde at 30 mg/kg (prophylactic CIN 30) or 60 mg/kg (prophylactic CIN 60) for 5 days before infection; finally, two therapeutic groups that received cinnamaldehyde at 30 mg/Kg (therapeutic CIN 30) or 60 mg/kg (therapeutic CIN 60) 2 h after infection. The animals were infected with suspensions containing 10^9 CFU/mL *E. coli* F5 (200 µL) inoculated intraperitoneally. All the treatments were performed by gavage.

The survival of the animals was evaluated every 12 h until the 5th day after infection [31], and after this period of time, the mice were euthanized with an overdose of the anesthetic (150 mg/kg ketamine hydrochloride and 120 mg/kg xylazine hydrochloride).

After survival analysis, it was possible to observe that the therapeutic CIN 60 group survived for longer periods. Then, another 24 animals were divided into 4 groups: PBS, CIN, *E. coli* F5, and therapeutic CIN 60, which were infected and treated, as described above.

4.3.2. Blood Sampling

After ten hours of infection, the animals were anesthetized to collect the blood with a solution of 2% xylazine chloride (20 mg/kg) and 5% ketamine chloride (25 mg/kg) at a 2:1 ratio [38] via an intramuscular injection. The blood was used to determine differential blood cell counts, nitric oxide (NO) and cytokine levels, as well as to quantify the presence of *E. coli* F5. Blood cell counts were performed using Bioclin 2.8 Vet (Bioclin, Belo Horizonte, Brazil).

4.3.3. Bone Marrow, Spleen, and Peritoneal Cell Counting

After euthanasia, animal peritoneal cells were collected by washing the peritoneal cavity with 5 mL sterile ice-cold PBS. Then, the femur was removed and perfused with 1 mL of PBS to obtain bone marrow cells. Finally, the spleen, liver, kidneys and the lungs were removed. The spleen was crushed with 5 mL PBS and passed through a silk sieve. Total cell number counting was performed according to Cruz et al. [39] using a hemocytometer with the aid of an optical microscope at 400× magnification.

4.3.4. Colony Forming Units (CFU) Counting

To quantify the number of bacteria in the organs, serial dilutions from the peritoneal lavage, blood, and macerated organs (liver, lungs, spleen, and kidneys), were cultivated on MacConkey agar at 37 °C. After overnight incubation, CFUs were counted and the results were expressed as Log of CFU/mL, as described by Dutra [7].

4.3.5. Histopathological Evaluation

The collected samples (kidney, liver, and lung) were fixed in 10% formaldehyde solution for 24 h. They were sectioned and processed for inclusion in paraffin. Sections with 5 mm were stained in hematoxylin-eosin for histopathological analysis. The following parameters were evaluated: edema, necrosis, cellular infiltrate, and hemorrhage. The classification (0 = absent; 1 = slight; 2 = moderate; 3 = intense) was performed, according to that described by Liberio et al. [40].

4.3.6. Cytokines Quantification

Dosage of MCP-1, IFN-γ, TNF-α, IL-12, IL-10, and IL-6 in serum and peritoneal lavage from animals was performed by flow cytometry, using the Mouse Inflammation Cytokine kit (Becton Dickinson Biosciences, San Jose, CA, USA), according to manufacturer's instructions.

4.4. Statistical Analysis

Statistical analyses were performed by using the GraphPad Prism software, version 8.0. The results were expressed as the mean ± standard deviation, and were subjected to ANOVA, followed by T-test and Tukey's multiple test, or Kruskal–Wallis and Mann–Whitney tests when the data normality assumption was not satisfied. The Kaplan–Meier curve and the log-rank test were used for survival analysis. The differences were considered to be significant when $p < 0.05$.

4.5. Ethics Approval

The experimental procedures were carried out under approved guidelines of the National Institutes of Health (NIH) Guide for the Care and Use of Laboratory Animals. All procedures were approved by the Committee of Ethics in Research at the Ceuma University (process no. 22/18).

5. Conclusions

Taken together, our data point to an important immunomodulator effect of cinnamaldehyde on the sepsis caused by extraintestinal pathogenic *E. coli*. The compound was also able to reduce the levels of cytokines IL-6, IL-10, IL-12, TNF-α, and IFN-γ, as well as the chemokine MCP-1 in serum and peritoneum. In addition, cinnamaldehyde increased the production of cells in both bone marrow and spleen, and the lymphocyte number at the infection site. Finally, it was able to reduce the inflammatory infiltrate in animals infected and treated in relation to non-treated infected group, decreasing the deleterious effects for the organism and contributing positively to the control of the sepsis and survival of animals.

The compound is efficient in the treatment of the sepsis caused by extraintestinal pathogenic *E. coli* and a promising candidate for the development of new drugs that can be used in the treatment of infections.

Author Contributions: Conceptualization, I.F.S.F. and A.G.A.; methodology, F.R.F.N. and A.G.A.; validation, J.R.N., F.R.F.N. and A.G.A.; formal analysis, J.R.N., W.P.E., F.R.F.N. and A.G.A.; investigation, I.F.S.F., L.G.A., R.G.A., I.L.D., J.R.N., F.S.R., C.S.R., L.S.R.A., M.A.F., S.F.C., J.C.d.S. and E.M.d.S.; resources, L.C.N.d.S., F.R.F.N. and A.G.A.; writing—original draft preparation, I.F.S.F.; writing—review and editing, L.C.N.d.S., W.P.E., F.R.F.N. and A.G.A.; supervision, F.R.F.N. and A.G.A.; project administration, A.G.A.; funding acquisition, F.R.F.N. and A.G.A. All authors have read and agreed to the published version of the manuscript.

Funding: This research was funded by Fundação de Amparo à Pesquisa e ao Desenvolvimento Científico e Tecnológico do Maranhão—FAPEMA (grants UNIVERSAL 01278/17, BEPP 02499/18, and IECT-02876/17), Conselho Nacional de Desenvolvimento em Pesquisa—CNPq (grants 426048/2018-0 and 306835/2018-5), and Coordenação de Aperfeiçoamento Pessoal de Nível Superior—CAPES (finance code 001).

Institutional Review Board Statement: The animal study protocol was approved by the Committee of Ethics in Research at the Ceuma University on 27 November 2018 (protocol code 22/18).

Informed Consent Statement: Not applicable.

Data Availability Statement: Data are contained within the article.

Acknowledgments: We thank the Ceuma University for technical and financial support.

Conflicts of Interest: The authors declare no conflict of interest.

References

1. Singer, M.; Deutschman, C.S.; Seymour, C.W.; Shankar-hari, M.; Annane, D.; Bauer, M.; Bellomo, R.; Bernard, G.R.; Chiche, J.D.; Coopersmith, C.M.; et al. The Third International Consensus Definitions for Sepsis and Septic Shock (Sepsis-3). *JAMA* **2016**, *315*, 801–810. [CrossRef] [PubMed]
2. Chalupka, A.N.; Talmor, D. The Economics of sepsis. *Crit. Care Clin.* **2012**, *28*, 57–76. [CrossRef] [PubMed]
3. Rulim, A.L.L.; Rulim, M.A.B.; Rolim Neto, M.L.; Pita, P.; Batista, I.L. Sepsis Epidemiology: Settings and Contexts. *Int. Arch. Med.* **2017**, *10*, 1–2. [CrossRef]
4. Gotts, J.E.; Matthay, M.A. Sepsis: Pathophysiology and clinical management. *BMJ* **2016**, *353*, i1585. [CrossRef] [PubMed]
5. Gomes, T.A.T.; Elias, W.P.; Scaletsky, I.C.A.; Guth, B.E.C.; Rodrigues, J.F.; Piazza, R.M.F.; Ferreira, L.C.S.; Martinez, M.B. Diarrheagenic *Escherichia coli*. *Braz. J. Microbiol.* **2016**, *47*, 3–30. [CrossRef] [PubMed]
6. Koga, V.L.; Tomazetto, G.; Cyoia, P.S.; Neves, M.S.; Vidotto, M.C.; Nakazato, G.; Kobayashi, R.K.T. Molecular screening of virulence genes in extraintestinal pathogenic *Escherichia coli* isolated from human blood culture in Brazil. *BioMed Res. Int.* **2014**, *2014*, 1–10. [CrossRef]
7. Dutra, I.L.; Araujo, L.G.; Assuncao, R.G.; Lima, Y.A.; Nascimento, J.R.; Vale, A.A.M.; Alves, P.C.S.; Trovao, L.O.; Santos, A.C.M.; Silva, R.M.; et al. Pic-Producing *Escherichia coli* Induces High Production of Proinflammatory Mediators by the Host Leading to Death by Sepsis. *Int. J. Mol. Sci.* **2020**, *21*, 2068. [CrossRef]
8. Abreu, A.G.; Abe, C.M.; Nunes, K.O.; Moraes, C.T.; Chavez-Duenas, L.; Navarro-Garcia, F.; Barbosa, A.S.; Piazza, R.M.F.; Elias, W.P. The serine protease Pic as a virulence factor of atypical enteropathogenic *Escherichia coli*. *Gut Microbes* **2016**, *7*, 115–125. [CrossRef]
9. Henderson, I.R.; Czeczulin, J.; Eslava, C.; Noriega, F.; Nataro, J.P. Characterization of Pic, a secreted protease of *Shigella flexneri* and enteroagregative *Escherichia coli*. *Infect. Immun.* **1999**, *67*, 5587–5596. [CrossRef]
10. Bhullar, K.; Zarepour, M.; Yu, H.; Yang, H.; Croxen, M.; Stahl, M.; Finlay, B.B.; Turvey, S.E.; Vallance, B.A. The serine protease autotransporter Pic modulates *Citrobacter rodentium* pathogenesis and its innate recognition by the host. *Infect. Immun.* **2015**, *83*, 2636–2650. [CrossRef]
11. Dutta, P.R.; Cappello, R.; Navarro-Garcia, F.; Nataro, J.P. Functional comparison of serine protease autotransporters of enterobacteriaceae. *Infect. Immun.* **2002**, *70*, 7105–7113. [CrossRef] [PubMed]
12. Ruiz-Perez, F.; Wahid, R.; Faherty, C.S.; Kolappaswamy, K.; Rodriguez, L.; Santiago, A.; Murphy, E.; Cross, A.; Sztein, M.B.; Nataro, J.P. Serine protease autotransporters from *Shigella flexneri* and pathogenic *Escherichia coli* target a broad range of leukocyte glycoproteins. *Proc. Natl. Acad. Sci. USA* **2011**, *108*, 12881–12886. [CrossRef] [PubMed]

13. Munera, D.; Ritchie, J.M.; Hatzios, S.K.; Bronson, R.; Fang, G.; Schadt, E.E.; Davis, B.M.; Waldor, M.K. Autotransporters but not pAA are critical for rabbit colonization by Shiga toxin-producing *Escherichia coli* O104:H4. *Nat. Commun.* **2014**, *5*, 1–9. [CrossRef] [PubMed]
14. Abreu, A.G.; Fraga, T.R.; Martinez, A.P.G.; Kondo, M.Y.; Juliano, M.A.; Juliano, L.; Navarro-Garcia, F.; Isaac, L.; Barbosa, A.S.; Elias, W.P. The serine protease Pic from enteroaggregative *Escherichia coli* mediates immune evasion by the direct cleavage of complement proteins. *J. Infect. Dis.* **2015**, *212*, 106–115. [CrossRef]
15. World Health Organization (WHO). *Antimicrobial Resistance: Global Report on Surveillance*; WHO: Geneva, Switzerland, 2014.
16. Doyle, A.A.; Stephens, J.C. A review of cinnamaldehyde and its derivatives as antibacterial agents. *Fitoterapia* **2019**, *139*, 104405. [CrossRef]
17. Sadeghi, S.; Davoodvandi, A.; Pourhanifeh, M.H.; Sharifi, N.; ArefNezhad, R.; Sahebnasagh, R.; Moghadam, S.A.; Sahebkar, A.; Mirzaei, H. Anti-cancer effects of cinnamon: Insights into its apoptosis effects. *Eur. J. Med. Chem.* **2019**, *178*, 131–140. [CrossRef]
18. Song, F.; Li, H.; Sun, J.; Wang, S. Protective effects of cinnamic acid and cinnamic aldehyde on isoproterenol-induced acute myocardial ischemia in rats. *J. Ethnopharmacol.* **2013**, *150*, 125–130. [CrossRef]
19. Lee, S.C.; Wang, S.Y.; Li, C.C.; Liu, C.T. Anti-inflammatory effect of cinnamaldehyde and linalool from the leaf essential oil of *Cinnamomum osmophloeum* Kanehira in endotoxin induced mice. *J. Food Drug Anal.* **2018**, *26*, 211–220. [CrossRef]
20. Liao, J.C.; Deng, J.S.; Chiu, C.S.; Hou, W.C.; Huang, S.S.; Shie, P.H.; Huang, G.J. Anti-inflammatory activities of *Cinnamomum cassia* constituents in vitro and in vivo. *Evid. Based Complement. Altern. Med.* **2012**, *2012*, 1–13. [CrossRef]
21. Sun, Q.; Shang, B.; Wang, L.; Lu, Z.; Liu, Y. Cinnamaldehyde inhibits fungal growth and aflatoxin B1 biosynthesis by modulating the oxidative stress response of *Aspergillus flavus*. *Appl. Microbiol. Biotechnol.* **2016**, *100*, 1355–1364. [CrossRef]
22. Pereira, W.A.; Pereira, C.D.S.; Assunção, R.G.; da Silva, I.S.C.; Rego, F.S.; Alves, L.S.R.; Santos, J.S.; Nogueira, F.J.R.; Zagmignan, A.; Thomsen, T.T.; et al. New insights into the antimicrobial action of Cinnamaldehyde towards *Escherichia coli* and its effects on intestinal colonization of mice. *Biomolecules* **2021**, *11*, 302. [CrossRef] [PubMed]
23. Yang, G.; Jin, T.; Yin, S.; Guo, D.; Zhang, C.; Xia, X.; Shi, C. trans-Cinnamaldehyde mitigated intestinal inflammation induced by *Cronobacter sakazakii* in newborn mice. *Food Funct.* **2019**, *10*, 2986–2996. [CrossRef] [PubMed]
24. Tung, Y.T.; Huang, C.C.; Ho, S.T.; Kuo, Y.; Lin, C.C.; Lin, C.T.; Wu, J. Bioactive phytochemicals of leaf essential oils of *Cinnamomum osmophloeum* prevent lipopolysaccharide/D-galactosamine (LPS/D-GalN)-induced acute hepatitis in mice. *J. Agric. Food Chem.* **2011**, *59*, 8117–8123. [CrossRef] [PubMed]
25. Almoiliqy, M.; Wen, J.; Xu, B.; Sun, Y.; Lian, M.; Li, Y.; Qaed, E.; Al-Azab, M.; Chen, D.; Shopit, A.; et al. Cinnamaldehyde protects against rat intestinal ischemia/reperfusion injuries by synergistic inhibition of NF-κB and p53. *Acta Pharmacol. Sin.* **2020**, *41*, 1208–1222. [CrossRef]
26. Chen, Y.F.; Wang, Y.W.; Huang, W.S.; Lee, M.; Wood, W.G.; Leung, Y.; Tsai, H. Trans-Cinnamaldehyde, n Essential Oil in Cinnamon Powder, Ameliorates Cerebral Ischemia-Induced Brain Injury via Inhibition of Neuroinflammation Through Attenuation of iNOS, COX-2 Expression and NFκ-B Signaling Pathway. *Neuromol. Med.* **2016**, *18*, 322–333. [CrossRef]
27. Le Tulzo, Y.; Pangault, C.; Gacouin, A.; Guilloux, V.; Tribut, O.; Amiot, L.; Tattevin, P.; Thomas, R.; Fauchet, R.; Drenou, B. Early circulating lymphocyte apoptosis in human septic shock is associated with poor outcome. *Shock* **2002**, *18*, 487–494. [CrossRef]
28. Hotchkiss, R.S.; Osmon, S.B.; Chang, K.C.; Wagner, T.H.; Coopersmith, C.M.; Karl, I.E. Accelerated lymphocyte death in sepsis occurs by both the death receptor and mitochondrial pathway. *J. Immunol.* **2005**, *174*, 5110–5118. [CrossRef]
29. Mendes, S.J.F.; Sousa, F.I.A.B.; Pereira, D.M.S.; Ferro, T.A.F.; Perreira, I.C.P.; Silva, B.L.R.; Pinheiros, A.J.M.C.R.; Mouchreka, A.Q.S.; Monteiro-Neto, V.; Costa, S.K.P.; et al. Cinnamaldehyde modulates LPS-induced systemic inflammatory response syndrome through TRPA1-dependent and independent mechanisms. *Int. Immunopharmacol.* **2016**, *34*, 60–70. [CrossRef]
30. Craciun, F.L.; Schuller, E.R.; Remick, D.G. Early enhanced local neutrophil recruitment in peritonitis-induced sepsis improves bacterial clearance and survival. *J. Immunol.* **2010**, *185*, 6930–6938. [CrossRef]
31. Maciel, M.C.G.; Fialho, E.M.S.; Guerra, R.N.M.; Borges, V.M.; Kwasniewski, F.H.; Nascimento, F.R.F. *Tityus serrulatus* scorpion venom improves survival and lung inflammation in lethal sepsis induced by CLP in mice. *Toxicon* **2014**, *89*, 1–8. [CrossRef]
32. Rios, C.E.; Abreu, A.G.; Braga Filho, J.A.; Nascimento, J.R.; Guerra, R.N.; Amaral, F.M.; Maciel, M.C.G.; Nascimento, F.R.F. *Chenopodium ambrosioides* L. Improves phagocytic activity and decreases bacterial growth and the systemic inflammatory response in sepsis induced by cecal ligation and puncture. *Front. Microbiol.* **2017**, *8*, 1–7. [CrossRef] [PubMed]
33. Braga Filho, J.A.F.; Abreu, A.G.; Rios, C.E.P.; Trovão, L.O.; Silva, D.L.F.; Cysne, D.N.; Nascimento, J.R.; Fortes, T.S.; Silva, L.A.; Guerra, R.N.M.; et al. Prophylactic treatment with simvastatin modulates the immune response and increases animal survival following lethal sepsis infection. *Front. Immunol.* **2018**, *21*, 2137. [CrossRef] [PubMed]
34. Pannee, C.; Chandhanee, I.; Wacharee, L. Antiinflammatory effects of essential oil from the leaves of *Cinnamomum cassia* and cinnamaldehyde on lipopolysaccharide-stimulated J774A.1 cells. *J. Adv. Pharm. Technol. Res.* **2014**, *5*, 164–170. [CrossRef] [PubMed]
35. Zhu, T.; Liao, X.; Feng, T.; Wu, Q.; Zhang, J.; Cao, X.; Li, H. Plasma Monocyte Chemoattractant Protein 1 as a Predictive Marker for Sepsis Prognosis: A Prospective Cohort Study. *Tohoku J. Exp. Med.* **2017**, *241*, 139–147. [CrossRef] [PubMed]
36. Mera, S.; Tatulescu, D.; Cismaru, C.; Bondor, C.; Slavcovici, A.; Zanc, V.; Carstina, D.; Oltean, M. Multiplex cytokine profiling in patients with sepsis. *APMIS* **2011**, *119*, 155–163. [CrossRef] [PubMed]
37. Clinical and Laboratory Standards Institute. *Performance Standards for Antimicrobial Susceptibility Testing*; Twenty-Seventh Informational Supplement; CLSI Document M100-S27; Clinical and Laboratory Standards Institute: Malvern, PA, USA, 2017.

38. Maciel, M.C.G.; Farias, J.C.; Maluf, M.J.; Gomes, E.A.; Pereira, P.V.S.; Aragão-Filho, J.W.C.; Frazão, B.; Costa, G.C.; Sousa, S.M.; Silva, L.A.; et al. *Syzygium jambolanum* treatment improves survival in lethal sepsis induced in mice. *BMC Complement. Altern. Med.* **2008**, *8*, 57–64. [CrossRef]
39. Cruz, G.V.B.; Pereira, P.V.S.; Patrício, F.J.; Costa, G.C.; Sousa, S.M.; Frazão, J.B.; Aragão-Filho, W.C.; Maciel, M.C.G.; Silva, L.A.; Amaral, F.M.M.; et al. Increase of cellular recruitment, phagocytosis ability and nitric oxide production induced by hydroalcoholic extract from *Chenopodium ambrosioides* leaves. *J. Ethnopharmacol.* **2007**, *111*, 148–154. [CrossRef]
40. Liberio, S.A.; Pereira, A.L.A.; Dutra, R.P.; Reis, A.S.; Araújo, M.J.A.M.; Mattar, N.S.; Silva, L.A.; Ribeiro, M.N.S.; Nascimento, F.R.F.; Guerra, R.N.M.; et al. Antimicrobial activity against oral pathogens and immunomodulatory effects and toxicity of geopropolis produced by the stingless bee *Melipona fasciculata* Smith. *BMC Complement. Altern. Med.* **2011**, *11*, 108. [CrossRef]

antibiotics MDPI

Article

Ethyl Acetate Fraction of *Punica granatum* and Its Galloyl-HHDP-Glucose Compound, Alone or in Combination with Fluconazole, Have Antifungal and Antivirulence Properties against *Candida* spp.

Aline Michelle Silva Mendonça [1], Cristina de Andrade Monteiro [2], Roberval Nascimento Moraes-Neto [3], Andrea Souza Monteiro [1], Renata Mondego-Oliveira [4], Camila Evangelista Carnib Nascimento [3], Luís Cláudio Nascimento da Silva [1,5], Lidio Gonçalves Lima-Neto [1], Rafael Cardoso Carvalho [3,*] and Eduardo Martins de Sousa [1,3]

1 Graduate Program in Microbial Biology, CEUMA University, UniCEUMA, São Luís 65055-310, MA, Brazil; alimichellemend@gmail.com (A.M.S.M.); andreasmont@gmail.com (A.S.M.); luiscn.silva@ceuma.br (L.C.N.d.S.); lidio.neto@ceuma.br (L.G.L.-N.); eduardo.martins@ceuma.br (E.M.d.S.)
2 Biology Laboratory, Maranhão Federal Institute, IFMA, São Luís 65030-005, MA, Brazil; cristinamonteiro@ifma.edu.br
3 Graduate Program in Health Sciences, Federal University of Maranhão, UFMA, São Luís 65080-805, MA, Brazil; robervalmoraes11@gmail.com (R.N.M.-N.); camila.carnib@ufma.br (C.E.C.N.)
4 Maurício de Nassau Faculty, UNINASSAU, São Luís 65040-840, MA, Brazil; re_mondego@hotmail.com
5 Graduate Program in Odontology, CEUMA University, UniCEUMA, São Luís 65075-120, MA, Brazil
* Correspondence: carvalho.rafael@ufma.br

Citation: Mendonça, A.M.S.; Monteiro, C.d.A.; Moraes-Neto, R.N.; Monteiro, A.S.; Mondego-Oliveira, R.; Nascimento, C.E.C.; da Silva, L.C.N.; Lima-Neto, L.G.; Carvalho, R.C.; de Sousa, E.M. Ethyl Acetate Fraction of *Punica granatum* and Its Galloyl-HHDP-Glucose Compound, Alone or in Combination with Fluconazole, Have Antifungal and Antivirulence Properties against *Candida* spp. *Antibiotics* **2022**, *11*, 265. https://doi.org/10.3390/antibiotics11020265

Academic Editor: Juraj Gregáň

Received: 13 January 2022
Accepted: 8 February 2022
Published: 18 February 2022

Abstract: Candidiasis is the most common fungal infection among immunocompromised patients. Its treatment includes the use of antifungals, which poses limitations such as toxicity and fungal resistance. Plant-derived extracts, such as *Punica granatum*, have been reported to have antimicrobial activity, but their antifungal effects are still unknown. We aimed to evaluate the antifungal and antiviral potential of the ethyl acetate fraction of *P. granatum* (PgEA) and its isolated compound galloyl-hexahydroxydiphenoyl-glucose (G-HHDP-G) against *Candida* spp. In silico analyses predicted the biological activity of G-HHDP-G. The minimum inhibitory concentrations (MIC) of PgEA and G-HHDP-G, and their effects on biofilm formation, preformed biofilms, and phospholipase production were determined. In silico analysis showed that G-HHDP-G has antifungal and hepatoprotective effects. An in vitro assay confirmed the antifungal effects of PgEA and G-HHDP-G, with MIC in the ranges of 31.25–250 µg/mL and 31.25 $\geq$ 500 µg/mL, respectively. G-HHDP-G and PgEA synergistically worked with fluconazole against planktonic cells. The substances showed antibiofilm action, alone or in combination with fluconazole, and interfered with phospholipase production. The antifungal and antibiofilm actions of PgEA and G-HHDP-G, alone or in combination with fluconazole, in addition to their effects on reducing *Candida* phospholipase production, identify them as promising candidates for therapeutics.

Keywords: *Punica granatum*; galloyl-HHDP-glucose; in silico analysis; in vitro analysis; candidiasis

1. Introduction

The incidence of fungal infections is increasing considerably in humans, due to the indiscriminate use of antibiotics, immunosuppressants, and increased use of invasive procedures, such as catheters. These infections can be debilitating, persistent, and result in costly treatments. Many of these microorganisms are natural colonizers of the human microbiota. However, they have an arsenal of factors and virulence properties that are associated with disorders in the host, such as immunodeficiency, trauma, and surgical procedures, which enable them to be opportunistic infections [1–3].

Candida spp. are commonly associated with diseases in humans, such as oral and vulvovaginal candidiasis, skin infections, and onychomycosis. These infections may occur even in immunocompetent hosts [4,5], when predisposing factors are present (age, antibiotic use, sexual activity, diabetes mellitus, and idiopathic causes), resulting in the possible need for prolonged treatments that may cause recurrences and generation of resistant species [6,7].

The main virulence factor of the *Candida* species, especially *C. albicans*, is its ability to form biofilms. Biofilms produced by pathogenic fungi are characterized by communities of filamentous fungi that adhere to biotic and abiotic surfaces, eventually expanding into highly organized communities that are resistant to antimicrobials and environmental conditions [8,9].

There are a few classes of antifungal agents available to treat *Candida* infections, but these have limitations in terms of their high cost and toxicity. Most infections caused by yeasts are preferably treated with Fluconazole (FCZ), an azole antifungal that acts by inhibiting ergosterol biosynthesis. FCZ has useful properties that make it the drug of choice, such as a wide spectrum of action, low cell toxicity compared to other antimycotics and high bioavailability. However, FCZ has a fungistatic action, inhibiting growth but not killing yeast cells, which can lead to the development of resistance [10].

Moreover, there is evidence of increased antifungal resistance to the available drugs, and most of them are poorly effective in treating diseases associated with biofilm formation [11–14]. Thus, there is a need to search for new antifungal compounds that are more effective, cheaper, and less toxic. Medicinal plants and their isolated compounds (e.g., cinnamaldehyde and eugenol) with antimicrobial properties are promising therapeutic alternatives for fungal infections [15].

Punica granatum is a plant belonging to the family Punicaceae. It is originally from Asia and is cultivated in several parts of the world, including Brazil [16]. Its fruits, roots, stems, and leaves are rich in tannins, flavonoids, ellagic acid, gallic acid, phenolic compounds, and other substances that have antioxidant, anti-inflammatory, antibacterial, and antifungal activities [17].

Our group previously characterized the ethyl acetate fraction obtained from the pomegranate leaf hydroalcoholic extract (PgEA), and identified a particularly interesting compound in it, galloyl-hexahydroxydiphenoil-glucose (G-HHDP-G), which is a hydrolyzable tannin whose pharmacological activities remain undetermined [18]. Later, Pinheiro et al. [19] showed that G-HHDP-G has anti-inflammatory properties and protects against acute lung injury in mice, and thus may be useful for the treatment of this condition and other inflammatory disorders. Nevertheless, there are no studies related to the antifungal effects of G-HHDP-G, and few have investigated the action of *P. granatum* leaf extract against *Candida* species. Therefore, we aimed to investigate the antifungal and antiviral actions of the PgEA fraction and G-HHDP-G against *Candida* spp. In addition, we also evaluated the antifungal activity of a combination of this fraction and compound with fluconazole (FCZ), to explore their synergistic interaction. This could help in the identification of compounds that can possibly serve as potential targets for the development of new herbal or drug formulations, in addition to providing a strategy for alternative therapies.

2. Results

2.1. *In Silico Analysis of the Biological Activities of G-HHDP-G and Its Hepatotoxic Action*

The biological activity spectra of G-HHDP-G were determined using an online version of Prediction of Activity Spectra for Substances (PASS) software. Table 1 shows the values obtained for the probable activity (Pa) and probable inactivity (Pi). Several activities were predicted for G-HHDP-G, including anti-infective, antioxidant, hepatoprotective, anti-inflammatory, immunostimulant, and antifungal. The highest Pa value was obtained for the anti-infective activity (0.962). The biological activities of FCZ were also determined for comparison. Table 2 shows the biological activity spectra of FCZ. The highest FCZ Pa value was for its antifungal activity (0.726).

Table 1. In silico analysis of the biological activities of G-HHDP-G.

Activities	PASS Predictions of G-HHDP-G	
	Pa	Pi
Anti-infective	0.962	0.003
Antioxidant	0.895	0.003
Hepatoprotective	0.883	0.002
Anti-inflammatory	0.775	0.008
Immunostimulant	0.752	0.011
Antifungal	0.692	0.015

Pa, probable activity; Pi, probable inactivity; PASS, prediction of activity spectra for substances; G-HHDP-G, galloyl-hexahydroxidifenoil-glucose.

Table 2. In silico analysis of the biological activities of Fluconazole.

Activities	PASS Predictions of FCZ	
	Pa	Pi
Antifungal	0.726	0.008
ATPase inhibitor of phospholipid translocation	0.480	0.069
Cell wall synthesis inhibitor	0.351	0.002
NADPH inhibitor-cytochrome-c2 reductase	0.366	0.134

Pa, probable activity; Pi, probable inactivity; PASS, prediction of activity spectra for substances; FCZ, fluconazole.

The in silico predictions of chemical toxicity are shown in Table 3. The G-HHDP-G did not show any possible damage to the analyzed cytochromes, but FCZ showed probable hepatotoxicity.

Table 3. In silico prediction of chemical toxicity in hepatic cytochromes.

Cytochrome	Predicted Values of Inhibitory Effect	
	G-HHDP-G	FCZ
CYP1A2	NT (0.8) *	T (0.606) **
CYP2C19	NT (0.872) *	NT (0.775) *
CYP2C9	NT (0.796) *	T (0.698) **
CYP2D6	NT (0.734) *	T (0.502) **
CYP3A4	NT (0.704) *	T (0.572) **

NT, non-toxic; T, toxic; * 0.7–0.9, no expected toxicity; ** 0.5–0.7, predicted toxicity.

2.2. Antifungal Activity

We evaluated the antifungal activity of PgEA and G-HHDP-G against a panel of two clinical and two reference strains of *Candida* spp. Both PgEA and G-HHDP-G exhibited antifungal activity against all the tested strains. However, the inhibitory concentrations varied among the isolates. The minimum inhibitory concentration (MIC) ranges were from 31.25 to 250 µg/mL for PgEA, from 31.25 to > 500 µg/mL for G-HHDP-G, and from 4 to 16 µg/mL for FCZ (Table 4).

Table 4. MICs of PgEA, G-HHDP-G, and FCZ against *Candida* spp. The readings of cell turbidity were recorded after a 48 h incubation, at 37 °C, in RPMI-1640 medium.

Strains	MIC (µg/mL)		
	PgEA	G-HHDP-G	FCZ
C. albicans ATCC 90028	125 ± 0	>500 ± 0	8 ± 0
C. albicans CAS	250 ± 0	>500 ± 0	8 ± 0
C. glabrata ATCC 2001	31.25 ± 0	125 ± 0	16 ± 0
C. glabrata FJF	31.25 ± 0	31.25 ± 0	4 ± 0

The assays were performed in triplicate. MIC, minimal inhibitory concentration; PgEA, ethyl acetate fraction of *P. granatum*; G-HHDP-G, galloyl-hexahydroxidifenoil-glucose; FCZ, fluconazole.

2.3. *In Vitro Interaction between PgEA/FCZ and G-HHDP-G/FCZ*

The interactions between PgEA/FCZ and G-HHDP-G/FCZ were evaluated using a checkerboard assay. The fractional inhibitory concentration index (FICI) showed a synergistic interaction between PgEA/FCZ (Figure 1A–D) and G-HHDP-G/FCZ (Figure 1E,F) against different *Candida* species. According to the assay, the combination of PgEA and FCZ showed synergistic effects against all *Candida* isolates. On the other hand, the combination of G-HHDP-G/FCZ had a synergistic effect against two *Candida* strains (Table 5). Furthermore, there was a drastic reduction in the MIC values of compounds when they were used in combination, compared to the MICs obtained for each compound alone. For example, the MIC values for PgEA against *C. albicans* ATCC 90028 reduced from 125 µg/mL to 3.9 µg/mL when the fraction was used in combination with FCZ. Against *C. albicans* CAS, the reduction was from 250 µg/mL to 7.8 µg/mL. In case of G-HHDP-G, the values reduced from 125 µg/mL and 31.25 µg/mL to 31.2 µg/mL and 7.8 µg/mL against *C. glabrata* ATCC 2001 and *C. glabrata* FJF, respectively. Even the MIC values for FCZ reduced for most strains when it was used in combination the fraction and compound (Table 5).

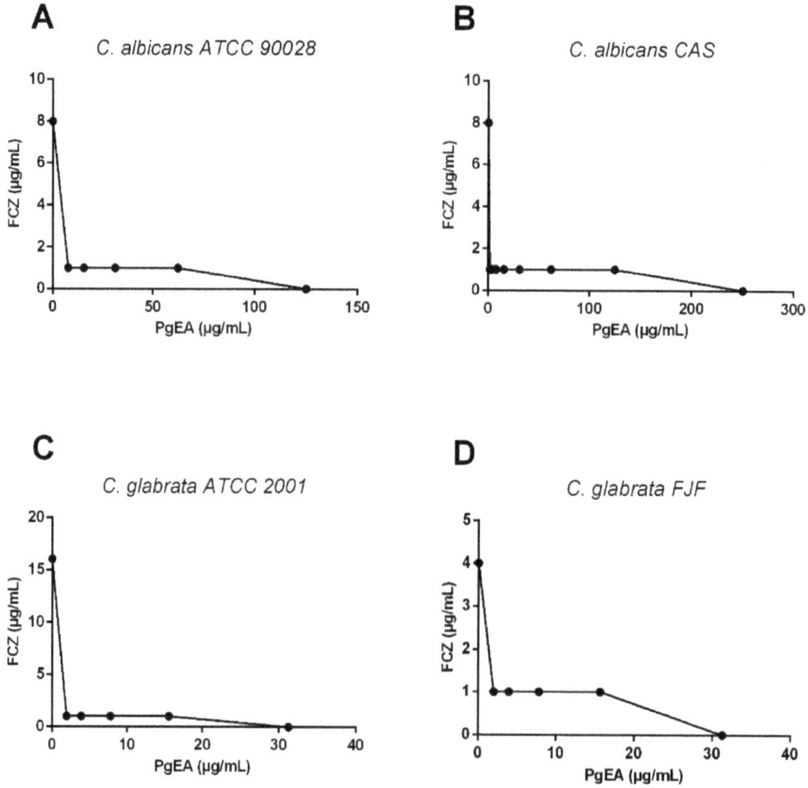

Figure 1. *Cont.*

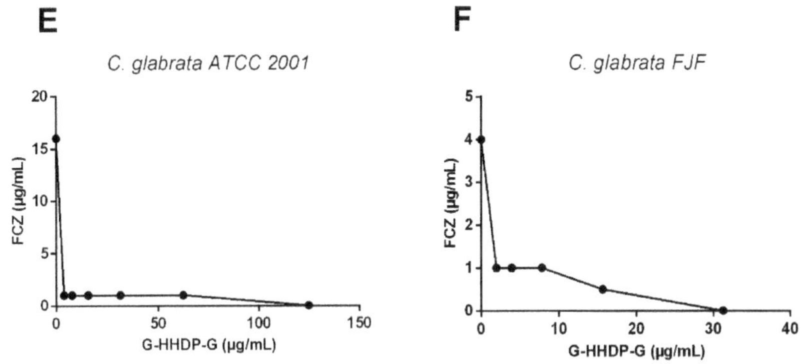

Figure 1. Interaction curves constructed for each pair of compounds when used in combination. Synergistic effect ethyl acetate fraction of *P. granatum* (PgEA) and fluconazole FCZ (**A–D**) against *C. albicans* (**A**,**B**) and *C. glabrata* (**C**,**D**) strains. Synergistic effect of galloyl-hexahydroxidifenoil-glucose (G-HHDP-G) and FCZ against *C. glabrata* strains (**E**,**F**).

Table 5. FICI and classification of the interaction between PgEA or G-HHDP-G and FCZ. The interaction was classified as synergism if FICI $\leq$ 0.5, non-interaction if 0.5 > FICI $\leq$ 4.0, and antagonism if FICI > 4.0.

Strain	MIC in Combination (µg/mL)			FICI			
	FCZ	PgEA	G-HHDP-G	$FICI_{FCZ+PgEA}$	It	$FICI_{FCZ+G\text{-}HHDP\text{-}G}$	It
C. albicans ATCC 90028	8	3.9	-	0.32	SYN	-	-
C. albicans CAS	1	7.8	-	0.36	SYN	-	-
C. glabrata ATCC 2001	4	15.6	31.2	0.45	SYN	0.47	SYN
C. glabrata FJF	0.5	7.8	7.8	0.49	SYN	0.37	SYN

MIC, minimum inhibitory concentration; FCZ, fluconazole; PgEA, ethyl acetate fraction; G-HHDP-G, galloyl-hexahydroxidifenoil-glucose; FICI, fraction inhibitory concentration index; It, interaction type; SYN, synergistic.

2.4. Antibiofilm Effect

Figure 2 shows the effects of PgEA, G-HHDP-G, and FCZ on biofilm formation and preformed biofilms. All tested strains formed biofilms. PgEA and G-HHDP-G reduced the biofilm formation and significantly interfered with the preformed biofilms of both *C. albicans* and *C. glabrata* ($p < 0.05$), both at sub-inhibitory and higher concentrations (Figures 2 and 3). FCZ was not able to interfere with the biofilm formation process of the strains, except at MIC concentrations. FCZ also did not interfere with the preformed biofilms of any strain. In contrast, all synergistic concentrations used were able to inhibit both stages of biofilm formation.

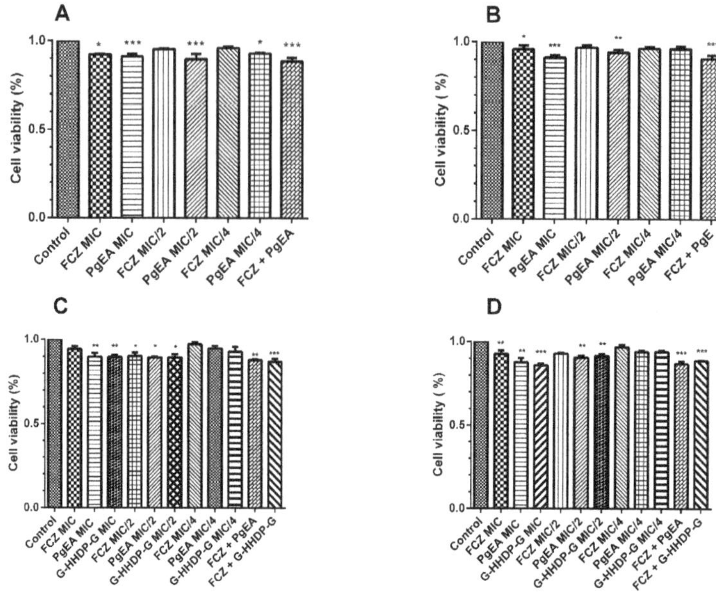

Figure 2. Effect of the ethyl acetate fraction of *P. granatum* (PgEA), galloyl-HHDP-glucose (G-HHDP-G), and Fluconazole (FCZ) on *Candida* biofilm formation. (**A**) Effect of PgEA and FCZ on the biofilm formation of the reference strain *C. albicans* 90028, using inhibitory, sub-inhibitory, and synergistic concentrations. (**B**) Effect of PgEA and FCZ on biofilm formation of the clinical isolate *C. albicans* CAS, using inhibitory, sub-inhibitory, and synergistic concentrations. (**C**) Effect of PgEA, G-HHDP-G, and FCZ on biofilm formation of the reference strain *C. glabrata* 2001, using inhibitory, sub-inhibitory, and synergistic concentrations. (**D**) Effect of PgEA, G-HHDP-G, and FCZ on the biofilm formation of the clinical isolate *C. glabrata* FJF, using inhibitory, sub-inhibitory, and synergistic concentrations. * $p < 0.05$, ** $p < 0.01$, and *** $p < 0.0001$. Data represent the mean ± SD of three independent experiments carried out in triplicate.

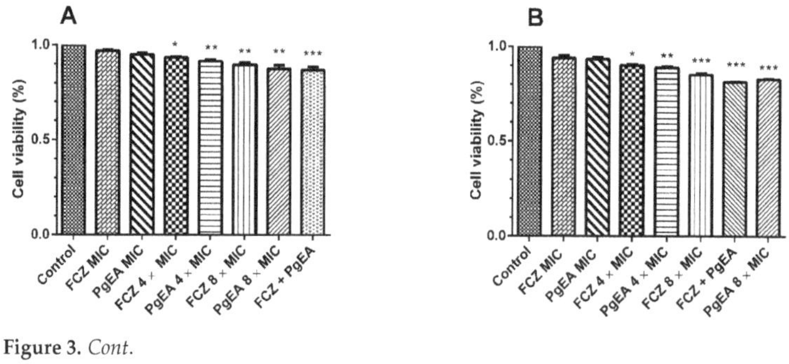

Figure 3. *Cont.*

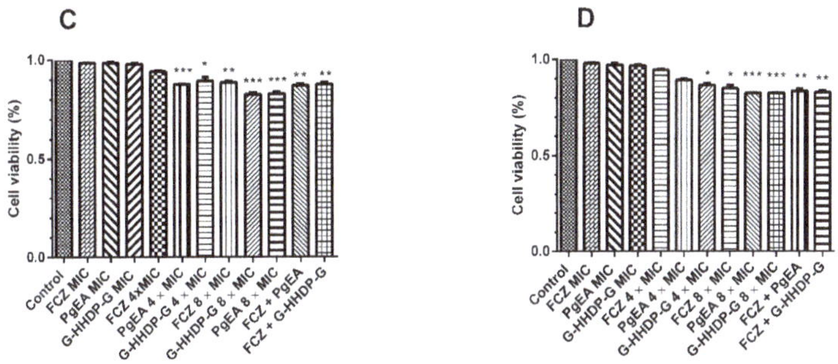

Figure 3. Effect of the ethyl acetate fraction of *P. granatum* (PgEA), galloyl-HHDP-glucose (G-HHDP-G), and Fluconazole (FCZ) on preformed biofilms of *Candida* spp. (**A**) Effect of PgEA and FCZ on the preformed biofilm of the reference strain *C. albicans* 90028, using inhibitory, higher than inhibitory, and synergistic concentrations. (**B**) Effect of PgEA and FCZ on the preformed biofilm of the isolate of clinical *C. albicans* CAS, using inhibitory, higher than inhibitory, and synergistic concentrations. (**C**) Effect of PgEA, G-HHDP-G, and FCZ on the preformed biofilm of the reference strain *C. glabrata* 2001, using inhibitory, higher than inhibitory, and synergistic concentrations. (**D**) Effect of PgEA, G-HHDP-G, and FCZ on the preformed biofilm of the clinical isolate *C. glabrata* FJF, using inhibitory, higher than inhibitory, and synergistic concentrations. * $p < 0.05$, ** $p < 0.01$, and *** $p < 0.0001$. Data represent the mean ± SD of three independent experiments carried out in triplicate.

2.5. Time Kill-Curve Assay

To evaluate the time-kill activity, this assay was performed over a period of 36 h with *C. glabrata* ATCC 2001, and *C. albicans* ATCC 90028 in the presence of G-HHDP-G. The G-HHDP-G concentration of 2 × MIC and 3 × MIC can inhibit *C. glabrata* ATCC 2001 cell viability from 24–30 h when compared to 1 × MIC and the negative control (4A). Interestingly, for *C. albicans* ATCC 90028, 3 × MIC (2000 µg/mL) was responsible for completely eliminating viable cells within 12 h of exposure ($_6$Log of cells/mL) (Figure 4B). The killing activity of G-HHDP-G appears to be dependent on the yeast species or strain; in turn, there is a degree of time and concentration dependence for microbial inhibition.

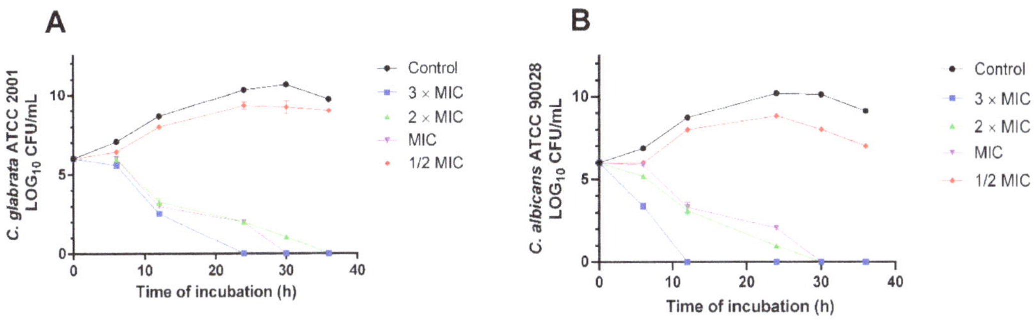

Figure 4. Time–kill curve for *C. glabrata* ATCC 2001 (**A**) and *C. albicans* ATCC 90028 (**B**). G-HHDP- G compound was tested at 3 × MIC. MIC: minimal inhibitory concentration. CFU: colony-forming unit. Time is expressed in hours. Negative control: no compound was added to the cell suspension.

2.6. Phospholipase Assay

We measured the extracellular phospholipase activity of *C. albicans* ATCC 90028, *C. albicans* CAS, *C. glabrata* ATCC 2001, and *C. glabrata* FJF. *C. glabrata* FJF did not produce phospholipases. The higher the phospholipase activity (as measured according to the calculated phospholipase precipitation zone (Pz)), the lower the Pz value. Thus, both PgEA and G-HHDP-G significantly reduced phospholipase production ($p < 0.05$), by interfering with the enzyme production levels (Figure 5 and Table 6).

Table 6. *Candida* spp. extracellular phospholipase activity when treated with PgEA, G-HHDP-G, and FCZ, as evaluated in egg yolk medium in terms of the precipitation zone.

Treatments	Precipitation *Zone*			Phospholipase Activity
	C. Albicans ATCC 90028	*C. Albicans CAS*	*C. Glabrata ATCC 2001*	
Control	0.76	0.67	0.68	H/VH/VH
PgEA MIC	0.91	0.70	0.73	VL/H/H
PgEA MIC/2	0.93	0.73	0.75	VL/H/H
G-HHDP-G MIC	—	—	0.76	H
G-HHDP-G MIC/2	—	—	0.84	L
FCZ MIC	0.82	0.71	0.72	L/H/H
FCZ MIC/2	0.81	0.70	0.70	L/H/H

The precipitation zone represents the ratio of the diameter of the colony to the cloudy zone and colony diameter. VL: very low (Pz = 0.90 to 0.99); L: low (Pz = 0.80 to 0.89); H: high (Pz = 0.70 to 0.79); VH: very high (Pz ≤ 0.69). PgEA—Ethyl acetate fraction of *P. granatum*; G-HHDP-G—Galloyl-Hexahydroxidifenoil-Glucose; FCZ—Fluconazole.

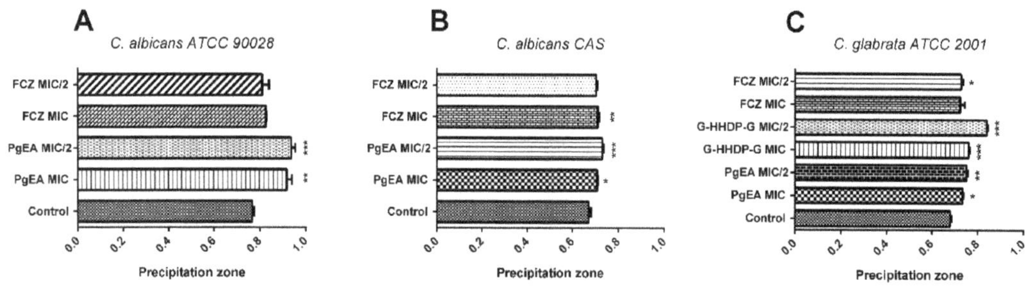

Figure 5. Evaluation of the degree of interference mediated by effect ethyl acetate fraction of *P. granatum* (PgEA), Galloyl-Hexahydroxidifenoil-Glucose (G-HHDP-G), and Fluconazole (FCZ) on *Candida* spp. phospholipase production. (**A**) *C. albicans* ATCC 90028, (**B**) *C. albicans* CAS, and (**C**) *C. glabrata* ATCC 2001. * $p < 0.05$, ** $p < 0.01$, and *** $p < 0.0001$.

3. Discussion

In the present study, we report the antifungal activities of PgEA and one of its phenolic compounds G-HHDP-G. *P. granatum* has attracted the interest of researchers due to its main biological activities, including antioxidant, anti-inflammatory, antibacterial, anticancer, and antiviral [17–21]. Phytochemical analysis performed previously by our research group showed that a richness of phytochemical compounds is present in the PgEA fraction, which corroborates its biological properties [18–20]. Among these compounds, we highlight G-HHDP-G in the present study.

In silico analysis of G-HHDP-G indicated a potential antifungal effect of the compound, with a Pa value of 0.692. Pa and Pi values range from 0.000 to 1.000. When Pa is greater than Pi, the compound is believed to be experimentally active. Pa values ranged from 0.5 to 0.7, indicating that the compound will likely show considerable pharmacological effects experimentally [22], which corroborates the data obtained herein. We also evaluated the in silico effects and toxicity of G-HHDP-G and compared the results obtained with those for FCZ, which is the drug of choice for the treatment of fungal infections. The results highlighted low hepatotoxicity of G-HHDP-G in comparison to that of FCZ. Furthermore, the analysis showed that the compound has a potential hepatoprotective effect (Pa = 0.883), which was not observed in case of FCZ. The examinations were based on the structure-activity ratio of approximately 200,000 compounds and 4000 types of pharmacological activities [23].

Some polyphenolic molecules derived from gallic acid have recently been studied to understand their biological properties. Zhang et al. [24] reviewed 1,2,3,4,6-penta-O-galloyl-β-D-glucose, a gallotanin derivative, and drew attention to its attractive pharmacological and physiological activities, such as anticancer, apoptosis-inducing, anti-inflammatory, and antioxidative. Antiviral, antibacterial, and antibiofilm activities have also been attributed to some galotannin derivatives [25,26]. Similarly, Al-Sayed and Esmat [22] verified the hepatoprotective and antioxidant effects of pentagalloyl glucose and other galloyl esters isolated from the extract of *Melaleuca styphelioides*. However, studies on the antifungal effect of PgEA are scarce, with no reports in the literature about the potential of G-HHDP-G against *Candida*. Therefore, we decided to demonstrate the promising antifungal activity of PgEA and G-HHDP-G in vitro.

We confirmed the results obtained in silico by estimating the MIC. MIC values for PgEA ranged from 31.25 μg/mL to 250 μg/mL, which was effective against all the tested strains. However, G-HHDP-G was only effective against *C. glabrata* strains, with MIC values ranging from 31.25 μg/mL to 125 μg/mL. These results are extremely relevant because *C. glabrata* is intrinsically resistant to azoles [26]. The highest G-HHDP-G value tested against *C. albicans* was 500 μg/mL, and this concentration was not growth inhibitory. One possibility is that G-HHDP-G has an antifungal effect against *C. albicans* when combined with one of the other compounds present in PgEA, since this fraction inhibited the growth of this species.

Most of the studies related to the antifungal effects of *P. granatum* refer to extracts from the fruit, bark, or peel. Lavaee et al. [17] verified that the methanolic and ethanolic extracts of the bark and root of *P. granatum* had anti-*Candida* activity. *P. granatum* peel ethanol extract also showed antifungal activity against oral *Candida* isolates when tested using the agar well diffusion method [27], and there are few reports on the activities of the leaf extract against *Candida*. In a recent study [28], the authors found that after fractionation of the hydroalcoholic extract of *P. granatum* leaves, the ethyl acetate fraction was the richest in polyphenols. However, this fraction did not inhibit the growth of *C. albicans*, which differs from the results obtained in the present study.

To the best of our knowledge, there are no studies in the scientific literature that verify the antifungal activity of G-HHDP-G from PgEA. It is well known that *P. granatum* extracts have antifungal activity; however, the compounds responsible for this effect have not yet been identified, and whether they have antivirulence activity is still poorly understood. An interesting investigation was conducted by Brighenti et al. [29], who verified the effect of phenolic compounds from *P. granatum*, such as punicalin, punicalagin, ellagic acid, and gallic acid, on clinical and reference strains of *C. albicans* and found punicalagin to be the most active. Other studies have also identified punicalagin as the bioactive compound responsible for the antimicrobial activity of pomegranate peel [30,31].

In this study, both PgEA and G-HHDP-G inhibited most *Candida* strains at very low MIC values. However, combination therapies are increasingly being used in clinical trials, with the aim of decreasing conventional antimycotic side effects or toxicity and selection of resistant isolates [32]. Therefore, we decided to evaluate whether PgEA and G-HHDP-G, in combination with FCZ, would present a synergistic interaction against *Candida* strains. The FICI values obtained were very low, and the combinations displayed increased antifungal efficacy against *C. albicans* and *C. glabrata*, over the compounds alone. The MIC values of all the compounds against the tested strains reduced drastically by 50%–97% and 75%–88% for PgEA and G-HHDP-G, respectively, when used in combination. Synergistic interactions overcome the limitations of traditional antifungals by reducing the associated side effects and increasing their spectrum of action [33].

Endo et al. [34], when evaluating the synergistic effect of *P. granatum* fruit extract + FCZ against *C. albicans* isolates, verified that the MIC values for FCZ decreased two-fold when combined with the fruit extract. Similar results were obtained by Silva et al. [35], who evaluated the effects of the combinations of nystatin and punicalagin against *C. albicans*. Combined concentrations increased the antifungal efficacy as compared to the compounds alone, and the two of them reduced punicalagin's MIC-50 by four- and eight-fold, increasing *Candida* inhibition and abrogating the cytotoxicity of punicalagin. These findings support our results since the MIC values for PgEA, G-HHDP-G, and FCZ decreased when they were combined with each other.

An alternative therapy for fungal infection treatment could be the use of compounds with action against. *Candida* spp. possesses several virulence factors that contribute to its high pathogenicity, including the ability to form biofilms. *Candida* species are capable of forming biofilms on both biotic (such as plant/animal cells and tissues) and abiotic surfaces (catheters, prosthetic devices, and dentures) [36]. Biofilms protect against cellular phagocytosis and make *Candida* cells resistant to antifungal drugs [37–39]. In our study, we evaluated the effects of PgEA, G-HHDP-G, and FCZ on biofilm formation and preformed biofilms.

PgEA and G-HHDP-G inhibited biofilm formation and reduced preformed biofilms of both *C. albicans* and *C. glabrata*, showing greater effectiveness than FCZ. In addition, the synergistic combinations of PgEA/FCZ and G-HHDP-G/FCZ were more efficient against *C. albicans* and *C. glabrata* biofilms than the substances alone. These results are extremely relevant because studies involving the effect of *P. granatum* on *Candida* biofilms are rare, and it is difficult to find effective compounds that efficiently inhibit biofilms. In a similar study, Bakkiyaraj et al. [40] also observed an antibiofilm action of the methanolic extract of *P. granatum* and its major compound ellagic acid, but at higher concentrations than those identified in our study. Almeida et al. [41] reported the antibiofilm activity of enriched fractions of *Equisetum giganteum* and *P. granatum* associated with and incorporated in a denture adhesive against *C. albicans*. The mixture was effective against the formation of biofilms on the surface of previously treated polymerized acrylic resin specimens. Villis et al. [42] used the same PgEA fraction as that in this study and verified that this fraction reduced the pre-formed biofilm of some *Cryptococcus* isolates, while showing better activity than FCZ. We highlight the importance of our findings in significantly reducing *Candida* biofilms, because these structures are generally associated with the majority of *Candida* infections and treatment failures, due to their drug-resistant biostructure [43–45]. Furthermore, this is the first study to assess the ability of PgEA, G-HHDP-G, and their combinations with FCZ to inhibit *Candida* biofilms.

Phospholipases are also relevant virulence factors produced by *Candida* spp. These are enzymes capable of breaking the phospholipid membranes or destroying proteins of the host immune system, and therefore, serve as relevant targets for antivirulence therapies [46]. In general, PgEA and G-HHDP-G significantly reduced phospholipase production ($p < 0.5$), as compared to FCZ, by interfering with the enzyme production levels. Liu et al. [32] showed that use of licofelone in combination with FCZ decreased the phospholipase activity at low concentrations, as compared to FCZ alone, with the inhibitory effect being positively correlated with the drug concentration. In turn, Nciki et al. [47] tested the effect of tannin-rich extracts in reducing phospholipase production in *Candida*, which required concentrations that were up to three times higher than those used in this study, suggesting that PgEA and G-HHDP-G have high antivirulence activity.

A limitation of our study is that the action of G-HHDP-G, alone and in association with fluconazole, was evaluated only against species of *C. albicans* and *C. glabrata*, since this was our main objective. Thus, there is a need to investigate a greater number of isolates of clinical origin from different anatomical sites and from different species in order to have a broader assessment of the findings of the present study. Additionally, we intend to carry out an evaluation of the effectiveness of the association of the compounds in experimental animal models.

4. Materials and Methods

4.1. Preparation of PgEA and Isolation/Identification of G-HHDP-G

The hydroalcoholic extract and ethyl acetate fraction of *P. granatum* leaves were obtained as described by Marques et al. [48] and Pinheiro et al. [18]. The ethyl acetate fraction was subjected to a silica gel chromatography column (230–400 mesh; 8 × 100 cm) and eluted with increasing polarities of mixtures of n-hexane/ethyl acetate and ethyl acetate/methanol, to obtain subfractions. The chromatographic separation resulted in 660 fractions. These fractions were grouped into 6 groups according to the similarity of the chromatographic profile. Group 6 was subjected to another round of chromatography and the compound was isolated in galloyl-HHDP-glucose of > 95% purity.

For Group 1, the following polarity gradient was used: Hexane (40%), Ethyl Acetate (60%) and Methanol (0%); Group 2, the following polarity gradient was used: Hexane (30%, 20% and 10%), Ethyl acetate (70%, 80% and 90%) 0% methanol; Group 3, the following polarity gradient was used: Hexane (0%), Ethyl Acetate (100% and 90%), and methanol (0% and 10%); Group 4, the following polarity gradient was used: Hexane (0%), Ethyl Acetate (80%) and Methanol (20%); Group 5, the following polarity gradient was used: Hexane (0%), Ethyl Acetate (70% and 60%) and Methanol (30% and 40%); Group 6 which contained galloyl-HHDP-glucose, the following polarity gradient was used: Hexane (0%), Ethyl Acetate (40%, 20% and 0%) and Methanol (60%, 80% and 100%).

The structure was determined using HPLC-DAD-ESI-IT/MS analysis, as previously described by Pinheiro et al. [18] (Figure 6). The compound G-HDP-G was characterized with data obtained by fragmentation by mass spectrometry and compared with an authentic standard.

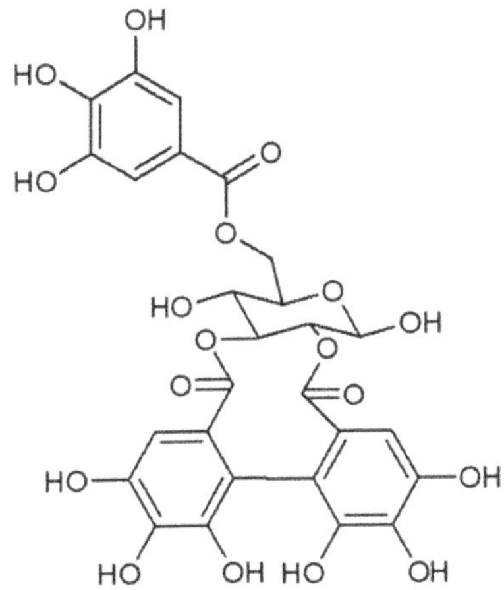

Figure 6. Chemical structure of Galloyl-Hexahydroxidifenoil-Glucose (G-HHDP-G), isolated from the ethyl acetate fraction of the *P. granatum* crude extract.

4.2. In Silico Analysis

4.2.1. Prediction of the Biological Activities of G-HHDP-G In Silico

The biological activities of G-HHDP-G and FCZ (standard drug) were evaluated using PASS Online [version 2.0, Way2Drug.com©2011–2022, Moscow, Russia] (www.way2drug.com/passonline/, accessed on 16 September 2021), which provides several characteristics of the biological action of a substance. The PASS program describes biological activity as "active" (Pa) or "inactive" (Pi), in which the estimated probability varies from zero to one. The chances of finding a particular activity increase when the Pa values are higher and Pi values are lower. The results of PASS prediction were interpreted as follows: (i) only biological activities with Pa > Pi were considered possible for a particular compound; (ii) if Pa > 0.7, the substance is likely to exhibit biological activity and the probability of the compound being an analog of a known pharmaceutical drug is also high; (iii) if 0.5 < Pa < 0.7, the compound is likely to present biological activity, but the substance is not similar to known drugs; (iv) if Pa < 0.5, the chance of finding a biological activity is lower, but the chance to find a structurally new compound is greater.

4.2.2. In Silico Analysis of G-HHDP-G Hepatic Toxicity

To assess the hepatic toxicity of G-HHDP-G and FCZ, we used the Super-CYPsPred [©Structural Bioinformatics Group 2019, Berlin, Germany] (http://insilico-cyp.charite.de/SuperCYPsPred/, accessed on 16 September 2021) web server, which includes machine learning models based on the random forest algorithm and different types of data sampling methods. The models presented in SuperCYSPred discriminate between inhibitors and non-inhibitors for the five main CYP450 isoforms. Fragment-based and structural similarity approaches were used to evaluate the applicability domain of the models, in addition to predicting a specific compound as active (inhibitor) or inactive (non-inhibitor) for a defined CYP isoform.

4.3. *In Vitro Analysis*

4.3.1. Candida Strains and Growth Conditions

For the in vitro assays, we used two clinical isolates from vaginal samples (*C. albicans* CAS and *C. glabrata* FJF 2001; CEP/UNICEUMA no.: 813.402/2014) and two reference strains from the American Type Culture Collection (ATCC; *C. albicans* ATCC 90028 and *C. glabrata* ATCC 2001). The reference strains were kindly donated by the São Paulo State University, Araraquara Dental School, São Paulo, Brazil. Strains were plated on Sabouraud dextrose agar (SDA, Merck, Darmstadt, Germany), incubated for 48 h at 37 °C, and maintained on SDA during the experiments.

4.3.2. MIC Determination

The MIC was determined using the broth dilution method, following the recommendations of the Clinical and Laboratory Standards Institute [49]. PgEA, G-HHDP-G, and FCZ solutions were diluted in RPMI-1640 (Sigma-Aldrich®, St. Louis, MO, USA) (pH 7.0) buffered with 0.165 M morpholinepropanesulfonic acid (MOPS; Sigma-Aldrich, St. Louis, MO, USA). Each substance was added to the first well of 96-well microplates (100 µL/well), with serial dilutions carried out in subsequent wells. The obtained and tested concentrations were 1000–1.95 µg/mL (PgEA), 500–0.07 µg/mL (G-HHDP-G), and 64–0.125 µg/mL (FCZ). Following that, 100 µL of *Candida* inoculum (1×10^3 CFU/mL) was added to each well and incubated at 37 °C for 48 h in RPMI-1640 medium. After the incubation period, MIC was defined as the lowest concentration that visibly inhibited fungal growth. FCZ was used as a positive control and RPMI (100 µL) plus standardized inoculum was used as a negative control. The results were obtained from three independent assays performed in triplicate.

4.3.3. In Vitro Interaction Assays between PgEA + FCZ and G-HHDP-G + FCZ

Interactions between PgEA/FCZ and G-HHDP-G/FCZ were evaluated using the checkerboard test [49]. The following concentrations of PgEA, G-HHDP-G, and FCZ were used for each *Candida* strain: Combination 1: PgEA (250–0.97 µg/mL) and FCZ (16–0.06 µg/mL), for *C. albicans* ATCC 90028; Combination 2: PgEA (500–1.95 µg/mL) and FCZ (16–0.06 µg/mL), for *C. albicans* CAS; Combination 3: PgEA (62.5–0.24 µg/mL) and FCZ (32–0.125 µg/mL), for *C. glabrata* ATCC 2001; Combination 4: PgEA (62.5–0.24 µg/mL) and FCZ (8–0.03 µg/mL), for *C. glabrata* FJF. All the substances were diluted in RPMI-1640/MOPS medium.

One hundred microliters of the inoculum (1×10^3 CFU/mL), 50 µL of PgEA or G-HHDP-G, and 50 µL of FCZ were added to 96-well plates. For sterility control, RPMI was used alone (100 µL), and growth was observed in RPMI (100 µL) plus standardized inoculum. Antimicrobial activity was assessed as described for MIC. After data normalization, the FICI was calculated for each compound, according to the general formula:

FICI = [MICFCZ in combination/MICFCZ] + [MICPgEA in combination/MICPgEA] or FICI = [MICFCZ in combination/MICFCZ] + [MICG-HHDP-G in combination/MICG-HHDP-G]. FICI was calculated for all possible combinations of different concentrations against the same strain, and the final result was expressed as the mean of the FICI values. In addition, interaction curves were also constructed. The interaction between compounds was classified as synergism if FICI $\leq$ 0.5, indifferent if 0.5 > FICI $\leq$ 4.0, and antagonism if FICI > 4.0 [50]. Three independent assays were performed in triplicate.

4.3.4. Effect of PgEA and G-HHDP-G on Candida Biofilms

Candida biofilms were developed using a slightly modified method [51,52]. To verify the interference of substances on biofilm formation, 200 µL of each substance at MIC, sub-inhibitory concentrations of MIC/4 and MIC/2, and established synergistic concentrations was used. The interference of substances on preformed biofilms was analyzed using the concentrations of MIC, 4× MIC, 8× MIC, and synergistic concentrations. PgEA and FCZ were tested against *C. albicans* (ATCC 90028 and CAS) and *C. glabrata* (ATCC 2001 and FJF). G-HHDP-G was tested against *C. glabrata* (ATCC 2001 and FJF).

Candida cells previously grown in SDA were transferred to Yeast Nitrogen Base Broth (YNB) (Sigma-Aldrich®, St. Louis, MO, USA) and incubated for 18 h at 37 °C. The cell pellet was washed three times with sterile phosphate-buffered saline (PBS). A standard cell suspension (1×10^6 CFU/mL, 200 µL) was added to 96-well plates and allowed to adhere for 90 min. After the adhesion phase, the microplates were gently washed three times with PBS to remove planktonic cells. To evaluate the interference on biofilm development, 200 µL of PgEA, G-HHDP-G, or FCZ, diluted in YNB + 100 mM glucose, were added to the corresponding wells, and the microplates were incubated for 24 h. For analysis of the interference on preformed biofilm, after the adhesion period, the wells were washed and each well was replaced with 200 µL of YNB. The microplates were then incubated for 24 h. Later, the wells were washed three times, 200 µL of PgEA, G-HHDP-G, and FCZ was added to the wells, and biofilms were incubated for a further 24 h. In all the experiments, biofilms without substances were used as controls.

After the final incubation, biofilms were evaluated for cell viability using 3-(4,5-dimethylthiazol-2-yl)-2,5-diphenyl-2H-tetrazolium bromide (MTT; Sigma-Aldrich) method [53]. Briefly, biofilms were washed with PBS, and 100 mL of MTT (5 mg/mL) was added to each sample and incubated for 4 h under light. Supernatants were then removed, 100 mL of Dimethyl Sulfoxide(DMSO) was added to each well, and the samples were incubated for another ten minutes. Readings were performed using a microplate reader (Softmax® Pro-Molecular Devices General Counsel, USA) at the wavelength of 490 nm. Each experiment was conducted three times in triplicate.

4.3.5. Time Kill-Curve Assay

The time-curve experiments were carried out in plastic tubes with screw caps in RPMI medium (Sigma-Aldrich), with a final volume of 500 µL at 37 °C for 36 h. The cells to the start of the experiment to obtain fungal cultures in early logarithmic phase growth. Cells were suspended in sterile distilled water to achieve a starting inoculum size of 1–5 $\times 10^6$ colony forming units (CFU)/mL and added to the tubes containing G-HHDP-G at concentrations 0.5, 1, 2, and 3 times the MIC. Growth control was also measured by adding the inoculum to tubes containing RPMI medium without drug. Sample for viable counts was taken at 0, 6, 12, 24, 30, and 36 h, plated in triplicate onto Sabouraud dextrose agar (SDA, Difco), and incubated for 24–48 h at 37 °C. After, incubation samples were first diluted in sterile saline (NaCl, 0.9%) and plated in the culture medium. Experiments were performed in duplicate for each isolate at different times. The results of the counts of the yeasts *C. albicans* ATCC 90028 *C. glabrata* ATCC 2001 were expressed in Log_{10} CFU/mL.

4.3.6. PgEA, G-HHDP-G, and FCZ Interference in Phospholipase Production

The phospholipase activity of *Candida* spp. was determined using egg yolk agar medium. Both *C. albicans* and *C. glabrata* cultures (1×10^3 CFU/mL) were treated with PgEA, G-HHDP-G, and FCZ at MIC, MIC/2, and MIC/4. A control group without substances was also included. The cultures were transferred into separate microtubes and incubated at 37 °C for 4 h. Subsequently, 10 µL of the suspension from each tube was inoculated into egg yolk agar medium and the plates were incubated at 37 °C for 72 h. After that, the diameters of the precipitation zones (a) and diameter of the precipitation zone plus diameter of the colony (b) were measured. The Pz was designated as Pz = a/b, as described by Price et al. [54] and Liu et al. [32]. According to this definition, the phospholipase production index was scored and categorized as follows: negative (Pz = 1), very low (Pz = 0.90 to 0.99), low (Pz = 0.80 to 0.89), high (Pz = 0.70 to 0.79), and very high (Pz $\leq$ 0.69) [50]. Each experiment was conducted three times in triplicate.

4.4. Statistical Analysis

All experiments were performed in triplicate, and the values have been expressed as mean ± standard deviation. The results were analyzed using one-way ANOVA, followed by Tukey's test. Statistical analyses were performed using Prism 7.00 software (GraphPad, San Diego, CA, USA), and differences were considered significant when $p < 0.05$.

5. Conclusions

The present study provides a substantial advance over recent studies on *P. granatum* and its compounds, and, to the best of our knowledge, is the first to discover the antifungal effects of G-HHDP-G. We are also pioneers in verifying the synergistic effect of PgEA and G-HHDP-G, in combination with FCZ, against *Candida* spp. planktonic cells and biofilms. These results indicate that both the PgEA fraction and the compound G-HHDP-G are potential candidates that could serve as antifungal agents and promising synergists with FCZ for the development of new drugs against *Candida*. However, more in-depth studies need to be conducted to uncover the mechanisms of action of these compounds.

Author Contributions: Conceptualization, R.C.C. and E.M.d.S.; Data curation, L.G.L.-N.; Formal analysis, A.M.S.M., C.d.A.M., R.N.M.-N., A.S.M., R.M.-O., C.E.C.N., L.C.N.d.S., L.G.L.-N. and E.M.d.S.; Funding acquisition, E.M.d.S.; Investigation, A.M.S.M., C.d.A.M., R.N.M.-N., A.S.M., L.C.N.d.S., L.G.L.-N., R.C.C. and E.M.d.S.; Methodology, A.M.S.M., C.d.A.M., R.N.M.-N., A.S.M., R.M.-O., C.E.C.N., L.C.N.d.S., L.G.L.-N., R.C.C. and E.M.d.S.; Project administration, R.C.C. and E.M.d.S.; Resources, E.M.d.S.; Supervision, R.C.C. and E.M.d.S.; Writing–original draft, E.M.d.S.; Writing–review & editing, R.C.C. and E.M.d.S. All authors have read and agreed to the published version of the manuscript.

Funding: The authors would like to thank the Foundation for the Support of Research and Scientific and Technological Development of Maranhão (BM-02332/19; INFRA-02012/21; BATI-05406/21) and CEUMA University for financial support.

Institutional Review Board Statement: This study was performed following the protocols approved by the Research Ethics Committee (REC/UniCEUMA no. 813.402/2014) of CEUMA University, São Luís, Maranhão, Brazil.

Informed Consent Statement: Not applicable.

Data Availability Statement: Not applicable.

Conflicts of Interest: The authors declare no conflict of interest.

References

1. Althaus, V.A.; Regginato, A.; Bossetti, V.; Schmidt, J.C. Espécies Se *Candida* spp. Em Isolados Clínicos e Suscetibilidade a Antifúngicos de Uso Hospitalar. *Saúde e Pesqui.* **2015**, *8*, 7. [CrossRef]
2. Bajpai, V.K.; Khan, I.; Shukla, S.; Kumar, P.; Rather, I.A.; Park, Y.H.; Huh, Y.S.; Han, Y.K. Invasive Fungal Infections and Their Epidemiology: Measures in the Clinical Scenario. *Biotechnol. Bioprocess Eng.* **2019**, *24*, 436–444. [CrossRef]

3. Bongomin, F.; Gago, S.; Oladele, R.O.; Denning, D.W. Global and Multi-National Prevalence of Fungal Diseases—Estimate Precision. *J. Fungi* **2017**, *3*, 57. [CrossRef]
4. Fakhim, H.; Vaezi, A.; Javidnia, J.; Nasri, E.; Mahdi, D.; Diba, K.; Badali, H. Candida Africana Vulvovaginitis: Prevalence and Geographical Distribution. *J. Mycol. Med.* **2020**, *30*, 100966. [CrossRef] [PubMed]
5. Huffnagle, G.B.; Noverr, M.C. The Emerging World of the Fungal Microbiome. *Trends Microbiol.* **2013**, *21*, 334–341. [CrossRef] [PubMed]
6. Rodríguez-Cerdeira, C.; Gregorio, M.C.; Molares-Vila, A.; López-Barcenas, A.; Fabbrocini, G.; Bardhi, B.; Sinani, A.; Sánchez-Blanco, E.; Arenas-Guzmán, R.; Hernandez-Castro, R. Biofilms and Vulvovaginal Candidiasis. *Colloids Surf. B Biointerfaces* **2019**, *174*, 110–125. [CrossRef] [PubMed]
7. Lima, J.S.; Braga, K.R.G.S.; Vieira, C.A.; Souza, W.W.R.; Chávez-Pavoni, J.H.; Araújo, C.d.; Goulart, L.S. Genotypic Analysis of Secreted Aspartyl Proteinases in Vaginal Candida Albicans Isolates. *J. Bras. Patol. e Med. Lab.* **2018**, *54*, 28–33. [CrossRef]
8. Costa-Orlandi, C.B.; Sardi, J.C.O.; Pitangui, N.S.; de Oliveira, H.C.; Scorzoni, L.; Galeane, M.C.; Medina-Alarcón, K.P.; Melo, W.C.M.A.; Marcelino, M.Y.; Braz, J.D.; et al. Fungal Biofilms and Polymicrobial Diseases. *J. Fungi* **2017**, *3*, 22. [CrossRef]
9. Roscetto, E.; Contursi, P.; Vollaro, A.; Fusco, S.; Notomista, E.; Catania, M.R. Antifungal and Anti-Biofilm Activity of the First Cryptic Antimicrobial Peptide from an Archaeal Protein against *Candida* spp. Clinical Isolates. *Sci. Rep.* **2018**, *8*, 17570. [CrossRef] [PubMed]
10. Lu, H.; Shrivastava, M.; Whiteway, M.; Jiang, Y. Candida Albicans Targets That Potentially Synergize with Fluconazole. *Crit. Rev. Microbiol.* **2021**, *47*, 323–337. [CrossRef]
11. Dantas-Medeiros, R.; Zanatta, A.C.; de Souza, L.B.F.C.; Fernandes, J.M.; Amorim-Carmo, B.; Torres-Rêgo, M.; Fernandes-Pedrosa, M.d.F.; Vilegas, W.; Araújo, T.A.d.S.; Michel, S.; et al. Antifungal and Antibiofilm Activities of B-Type Oligomeric Procyanidins From Commiphora Leptophloeos Used Alone or in Combination with Fluconazole against *Candida* spp. *Front. Microbiol.* **2021**, *12*, 613155. [CrossRef] [PubMed]
12. Fuentefria, A.M.; Pippi, B.; Dalla Lana, D.F.; Donato, K.K.; de Andrade, S.F. Antifungals Discovery: An Insight into New Strategies to Combat Antifungal Resistance. *Lett. Appl. Microbiol.* **2018**, *66*, 2–13. [CrossRef] [PubMed]
13. Perlin, D.S.; Rautemaa-Richardson, R.; Alastruey-Izquierdo, A. The Global Problem of Antifungal Resistance: Prevalence, Mechanisms, and Management. *Lancet Infect. Dis.* **2017**, *17*, e383–e392. [CrossRef]
14. Wiederhold, N.P. Antifungal Resistance: Current Trends and Future Strategies to Combat. *Infect. Drug Resist.* **2017**, *10*, 249–259. [CrossRef] [PubMed]
15. Saracino, I.M.; Foschi, C.; Pavoni, M.; Spigarelli, R.; Valerii, M.C.; Spisni, E. Antifungal Activity of Natural Compounds vs. *Candida* spp.: A Mixture of Cinnamaldehyde and Eugenol Shows Promising In Vitro Results. *Antibiotics* **2022**, *11*, 73. [CrossRef]
16. Degaspari, C.H.; Dutra, A.P.C. Propriedades fitoterápicas da romã (*Punica granatum* L.). *Visão Acadêmica* **2011**, *12*, 36–46. [CrossRef]
17. Lavaee, F.; Motaghi, D.; Jassbi, A.R.; Jafarian, H.; Ghasemi, F.; Badiee, P. Antifungal Effect of the Bark and Root Extracts of Punica Granatum on Oral Candida Isolates. *Curr. Med. Mycol.* **2018**, *4*, 20–24. [CrossRef] [PubMed]
18. Pinheiro, A.J.M.C.R.; Gonçalves, J.S.; Dourado, Á.W.A.; De Sousa, E.M.; Brito, N.M.; Silva, L.K.; Batista, M.C.A.; De Sá, J.C.; Monteiro, C.R.A.V.; Fernandes, E.S.; et al. *Punica granatum* L. Leaf Extract Attenuates Lung Inflammation in Mice with Acute Lung Injury. *J. Immunol. Res.* **2018**, *2018*, 6879183. [CrossRef]
19. Pinheiro, A.J.M.C.R.; Mendes, A.R.S.; Neves, M.D.F.d.J.; Prado, C.M.; Bittencourt-Mernak, M.I.; Santana, F.P.R.; Lago, J.H.G.; de Sá, J.C.; da Rocha, C.Q.; de Sousa, E.M.; et al. Galloyl-Hexahydroxydiphenoyl (HHDP)-Glucose Isolated from *Punica granatum* L. Leaves Protects against Lipopolysaccharide (LPS)-Induced Acute Lung Injury in BALB/c Mice. *Front. Immunol.* **2019**, *10*, 1978. [CrossRef]
20. Lansky, E.P.; Newman, R.A. Punica Granatum (Pomegranate) and Its Potential for Prevention and Treatment of Inflammation and Cancer. *J. Ethnopharmacol.* **2007**, *109*, 177–206. [CrossRef]
21. Reddy, M.K.; Gupta, S.K.; Jacob, M.R.; Khan, S.I.; Ferreira, D. Antioxidant, Antimalarial and Antimicrobial Activities of Tannin-Rich Fractions, Ellagitannins and Phenolic Acids from *Punica granatum* L. *Planta Med.* **2007**, *73*, 461–467. [CrossRef] [PubMed]
22. Al-Sayed, E.; Esmat, A. Hepatoprotective and Antioxidant Effect of Ellagitannins and Galloyl Esters Isolated from Melaleuca Styphelioides on Carbon Tetrachloride-Induced Hepatotoxicity in HepG2 Cells. *Pharm. Biol.* **2016**, *54*, 1727–1735. [CrossRef] [PubMed]
23. Rim, K.T. In Silico Prediction of Toxicity and Its Applications for Chemicals at Work. *Toxicol. Environ. Health Sci.* **2020**, *12*, 191–202. [CrossRef] [PubMed]
24. Zhang, J.; Li, L.; Kim, S.H.; Hagerman, A.E.; Lü, J. Anti-Cancer, Anti-Diabetic and Other Pharmacologic and Biological Activities of Penta-Galloyl-Glucose. *Pharm. Res.* **2009**, *26*, 2066–2080. [CrossRef] [PubMed]
25. Bag, A.; Chattopadhyay, R.R. Synergistic Antibiofilm Efficacy of a Gallotannin 1,2,6-Tri-O-Galloyl-β-D-Glucopyranose from Terminalia Chebula Fruit in Combination with Gentamicin and Trimethoprim against Multidrug Resistant Uropathogenic Escherichia Coli Biofilms. *PLoS ONE* **2017**, *12*, e0178712. [CrossRef] [PubMed]
26. Santos, G.C.d.O.; Vasconcelos, C.C.; Lopes, A.J.O.; Cartágenes, M.d.S.d.S.; Filho, A.K.D.B.; do Nascimento, F.R.F.; Ramos, R.M.; Pires, E.R.R.B.; de Andrade, M.S.; Rocha, F.M.G.; et al. Candida Infections and Therapeutic Strategies: Mechanisms of Action for Traditional and Alternative Agents. *Front. Microbiol.* **2018**, *9*, 1351. [CrossRef] [PubMed]
27. Prem Kumar, K.; Samlin, S.; Siva, B.; Sudharshan, R.; Vignesswary, A.; Divya, K. Punica Granatum as a Salutiferous Superfruit in the Treatment of Oral Candidiasis—An in-Vitro Study. *J. Oral Maxillofac. Pathol.* **2020**, *24*, 188. [CrossRef]

28. Swilam, N.; Nematallah, K.A. Polyphenols Profile of Pomegranate Leaves and Their Role in Green Synthesis of Silver Nanoparticles. *Sci. Rep.* **2020**, *10*, 14851. [CrossRef]
29. Brighenti, V.; Iseppi, R.; Pinzi, L.; Mincuzzi, A.; Ippolito, A.; Messi, P.; Sanzani, S.M.; Rastelli, G.; Pellati, F. Antifungal Activity and DNA Topoisomerase Inhibition of Hydrolysable Tannins from *Punica granatum* L. *Int. J. Mol. Sci.* **2021**, *22*, 4175. [CrossRef] [PubMed]
30. Silva, T.C.; de Ávila, R.I.; Zara, A.L.S.A.; Santos, A.S.; Ataídes, F.; Freitas, V.A.Q.; Costa, C.R.; Valadares, M.C.; Silva, M.d.R.R. Punicalagin Triggers Ergosterol Biosynthesis Disruption and Cell Cycle Arrest in Cryptococcus Gattii and Candida Albicans: Action Mechanisms of Punicalagin against Yeasts. *Braz. J. Microbiol.* **2020**, *51*, 1719–1727. [CrossRef] [PubMed]
31. Gosset-Erard, C.; Zhao, M.; Lordel-Madeleine, S.; Ennahar, S. Identification of Punicalagin as the Bioactive Compound behind the Antimicrobial Activity of Pomegranate (*Punica granatum* L.) Peels. *Food Chem.* **2021**, *352*, 129396. [CrossRef] [PubMed]
32. Liu, X.; Li, T.; Wang, D.; Yang, Y.; Sun, W.; Liu, J.; Sun, S. Synergistic Antifungal Effect of Fluconazole Combined with Licofelone against Resistant Candida Albicans. *Front. Microbiol.* **2017**, *8*, 2101. [CrossRef] [PubMed]
33. Wagner, H.; Ulrich-Merzenich, G. Synergy Research: Approaching a New Generation of Phytopharmaceuticals. *Phytomedicine* **2009**, *16*, 97–110. [CrossRef] [PubMed]
34. Endo, E.H.; Garcia Cortez, D.A.; Ueda-Nakamura, T.; Nakamura, C.V.; Dias Filho, B.P. Potent Antifungal Activity of Extracts and Pure Compound Isolated from Pomegranate Peels and Synergism with Fluconazole against Candida Albicans. *Res. Microbiol.* **2010**, *161*, 534–540. [CrossRef] [PubMed]
35. Silva, R.A.; Ishikiriama, B.L.C.; Ribeiro Lopes, M.M.; de Castro, R.D.; Garcia, C.R.; Porto, V.C.; Santos, C.F.; Neppelenbroek, K.H.; Lara, V.S. Antifungal Activity of Punicalagin–Nystatin Combinations against Candida Albicans. *Oral Dis.* **2020**, *26*, 1810–1819. [CrossRef] [PubMed]
36. Silva, P.M.; de Moura, M.C.; Gomes, F.S.; da Silva Trentin, D.; Silva de Oliveira, A.P.; de Mello, G.S.V.; da Rocha Pitta, M.G.; de Melo Rego, M.J.B.; Coelho, L.C.B.B.; Macedo, A.J.; et al. PgTeL, the Lectin Found in Punica Granatum Juice, Is an Antifungal Agent against Candida Albicans and Candida Krusei. *Int. J. Biol. Macromol.* **2018**, *108*, 391–400. [CrossRef] [PubMed]
37. Dadar, M.; Tiwari, R.; Karthik, K.; Chakraborty, S.; Shahali, Y.; Dhama, K. Candida Albicans-Biology, Molecular Characterization, Pathogenicity, and Advances in Diagnosis and Control—An Update. *Microb. Pathog.* **2018**, *117*, 128–138. [CrossRef] [PubMed]
38. Silva, S.; Negri, M.; Henriques, M.; Oliveira, R.; Williams, D.W.; Azeredo, J. Candida Glabrata, Candida Parapsilosis and Candida Tropicalis: Biology, Epidemiology, Pathogenicity and Antifungal Resistance. *FEMS Microbiol. Rev.* **2012**, *36*, 288–305. [CrossRef] [PubMed]
39. Barros, P.P.; Rossoni, R.D.; de Souza, C.M.; Scorzoni, L.; Fenley, J.D.C.; Junqueira, J.C. Candida Biofilms: An Update on Developmental Mechanisms and Therapeutic Challenges. *Mycopathologia* **2020**, *185*, 415–424. [CrossRef] [PubMed]
40. Bakkiyaraj, D.; Nandhini, J.R.; Malathy, B.; Pandian, S.K. The Anti-Biofilm Potential of Pomegranate (*Punica granatum* L.) Extract against Human Bacterial and Fungal Pathogens. *Biofouling* **2013**, *29*, 929–937. [CrossRef] [PubMed]
41. Almeida, N.L.M.; Saldanha, L.L.; da Silva, R.A.; Pinke, K.H.; da Costa, E.F.; Porto, V.C.; Dokkedal, A.L.; Lara, V.S. Antimicrobial Activity of Denture Adhesive Associated with Equisetum Giganteumand Punica Granatum-Enriched Fractions against Candida Albicans Biofilms on Acrylic Resin Surfaces. *Biofouling* **2018**, *34*, 62–73. [CrossRef] [PubMed]
42. Villis, P.C.M.; de Macedo, A.T.; Furtado, H.L.A.; Fontenelle, P.H.C.; Gonçalves, I.S.; Mendes, T.L.; Motta, B.L.A.; Marinho, P.L.L.; Pinheiro, A.J.M.C.R.; Lima-Neto, L.G.; et al. A Study of the Disruptive Effect of the Acetate Fraction of Punica Granatum Extract on Cryptococcus Biofilms. *Front. Microbiol.* **2021**, *11*, 568258. [CrossRef]
43. Nobile, C.J.; Johnson, A.D. Candida Albicans Biofilms and Human Disease. *Annu. Rev. Microbiol.* **2015**, *69*, 71–92. [CrossRef] [PubMed]
44. Tsui, C.; Kong, E.F.; Jabra-Rizk, M.A. Pathogenesis of Candida Albicans Biofilm. *Pathog. Dis.* **2016**, *74*, ftw018. [CrossRef] [PubMed]
45. Zago, P.M.W.; Dos Santos Castelo Branco, S.J.; De Albuquerque Bogéa Fecury, L.; Carvalho, L.T.; Rocha, C.Q.; Madeira, P.L.B.; De Sousa, E.M.; De Siqueira, F.S.F.; Paschoal, M.A.B.; Diniz, R.S.; et al. Anti-Biofilm Action of Chenopodium Ambrosioides Extract, Cytotoxic Potential and Effects on Acrylic Denture Surface. *Front. Microbiol.* **2019**, *10*, 1724. [CrossRef] [PubMed]
46. Sriphannam, C.; Nuanmuang, N.; Saengsawang, K.; Amornthipayawong, D.; Kummasook, A. Anti-Fungal Susceptibility and Virulence Factors of *Candida* spp. Isolated from Blood Cultures. *J. Mycol. Med.* **2019**, *29*, 325–330. [CrossRef] [PubMed]
47. Nciki, S.; Oderinlo, O.O.; Gulube, Z.; Osamudiamen, P.M.; Idahosa, K.C.; Patel, M. Mezoneuron Benthamianum Inhibits Cell Adherence, Hyphae Formation, and Phospholipase Production in Candida Albicans. *Arch. Microbiol.* **2020**, *202*, 2533–2542. [CrossRef] [PubMed]
48. Marques, L.C.F.; Pinheiro, A.J.M.C.R.; Araújo, J.G.G.; De Oliveira, R.A.G.; Silva, S.N.; Abreu, I.C.; De Sousa, E.M.; Fernandes, E.S.; Luchessi, A.D.; Silbiger, V.N.; et al. Anti-Inflammatory Effects of a Pomegranate Leaf Extract in LPS-Induced Peritonitis. *Planta Med.* **2016**, *82*, 1463–1467. [CrossRef] [PubMed]
49. *CLSI Standard M27*; Reference Method for Broth Dilution Antifungal Susceptibility Testing of Yeasts. 4th ed. Clinical and Laboratory Standards Institute: Wayne, UK, 2017.
50. Odds, F.C. Synergy, Antagonism, and What the Chequerboard Puts between Them. *J. Antimicrob. Chemother.* **2003**, *52*, 1. [CrossRef]
51. Gu, W.; Guo, D.; Zhang, L.; Xu, D.; Sun, S. The Synergistic Effect of Azoles and Fluoxetine against Resistant Candida Albicans Strains Is Attributed to Attenuating Fungal Virulence. *Antimicrob. Agents Chemother.* **2016**, *60*, 6179–6188. [CrossRef]

52. Sun, W.; Wang, D.; Yu, C.; Huang, X.; Li, X.; Sun, S. Strong Synergism of Dexamethasone in Combination with Fluconazole against Resistant Candida Albicans Mediated by Inhibiting Drug Efflux and Reducing Virulence. *Int. J. Antimicrob. Agents* **2017**, *50*, 399–405. [CrossRef] [PubMed]
53. Santos, J.R.A.; Gouveia, L.F.; Taylor, E.L.S.; Resende-Stoianoff, M.A.; Pianetti, G.A.; César, I.C.; Santos, D.A. Dynamic Interaction between Fluconazole and Amphotericin B against Cryptococcus Gattii. *Antimicrob. Agents Chemother.* **2012**, *56*, 2553–2558. [CrossRef]
54. Price, M.F.; Wilkinson, A.N.D.; Gentry, L.O. Plate method for detection of phospholipase activity in Candida albicans. *Sabouraudia* **1982**, *20*, 7–14. [CrossRef] [PubMed]

MDPI
St. Alban-Anlage 66
4052 Basel
Switzerland
www.mdpi.com

Antibiotics Editorial Office
E-mail: antibiotics@mdpi.com
www.mdpi.com/journal/antibiotics

www.ingramcontent.com/pod-product-compliance
Lightning Source LLC
Chambersburg PA
CBHW061606220926
51850CB00005B/435